Epigenetics and Chromatin: Advanced Concepts of Cell Biology

Epigenetics and Chromatin: Advanced Concepts of Cell Biology

Editor: Lily Kingsley

R CALLISTO REFERENCE

www.callistoreference.com

Callisto Reference,
118-35 Queens Blvd., Suite 400,
Forest Hills, NY 11375, USA

Visit us on the World Wide Web at:
www.callistoreference.com

ISBN: 978-1-64116-126-8 (Hardback)

Cataloging-in-Publication Data

Epigenetics and chromatin : advanced concepts of cell biology / edited by Lily Kingsley.
    p. cm.
Includes bibliographical references and index.
ISBN 978-1-64116-126-8
1. Cytology. 2. Epigenetics. 3. Chromatin. 4. Genetics. I. Kingsley, Lily.
QH581.2 .E65 2019
571.6--dc23

Table of Contents

Preface

This book has been an outcome of determined endeavour from a group of educationists in the field. The primary objective was to involve a broad spectrum of professionals from diverse cultural background involved in the field for developing new researches. The book not only targets students but also scholars pursuing higher research for further enhancement of the theoretical and practical applications of the subject.

Epigenetics studies heritable genetic changes generated by a change in gene function, generated through processes like DNA methylation, histone modification, etc. Epigenetic changes alter the microstructure of DNA or the associated chromatin proteins leading to gene activation or silencing. Chromatin is a macromolecular complex, consisting of DNA, RNA and protein found in cells. Its primary functions are packaging DNA into more compact and denser shape, preventing DNA damage, controlling gene expression and DNA replication. The structure of chromatin is subject to change depending on the stage of the cell cycle. The aim of this book is to present researches that have transformed the understanding of epigenetics and chromatin. It is a compilation of chapters that discuss the most vital concepts and emerging trends in these fields. It will be a valuable source of reference for students and researchers alike.

It was an honour to edit such a profound book and also a challenging task to compile and examine all the relevant data for accuracy and originality. I wish to acknowledge the efforts of the contributors for submitting such brilliant and diverse chapters in the field and for endlessly working for the completion of the book. Last, but not the least; I thank my family for being a constant source of support in all my research endeavours.

Editor

Sex-specific chromatin landscapes in an ultra-compact chordate genome

Pavla Navratilova[1†], Gemma Barbara Danks[1†], Abby Long[2], Stephen Butcher[2], John Robert Manak[2] and Eric M. Thompson[1,3*] (ID)

Abstract

Background: In multicellular organisms, epigenome dynamics are associated with transitions in the cell cycle, development, germline specification, gametogenesis and inheritance. Evolutionarily, regulatory space has increased in complex metazoans to accommodate these functions. In tunicates, the sister lineage to vertebrates, we examine epigenome adaptations to strong secondary genome compaction, sex chromosome evolution and cell cycle modes.

Results: Across the 70 MB *Oikopleura dioica* genome, we profiled 19 histone modifications, and RNA polymerase II, CTCF and p300 occupancies, to define chromatin states within two homogeneous tissues with distinct cell cycle modes: ovarian endocycling nurse nuclei and mitotically proliferating germ nuclei in testes. Nurse nuclei had active chromatin states similar to other metazoan epigenomes, with large domains of operon-associated transcription, a general lack of heterochromatin, and a possible role of Polycomb PRC2 in dosage compensation. Testis chromatin states reflected transcriptional activity linked to spermatogenesis and epigenetic marks that have been associated with establishment of transgenerational inheritance in other organisms. We also uncovered an unusual chromatin state specific to the Y-chromosome, which combined active and heterochromatic histone modifications on specific transposable elements classes, perhaps involved in regulating their activity.

Conclusions: Compacted regulatory space in this tunicate genome is accompanied by reduced heterochromatin and chromatin state domain widths. Enhancers, promoters and protein-coding genes have conserved epigenomic features, with adaptations to the organization of a proportion of genes in operon units. We further identified features specific to sex chromosomes, cell cycle modes, germline identity and dosage compensation, and unusual combinations of histone PTMs with opposing consensus functions.

Keywords: Histone, Enhancer, Spermatogenesis, Polycomb, Dosage compensation, Heterochromatin, Transposable elements, Endocycle

Background

Histone proteins, which package genomic DNA, provide multiple sites for covalent posttranslational modifications (PTMs) by evolutionarily conserved histone modifiers that form multimeric complexes and cooperate with non-histone proteins [1, 2]. Histone PTMs are associated with chromatin dynamics linked to transcription, replication, DNA repair, recombination, chromosome segregation and other mitotic and meiotic processes [3, 4]. Importantly, in the germ line, they help to secure correct transgenerational inheritance, setting the stage for early embryonic development [5, 6]. Combinatorics of histone PTMs, proposed to constitute a "histone code" [7], are part of the mechanism through which a single genome generates a variety of cell types and states that respond to developmental and environmental cues. Prevalent combinations of modifications, referred to as "chromatin states," correlate with specific functional regions of the genome, and many appear to be conserved among eukaryotes [8–12]. To date, however, only a few metazoan

*Correspondence: Eric.Thompson@uib.no
†Pavla Navratilova and Gemma Barbara Danks contributed equally to this work
1 Sars International Centre for Marine Molecular Biology, University of Bergen, 5008 Bergen, Norway
Full list of author information is available at the end of the article

epigenomes have been studied in detail [10, 13–15], often using cell lines or heterogeneous cell/tissue populations from organisms, with the exception of in vivo cell population studies that focused on a few histone PTMs [16–19]. Here, we present the germline epigenomes of the chordate *Oikopleura dioica*, a member of the lineage that comprises the closest living relatives to vertebrates [20].

Oikopleura dioica is a semelparous pelagic tunicate (Urochordate, Appendicularian) with a simple chordate body plan and short, 6-day life cycle [21]. Several major developmental transitions are accompanied by switches between mitotic and endocycling cell cycle modes in both somatic tissues and the ovary [22–24]. *O. dioica* has the smallest metazoan genome sequenced to date, organized in a haploid complement of 3 chromosomes. At 70 Mb, it is ~44-fold smaller than the human genome despite maintaining >18,000 protein-coding genes [25] compared to ~20,000 in humans [26, 27]. Introns are frequently very small (peak at 47 bp; only 2.4% >1 kb), as are intergenic spaces (53% <1 kb). One quarter of the gene complement is organized into operons [28], and *trans*-splicing of a short spliced-leader (SL) RNA occurs at the 5' ends of 39% of protein-coding genes [29]. Transposable elements (TEs) form a significant proportion of vertebrate genomes, but most vertebrate TE families are absent in *O. dioica* and the density of TEs is low, with most concentrated on the gene-poor Y-chromosome [25]. Major clades of non-LTR (long terminal repeat) retrotransposons are missing from the *O. dioica* genome, but it has variety of LTR retrotransposons from the *Ty3/gypsy* group, divergent from those found in other organisms, as well as *Dictyostelium* intermediate repeat sequence 1 (DIRS1) and Penelope-like elements [30]. These autonomous elements carry an *env* gene and are expressed in a variety of *Oikopleura* tissues including germline-associated cells [31]. A comprehensive developmental transcriptome for *O. dioica* has been assembled and includes ovary and testes samples [32]. The full histone complement and associated PTMs have also been characterized, showing conservation of histone variants and a histone modification repertoire comparable to vertebrates [33].

Regions of the *O. dioica* genome that have potential regulatory function (introns and intergenic regions) have been reduced, often to the order of one nucleosome in size. How this compaction affects long-range enhancer-mediated gene regulation [34] and the epigenetic inheritance of chromatin domains through replication, mitosis and trans-generationally [35] is unknown. Polycomb complexes (PRC 1 and 2), via the tri-methylation of histone 3 on lysine 27 (H3K27me3), govern core mechanisms of metazoan epigenome heritability, organization and developmental dynamics [36]. Polycomb

complexes also function in sex chromosome inactivation during dosage compensation [37]. PRC1 is an ancient complex and a determinant of cellular stemness [38], but its canonical composition has been reduced in *O. dioica* and nematodes, possibly correlating with limited cellular plasticity and lack of regeneration [39]. A number of Polycomb complexes and modes of recruitment and function exist, but these vary in the extent to which they have been characterized [40, 41]. *O. dioica* is an interesting model in which to investigate non-canonical Polycomb complexes and their functions in a rapidly evolving lineage.

Oikopleura dioica is unusual among tunicates in that it has genetically determined separate sexes with heterogametic (XY) males and homogametic (XX) females. Organisms with heterogametic sex chromosomes have evolved dosage compensation mechanisms to equalize the abundance of transcripts produced by the single X-chromosome in males and the double X-chromosome in females [42, 43]. In male mammals, flies and worms, transcription of genes on the X-chromosome is upregulated. In female mammals one X-chromosome is inactivated, and in hermaphrodite worms, expression from X-chromosomes is downregulated. Different underlying components and molecular mechanisms behind the recruitment and targeting of dosage compensation complexes as well as the resulting changes in chromatin have been well documented [44]. A common feature is that complexes with other functions in the organism, such as Polycomb in mammals, DCC in worms, or fly MSL, have been recruited for domain regulation of X-linked genes. It has thus far not been established that dosage compensation occurs in *O. dioica*, nor through what mechanism it might be achieved, if it does occur.

Here, we sampled *O. dioica* testis, at early day 6 (a few hours before germ-cell release), when it is a syncytium of mitotically proliferating, transcriptionally active, spermatogonia nuclei (Additional file 2: Fig. S1). A proportion of active somatic genes including housekeeping, self-renewal and proliferation genes are required for mitosis and germline reprogramming. At the same time, a testis-specific transcriptional program is required for initiating spermatogenic gene transcription, setting up the transmission of epigenetic memory and poising developmental genes for expression following fertilization. Transitioning between these processes is rapid in the *O. dioica* male germ line, but meiosis itself occurs only in late day 6, about 2 h before spawning. The day 6 *O. dioica* ovary consists of one single giant cell (the coenocyst), where endocycling nurse nuclei share a common cytoplasm with meiotic nuclei arrested in prophase I [24, 45] (Additional file 2: Fig. S1). These two populations of nuclei occur in equivalent numbers, but the ploidy of nurse nuclei (200C) compared to that of the prophase I meiotic nuclei (4C) [22] means that the nurse nuclei dominate (98% contribution) the chromatin content of the ovary. Nurse nuclei are terminally differentiated and help direct oocyte maturation and cellularization. A large portion of their transcriptional output is maternal mRNA that is subsequently stocked in the oocytes. Unlike testis or oocyte meiotic nuclei, nurse nuclei do not traverse mitosis and do not need to re-establish post-mitotic epigenetic landscapes or undergo germline-specific gene repression.

We extracted homogeneous nuclear populations from testes and ovaries and profiled key histone PTMs and nonhistone chromatin-associated proteins to explore chromatin state landscapes in *O. dioica* germ lines and their relationship to genome compaction, sex chromosomes and autosomes. We found RNAPII activity-linked signatures known from other metazoans. Chromatin domains were generally reduced in size, but we did identify regions with histone PTMs typical of enhancers. The ovarian, nurse nuclear epigenome consisted of large domains of active transcription and a general lack of repressive heterochromatin. The male germline epigenome contained chromatin states specific to the spermatogenic program and the X-chromosome and included an intriguing combination of histone PTMs on the Y-chromosome, which may be involved in regulating the activity of transposable elements. This work provides the first comprehensive view of a protochordate epigenome, providing insight into its organization in two sex-specific tissue samples.

Results

Oikopleura histone PTMs and their combinations

We profiled the following in maturing *O. dioica* testes and ovaries: 19 histone PTMs (Additional file 3: Table S1), using native ChIP-chip; CTCF, p300 and RNA polymerase II (RNAPII) occupancy, using cross-linked ChIP-chip; and 5-methylcytosine DNA methylation (5 mC), using meDIP-chip. Sampled testes were in the mitotically dividing pre-meiotic (spermatogonia) stage, whereas ovaries were dominated by endocycling, transcriptionally active nurse nuclei. We focused on histone H3 and H4 PTMs and related these profiles to gene expression levels [32], *trans*-splicing status [28], chromosomal location, and GC content of promoters (Fig. 1; Additional file 1: Supplemental Results; Additional file 2: Fig. S2). We compared our results to those in human cells, *Saccharomyces cerevisiae*, *Drosophila melanogaster* and *C. elegans* (Table 1) [8, 10, 12–15, 46–52].

Combinatorial deposition of chromatin marks was analyzed by classifying testis and ovary chromatin into 15 states (Fig. 2a; Additional file 3: Tables S2 and S3), learnt jointly across both cell types, using a multivariate hidden Markov model (chromHMM) [53]. These 15 states were reproducible when processing ovary and testes datasets independently (Additional file 2: Fig. S3). Functions were assigned to jointly learned states according to their enrichments in an array of transcriptionally repressed and active genomic features (Fig. 2c). Feature annotation of the separately learnt 15-state models underscored some different uses of individual modifications in the testis versus ovary (Additional file 2: Figs. S3 and S4). We were also able to resolve 50 biologically meaningful chromatin sub-states (Additional file 1: Supplemental Results; Additional file 2: Fig. S5; and Additional file 3: Table S4) including a Polycomb-repressed state (state 46: enriched for H3K27me3 and marks promoters of silent developmental genes) that was less pronounced in the 15-state models. Unless stated otherwise, all subsequent analyses were based on the main functional chromatin states captured by the jointly learnt 15-state model.

Four chromatin states were specific to the ovary (1, 3, 4, 5), and four were specific to the testis (7, 8, 11, 12). Specific chromatin states were associated with active promoters (states 5, 8 and 9), transcription elongation (states 3, 4, 6, 7) and silent regions (states 1, 10–12, 14–15), which included states specific to the Y-chromosome (state 11) and silent transcription factors (TFs) (state 1). State 2, which had no enrichment of any profiled modifications, covered 54% of the ovary and 40% of the testis genomes (Fig. 2b), similar to the sum of "weak signal" states calculated for human (45%), fly (35%) and worm (45%) [12]. We grouped active and silent genes by GO terms and calculated chromatin state enrichments on their promoters and gene bodies to reveal differential use of chromatin states on genes with different biological functions (Fig. 3; Additional file 3: Table S5).

We compared chromatin state domain widths in *O. dioica* to those found in nine human cell lines [9] and found that both activating and repressive domains were significantly narrower in *O. dioica* (Mann–Whitney test: $W = 1.189401e+12$, p value <2.2e−16; Additional file 2: Fig. S6). Domains spanning over 7 nucleosomes were largely absent (Fig. 4a). The absence of large repressive regions in *O. dioica* was notable (Fig. 4b; Additional file 2: Fig. S6) and supports previous observations of a decline in heterochromatin coverage with decreasing genome size [12].

We found homologs of human histone modifier proteins and extracted gene expression values for these homologs from a previously published transcriptomic dataset [32]. The expression patterns of the complement of *O. dioica* histone modifiers in testes and ovaries corresponded to the presence or absence of their associated histone PTMs. Gene duplications of some modifiers and

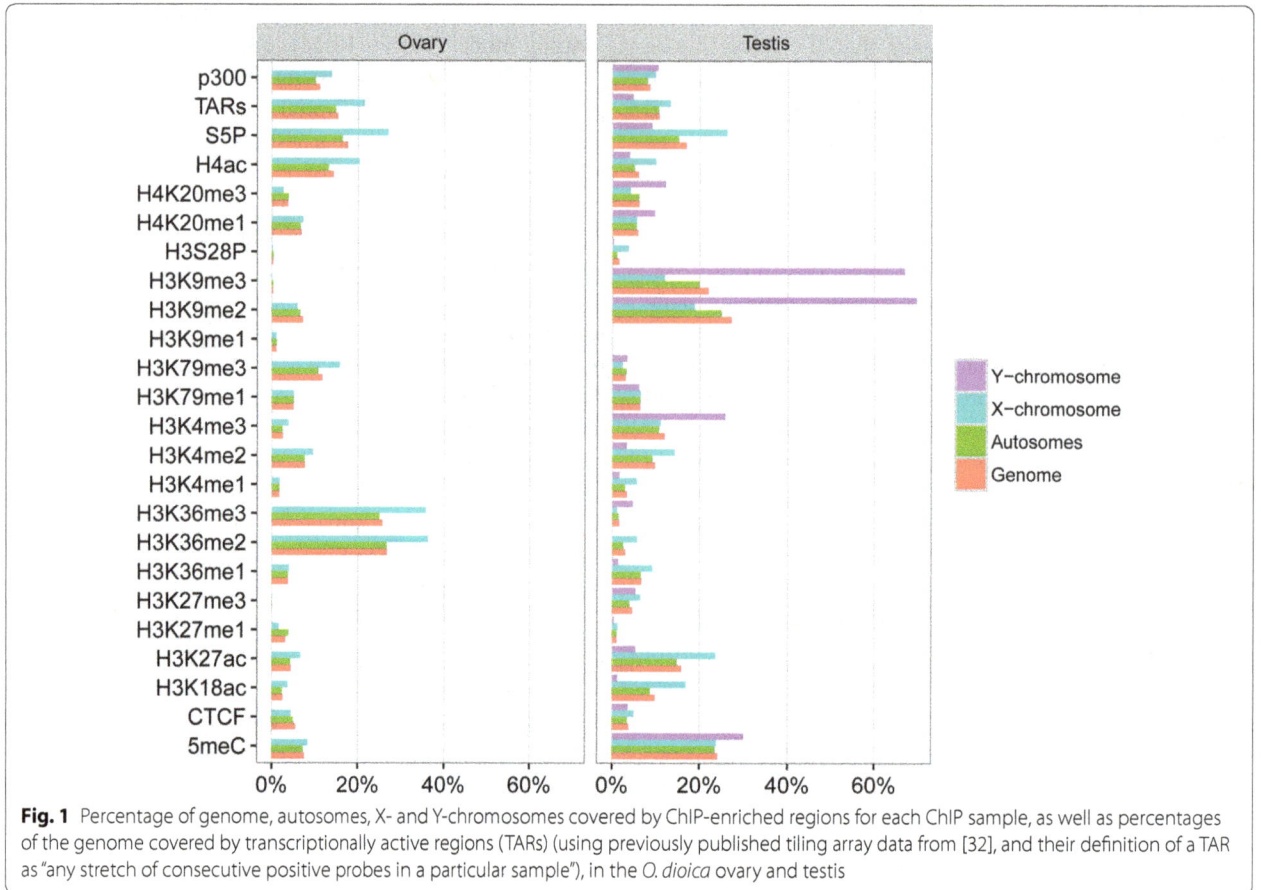

Fig. 1 Percentage of genome, autosomes, X- and Y-chromosomes covered by ChIP-enriched regions for each ChIP sample, as well as percentages of the genome covered by transcriptionally active regions (TARs) (using previously published tiling array data from [32], and their definition of a TAR as "any stretch of consecutive positive probes in a particular sample"), in the *O. dioica* ovary and testis

losses of others, particularly those related to DNA repair, hormone response, Hox gene activating, RA response and histone methyltransferases, reflected *O. dioica's* reduced NHEJ DNA repair toolkit, altered Polycomb complex complement and dispersion of developmental gene clusters (Fig. 5; Additional file 3: Table S6; examples and details in Additional file 1).

Transcriptionally permissive chromatin in the ovary

The ovarian chromatin landscape was dominated by ovary-specific states 3–5 (Fig. 2a, b). These states were enriched for H3K36me2 and H3K36me3 (typical of transcribed gene bodies in metazoans [46]), covered 25% of the genome (Fig. 2b), overlapped RNAPII-occupied regions and correlated with high levels of transcription. Active genes in the ovary were mostly related to housekeeping functions, maternal transcription and oogenesis (Fig. 3; Additional file 3: Table S5). *Trans*-spliced genes, despite no significant differences in mean expression levels compared to non-*trans*-spliced genes, had higher enrichment for these two methylation marks (Additional file 2: Fig. S2A). States 3–5 were also more enriched on operons (a subset of *trans*-spliced genes) (Fig. 2c). State

3 was distinct from state 4 in its higher enrichment of H4ac, higher prevalence in UTRs compared to gene bodies and higher enrichment in regions of RNAPII occupancy and transcribed operons. Interestingly, state 3 was also enriched at the TSS in a subset of silent genes annotated with GO terms related to nutrient response (cluster 17, Fig. 3). This may reflect RNAPII pausing in genes that regulate oocyte production in a nutrient-dependent manner.

Promoters of operon genes were enriched for promoter state 5, characterized by H4ac, H3K27ac, H3K4me2/3 and H3K36 methylations. The typically active gene body marks (H3K36me and H4ac) in these promoters may reflect the uncertainty of TSS annotations for a subset of *trans*-spliced operon transcripts (since a stretch of 5' sequence is removed from mRNAs and replaced by the SL RNA). Enrichment of H3K4me3, the hallmark of active promoters [54], was overrepresented at active promoters of operons compared to silent operon promoters (Fisher's test *p* value = 2.044×10^{-10}) but not at regions surrounding the 5' ends of expressed downstream operon genes. This provides evidence that these are indeed co-transcribed

Table 1 Comparative residency of histone PTMs on genomic features of diverse eukaryotes

Histone PTM	Human	Saccharomyces cerevisiae	Caenorhabditis elegans	Drosophila melanogaster	Oikopleura dioica ovary	Oikopleura dioica testis
H3K18ac	Promoters, enhancers	Promoters and 5' end of genes	Promoters and 5' end of genes	Promoters, enhancers	Promoters and 5' end of genes	Promoters and 5' end of genes
H3K27ac	Promoters, enhancers	Promoters and gene bodies	Promoters, enhancers	Promoters, enhancers	Promoters, enhancers	Promoters and gene bodies
H4ac	Promoters and gene bodies	Transcription start sites	Promoters and gene bodies	Active and silent gene bodies, enriched on X	Promoters and gene bodies	Silent and active HGP genes, active genes on X
H3K4me1	Promoters, enhancers	Transcription end sites	Promoters, enhancers	Promoters, enhancers	Promoters	Promoters
H3K4me2	Promoters, enhancers	Gene bodies	Promoters, enhancers	Promoters, enhancers	Active and silent LGP, active HGP	Active and silent LGP, active HGP
H3K4me3	Active and bivalent promoters	Promoters	Promoters	Promoters	LGP	Active and silent LGP, active HGP, silent broad regions on Y
H3K9me1	Gene bodies, esp. active but also silent, enhancers	not present	Gene bodies enriched on X, enhancers	Gene bodies, esp. active but also silent	Gene bodies, esp. active but also silent	Gene bodies enriched on X
H3K9me2	Gene bodies, mobile DNA	not present	Silent and active genes, mobile DNA	Silent and active genes, mobile DNA	Both active and silent promoters, esp. LGP	Active and silent promoters, esp. LGP, enriched on Y and mobile DNA
H3K9me3	Silent regions	not present	Silent regions, mobile DNA	Silent regions, mobile DNA	Silent regions	Silent regions, enriched on Y and mobile DNA
H3K27me1	Gene bodies	not present	Gene bodies, enriched on X	Gene bodies	Gene bodies, enriched on SL	Gene bodies, enriched on X
H3K27me3	Silent, Polycomb-repressed regions	not present	Gene bodies	Gene bodies	Active and silent genes, enriched on X	Active and silent genes, enriched on X
H3K36me1	5' end of highly active HGP genes	Gene bodies	Gene bodies enriched on X	Gene bodies	5' end of HGP genes	5' end of HGP genes
H3K36me2	Gene bodies	Gene bodies	Gene bodies	Gene bodies	Gene bodies, enriched on SL	Gene bodies
H3K36me3	Gene bodies	Gene bodies	Gene bodies	Gene bodies	Gene bodies	Active and silent gene bodies, peaks at silent LGC promoters, mobile DNA, Y
H3K79me1	Gene bodies	Gene bodies	Gene bodies, peaks towards 3'	Gene bodies	Gene bodies	Active and silent gene bodies
H3K79me3	Active and silent promoters and gene bodies	Active gene bodies; Telomeres	Gene bodies, peaks towards 5'	Active gene bodies, peak downstream of promoter	Active and silent gene bodies	Gene bodies
H3S28P	Stress response gene promoters	NA	NA	NA	Active gene bodies, esp. SL, both silent and expressed LGP	Active and silent genes, enriched on X
H4K20me1	5' of active and silent genes. X	not present	Gene bodies enriched on X	Gene bodies	Gene bodies	Active and silent gene bodies
H4K20me3	Silent regions	not present	Silent regions	Silent regions	Silent regions, enriched on HGP genes	Silent regions, enriched on Y

Assignments of features marked by histone PTMs for species other than *Oikopleura dioica* were based on reference literature as follows: human [12, 13, 46–49]; *Saccharomyces cerevisiae* [50–52]; *Caenorhabditis elegans* [12, 14, 15]; *Drosophila melanogaster* [8, 10, 12]. Associations of individual histone modification *Oikopleura dioica* were assigned based on assessment of enrichment plots shown in Supplemental Figure S1. These plots show the mean signal intensity around gene start and end sites with genes grouped according to their expression level, the GC content of their promoter, their trans-splicing status and chromosomal locations. We used the error bars indicating 95% confidence intervals on the mean to define (by the lack of their overlap) relative enrichments between compared gene sets as significant. Colors indicate transcriptional activity of features marked by PTMs: green = transcribed; red = repressed; yellow = both active and silent. Esp. = especially; X = X-chromosome; Y = Y-chromosome; HGP/LGP = high/low GC content promoter; SL genes = genes whose transcripts are subject to trans-splicing of the splice leader (SL) sequence; N = not analyzed

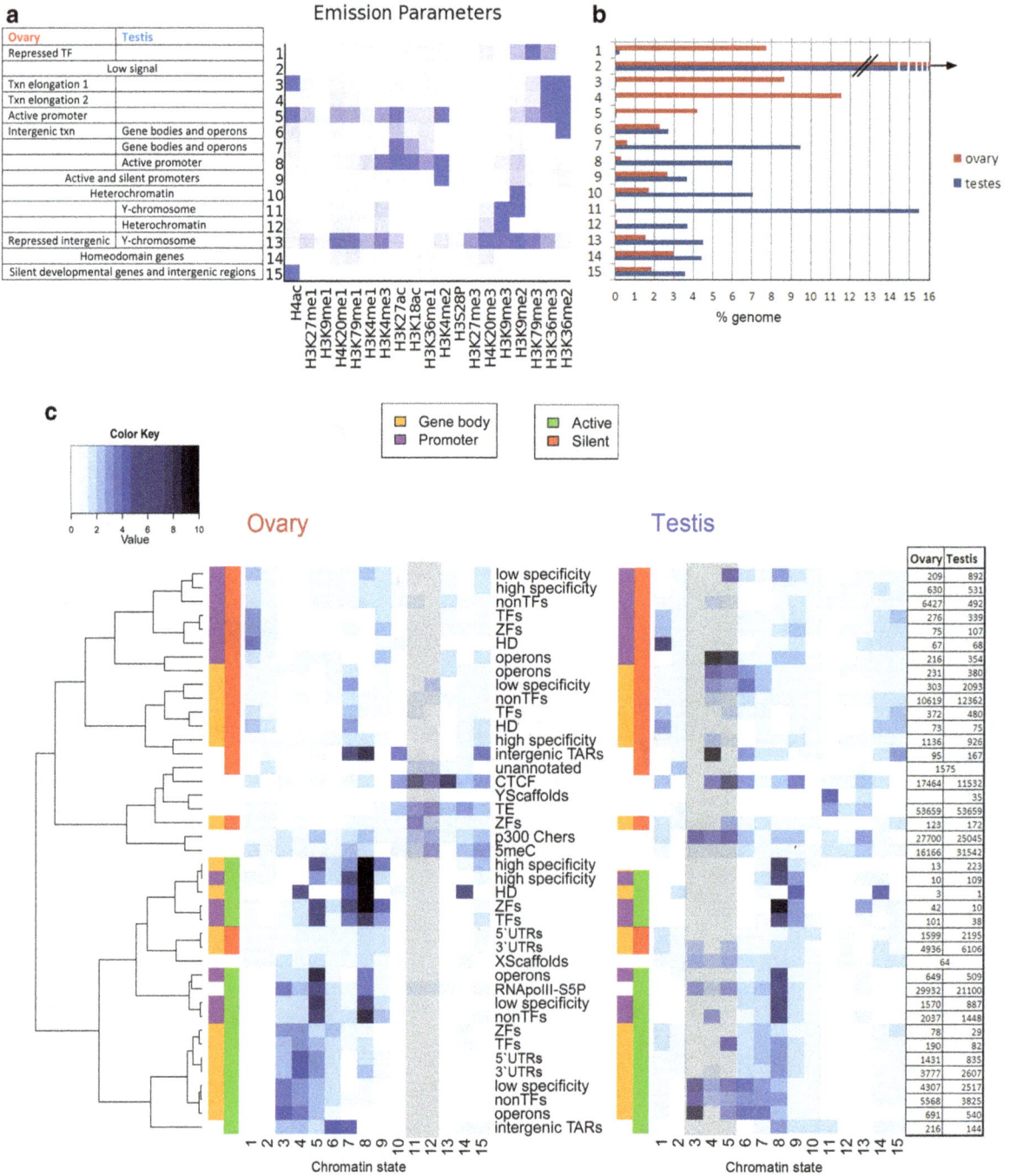

Fig. 2 Chromatin states in the *O. dioica* ovary and testis reveal two distinct epigenetic landscapes. Heatmap of model emission parameters (**a**) for 15 chromatin states learned across the genome, using both samples (see also Additional file 3: Table S3). Proportions of the genome in the ovary and testis covered by each chromatin state (**b**) show sex/tissue specificity of chromatin states. Heatmaps (**c**) visualize the fold enrichments of each chromatin state (*columns*) for a set of genomic features (*rows*), as listed, in the ovary and testis. This facilitates assignment of putative biological function(s) of each chromatin state in each tissue. *Gray shading* indicates states that cover below 0.1% of the genome (70,000 bp) in each tissue. These low-coverage states nevertheless have large fold enrichments over certain genomic features, e.g., state 3 has low coverage in the testis but is found more often than expected by chance at active operons. Features are clustered according to their correlation in the ovary, and this ordering is used for the testis heatmap to allow comparison. The table gives the numbers of each feature in each sample. Gene bodies (*orange*) and promoter regions (*purple*) are split into active (*green*) and silent (*red*) states in the respective ovary and testis tissues. Txn, transcription; TF, transcription factor; ZF, zinc finger protein; HD, homeodomain protein, high- and low-specificity genes according to breadth of expression across development (see "Methods" section); TAR, transcriptionally active region; TE, transposable element; unannotated, regions lacking annotation or transcription; 5meC, methylcytosine; RNApolII-S5P, serine 5 phosphorylated RNA polymerase II

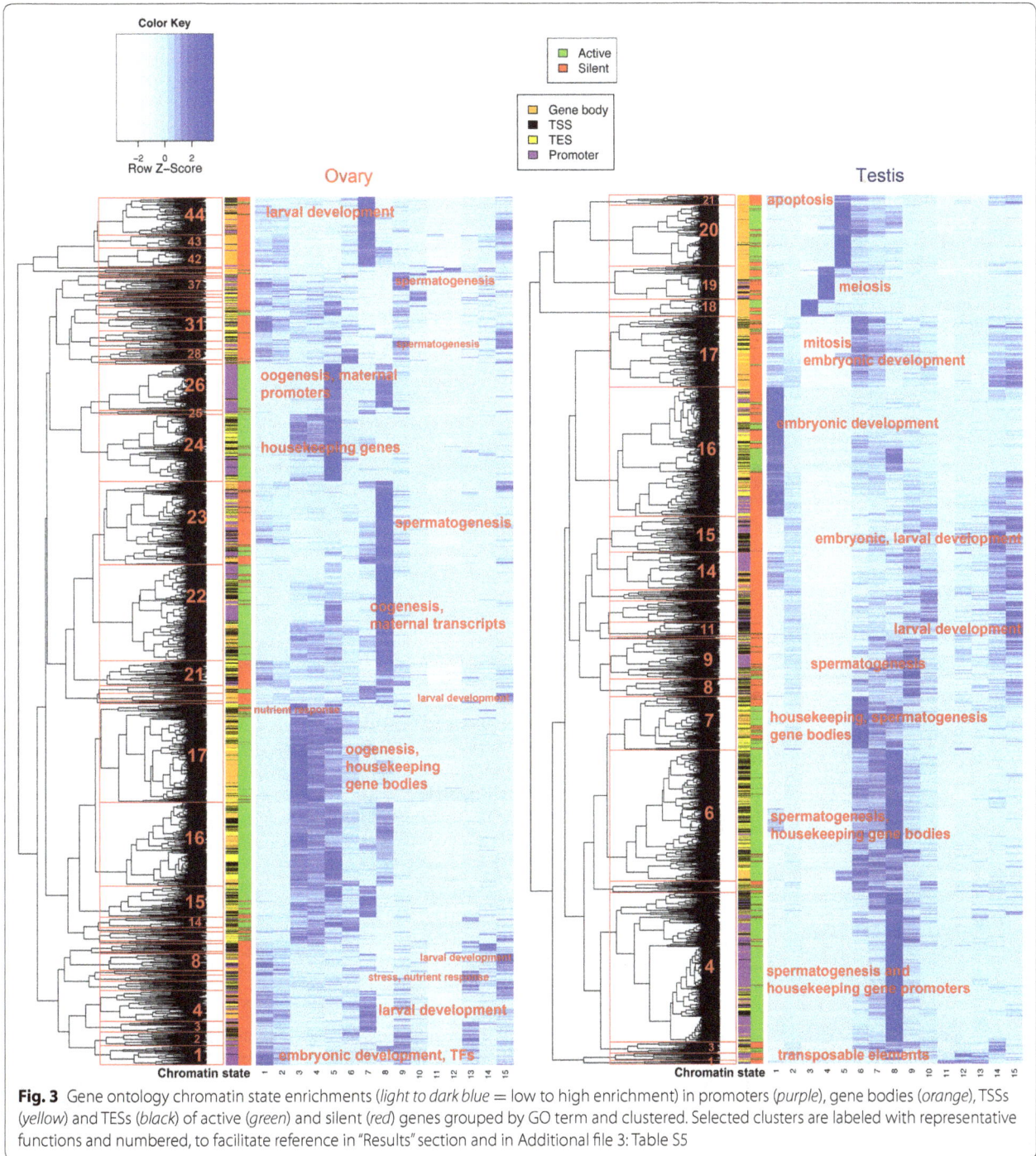

Fig. 3 Gene ontology chromatin state enrichments (*light to dark blue* = low to high enrichment) in promoters (*purple*), gene bodies (*orange*), TSSs (*yellow*) and TESs (*black*) of active (*green*) and silent (*red*) genes grouped by GO term and clustered. Selected clusters are labeled with representative functions and numbered, to facilitate reference in "Results" section and in Additional file 3: Table S5

from a single upstream promoter. Active promoter state 8 (H3K27ac, H3K18ac, H3K4me2/3) was enriched in active genes related to oogenesis and embryonic development (presumably maternal transcripts stocked in oocytes: cluster 22, Fig. 3), but also marked a group of silent spermatogenesis genes in the ovary (cluster 23, Fig. 3).

RNAPII pausing

Pausing of RNAPII is a pervasive feature of promoters in metazoans that facilitates integration of multiple cellular signals, poising genes for rapid expression and/or synchronous activation [55]. In *O. dioica*, 15% of the ovary genome and 8% of the testis genome are transcribed [32]. Serine-5 phosphorylated RNAPII ChIP-chip, however,

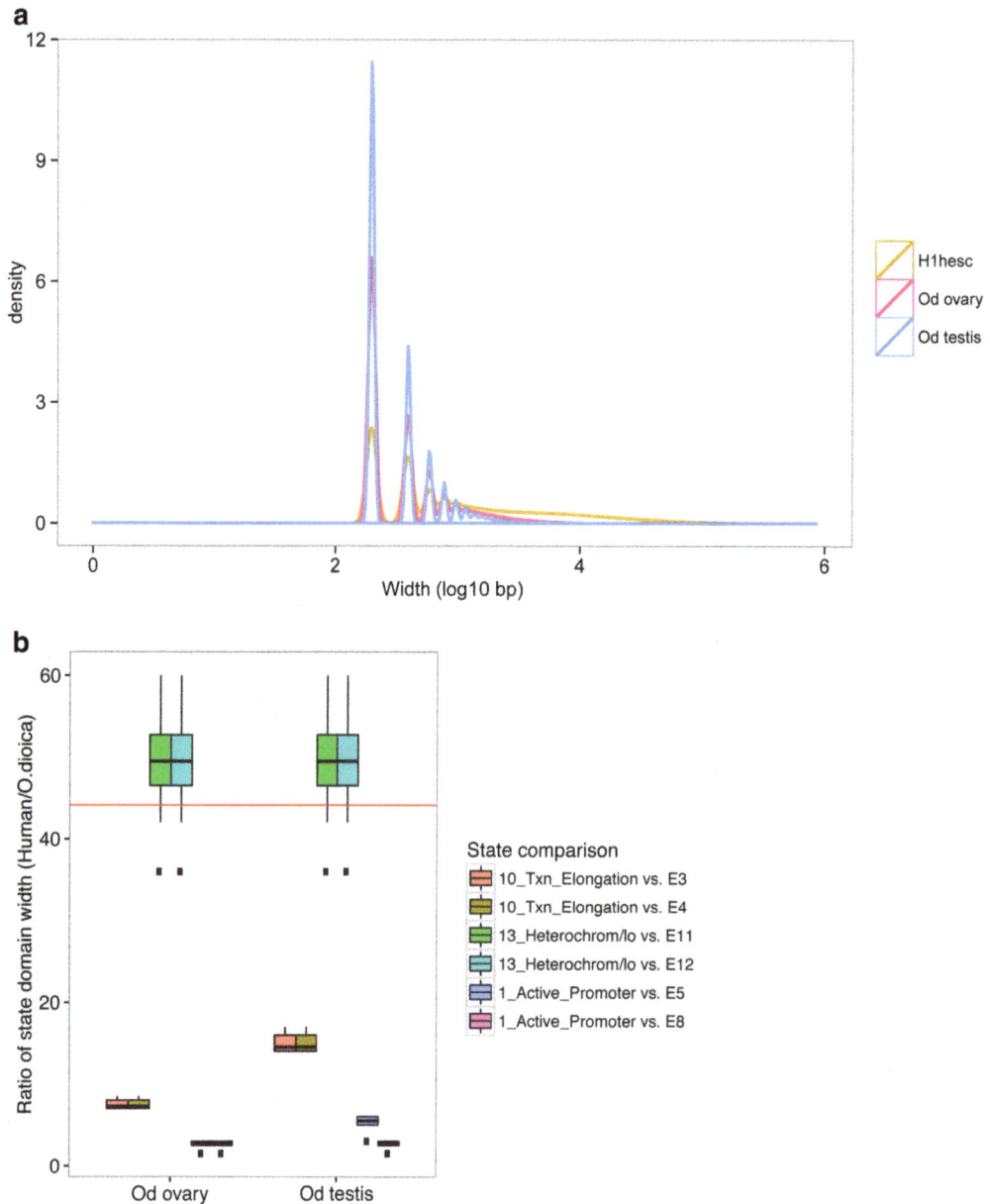

Fig. 4 Compact chromatin state domains in the *O. dioica* epigenome. **a** Distributions of all chromatin state domain widths for each cell type in *O. dioica* compared to those in human embryonic stem cells (H1-hESC) [92]. *O. dioica* domain widths were adjusted for the difference in resolution (50 vs. 200 bp) by rounding *O. dioica* widths up to the nearest 200 bp. **b** Comparison of the ratios of human state domain widths for active promoters, transcriptional elongation and heterochromatin (see Additional file 2: Fig. S6) to corresponding state domains in *O. dioica*. *Each box* summarizes the ratios of median domain widths (*y*-axis) for the human cell lines relative to the similar domains in each *O. dioica* tissue (*x*-axis). The *red line* indicates the ratio of the human genome size to that of *O. dioica* (44-fold smaller). Relative to the genome compaction, heterochromatin domains in *O. dioica* are disproportionately more compact

showed near-equal occupancy across the genome: 17.6% of the genome in the ovary and 17.0% in the testis were enriched for RNAPII S5P (Fig. 1), suggesting that paused RNAPII is frequent in *O. dioica*. We identified 1106 genes in the ovary and 1389 genes in the testis, which have proximal promoter regions that are repressed but associated with RNAPII (16 and 18% of silent genes in the ovary and testis, respectively, which is higher than expected in the testis $\chi^2 = 12.17$, *df* 1, *p* value $= 4.85 \times 10^{-4}$). Most of these genes encode proteins for RNA-processing functions, cell cycle and to a lesser extent, development, in both samples, and for

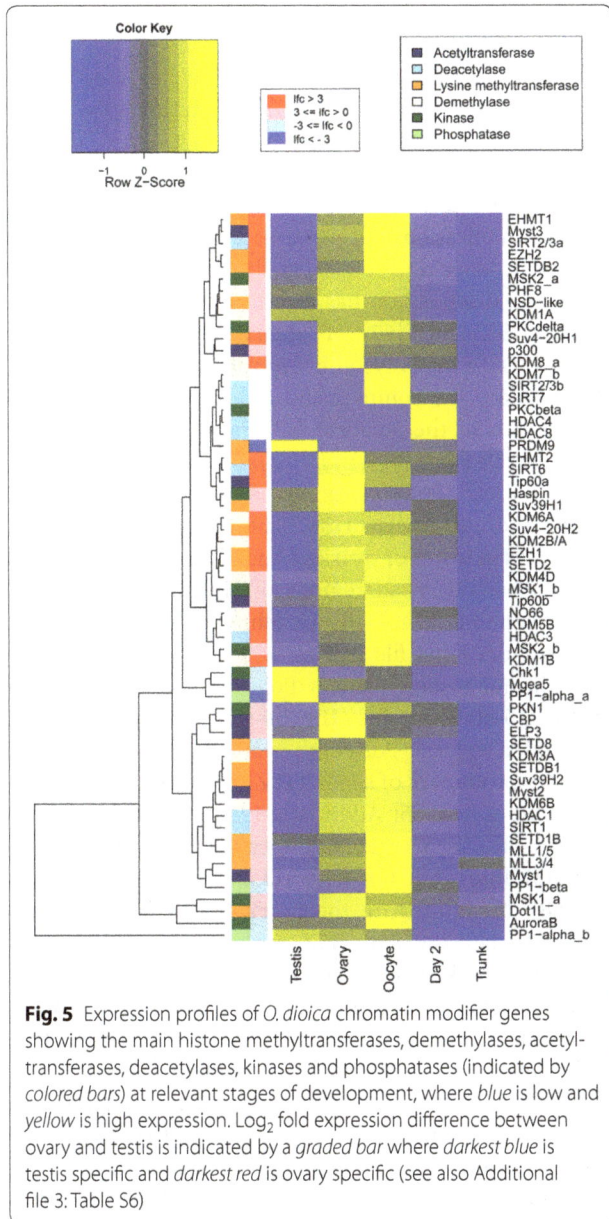

Fig. 5 Expression profiles of *O. dioica* chromatin modifier genes showing the main histone methyltransferases, demethylases, acetyltransferases, deacetylases, kinases and phosphatases (indicated by *colored bars*) at relevant stages of development, where *blue* is low and *yellow* is high expression. Log$_2$ fold expression difference between ovary and testis is indicated by a *graded bar* where *darkest blue* is testis specific and *darkest red* is ovary specific (see also Additional file 3: Table S6)

the co-activator p300, a tissue-specific enhancer-binding histone acetyltransferase [58, 59], and found 353 and 2281 strong candidates of active (p300-bound) enhancers in the ovary and testis, respectively. Sites bound by p300 frequently co-localized with the E2F transcription factor and CTCF (Additional file 2: Fig. S8). Chromatin states enriched at p300-binding sites (Fig. 2c) included states 5 and 8, which contained typical enhancer PTMs. Enhancer marks were present to a lesser extent in state 13, which comprises a complex combination of histone PTMs, and heterochromatic state 12, which may represent repressed regulatory elements. Highly conserved elements (HCNEs), indicative of regions with conserved regulatory function [60], were identified around genes encoding developmental TFs [25]. HCNEs were enriched in states 1 and 15 in the testis and 1, 6 and 7 in the ovary; states also associated with developmental TFs (Fig. 2c; Additional file 2: Fig. S9). As expected, given that developmental TFs are primarily active during embryogenesis, we did not find any overlap of HCNEs with our candidate enhancers in the gonads. Together the data indicate that enhancers are present in the *O. dioica* genome, despite its high compaction, reshuffling and loss of some regions of evolutionarily conserved synteny. More enhancers would likely be detected in early embryonic and larval samples when developmental regulatory networks are activated.

Marking of repressed developmental regulatory genes differs in ovary and testis

Repressed TF and zinc finger (ZF) genes in the ovary were associated with chromatin state 1 (enriched for H3K79me3, together with H3K36me3, H3K9me2 and H4K20me3) (Fig. 2a, b; Additional file 3: Table S3). State 1, and H3K79me3 in general, was found almost exclusively in the ovary (Figs. 1, 2a; Additional file 2: Fig. S3) and was enriched in the long introns of silent genes (Additional file 2: Fig. S10). It is interesting, therefore, to note that 3 testis-specific H3t variants have substantial modification of the Dot1 recognition site for H3K79 methylation [33]. Repressed TF, ZF and homeodomain (HD) genes, particularly in the testis, had an unusual signature: strong enrichment for H4ac that resulted in assignment to state #15 (Figs. 2a, c, 3; Additional file 3: Tables S3 and S4). Extended analysis to the 50-state model confirmed these findings and revealed additional marks such as H4K20me1 in addition to H4ac (state 14; Additional file 2: Fig. S5A, C), H3K4me1, H3K27me3 or H3K36me1 (states 43–46). These loci correspond to typical Polycomb targets known from metazoan somatic cells [12]. Interestingly, a group of silent TFs, HD and ZF genes with functions in larval development had gene bodies enriched for state 7 in the ovary (Fig. 3, clusters, 3, 4, 42, 43, 44) and both states 6 and 7 in the testis (Fig. 3, cluster

nutrient response functions, particularly in the ovary (Additional file 2: Fig. S7). Such a mechanism is consistent with the observation that *O. dioica* can rapidly adjust its gamete output in response to shifting nutrient conditions (e.g., algal blooms) [28, 56, 57].

Enhancers in the compact *Oikopleura* genome
We identified 8070 regions of H3K4me1 enrichment that did not overlap annotated gene start sites in the ovary and 12,272 in the testis. Of these, 1087 and 6829 overlapped regions of H3K27ac enrichment in the ovary and testis, respectively, and constituted candidate enhancers. We intersected these with regions associated with

17). These data indicate that *O. dioica* testes deploy distinct epigenomic states (and possibly histone variants) in the control of important developmental regulatory loci.

Tissue-specific spermatogenesis-related chromatin states and a switch from SETD2- to NSD-mediated H3K36 methylation in the testis

State 8 was the most prevalent active promoter state in the testis, driving spermatogenesis-related genes (Fig. 3, testis clusters 4 and 6). This profile, to a lesser extent, corresponded to a general signature of tissue-specific gene promoters for the ovary as well. Transcriptional elongation on less tissue-specific expressed genes and operon genes in testis mapped to chromatin states 6 and 7 (Figs. 2a, c, 3b), which consisted of H3K27 and H3K18 acetylations but differed in the presence (state 6) or absence (state 7) of H3K36me2 (Fig. 2a). Remarkably, 9.5% of the testis genome was transcriptionally elongated in the absence or with low enrichment of H3K36me2/3 (Fig. 2a, b). Expression data showed that the SETD2 ortholog (the main histone methyltransferase generating H3K36me3 from H3K36me2) was repressed in the testis and RT-PCR revealed alternative splicing of the SET domain of this gene (not shown). The presence of H3K36me2, independent of gene activity, in state 6 (Fig. 2a) suggests a developmental switch where the SETD2 co-transcriptional deposition of H3K36me2/3 is replaced by an NSD-like H3K36-methyltransferase that is highly transcribed in the testis (Fig. 5; Additional file 3: Table S6). This resembles the switch from somatic met1 (worm SETD2 ortholog) to the germline mes-4 (worm NSD1 ortholog) in *C. elegans*. This switch is essential for production and maintenance of H3K36 methylation patterns during transcription in the parental germ line as well as for proper post-embryonic development of germ cells in the offspring [61].

A modified chordate Polycomb system

Homologs of core components of the canonical Polycomb complexes were not identified (RNF1, SUZ12, PCGF, SCMH1), and RA signaling is absent *O. dioica* [39, 62]. Several trithorax group proteins (MLL1, Ash1 and PRC-recruiting KDM2B) that activate Hox genes in a cell-specific manner in response to RA [41] were also not found in the *O. dioica* genome (Additional file 3: Table S6). We found, however, duplications for EZH2, which is responsible for PRC2 methyltransferase activity, and the EED subunit, which recognizes H3K27me3 and stimulates an activity-based feedback loop. Both EZH2 paralogs were abundantly transcribed in the endocycling ovary, but only weakly expressed or silent in the testis (Fig. 5; Additional file 3: Table S6). Taken together, these observations raise interesting questions on the

mechanism of Polycomb-mediated repression in this chordate lineage.

Prior to H3K27me3 deposition, PRC2 can be recruited in a number of ways that depend on composition of the complex, its interacting partners, target genes and the cell cycle and developmental stage [41, 63, 64]. In nematodes, H3K27me3 can remain on daughter chromatin through replication and serve as a template for PRC2-mediated deposition [65]. G9a- [66] or EZH2-mediated H3K27me1 deposition has also been reported as a step toward H3K27me3 [67]. Mitotic H3S28P creates an optimal substrate for Polycomb deposition [68, 69], but can also displace the complex and induce a methyl-acetylation switch of the adjacent K27 residue, depending on the cell cycle stage [70]. In *Drosophila* [71] and mammals [72], PRC exhibits a preference for CG-rich sequences, but we did not observe H3K27me3 enrichment on such sequences in testes or ovaries (Additional file 2: Fig. S2B).

We observed broad regions of H3K27me3 in the *O. dioica* genome at a lower enrichment level than other marks (Additional file 2: Figs. S11 and S12) and therefore did not resolve distinct Polycomb repressive states in the 15-state model. The 50-state model, however, resolved a typical Polycomb state 46, unambiguously marking promoters of silent TF, ZF and HD genes (Additional file 2: Fig. S5; Additional file 3: Table S5). In the testis, H3K27me3 regions co-localized with H3K27me1 and H3K4me1, but not with H3S28P (Additional file 2: Fig. S8). In contrast, in the ovary, H3K27me3 co-localized with H3S28P and H3K4me1 but not H3K27me1, the latter mark instead being strongly associated with transcriptional elongation and H3K36me2/3 (Additional file 2: Figs. S8 and S11). We also observed that H3K27me3 blocks were largely complementary to those of H3K36me2 in *O. dioica* (Additional file 2: Fig. S11A). It is known that mes-4 (NSD1)-dependent H3K36 methylation not only marks germ-cell-expressed genes in the germ line of adult nematodes, but also leads to exclusion of PRC2 (mes2/3/6) from germline-activated genes and repression of genes normally expressed in somatic cells, as well as genes on the X-chromosome [61, 73]. This mechanism would explain X-chromosome bias of H3K27me3 and complementarity of the two marks in the *O. dioica* testis although besides NSD1, H3K36me3 could in addition be deposited by the activity of male-specific PRDM9 [74]. Testis X-chromosome genes are less transcriptionally active than those in the ovary (cf. TARs in Fig. 1) despite similar RNAPII occupancy.

PRC2 as a central component of dosage compensation in *Oikopleura*

The *O. dioica* X-chromosome is gene rich and was transcriptionally active in both day 6 heterogametic (XY) males and homogametic (XX) females. Autosomal and X-chromosome genes have similar expression levels in both ovary and testis (mean X:A ratio: 1.1 in the ovary, 1.12 for constitutively expressed genes; and 0.88 in the testis, 0.99 for constitutively expressed genes; see also Additional file 2: Fig. S13), indicating dosage compensation. Several histone PTMs showed differential enrichment on autosomes versus the X-chromosome (Additional file 2: Fig. S2C). Ovarian X-chromosomes showed depleted levels of H3K27me1, the first step in Polycomb repression and memory, but were enriched in H3K27me3, the final Polycomb-repressed state. We interpret this as a possible dosage compensation mechanism.

In the testis, transcriptionally active promoters on the X-chromosome had a higher enrichment of H4ac, H3K27me1 and H3K4me1 compared to transcriptionally active genes on autosomes. Repressive H4K20me3, H3K9me2 and 3, strongly enriched on the Y-chromosome, were depleted from the X-chromosome as compared to autosomes (Additional file 2: Fig. S2C, D). On the other hand, the active promoter-associated H3K9me1 [12, 13] was enriched on male X-chromosome genes, similar to observations in *C. elegans* [14]. Enrichment of H3S28P in the testis was higher around X-chromosome genes compared to autosomes (Additional file 2: Fig. S2C). This may be related to Polycomb-mediated higher-order chromatin structures and/or higher residence/recruitment of PRC on this chromosome during mitotic divisions.

Together, our data revealed specific distributions of marks on sex chromosomes and are consistent with the Polycomb system being a central component of the *O. dioica* dosage compensation system.

H3K4me3 and H3K36me3 enriched heterochromatic states linked to mobile DNA elements

We resolved several distinct repressive chromatin states (11–13). While we found regions of mutual exclusion of the repressive marks H3K27me3 and H3K9me3 in *O. dioica* (Additional file 2: Fig. S12), we also found that they overlap in heterochromatin (Fig. 2a, state 13 in the 15-state model and Additional file 2: Fig. S5; states 18, 26, 29 in the 50-state model, Additional file 2: Fig. S8). This overlap was previously thought to be specific to worms [12], although cross talk between PRC and G9a/GLP in regulating a subset of genes has been found [66] and these H3K27 and H3K9-, mono- and di-methyltransferases are known to form multimeric complexes with tri-methyltransferases in human cells [2].

The *O. dioica* Y-chromosome was enriched for repressive states characterized by H3K9me3, H3K9me2 and H4K20me3 (Fig. 2a, states 11 and 13; Additional file 2: Fig. S12). In the testis, mobile DNA elements were also

enriched for these states. State 11 (H3K9me2/3 heterochromatic marks) was enriched at regions of 5mC despite limited DNA methylation potential in *O. dioica* (DNMT1 and 3 absence) (Fig. 2a, c).

Beside the conserved role of H3K4me3 at active promoters in *O. dioica* (Fig. 2c, chromatin states 5 and 8), we also found broad regions of this mark within heterochromatin states 11, 12 and 13 (and states 16, 19 and 18 in the 50-state model, Additional file 2: Fig. S5). Independent support for this observation comes from immunolocalizations on both *O. dioica* diploid and endocycling nuclei, showing H3K4me3 signals in RNAPII-depleted heterochromatic knobs together with other typical heterochromatic marks and DNA methylation [75]. Intriguingly, we also found that regions containing the active transcription mark, H3K36me3, coincided with these heterochromatic regions in the testis (Fig. 2c; Additional file 2: Fig. S5, state 18 in the 50-state model; Fig. S12).

We found a significant relationship between the type of TE on the Y-chromosome and whether or not it overlapped the heterochromatic mark H3K9me3 and either of the active marks H3K4me3 or H3K36me3 ($\chi^2 = 97.454$, $df 8$, p value <2.2e−16). The TE orders DIRS (102 regions out of 519) and LTR (183 out of 1729) both had an unusually high frequency of this combination of heterochromatin and active marks, whereas Mariner, MITEs and repeats annotated as possible Env had lower than expected frequencies. We tested for the combination of H3K9me3 with each active mark separately and obtained similar results when considering H3K9me3 with H3K4me3 ($\chi^2 = 95.685$, $df 8$, p value <2.2e−16) (Fig. 6a). When considering only H3K9me3 and H3K36me3, however, the DIRS order was the only one that had a higher than expected frequency of the two marks together (X-squared = 109.76, $df 8$, p value <2.2e−16) (Fig. 6b). These retrotransposons carrying tyrosine recombinases are widespread in eukaryotes. Their sequences are bordered by terminal repeats related to their replication via free circular dsDNA intermediates that are integrated without duplications in the site of integration [76].

Discussion

Compared to vertebrates, the protochordate *Oikopleura dioica* has undergone strong secondary genome compaction and morphological simplification, characteristic of the rapidly evolving larvacean lineage. *O. dioica* is the only known larvacean with separate sexes, and this is rare in tunicates in general [77] indicating an ancestral hermaphroditic state from which separate sexes and heteromorphic sex chromosomes have evolved more recently in only a few species. In this study, we compared the chromatin landscapes of the *O. dioica* ovary and testis and assessed differences between the autosomes and

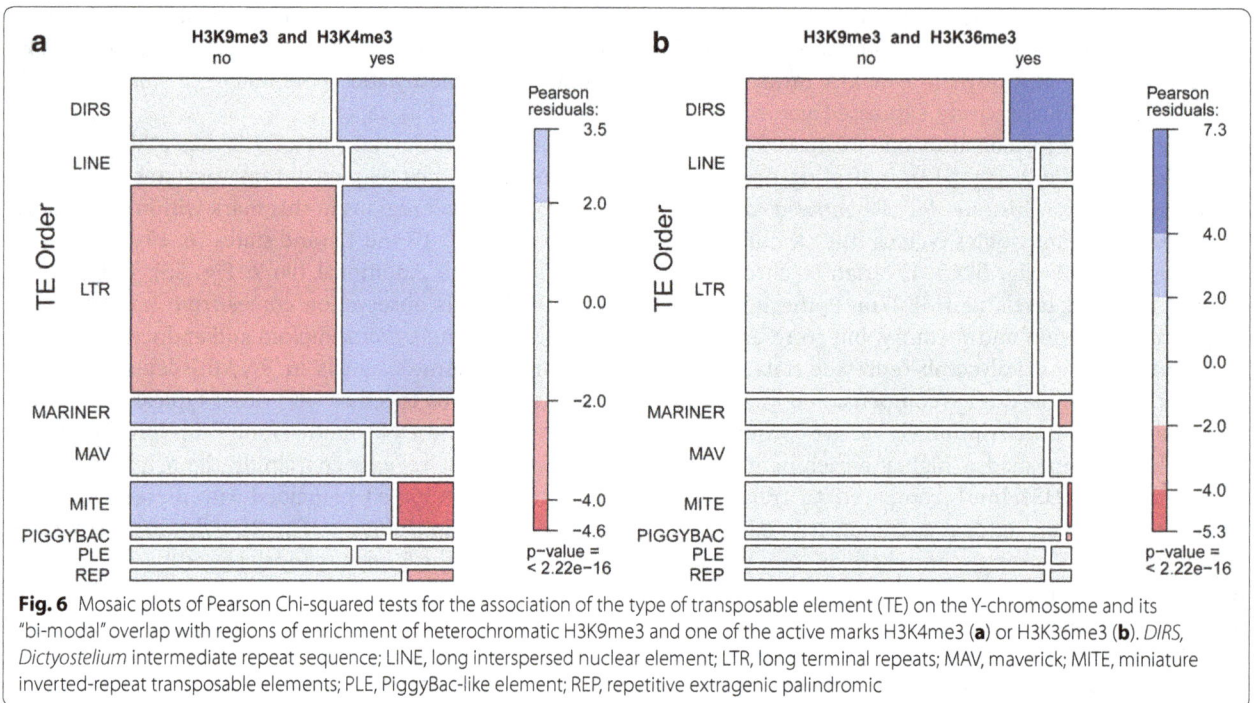

Fig. 6 Mosaic plots of Pearson Chi-squared tests for the association of the type of transposable element (TE) on the Y-chromosome and its "bi-modal" overlap with regions of enrichment of heterochromatic H3K9me3 and one of the active marks H3K4me3 (**a**) or H3K36me3 (**b**). *DIRS, Dictyostelium* intermediate repeat sequence; LINE, long interspersed nuclear element; LTR, long terminal repeats; MAV, maverick; MITE, miniature inverted-repeat transposable elements; PLE, PiggyBac-like element; REP, repetitive extragenic palindromic

sex chromosomes. A greater diversity of chromatin states was deployed in the testes, compared to the ovary, paralleling the previously observed amplification of testes-specific histone variants [33].

Chromatin states in *O. dioica* included the typical epigenetic signature of active promoters conserved among other metazoans (H3K4me3, H3K27ac, H3K18ac, weak H3K4me1 and H3K79me2 [12]). Tissue-specific promoter states also often marked gene bodies. This has also been observed in other species on frequently transcribed genes [78] and may be a consequence of very short coding regions and introns and/or high RNAPII processivity in *O. dioica*.

Histone acetylations are generally associated with accessible chromatin and are related to transcriptional activation. We found, however, silent developmental genes marked by H4ac, particularly in the testis (state 15). In endocycling nurse nuclei, silent developmental genes were enriched in H3K79me3 (state 1) (e.g., sox2 locus; Additional file 2: Fig. S12) though these two marks also co-occurred in both sexes to some extent, as resolved by the 50-state model (states 43–45, Additional file 2: Fig. S5). All three methylation states of H3K79 are mediated by Dot1 [79]. This mark acts to regulate endocycle progression, and its peak at G1/S prohibits re-replication in mammalian cells [3]. H4 acetylation and H4K20me1 peak during M and early G1 and decrease during S-phase [80] to ensure replication origin licensing and a chromatin state accessible to replication factors [81]. Thus,

one potential reason for differential H4ac (testes) and H3K79me3 (ovaries) marking could be related to the different mitotic versus endocycling cell cycle modes. Alternatively, these nucleosomes may be subject to histone variant exchange in the testis. *O. dioica* sperm chromatin does not undergo histone-to-protamine replacement that is typically preceded by massive H4 hyperacetylation in both human and *Drosophila* [82]. Instead, histone H3.3 is replaced by three isoforms of the *O. dioica* testis-specific variant H3t [33], and nucleosomes are retained. Amino acid substitutions surrounding K79 within the H3ts most likely preclude Dot1 binding and its ability to methylate these histones. H4 is also replaced by the male-specific H4t that has two residue changes adjacent to K20 that may hinder the methylation of this residue. Together with the retention of histones in *O. dioica* sperm, the association of generally activating H4ac with silent developmental genes in the testes could be viewed as a potentiating transgenerational mark. The lack of a transition to protamines, and extremely rapid embryonic and larval development, may have increased weighting on the intergenerational transmission of potentiated epigenetic states. Resolving individual lysine acetylations on H4 would be a step toward better understanding this unusual marking.

Along with promoter structure, enhancers determine cell-type specificity of gene expression and are important in developmental switching of metazoan genes. *O. dioica* intergenic and intronic spaces are very limited compared to vertebrate genomes, and the presence of enhancers is as yet poorly defined. We identified a number of candidate enhancer regions in this compact genome, in both the ovary and the testis, using intersections of typical enhancer PTMs and p300-bound regions. Histone PTMs marking enhancers in human, fly, cnidarians and worm (H3K4me1 and 2, H3K27ac, H3K79me2 and 3 [12, 16, 83]) were found in actively differentiating embryonic cells and tissue cell cultures. The O. dioica ovary and testis comprise terminally differentiated, highly specialized cell types, and the lower abundance of typical enhancer marks is in accordance with recent models of gene regulation during differentiation [84]. Moreover, genes organized in operons, particularly those devoted to maternal transcripts in the ovary, might have high transcriptional rates by default and be more subject to translational regulation via mTOR signaling [57]. The longer introns of O. dioica developmental genes exhibit more conserved intron positioning and also contain HCNEs [25]. We observed enrichment of chromatin states 1 and 15 on HCNEs. State 15 was under-represented in regions bound by p300 in the testis, but in the ovary, p300 co-localized with actively transcribed gene promoter state 5 or with 5mC in silent intergenic regions. The detection of typical active developmental enhancer activity would be further addressed through analysis of embryonic stages.

We found evidence of Polycomb interactions on autosomes and X-chromosomes in both sexes and propose roles of PRC2 in dosage compensation. We also observed in the O. dioica male germ line that the somatic H3K36 methyltransferase SETD2 is repressed, leaving NSD-(mes-4)-like and PRDM9-like to deposit H3K36-methylation. Complementary patterns of H3K36me and H3K27me3 in testis indicate regulation of transcript levels in the germline analogous to mes-4 (NSD-like) catalyzed H3K36me in C. elegans, which antagonizes PRC on germline genes such that PRC is excluded from autosomes but remains on (and represses) the X-chromosome. The mutual antagonism between PRC2 and mes-4 (NSD-like) is thought to be important in the transgenerational inheritance of germline-specific transcriptional programs in C. elegans [61, 85–87]. Thus, similar mechanisms may be operating in O. dioica, and it would be of future interest to determine the terminal chromatin states of mature sperm in this regard. In C. elegans, this proposed mechanism operates in the absence of DNA methylation or histone replacement by protamines during spermatogenesis. Both of these latter mechanisms are features of mammalian spermatogenic programs, though protamine replacement is not complete and some histones remain. Interestingly, O. dioica is intermediate in that DNA methylation is present whereas histone replacement by protamines is absent [33], offering

perhaps a useful comparative reference perspective in the evolution of mechanisms assuring transgenerational inheritance of germline transcriptional programs.

Given the rarity of separate sexes among tunicates, O. dioica offers an interesting model of more recent independent evolution of heteromorphic sex chromosomes. The O. dioica Y-chromosome contains a pseudoautosomal region, accumulated mobile elements, and a few male-specific genes, indicating that Y-chromosomal degeneration progressed rapidly [25]. Evolutionary forces driving Y-chromosomal sequence decay are well studied [88], but little is known about autosomal epigenome transitions toward the largely heterochromatic nature of ancient Y-chromosomes. We observed strong co-localization of heterochromatic marks and 5mC on this chromosome. This cooperation is an ancient feature based on HP1-mediated DNA methyltransferase recruitment [89, 90]. Patterns of histone PTMs on the Y-chromosome reflect the functional state and evolutionary history of the sequences [88]. The combination of histone and DNA modifications on the O. dioica Y-chromosome appears to have adapted to repress the activity of accumulated mobile DNA elements [25].

Our finding of unusual chromatin states containing typically heterochromatic marks (H3K9me2, H3K9me3, H4K20me3) combined with active transcription-related ones (H3K4me3, H3K36me3, H3K27ac, H4K20me1, H3K79me1) might represent a transition state in the time course of heteromorphic chromosome evolution. To our knowledge, such patterns of H3K4me3 and H3K36me3 overlapping heterochromatin signatures have been observed only in mammalian imprinted loci [91] and C. elegans mobile elements [14]. In the analysis of the epigenomes in nine human cell types, a group of endogenous retroviruses was enriched in a complex chromatin state that consisted of a number of histone modifications including H3K36me3, H3K4me3 and H3K9me3 [92]. However, these states have not been reported specifically on the Y-chromosome. O. dioica has little heterochromatin, consistent with a scarcity of TE elements [30] and generally reduced noncoding regions. Most TEs are present on the largely non-recombining Y-chromosome. The genome contains active Tor-family retrotransposons [30] that are transcribed in primordial germ-cell-adjacent somatic cells during embryonic development and in the adult testis [31]. These elements are specifically activated, often from their own non-LTR internal promoters and not genome-wide de-repressed. The presence of H3K4me3 and H3K36me3 on TEs could be a consequence of TE transcription with the co-occurrence of H3K9me3 due to the location of some copies of these elements in repressive chromatin environments, whereas others are active, either within one animal, or in different individuals. Notably, these unusual states were restricted to marking certain classes of TEs in *O. dioica*.

The short life cycle and ability to rapidly regulate gamete output over 3 orders of magnitude [45, 56, 93] may also be relevant to the extent of paused RNAPII signatures we observed in the gonads. Such a strategy would be compatible with more rapid transcriptional response to nutrient availability in adjusting gamete number. Finally, it is suspected that the width of broad chromatin domains contributes to the heritability of epigenetic states because of random segregation of nucleosomes to daughter cells during genome replication [35]. Thus, the probability of loss of an epigenetic state will increase as a function of decreasing breadth of chromatin domains. This poses a challenge to epigenetic regulation and inheritance on the compact *O. dioica genome*, where intergenic regulatory regions are frequently on the order of one nucleosome. Indeed, we found that chromatin state domain widths in *O. dioica* were generally smaller than those in their chordate relatives, the vertebrates, and rarely exceeded 7 nucleosomes in size. Nonetheless, this is evidently compatible with fidelity of heritable transmission of epigenetic states in this species.

Conclusions

Our work provides the first comprehensive view of a protochordate epigenome. In the transcriptionally active endocycling, ovarian, nurse nuclei, histone modifications linked to RNAPII activity were conserved compared to those in other metazoans. We found evidence that RNAPII pausing is frequently involved in controlling gene transcription during oogenesis. We also identified candidate enhancers that have enhancer PTMs typical of other metazoan enhancer regions. Large heterochromatic domains of H3K9me3 were largely absent in the ovary.

The epigenome of the male germ line reflected a shift from tissue-specific gene expression toward establishment of transgenerational inheritance and showed features related to the mitotic proliferation ongoing in this tissue. Our data support the involvement of a modified Polycomb complex in sex chromosome dosage compensation in homogametic females. We also identified a novel chromatin states that combines both euchromatic and heterochromatic histone PTMs that are associated with specific classes of TEs and the Y-chromosome.

Strong secondary genome compaction and disruption of evolutionarily conserved linear genome architecture in *O. dioica* have been epigenetically accompanied by reduced chromatin state domain widths and a general reduction of heterochromatin. It will be of interest to build on the foundation of the current study to determine how the three-dimensional genomic architecture of topologically associated domains (TADs) has been affected by these processes.

Methods
Animal culture and collection
Oikopleura dioica were cultured at 15 °C [21]. At late day 5/early day 6, males and females were separated from each other, and gonads were dissected in cold N-ChIP collection buffer (0.32 M sucrose, 1 mM $CaCl_2$, 4 mM $MgCl_2$, 50 mM Tris/HCl pH 7.7, 0.5 mM PMSF and 1× protease inhibitor. Sampled testes were in the mitotically dividing pre-meiotic (spermatogonia) stage, using H3S28P immunostaining as a reference mark of meiotic versus mitotic events. For histone acetylation samples, 5 mM sodium butyrate was also added, and for histone phosphorylation samples, phosphatase inhibitor (phosphatase inhibitor cocktails 2 and 3, Sigma) was included. Samples were sedimented at 8000 rpm and snap-frozen in liquid nitrogen.

Native and cross-linked ChIP-chip
We performed native ChIP [94], with minor modifications, using 10 male or 40 female gonads in 300-μl buffer per ChIP. Mammalian and *O. dioica* histone H3 and H4 sequences are identical [33], and we used antibodies (10 μg per ChIP) validated as ChIP grade for human cells (Additional file 3: Table S1). Antibodies were bound to 50 μl Dynabeads® Protein G (Invitrogen) and incubated with chromatin overnight (O/N). Immunoprecipitated DNA was purified by phenol–chloroform–isoamylalcohol extraction and EtOH precipitation followed by Min Elute PCR purification (Qiagen). Resulting DNA was amplified using Whole Genome Amplification kit 2 or 4 (Sigma) according to manufacturer's instructions (the chemical fragmentation step was omitted). With the exception of H3K4me2 and H3K9me2, histone PTM profiles were replicated at least twice in both ovary and testis samples. Amplified material was labeled and hybridized to *O. dioica* genomic microarrays (Nimblegen) [32].

Animals were fixed in 1% formaldehyde in PBS for 12 min. Cross-linking was stopped by adding glycine to 0.13 M for 5 min followed by 3 washes with cold PBS. Fixed animals were kept in PBS containing 0.5 mM PMSF, 1× protease inhibitor and phosphatase inhibitor for dissection of gonads. Gonads were lysed in RIPA buffer (150 mM NaCl, 1% NP-40, 0.5% Na deoxycholate, 0.1% SDS, 50 mM Tris pH 8, 5 mM EDTA), supplemented with fresh PMSF, protease and phosphatase inhibitor for 15 min. Chromatin was fragmented to a size range of 200–600 bp using a Bioruptor UCD-200 (Diagenode) for 8 × 30 s on/1 min off. ChIP was performed using 300 μl of centrifuged chromatin. We used custom antibodies (twentieth-century biochemicals) against *O. dioica* p300 and CTCF. Antibodies were bound to 50 µl Dynabeads® Protein G and incubated with chromatin O/N. Immune complexes were washed twice with 800 µl RIPA, twice with 500 mM NaCl/RIPA and twice with LiCl buffer (500 mM LiCl, 10 mM Tris–HCl pH 8, 1% sodium deoxycholate, 1% NP-40) and once with TE. Enriched DNA was recovered by 15-min elution at 65 °C in 1% SDS, 50 mM Tris–HCl pH 8, 1 mM EDTA and subjected to cross-link reversal for 5 h at 65 °C followed by RNase A and proteinase K treatments. DNA was then purified as for native ChIP. Immunoprecipitated DNA was amplified using the Whole Genome Amplification 4 kit (WGA4 Sigma), labeled and hybridized to *Oikopleura* genomic microarrays (Nimblegen).

Methylated DNA immunoprecipitation on chip (MeDIP-chip)

Oikopleura dioica genomic DNA from day 6 immature ovaries and testes was isolated using proteinase K lysis followed by RNAse A treatment and phenol–chloroform extraction. Ethanol-precipitated DNA was diluted in 450 µl TE, randomly fragmented by sonication using the Bioruptor UCD-200 (Diagenode). Fragments ranged in size from 200 to 600 bp and were denatured at 95 °C for 5 min. Protein G Dynabeads (50 µl) were coupled to the monoclonal antibody against 5-methylcytidine (5mC) (Eurogentec #BI-MECY-1000). After three washes with IP buffer (10 mM Na-phosphate pH 7.0, 140 mM NaCl, 0.05% Triton X-100), DNA fragments together with 51 µl of $10 \times$ IP buffer were added to the beads and incubated 16 h at 4 °C with constant rotation. Beads were collected with a magnetic rack and washed with 700 µl $1\times$ IP buffer for 10 min with rotation. Washing was repeated twice, followed by elution in proteinase K buffer (50 mM Tris pH 8.0, 10 mM EDTA, 0.5% SDS, 50 µg proteinase K (NEB)) and incubation for 2 h at 55° C with shaking. Immunoprecipitated DNA was purified by phenol–chloroform–isoamylalcohol extraction and EtOH precipitation followed by Min Elute PCR purification (Qiagen). Resulting DNA was amplified using Whole Genome Amplification kit 4 (Sigma) according to manufacturer's instructions (the chemical fragmentation step was omitted).

Preprocessing ChIP-chip data

ChIP-chip data were preprocessed using the Bioconductor R package *Ringo* [95]. We normalized raw probe intensities from each sample (Cy5 channel) to corresponding input DNA probe intensities (Cy3 channel) by computing $\log_2(\text{Cy5/Cy3})$. We applied the NimbleGen normalization method, which adjusts for systematic dye and labeling biases by subtracting from individual $\log_2$ ratios, the Tukey's biweight mean, computed across each sample's $\log_2$ ratios. To reduce noise, we smoothened the normalized $\log_2$ ratios using a running median across a 150-bp window (approximate nucleosome size) with a minimum threshold of three nonzero probes. We used the resulting $\log_2$ ratios in all further analyses and browser visualizations.

Histone PTM distributions and defining ChIP-enriched regions

We defined transcription start sites (TSSs) and transcription end sites (TESs) using gene models on the *O. dioica* reference genome [25]. We categorized genes into highly expressed (top quartile) and silent using ovary and testis transcriptome profiling data [32]. We further categorized genes according to whether or not they are *trans*-spliced with the spliced leader (SL) using an SL CAGE dataset [28]. We also categorized genes according to GC content 500 bp upstream of TSSs and defined the top 10% as high GC (HGC) and the bottom 10% as low GC (LGC). We also categorized genes according to their localization on X- and Y-chromosomes or autosomes. For each gene category, we calculated the mean $\log_2$ ratio at each probe position in a 1000-bp window centered on TSSs and TESs. Genes shorter than 1000-bp and with less than 1100-bp intergenic space were excluded, as were genes with no expression data. A total of 5444 gene models met our criteria for inclusion.

We used the R package *Ringo* [95] to identify ChIP-enriched regions (chers). This models the distribution of smoothened probe intensities (y) as two underlying distributions: a null distribution from non-enriched regions and an alternative distribution from enriched regions. The null distribution is estimated for each sample individually using the empirical distribution of probe intensities: The mode (m_0) is calculated and probe intensities $y < m_0$ are reflected onto $y > m_0$. A threshold y_0 is then computed on the null distribution to define enrichment. We computed two enrichment thresholds using the 99 and 95% quantiles of the null distribution estimated for each sample individually. A region was called ChIP-enriched if a minimum of two probes exceeded the enrichment threshold. Regions were merged if the width between them was less than 50 bp. For samples with replicates, we defined final chers by intersecting the chers of each replicate: A region was only classed as ChIP-enriched if it was present in at least two replicates at the 95% threshold and present at the 99% threshold in at least one replicate. For H3K27me3, we also lowered the enrichment threshold to 75% in order to identify broader domains.

Chromatin states and related genomic features

The genome was segmented into 15 or 50 chromatin states using chromHMM [9]. We created a binary matrix indicating the presence or absence of ChIP-enriched regions for each histone PTM in the ovary and testis, using 50-bp intervals across the genome. A PTM was defined as present if the 50-bp window overlapped a cher. States were learned using this matrix as input to chromHMM. We used models learnt jointly across both ovary and testis for all downstream analyses. We also used chromHMM to learn 15-state models for the ovary and testis data separately. We used chromHMM to calculate fold enrichments of states for various genomic features to functionally classify chromatin states. We used O. dioica annotations [25] to define TSS, TES, exon/intron boundaries, TEs and operons. A promoter region was defined as the 400 bp region upstream of the annotated TSS. We used InterPro domains, together with GO terms and manual curation, to identify transcription factors, zinc fingers and homeodomain protein genes. O. dioica homeodomain proteins were classified according to whether or not they had DNA-binding specificity PWM predictions using PreMoTF [96] (http://stormo.wustl.edu/PreMoTF/). We calculated the specificity of each O. dioica gene across the developmental transcriptome using Shannon entropy, normalized to range between 0 and 1, following [97]. The specificity of a gene ranged from 1 (specific to a single stage) to 0 (equally expressed in all stages). We defined low-specificity and high-specificity genes using thresholds of 0.2 and 0.9, respectively. In order to assign genomic features to chromatin states, we used a threshold of 2 on the enrichments calculated by chromHMM. The main histone modifications for each state were extracted using a threshold of 0.5 on emission parameters.

GO term analyses

We used published O. dioica GO term annotations [32]. A subset of genes was created for each GO term that was associated with 50 or more genes (9665 in total). This subset was further subdivided into active and silent genes for the ovary and testis, separately. Only subsets with 10 or more genes were analyzed further. For each subset, we computed the enrichment of chromatin states at different genomic features (TSS, TES, promoter region and gene body) using chromHMM. Enrichments were clustered hierarchically using (1—the Pearson's correlation coefficient). Clusters were defined by cutting the resulting dendrogram at a height of 1. Clusters were further defined and functions summarized manually by visual inspection of the respective data files.

Abbreviations

bp: base pair; ChIP-seq: chromatin immunoprecipitation followed by sequencing; CTCF: CCCTC-binding factor; GO term: Gene Ontology term; HCNE: highly conserved element; meDIP: methylated DNA immunoprecipitation; MNase: micrococcal nuclease; PRC: Polycomb repressive complex; RNAPII: RNA polymerase II; TAR: transcriptionally active region; TE: transposable element; TSS: transcription start site.

Authors' contributions

EMT and PN conceived and designed the study. PN performed experiments and SB, AL and JRM carried out hybridizations. GBD processed the data. GBD, PN and EMT analyzed results. PN, GBD and EMT wrote the manuscript. All authors read and approved the final manuscript.

Author details

[1] Sars International Centre for Marine Molecular Biology, University of Bergen, 5008 Bergen, Norway. [2] Departments of Biology and Pediatrics and the Roy J. Carver Center for Genomics, 459 Biology Building, University of Iowa, Iowa City, IA 52242, USA. [3] Department of Biology, University of Bergen, 5020 Bergen, Norway.

Acknowledgements

We thank the staff of the Appendicularia culture facility for supplying animals.

Competing interests

The authors declare that they have no competing interests.

Funding

This work was supported by Grants 204891 and 133335/V40 from the Norwegian Research Council (E.M.T.).

References

1. Fang R, Barbera AJ, Xu Y, Rutenberg M, Leonor T, Bi Q, et al. Human LSD2/KDM1b/AOF1 regulates gene transcription by modulating intragenic H3K4me2 methylation. Mol Cell. 2010;39:222–33.
2. Fritsch L, Robin P, Mathieu JRR, Souidi M, Hinaux H, Rougeulle C, et al. A subset of the histone H3 lysine 9 methyltransferases Suv39h1, G9a, GLP, and SETDB1 participate in a multimeric complex. Mol Cell. 2010;37:46–56.
3. Fu H, Maunakea AK, Martin MM, Huang L, Zhang Y, Ryan M, et al. Methylation of histone H3 on lysine 79 associates with a group of replication origins and helps limit DNA replication once per cell cycle. PLoS Genet. 2013;9:e1003542.
4. Acquaviva L, Székvölgyi L, Dichtl B, Dichtl BS, de La Roche Saint Andrède C, Nicolas A, et al. The COMPASS subunit Spp1 links histone methylation to initiation of meiotic recombination. Science. 2013;339:215–8.
5. Kelly WG. Multigenerational chromatin marks: no enzymes need apply. Dev Cell. 2014;31:142–4.
6. Heard E, Martienssen RA. Transgenerational epigenetic inheritance: myths and mechanisms. Cell. 2014;157:95–109.
7. Jenuwein T, Allis CD. Translating the histone code. Science. 2001;293:1074–80.
8. Filion GJ, van Bemmel JG, Braunschweig U, Talhout W, Kind J, Ward LD, et al. Systematic protein location mapping reveals five principal chromatin types in Drosophila cells. Cell. 2010;143:212–24.
9. Ernst J, Kellis M. ChromHMM: automating chromatin-state discovery and characterization. Nat Methods. 2012;9:215–6.

10. Kharchenko PV, Alekseyenko AA, Schwartz YB, Minoda A, Riddle NC, Ernst J, et al. Comprehensive analysis of the chromatin landscape in Drosophila melanogaster. Nature. 2011;471:480–5.

11. Roudier F, Ahmed I, Bérard C, Sarazin A, Mary-Huard T, Cortijo S, et al. Integrative epigenomic mapping defines four main chromatin states in Arabidopsis. EMBO J. 2011;30:1928–38.

12. Ho JWK, Jung YL, Liu T, Alver BH, Lee S, Ikegami K, et al. Comparative analysis of metazoan chromatin organization. Nature. 2014;512:449–52.

13. Barski A, Cuddapah S, Cui K, Roh T-Y, Schones DE, Wang Z, et al. High-resolution profiling of histone methylations in the human genome. Cell. 2007;129:823–37.

14. Gerstein MB, Lu ZJ, Van Nostrand EL, Cheng C, Arshinoff BI, Liu T, et al. Integrative analysis of the Caenorhabditis elegans genome by the modEN-CODE project. Science. 2010;330:1775–87.

15. Liu T, Rechtsteiner A, Egelhofer TA, Vielle A, Latorre I, Cheung M-S, et al. Broad chromosomal domains of histone modification patterns in C. elegans. Genome Res. 2011;21:227–36.

16. Bonn S, Zinzen RP, Girardot C, Gustafson EH, Perez-Gonzalez A, Delhomme N, et al. Tissue-specific analysis of chromatin state identifies temporal signatures of enhancer activity during embryonic development. Nat Genet. 2012;44:148–56.

17. Lara-Astiaso D, Weiner A, Lorenzo-Vivas E, Zaretsky I, Jaitin DA, David E, et al. Chromatin state dynamics during blood formation. Science. 2014;345:943–9.

18. Ng J-H, Kumar V, Muratani M, Kraus P, Yeo J-C, Yaw L-P, et al. In vivo epigenomic profiling of germ cells reveals germ cell molecular signatures. Dev Cell. 2013;24:324–33.

19. Consortium RE, Kundaje A, Meuleman W, Ernst J, Bilenky M, Yen A, et al. Integrative analysis of 111 reference human epigenomes. Nature. 2015;518:317–30.

20. Delsuc F, Brinkmann H, Chourrout D, Philippe H. Tunicates and not cephalochordates are the closest living relatives of vertebrates. Nature. 2006;439:965–8.

21. Bouquet J-M, Spriet E, Troedsson C, Otterå H, Chourrout D, Thompson EM. Culture optimization for the emergent zooplanktonic model organism Oikopleura dioica. J Plankton Res. 2009;31:359–70.

22. Ganot P, Thompson EM. Patterning through differential endoreduplication in epithelial organogenesis of the chordate, Oikopleura dioica. Dev Biol. 2002;252:59–71.

23. Campsteijn C, Ovrebø JI, Karlsen BO, Thompson EM. Expansion of cyclin D and CDK1 paralogs in Oikopleura dioica, a chordate employing diverse cell cycle variants. Mol Biol Evol. 2012;29:487–502.

24. Ganot P, Bouquet J-M, Kallesøe T, Thompson EM. The Oikopleura coenocyst, a unique chordate germ cell permitting rapid, extensive modulation of oocyte production. Dev Biol. 2007;302:591–600.

25. Denoeud F, Henriet S, Mungpakdee S, Aury J-M, Da Silva C, Brinkmann H, et al. Plasticity of animal genome architecture unmasked by rapid evolution of a pelagic tunicate. Science. 2010;330:1381–5.

26. Kim M-S, Pinto SM, Getnet D, Nirujogi RS, Manda SS, Chaerkady R, et al. A draft map of the human proteome. Nature. 2014;509:575–81.

27. Wilhelm M, Schlegl J, Hahne H, Moghaddas Gholami A, Lieberenz M, Savitski MM, et al. Mass-spectrometry-based draft of the human proteome. Nature. 2014;509:582–7.

28. Danks GB, Raasholm M, Campsteijn C, Long AM, Manak JR, Lenhard B, et al. Trans-splicing and operons in metazoans: translational control in maternally regulated development and recovery from growth arrest. Mol Biol Evol. 2015;32:585–99.

29. Ganot P, Kallesøe T, Reinhardt R, Chourrout D, Thompson EM. Spliced-leader RNA trans splicing in a chordate, Oikopleura dioica, with a compact genome. Mol Cell Biol. 2004;24:7795–805.

30. Volff J-N, Lehrach H, Reinhardt R, Chourrout D. Retroelement dynamics and a novel type of chordate retrovirus-like element in the miniature genome of the tunicate Oikopleura dioica. Mol Biol Evol. 2004;21:2022–33.

31. Henriet S, Sumic S, Doufoundou-Guilengui C, Jensen MF, Grandmougin C, Fal K, et al. Embryonic expression of endogenous retroviral RNAs in somatic tissues adjacent to the Oikopleura germline. Nucleic Acids Res. 2015;43:3701–11.

32. Danks G, Campsteijn C, Parida M, Butcher S, Doddapaneni H, Fu B, et al. OikoBase: a genomics and developmental transcriptomics resource for the urochordate Oikopleura dioica. Nucleic Acids Res. 2013;41:D845–53.

33. Moosmann A, Campsteijn C, Jansen PW, Nasrallah C, Raasholm M, Stunnenberg HG, et al. Histone variant innovation in a rapidly evolving chordate lineage. BMC Evol Biol. 2011;11:208.

34. Harmston N, Lenhard B. Chromatin and epigenetic features of long-range gene regulation. Nucleic Acids Res. 2013;41(15):7185–99. doi:10.1093/nar/gkt499.

35. Rando OJ. Global patterns of histone modifications. Curr Opin Genet Dev. 2007;17:94–9.

36. Bantignies F, Cavalli G. Polycomb group proteins: repression in 3D. Trends Genet. 2011;27:454–64.

37. Wang J, Mager J, Chen Y, Schneider E, Cross JC, Nagy A, et al. Imprinted X inactivation maintained by a mouse Polycomb group gene. Nat Genet. 2001;28:371–5.

38. Klauke K, Radulović V, Broekhuis M, Weersing E, Zwart E, Olthof S, et al. Polycomb Cbx family members mediate the balance between haematopoietic stem cell self-renewal and differentiation. Nat Cell Biol. 2013;15:353–62.

39. Schuettengruber B, Chourrout D, Vervoort M, Leblanc B, Cavalli G. Genome regulation by polycomb and trithorax proteins. Cell. 2007;128:735–45.

40. Gil J, O'Loghlen A. PRC1 complex diversity: where is it taking us? Trends Cell Biol. 2014;24:632–41.

41. Blackledge NP, Farcas AM, Kondo T, King HW, McGouran JF, Hanssen LLP, et al. Variant PRC1 complex-dependent H2A ubiquitylation drives PRC2 recruitment and polycomb domain formation. Cell. 2014;157:1445–59.

42. Muyle A, Zemp N, Deschamps C, Mousset S, Widmer A, Marais GAB. Rapid de novo evolution of X chromosome dosage compensation in Silene latifolia, a plant with young sex chromosomes. PLoS Biol. 2012;10:e1001308.

43. Disteche CM. Dosage compensation of the sex chromosomes. Annu Rev Genet. 2012;46:537–60.

44. Ferrari F, Alekseyenko AA, Park PJ, Kuroda MI. Transcriptional control of a whole chromosome: emerging models for dosage compensation. Nat Struct Mol Biol. 2014;21:118–25.

45. Ganot P, Moosmann-Schulmeister A, Thompson EM. Oocyte selection is concurrent with meiosis resumption in the coenocystic oogenesis of Oikopleura. Dev Biol. 2008;324:266–76.

46. Wagner EJ, Carpenter PB. Understanding the language of Lys36 methylation at histone H3. Nat Rev Mol Cell Biol. 2012;13:115–26.

47. Wang Z, Zang C, Rosenfeld JA, Schones DE, Barski A, Cuddapah S, et al. Combinatorial patterns of histone acetylations and methylations in the human genome. Nat Genet. 2008;40:897–903.

48. Vavouri T, Lehner B. Human genes with CpG island promoters have a distinct transcription-associated chromatin organization. Genome Biol. 2012;13:R110.

49. Sawicka A, Hartl D, Goiser M, Pusch O, Stocsits RR, Tamir IM, et al. H3S28 phosphorylation is a hallmark of the transcriptional response to cellular stress. Genome Res. 2014;24:1808–20.

50. Pokholok DK, Harbison CT, Levine S, Cole M, Hannett NM, Lee TI, et al. Genome-wide map of nucleosome acetylation and methylation in yeast. Cell. 2005;122:517–27.

51. Liu CL, Kaplan T, Kim M, Buratowski S, Schreiber SL, Friedman N, et al. Single-nucleosome mapping of histone modifications in S. cerevisiae. PLoS Biol. 2005;3:e328.

52. Frederiks F, Tzouros M, Oudgenoeg G, van Welsem T, Fornerod M, Krijgsveld J, et al. Nonprocessive methylation by Dot1 leads to functional redundancy of histone H3K79 methylation states. Nat Struct Mol Biol. 2008;15:550–7.

53. Ernst J, Kellis M. Discovery and characterization of chromatin states for systematic annotation of the human genome. Nat Biotechnol. 2010;28:817–25.

54. Bernstein BE, Kamal M, Lindblad-Toh K, Bekiranov S, Bailey DK, Huebert DJ, et al. Genomic maps and comparative analysis of histone modifications in human and mouse. Cell. 2005;120:169–81.

55. Adelman K, Lis JT. Promoter-proximal pausing of RNA polymerase II: emerging roles in metazoans. Nat Rev Genet. 2012;13:720–31.

56. Troedsson C, Bouquet JM, Aksnes DL, Thompson EM. Resource allocation between somatic growth and reproductive output in the pelagic chordate Oikopleura dioica allows opportunistic response to nutritional variation. Mar Ecol Prog Ser. 2002;243:83–91.

57. Danks G, Thompson EM. *Trans*-splicing in metazoans: a link to translational control? Worm. 2015;4(3):e1046030. doi:10.1080/21624054.2015.1046030.

58. Ogryzko VV, Schiltz RL, Russanova V, Howard BH, Nakatani Y. The transcriptional coactivators p300 and CBP are histone acetyltransferases. Cell. 1996;87:953–9.

59. Nord AS, Blow MJ, Attanasio C, Akiyama JA, Holt A, Hosseini R, et al. Rapid and pervasive changes in genome-wide enhancer usage during mammalian development. Cell. 2013;155:1521–31.

60. Kikuta H, Laplante M, Navratilova P, Komisarczuk AZ, Engström PG, Fredman D, et al. Genomic regulatory blocks encompass multiple neighboring genes and maintain conserved synteny in vertebrates. Genome Res. 2007;17:545–55.

61. Gaydos LJ, Rechtsteiner A, Egelhofer TA, Carroll CR, Strome S. Antagonism between MES-4 and Polycomb repressive complex 2 promotes appropriate gene expression in *C. elegans* germ cells. Cell Rep. 2012;2:1169–77.

62. Cañestro C, Postlethwait JH. Development of a chordate anterior-posterior axis without classical retinoic acid signaling. Dev Biol. 2007;305:522–38.

63. Cao R, Wang L, Wang H, Xia L, Erdjument-Bromage H, Tempst P, et al. Role of histone H3 lysine 27 methylation in Polycomb-group silencing. Science. 2002;298:1039–43.

64. Riising EM, Comet I, Leblanc B, Wu X, Johansen JV, Helin K. Gene silencing triggers polycomb repressive complex 2 recruitment to CpG islands genome wide. Mol Cell. 2014;55:347–60.

65. Gaydos LJ, Wang W, Strome S. H3K27me and PRC2 transmit a memory of repression across generations and during development. Science. 2014;345:1515–8.

66. Mozzetta C, Pontis J, Fritsch L, Robin P, Portoso M, Proux C, et al. The histone H3 lysine 9 methyltransferases G9a and GLP regulate polycomb repressive complex 2-mediated gene silencing. Mol Cell. 2014;53:277–89.

67. Ferrari KJ, Scelfo A, Jammula S, Cuomo A, Barozzi I, Stützer A, et al. Polycomb-dependent H3K27me1 and H3K27me2 regulate active transcription and enhancer fidelity. Mol Cell. 2014;53:49–62.

68. Yung PYK, Stuetzer A, Fischle W, Martinez A-M, Cavalli G. Histone H3 serine 28 is essential for efficient polycomb-mediated gene repression in *Drosophila*. Cell Rep. 2015;11:1437–45.

69. Fonseca JP, Steffen PA, Müller S, Lu J, Sawicka A, Seiser C, et al. In vivo Polycomb kinetics and mitotic chromatin binding distinguish stem cells from differentiated cells. Genes Dev. 2012;26:857–71.

70. Lau PNI, Cheung P. Histone code pathway involving H3 S28 phosphorylation and K27 acetylation activates transcription and antagonizes polycomb silencing. Proc Natl Acad Sci USA. 2011;108:2801–6.

71. Petruk S, Sedkov Y, Johnston DM, Hodgson JW, Black KL, Kovermann SK, et al. TrxG and PcG proteins but not methylated histones remain associated with DNA through replication. Cell. 2012;150:922–33.

72. Ku M, Koche RP, Rheinbay E, Mendenhall EM, Endoh M, Mikkelsen TS, et al. Genomewide analysis of PRC1 and PRC2 occupancy identifies two classes of bivalent domains. PLoS Genet. 2008;4:e1000242.

73. Bender LB, Suh J, Carroll CR, Fong Y, Fingerman IM, Briggs SD, et al. MES-4: an autosome-associated histone methyltransferase that participates in silencing the X chromosomes in the *C. elegans* germ line. Development. 2006;133:3907–17.

74. Eram MS, Bustos SP, Lima-Fernandes E, Siarheyeva A, Senisterra G, Hajian T, et al. Trimethylation of histone H3 lysine 36 by human methyltransferase PRDM9 protein. J Biol Chem. 2014;289:12177–88.

75. Spada F, Vincent M, Thompson EM. Plasticity of histone modifications across the invertebrate to vertebrate transition: histone H3 lysine 4 trimethylation in heterochromatin. Chromosome Res. 2005;13(1):57–72.

76. Cappello J, Handelsman K, Lodish HF, Biggin MD, Gibson TJ, Hong GF, et al. Sequence of *Dictyostelium* DIRS-1: an apparent retrotransposon with inverted terminal repeats and an internal circle junction sequence. Cell. 1985;43:105–15.

77. Huus J. Handbuch der Zoologie: Tunicata/bearb. v. Johan Huus. de Gruyter; 1933.

78. Benayoun BA, Pollina EA, Ucar D, Mahmoudi S, Karra K, Wong ED, et al. H3K4me3 breadth is linked to cell identity and transcriptional consistency. Cell. 2014;158:673–88.

79. van Leeuwen F, Gafken PR, Gottschling DE, Brachmann CB, Davies A, Cost GJ, et al. Dot1p modulates silencing in yeast by methylation of the nucleosome core. Cell. 2002;109:745–56.

80. Tardat M, Brustel J, Kirsh O, Lefevbre C, Callanan M, Sardet C, et al. The histone H4 Lys 20 methyltransferase PR-Set7 regulates replication origins in mammalian cells. Nat Cell Biol. 2010;12:1086–93.

81. Unnikrishnan A, Gafken PR, Tsukiyama T. Dynamic changes in histone acetylation regulate origins of DNA replication. Nat Struct Mol Biol. 2010;17:430–7.

82. Rathke C, Baarends WM, Awe S, Renkawitz-Pohl R. Chromatin dynamics during spermiogenesis. Biochim Biophys Acta. 2014;1839:155–68.

83. Schwaiger M, Schönauer A, Rendeiro AF, Pribitzer C, Schauer A, Gilles AF, et al. Evolutionary conservation of the eumetazoan gene regulatory landscape. Genome Res. 2014;24:639–50.

84. Arner E, Daub CO, Vitting-Seerup K, Andersson R, Lilje B, Drablos F, et al. Transcribed enhancers lead waves of coordinated transcription in transitioning mammalian cells. Science. 2015;347:1010–4.

85. Fong Y, Bender L, Wang W, Strome S. Regulation of the different chromatin states of autosomes and X chromosomes in the germ line of *C. elegans*. Science. 2002;296:2235–8.

86. Evans KJ, Huang N, Stempor P, Chesney MA, Down TA, Ahringer J. Stable *Caenorhabditis elegans* chromatin domains separate broadly expressed and developmentally regulated genes. Proc Natl Acad Sci USA 2016;113:E7020–29.

87. Kelly WG, Berger S, Kouzarides T, Shiekhattar R, Shilatifard A, Strahl B, et al. Transgenerational epigenetics in the germline cycle of *Caenorhabditis elegans*. Epigenetics Chromatin. 2014;7:6.

88. Bachtrog D. Y-chromosome evolution: emerging insights into processes of Y-chromosome degeneration. Nat Rev Genet. 2013;14:113–24.

89. Lehnertz B, Ueda Y, Derijck AAHA, Braunschweig U, Perez-Burgos L, Kubicek S, et al. Suv39 h-mediated histone H3 lysine 9 methylation directs DNA methylation to major satellite repeats at pericentric heterochromatin. Curr Biol. 2003;13:1192–200.

90. Rountree MR, Selker EU. DNA methylation and the formation of heterochromatin in *Neurospora crassa*. Heredity. 2010;105:38–44.

91. Mikkelsen TS, Ku M, Jaffe DB, Issac B, Lieberman E, Giannoukos G, et al. Genome-wide maps of chromatin state in pluripotent and lineage-committed cells. Nature. 2007;448:553–60.

92. Ernst J, Kheradpour P, Mikkelsen TS, Shoresh N, Ward LD, Epstein CB, et al. Mapping and analysis of chromatin state dynamics in nine human cell types. Nature. 2011;473:43–9.

93. Ganot P, Kallesøe T, Thompson EM. The cytoskeleton organizes germ nuclei with divergent fates and asynchronous cycles in a common cytoplasm during oogenesis in the chordate *Oikopleura*. Dev Biol. 2007;302:577–90.

94. Cuddapah S, Barski A, Cui K, Schones DE, Wang Z, Wei G, et al. Native chromatin preparation and Illumina/Solexa library construction. Cold Spring Harb Protoc. 2009;2009:pdb.prot5237.

95. Toedling J, Skylar O, Sklyar O, Krueger T, Fischer JJ, Sperling S, et al. Ringo—an R/Bioconductor package for analyzing ChIP-chip readouts. BMC Bioinform. 2007;8:221.

96. Christensen RG, Enuameh MS, Noyes MB, Brodsky MH, Wolfe SA, Stormo GD. Recognition models to predict DNA-binding specificities of homeodomain proteins. Bioinformatics. 2012;28:i84–9.

97. Schug J, Schuller W-P, Kappen C, Salbaum JM, Bucan M, Stoeckert CJ, et al. Promoter features related to tissue specificity as measured by Shannon entropy. Genome Biol. 2005;6:R33.

2

Flightless-I governs cell fate by recruiting the SUMO isopeptidase SENP3 to distinct *HOX* genes

Arnab Nayak[1][*], Anja Reck[2], Christian Morsczeck[2] and Stefan Müller[1][*]

Abstract

Background: Despite recent studies on the role of ubiquitin-related SUMO modifier in cell fate decisions, our understanding on precise molecular mechanisms of these processes is limited. Previously, we established that the SUMO isopeptidase SENP3 regulates chromatin assembly of the MLL1/2 histone methyltransferase complex at distinct *HOX* genes, including the osteogenic master regulator *DLX3*. A comprehensive mechanism that regulates SENP3 transcriptional function was not understood.

Results: Here, we identified flightless-I homolog (FLII), a member of the gelsolin family of actin-remodeling proteins, as a novel regulator of SENP3. We demonstrate that FLII is associated with SENP3 and the MLL1/2 complex. We further show that FLII determines SENP3 recruitment and MLL1/2 complex assembly on the *DLX3* gene. Consequently, FLII is indispensible for H3K4 methylation and proper loading of active RNA polymerase II at this gene locus. Most importantly, FLII-mediated SENP3 regulation governs osteogenic differentiation of human mesenchymal stem cells.

Conclusion: Altogether, these data reveal a crucial functional interconnection of FLII with the sumoylation machinery that converges on epigenetic regulation and cell fate determination.

Keywords: Sumoylation, MLL1/2, *HOX* gene, Mesenchymal stem cells, Flightless-I, Chromatin, SILAC-based proteomics

Background

The ubiquitin-like SUMO (small ubiquitin-like modifier) system has emerged as a key regulator of cell function [1–5]. SUMO covalently modifies its target proteins and has important implications not only in adult life but also during early development [6, 7]. The multistep, ATP-dependent SUMO conjugation pathway requires an enzymatic (E1–E2–E3) cascade for the formation of an isopeptide bond between the carboxy-terminal di-glycine motif of SUMO paralogs (SUMO1 and the highly related SUMO2/3) and lysine residues in a target protein. The modification is reversed by specific cysteine proteases,

commonly termed SUMO-specific isopeptidases or SUMO proteases. These enzymes cleave the isopeptide bond between the SUMO moiety and substrates. SUMO-specific isopeptidases belong to three distinct families, viz. the Ulp/SENP, the Desi and the USPL1 family [8–10]. In humans, six different SENP isoforms (SENP1, 2, 3, 5, 6 and 7) with distinct subcellular localizations, substrate specificities and functions have been identified. For example, SENP3 exerts preferential cleavage activity toward SUMO2-/3-modified substrates. It is distributed in the nucleolus and nucleoplasm and shuttles between these two compartments in an mTOR-controlled process [11]. Work from our group and others has initially established a role of SENP3 in ribosome maturation [11–14]. Subsequently, we and others could also unravel a crucial function of SENP3 in the control of gene expression [15–17], in particular in the regulation of homeobox (*HOX*) genes [18].

*Correspondence: nayak@med.uni-frankfurt.de;
ste.mueller@em.uni-frankfurt.de; nayak@med.uni-frankfurt.de
[1] Institute of Biochemistry II, Goethe University Medical School, University Hospital Building 75, Theodor-Stern-Kai 7, 60590 Frankfurt am Main, Germany
Full list of author information is available at the end of the article

The coordinated expression of *HOX* genes that encode homeodomain-containing transcription factors is critical for embryonic development [19–21]. In adult tissues, *HOX* genes are involved in regulating cell commitment pathways such as hematopoietic differentiation [22]. The distal-less (Dll) family of *HOX* genes, which encodes DLX transcription factors, has a critical role in limb development, brain patterning and craniofacial morphogenesis [23]. *DLX3* exerts key functions in osteogenic differentiation, and accordingly loss of *DLX3* expression is a signature of defective osteogenic differentiation processes [24–27].

The expression of *HOX* genes is exquisitely regulated by the epigenetic modifiers, trithorax (*trxG*)- and Polycomb (*PcG*)-group proteins, which activate or repress *HOX* genes, respectively [28–32]. The mixed lineage leukemia (MLL)–histone methyltransferase (HMT) complex proteins belong to the trxG group and catalyze trimethylation of lysine 4 on histone 3 (H3K4me3) [33, 34]. This is typically associated with an active state of chromatin and thus found on transcriptionally active genes or developmentally regulated bivalent promoters [35–40]. In mammals, distinct subtypes of MLL complexes are found. They all consist of a SET1/MLL methyltransferase (SET1A/B, MLL1, 2, 3 or 4) as the catalytic core [41–43]. However, for optimum catalytic activity the core requires the association with a multiprotein module minimally composed of WDR5, RbBP5, Ash2L and DPY-30, known as WRAD module [44]. The ordered and timed assembly of the WRAD module at chromatin is one critical point in the control of SET1/MLL function [41, 45]. Our earlier work established a SUMO-dependent mechanism of this assembly process at a subset of *HOX* genes. We had shown that the MLL1/2 complex subunit RbBP5 is covalently modified by SUMO2. We further demonstrated that the SUMO isopeptidase SENP3 catalyzes the desumoylation of RbBP5, which is a prerequisite for the recruitment of Ash2L and menin into functional MLL1/2 complexes at a subset of *HOX* genes, including the *DLX3* gene [18]. However, how SENP3 itself is regulated and targeted to perform its transcription regulatory role remained unclear.

Here, we identify flightless-I homolog (FLII) as a central player in this process. FLII has initially been described as an actin-remodeling protein that belongs to the evolutionary conserved gelsolin protein superfamily [46–48]. The FLII protein consists of an N-terminal leucine-rich repeat (LRR) domain involved in protein–protein interaction. The C-terminus of FLII is made up of the gelsolin-like domain, which mediates actin binding and protein–protein interaction. FLII function has been assigned to cytoskeleton organization during cell migration and negative regulation of wound repair [49]. Apart from cytoskeleton-related functions, FLII has also been shown to act as a transcriptional co-activator in nuclear receptor signaling [50].

In this study, we identify FLII as a major regulator of SENP3. We demonstrate that FLII regulates the chromatin association of SENP3, thereby modulating MLL1/2 complex activity. Through this pathway, FLII governs *DLX3* gene expression and determines osteogenic differentiation process of human mesenchymal stem cells.

Result

FLII is a major interactor of SENP3

Our earlier work unraveled a functional and physical association of SENP3 with pre-60S ribosomes and MLL1/2 histone methyltransferase complexes [13, 14, 18]. A major interaction platform for SENP3 at pre-60S ribosomes is the PELP1 complex, but how SENP3 is targeted to chromatin and the MLL1/2 complex is currently unknown. In order to investigate these issues, we performed a system-wide interactome study for endogenous SENP3 using SILAC (stable isotope labeling in cell culture)-based mass spectrometry (MS). Cell lysates from HeLa cells labeled with light or heavy amino acids were incubated with either control or anti-SENP3 antibodies for immunoprecipitation. Following separation by SDS-PAGE, proteins were excised from gels, trypsinized and peptides measured by MS. After applying a number of stringent filtering criteria (as detailed in the legends to Additional file 1: Figs. S1 and S2), a cytoscape (version 3.3.0) map of the SENP3 interactome was generated. The map is comprised of 27 proteins that satisfied multilayered filtering criteria. Among the SENP3-associated proteins with the highest SILAC ratio, highest peptide count and lowest PEP score, we consistently found FLII (Additional file 1: Fig. S1, Additional file 2: Table S1).

When these 27 high-confident proteins were analyzed in STRING database [51], a high clustering coefficient of 0.863 was observed (Additional file 1: Fig. S2). This indicates that the proteins enriched by SENP3 IP are high probability clusters of functionally related protein groups. Accordingly, when the SENP3 cytoscape cluster was integrated in the known FLII network extracted from STRING database (version 10.0), we found another FLII interacting protein LRRFIP1 as a common high confidence candidate identified in the SENP3 interactome (Additional file 1: Fig. S2b). This further indicates that SENP3 is connected to FLII.

To validate the MS data, directed co-immunoprecipitation assays were performed. To this end, we first expressed a Flag-tagged SENP3 version in HeLa cells by transient transfection and performed anti-Flag IPs. In this setup, endogenous FLII was specifically captured on anti-Flag beads (Additional file 1: Fig. S3a). When immunoprecipitating endogenous SENP3, we detected the known SENP3 interactors PELP1 and LAS1L upon immunoblotting with the respective antibodies. Moreover, we strongly enriched for endogenous FLII and LRRFIP1 in the anti-SENP3 IP, but not in the IgG control IP (Fig. 1a). Notably, upon depletion of SENP3 by siRNA, FLII was lost in the anti-SENP3 IPs, ruling out that the SENP3–FLII co-IP results from an antibody cross-reaction (Fig. 1b). Moreover, in a reverse anti-FLII-IP, endogenous SENP3 as well as LRRFIP1 was detected by immunoblotting with the respective antibodies (Fig. 1c).

To probe for direct interaction between FLII and SENP3 and to map the binding domain of SENP3 in FLII, we performed GST pull-down assays using recombinantly expressed fragments of FLII. The LRR, GelA and GelB domains were fused to GST and used as baits in pull-down assays to test for binding to SENP3, which was generated by in vitro transcription/translation in a rabbit reticulocyte-based system (Fig. 1d). The experiment revealed a physical interaction of SENP3 predominantly with the GelA domain of FLII.

Previous reports have described both nuclear and cytosolic localizations of FLII [47, 48, 52]. To see in which compartment the SENP3–FLII interaction occurs, cellular proteins were separated in a nuclear and cytosolic fraction and IPs were performed. Anti-FLII immunoprecipitation from the different fractions indicates that the FLII–SENP3 interaction is mostly nuclear (Additional file 1: Fig. S3b). In accordance with this, immunofluorescence assays revealed the distribution of both proteins in the nucleoplasm (Additional file 1: Fig. S3c). Taken together, these observations establish a robust and specific nuclear interaction of SENP3 with FLII via its GelA domain.

FLII is involved in the regulation of SENP3-controlled *DLX3* gene expression

Next, we addressed the physiological relevance of FLII binding to SENP3. Because we had defined *DLX3* as a SENP3-regulated gene, *DLX3* expression was chosen as a model system [18]. We initially monitored *DLX3* mRNA levels by RT-qPCR in control cells or in cells

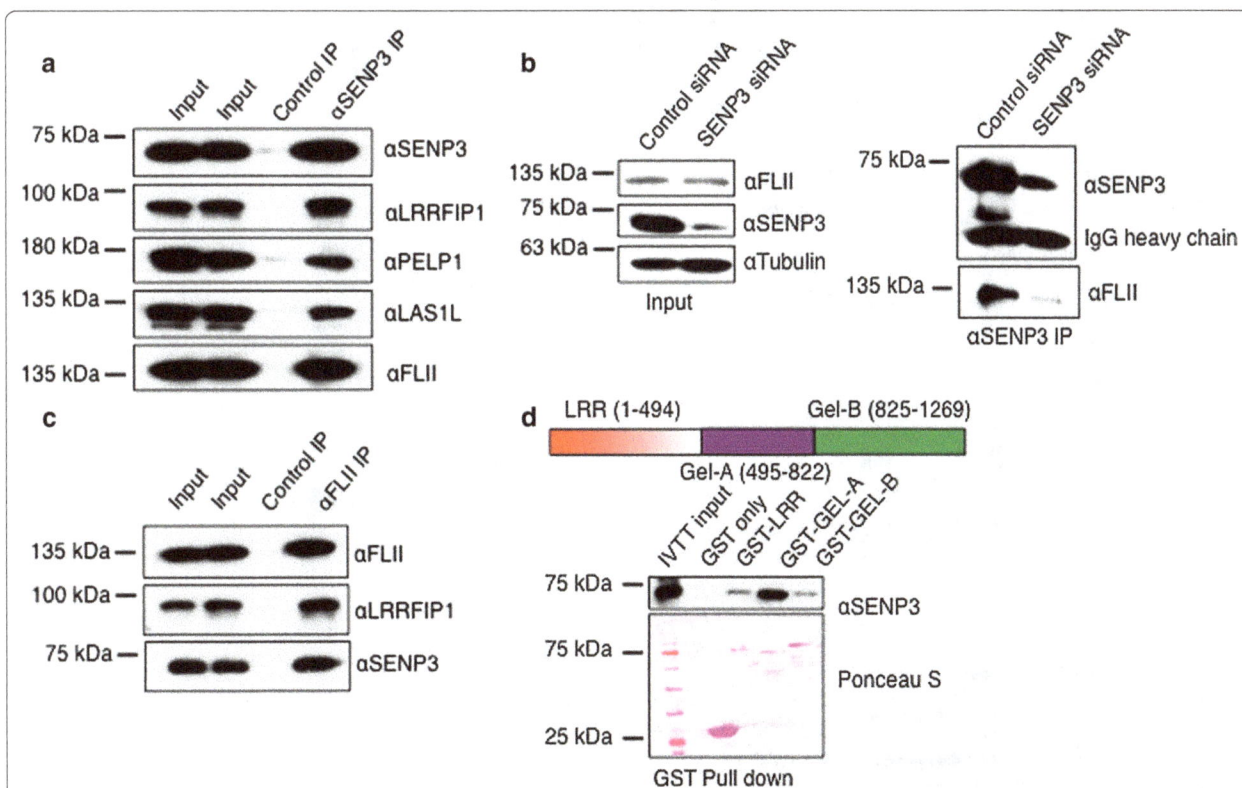

Fig. 1 FLII is a major novel interactor of SENP3. **a** Immunoprecipitation using anti-SENP3 antibody was performed on extracts from HeLa cells. Immunoprecipitates were analyzed by SDS-PAGE and probed with antibodies as indicated. **b** Similar to **a** except, SENP3 was depleted from HeLa cells by siRNA. After 72 h of siRNA transfection, SENP3 IP was performed. SDS-PAGE and immunoblotting with selected antibodies were performed as indicated. **c** Same as **b** but antibody directed against FLII was used for immunoprecipitation. **d** Schematic representation of FLII domain structure. The *numbers in brackets* indicate the amino acids. Equal amounts of in vitro transcribed/translated SENP3 were mixed with either GST only as a control or with GST-tagged FLII fragments (GST-LRR, GST-GelA and GST-GelB) and GST pull-down was performed. After intensive washing, bound proteins were eluted in SDS-sample buffer, separated by SDS-PAGE and probed for the presence of SENP3

depleted from FLII by two different siRNAs. Intriguingly, when compared to controls, depletion of FLII with either siRNA results in a fivefold–tenfold down-regulation of *DLX3* mRNA expression (Fig. 2a). Expression of the siRNA-resistant Gel A or GelA/B domains of FLII restores the *DLX3* gene expression to nearly 65 and 80%, respectively (Fig. 2b). Notably, FLII protein depletion had almost no effect on expression of the homeobox gene *HOXC8* (Fig. 2a), which is also not sensitive to SENP3 depletion [18]. To get a first clue whether FLII and SENP3 dictate DLX3 expression via a common pathway, we co-depleted both proteins by siRNA. Importantly, when compared to individual depletion of either SENP3 or FLII, co-depletion of both proteins did not further decrease DLX3 mRNA levels indeed supporting the idea that SENP3 and FLII act in a common pathway in controlling DLX3 expression (Fig. 2c).

In our previous work, we demonstrated that SENP3 is associated with MLL1/2 methyltransferase complexes and exerts control on DLX3 expression by regulating the assembly of these complexes. Based the above physical association of SENP3 and FLII, we therefore asked whether FLII is also associated with MLL1/2 components. To address this, we performed endogenous anti-FLII IPs and tested for the presence of components of MLL1/2 complexes in the immunoprecipitated material. We detected RbBP5, menin, WDR5 and the catalytic core subunits MLL1 and MLL2, but not the related Set1A protein in the FLII immunoprecipitate (Fig. 3a; Additional file 1: Fig. S4a–c). Moreover, by employing GST pull-down assays we observed a physical interaction between RbBP5 and FLII mediated mostly through the GelA domain (Additional file 1: Fig. S4d and e).

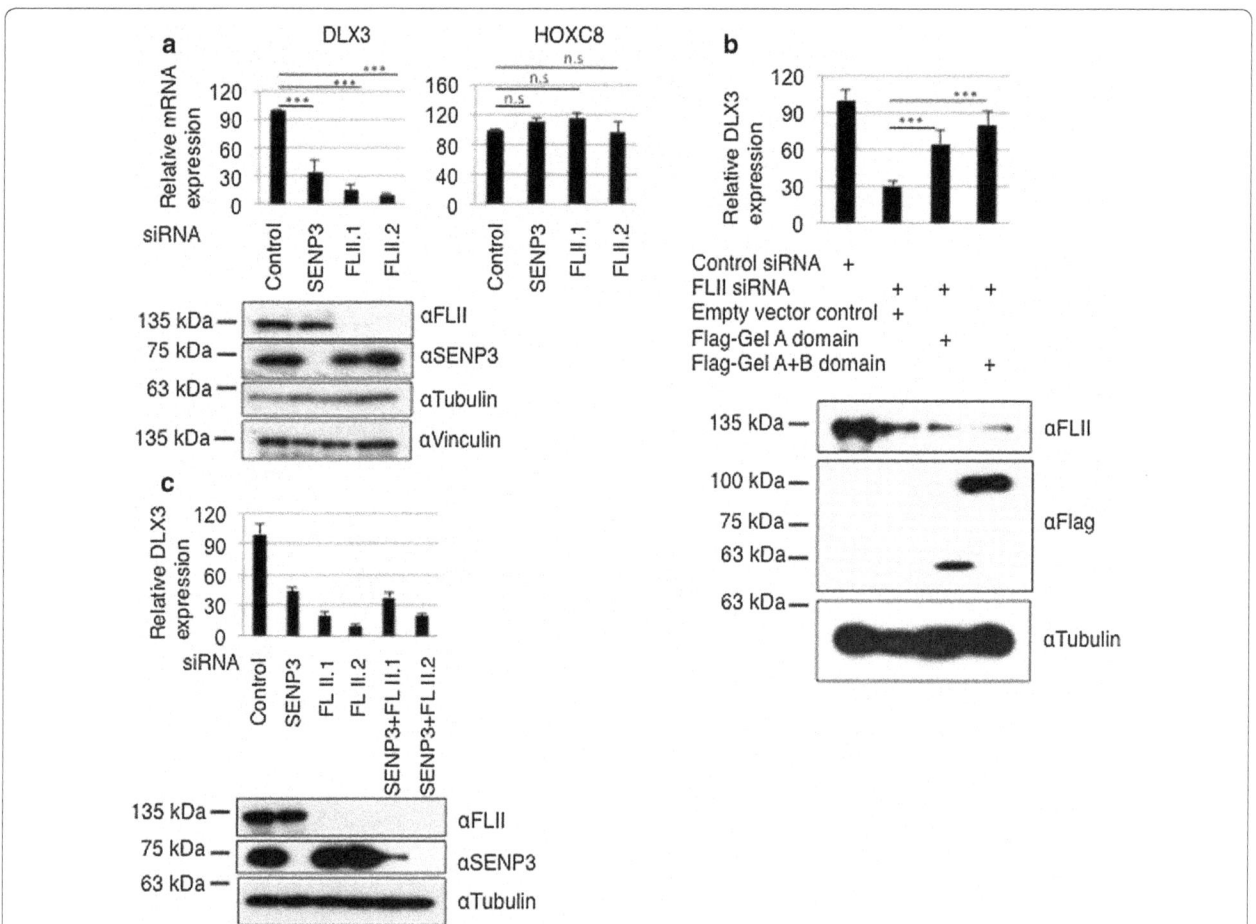

Fig. 2 FLII controls *DLX3* gene expression. **a** FLII was depleted from HeLa by two independent siRNAs. Total RNA was prepared from control cells or cells transfected with either SENP3 or FLII siRNA. RNA was reverse transcribed and cDNA was used in real-time quantitative PCR (RT-qPCR) to monitor the expression of *DLX3* and *HOXC8*. Normalized values (against GAPDH mRNA level) represent the average of three independent experiments performed in triplicate ± SEM ($n = 3$, T test, ***$p < 0.001$, n.s. not significant). The *lower panel* demonstrates efficient depletion of SENP3 or FLII. **b** Same as in **a** except flag-FLII domains were expressed. Data represent the average values from three independent experiments in duplicate ± SEM. **c** Same as in **a** except co-depletion of FLII together with SENP3 was performed. The *lower panel* is a representative western blot, demonstrating efficient depletion of SENP3 and FLII

To reveal a functional connection of FLII to the MLL1/2 methyltransferase complex, we co-depleted cells from FLII together with distinct components of the MLL1/2 complexes. Similar to what was observed upon co-depleting SENP3 and FLII, co-depletion of RbBP5 or WDR5 together with FLII did not further reduce the level of DLX3 mRNA when compared to their single depletion (Fig. 3b). These observations support the idea of a common signaling pathway involving SENP3, FLII and MLL1/2 complexes in the execution of *DLX3* expression.

Next, we asked whether FLII regulates a broader spectrum of *HOX* genes. To this end, we investigated the effect of FLII depletion on the expression of a set of selected *HOX* genes—*HOXA9*, *HOXB3*, *MEIS1* and *MEOX1*—that were previously defined as SENP3-responsive genes [18]. Importantly, depletion of FLII by two different siRNAs resulted in down-regulation of mRNA transcripts of all four genes (Fig. 3c) and further strengthened the idea that FLII is involved in regulating the expression of *HOX* genes in conjunction with SENP3.

Noteworthy, the transcriptional function of FLII has previously connected to the estrogen receptorα (ERα) pathway (50). However, since ERα is not expressed in HeLa cells, the effects described here cannot be assigned to an interplay of FLII with ERα. Moreover, in the ERα-positive MCF7 cell line, SENP3 does not act in conjunction with FLII in ERα-mediated gene regulation (Additional file 1: Fig. S5a).

SENP3 recruitment to its target gene *DLX3* is dependent on FLII

We next aimed to determine the molecular mechanism of FLII-controlled SENP3 function. As the catalytic activity of SENP3 is indispensible for *DLX3* gene expression, we first checked whether FLII has any role in modulating SENP3 catalytic activity. However, a SENP3 activity assay in HeLa cell lysates from control cells or cells depleted from FLII did not reveal differences in catalytic activity (Additional file 1: Fig. S5b and c).

We next focused on a possible direct role of FLII at chromatin and therefore asked whether FLII is present

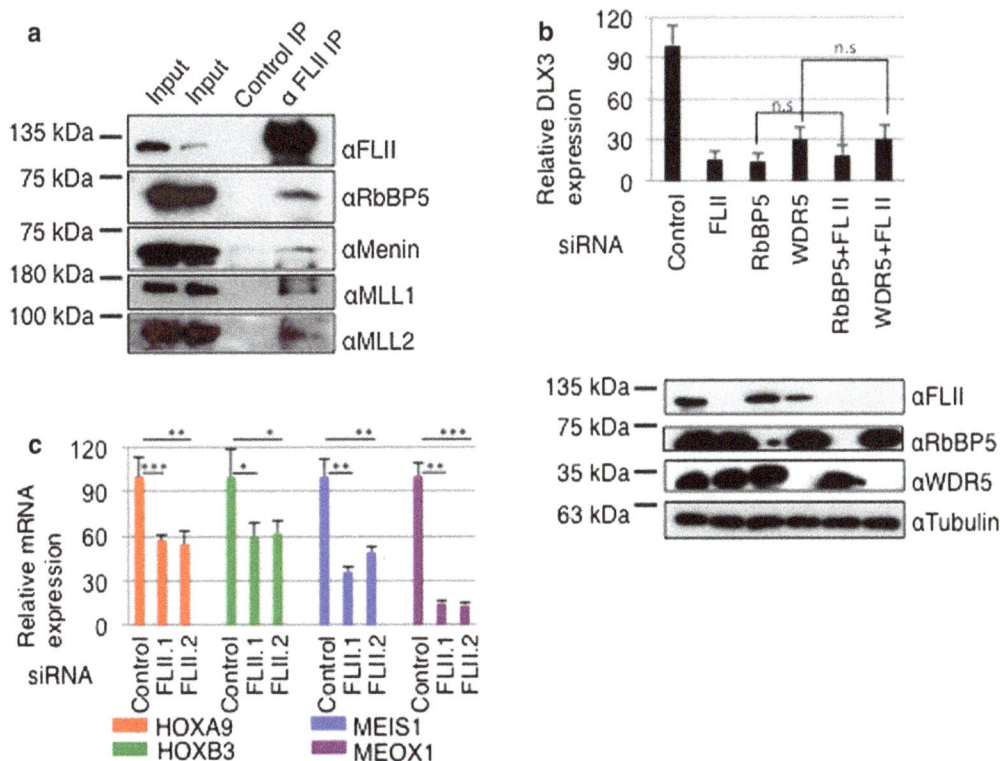

Fig. 3 FLII is functionally associated with MLL1/2 complexes and governs *HOX* gene expression. **a** Endogenous FLII was immunoprecipitated from HeLa cell extracts, and the immunoprecipitates were probed for the presence of RbBP5, menin, MLL1 or MLL2 as indicated. **b** Same as Fig. 2a except total RNA was prepared from control cells or cells individually depleted for FLII, RbBP5 and WDR5 (SET1/MLL component subunit) by siRNAs or co-depleted for FLII and the indicated SET1/MLL components. RNA was reverse transcribed and cDNA was used in RT-qPCR to monitor *DLX3* gene expression. Data represent as the average of three independent experiments performed in triplicate ± SEM. **c** FLII was depleted from HeLa by two independent siRNAs, and RT-qPCR was performed to monitor the expression of *HOXA9*, *HOXB3*, *MEIS1* and *MEOX1* ($n = 3$, T test, $*p < 0.05$; $**p < 0.01$; $***p < 0.001$, n.s. not significant)

at the *DLX3* gene. To address this issue, ChIP (chromatin immunoprecipitation) assays were performed in order to monitor the binding of SENP3 and FLII in the promoter region or exon 3 of *DLX3* gene as we had previously detected SENP3 at these sites (Fig. 4a) [18]. Cross-linked chromatin–protein complexes from HeLa cells indeed revealed a strong enrichment of exogenous Flag-tagged FLII as well as endogenous FLII on the *DLX3* gene locus (Additional file 1: Fig. S6a; Fig. 4b). Importantly, enrichment of FLII on chromatin was not influenced by depletion of SENP3 (Fig. 4b), indicating that FLII was recruited to chromatin independently from SENP3. These results

led us to the idea that FLII may in turn target SENP3 to chromatin and in particular to the *DLX3* gene locus. To test this idea, ChIP assays with an anti-SENP3 antibody were performed as above, and immunopurified *DLX3* gene fragments were quantified by qPCR. Importantly, depletion of FLII leads to a more than threefold reduction of SENP3 at the *DLX3* gene locus, both in the promoter and in the exon regions (Fig. 4c). Noteworthy, this reduction was not due to an altered stability or expression of SENP3, as SENP3 levels remained unchanged upon FLII depletion (Fig. 4d; Additional file 1: Fig. S6b). As SENP3 is required for proper chromatin recruitment

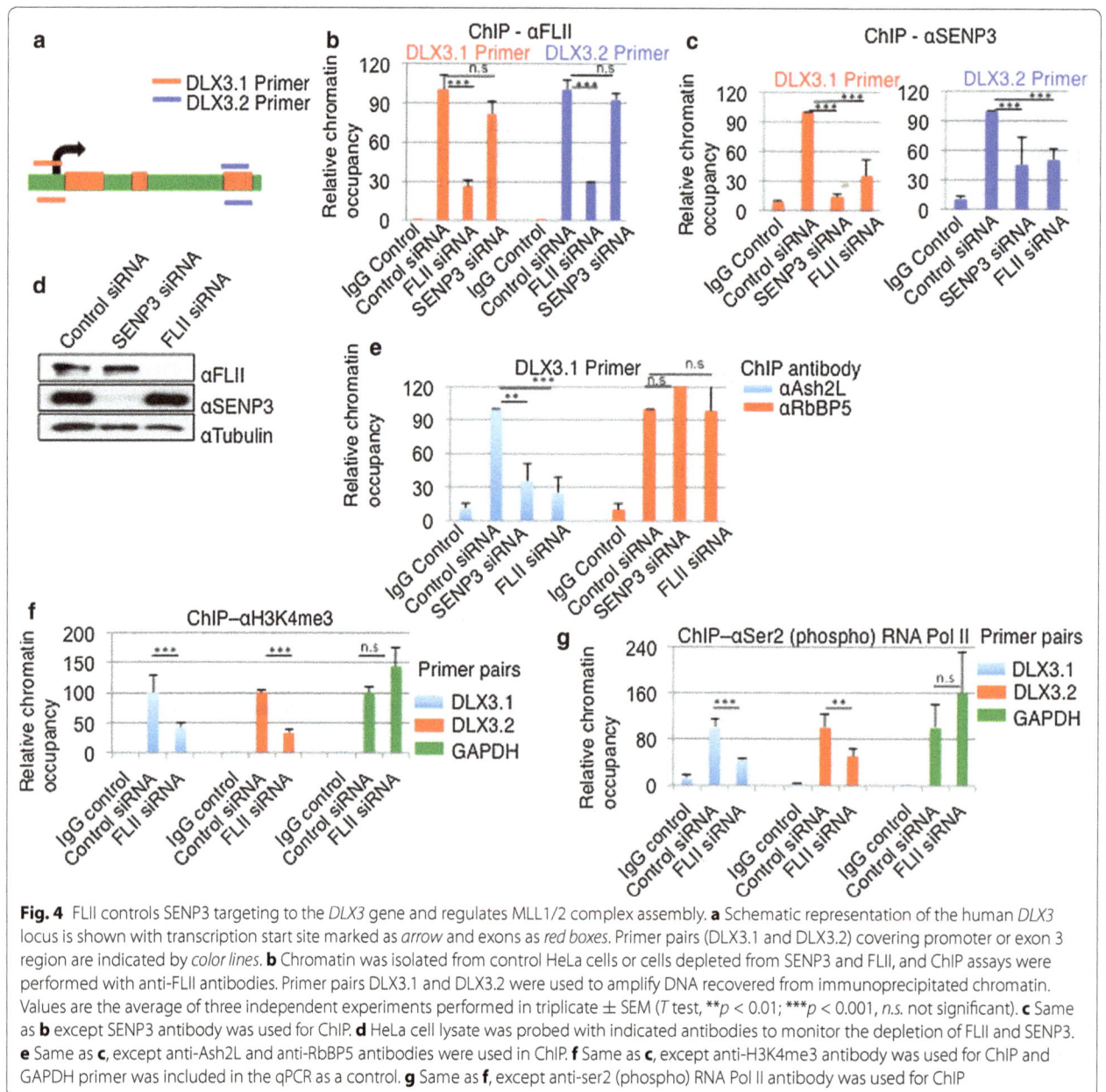

Fig. 4 FLII controls SENP3 targeting to the *DLX3* gene and regulates MLL1/2 complex assembly. **a** Schematic representation of the human *DLX3* locus is shown with transcription start site marked as *arrow* and exons as *red boxes*. Primer pairs (DLX3.1 and DLX3.2) covering promoter or exon 3 region are indicated by *color lines*. **b** Chromatin was isolated from control HeLa cells or cells depleted from SENP3 and FLII, and ChIP assays were performed with anti-FLII antibodies. Primer pairs DLX3.1 and DLX3.2 were used to amplify DNA recovered from immunoprecipitated chromatin. Values are the average of three independent experiments performed in triplicate ± SEM (*T* test, **$p < 0.01$; ***$p < 0.001$, *n.s.* not significant). **c** Same as **b** except SENP3 antibody was used for ChIP. **d** HeLa cell lysate was probed with indicated antibodies to monitor the depletion of FLII and SENP3. **e** Same as **c**, except anti-Ash2L and anti-RbBP5 antibodies were used in ChIP. **f** Same as **c**, except anti-H3K4me3 antibody was used for ChIP and GAPDH primer was included in the qPCR as a control. **g** Same as **f**, except anti-ser2 (phospho) RNA Pol II antibody was used for ChIP

of Ash2L and menin [18], we wondered whether defective chromatin residency of SENP3 in the absence of FLII might also affect MLL1/2 complex assembly. Therefore, we followed the chromatin association of Ash2L on the *DLX3* gene by ChIP as described above. Importantly, FLII depletion perfectly phenocopied the loss of SENP3 with respect to Ash2L residency at the *DLX3* gene. Very similar to what was observed upon depletion of SENP3, Ash2L occupancy at the *DLX3* promoter is reduced to 20–30%. However, RbBP5, another core subunit of the MLL1/2 complexes, as well as MLL1 and MLL2 itself remains associated with the *DLX3* gene in the absence of either FLII or SENP3 (Fig. 4e; Additional file 1: Fig. S6c–e). Noteworthy, FLII depletion did not affect the protein level of any of the core components of MLL1/2 complexes, ruling out that the reduced chromatin occupancy in the absence of FLL is due to changes in protein stability (Additional file 1: Fig. S7a, b). Importantly, FLII depletion had no effect on Ash2L chromatin association at the *HOXC8* gene, which is consistent with the observation that expression of HOXC8 was not sensitive to either FLII or SENP3 depletion (Additional file 1: Fig. S7c). Altogether, this indicates that FLII is involved in proper SENP3 targeting to distinct *HOX* genes, thereby controlling the assembly and function of MLL1/2 complexes.

To further validate this idea, we monitored H3K4 trimethylation (H3K4me3) at the *DLX3* gene, because this epigenetic mark is catalyzed by MLL1/2 complexes. H3K4me3 can be assayed by ChIP experiments using an antibody directed against this modification. In FLII-depleted cells, H3K4me3 marks at the *DLX3* gene were reduced to 30–40%, when compared to control cells. It is worth noting that no reduction of H3K4 trimethylation was detected on the unrelated control gene *GAPDH* (Fig. 4f) or *HOXC8* (Additional file 1: Fig. S7d). H3K4me3 is typically considered as a chromatin mark for active genes, and accordingly active RNA Pol II is found on genes that harbor H3K4-trimethyl marks. In particular, the serine 2 phosphorylated form of RNA Pol II C-terminal domain (CTD) is a signature of productive transcription, and it overlaps with H3K4-trimethyl-enriched chromatin domains [53]. We therefore monitored the presence of ser2P-RNA Pol II at the *DLX3* gene locus in the presence or absence of FLII. Again, we observed more than twofold reduction of Pol II residency at the *DLX3* gene, but not on the control *GAPDH* or *HOXC8* gene, when endogenous FLII is knocked down (Fig. 4g; Additional file 1: Fig. S7e). Collectively, these data suggest that FLII governs chromatin association of SENP3 on specific *HOX* genes, thereby regulating MLL1/2 complex assembly, which ultimately impinges on gene activation.

To further strengthen the idea that the FLII recruits SENP3 to the MLL complex assembly on distinct gene

loci at chromatin, we performed FLII siRNA followed by immunoprecipitation of endogenous SENP3 and probed for co-precipitation of RbBP5 and vice versa. When compared to control cells, RBBP5 is much reduced in the anti-SENP3 IP upon loss of FLII. Even more strikingly SENP3 does not co-immunoprecipitate with RbBP5 in the absence of FLII (Additional file 1: Fig. S7f). Importantly, however, SENP3 depletion did not affect Ash2L interaction with RbBP5 further supporting our notion that SENP3 does not generally affect the assembly of MLL complexes in solution, but acts at chromatin on distinct *HOX* genes. (Additional file 1: Fig. S7g).

FLII regulates sumoylation of RbBP5 and is critical for differentiation of human dental follicle stem cells

In our previous work, we had shown that the MLL1/2 complex subunit RbBP5 is covalently modified by SUMO2. We further demonstrated that SENP3 catalyzes the desumoylation of RbBP5, which is a prerequisite for the recruitment of Ash2L and menin into functional MLL1/2 complexes at the *DLX3* gene [18]. Since FLII is needed for SENP3 recruitment to the *DLX3* gene, we reasoned that loss of FLII would translate into an altered sumoylation status of RbBP5. To test this possibility, we introduced His-SUMO2 together with Flag-RbBP5 in HeLa cell and performed anti-Flag immunoprecipitation followed by anti-SUMO2/3 immunoblotting. Cells were either mock depleted or depleted from FLII by siRNA. Importantly, in anti-FLAG IPs, RbBP5–SUMO2 conjugates where enriched upon depletion of FLII (Fig. 5a) indeed suggesting that FLII is needed to channel SENP3 activity to RbBP5.

Since loss of FLII phenocopied the loss of SENP3 with respect to DLX3 expression, we asked whether FLII is also functionally connected to osteogenic differentiation. Primary human dental follicle stem cells (DFCs) were employed as an experimental system for the investigation of FLII function in this process. In the absence of SENP3, osteogenic differentiation is prevented due to improper induction of the DLX3, which is needed for the activation of downstream targets, such as RUNX2 and alkaline phosphatase (ALP). To investigate whether FLII is also involved in this pathway, we depleted endogenous FLII from DFCs and followed the expression of DLX3 upon induction of differentiation. In this experimental setup, loss of FLII resulted in a threefold to fourfold reduction of DLX3 mRNA and protein levels when compared to control cells (Fig. 5b–d). Accordingly, the downstream targets of DLX3, RUNX2 and ALP were also down-regulated (Additional file 1: Fig. S8a). Moreover, a significant decrease in ALP activity (Fig. 5e), which is indicative of defective osteogenic differentiation, was observed. Proper osteogenic differentiation of DFCs

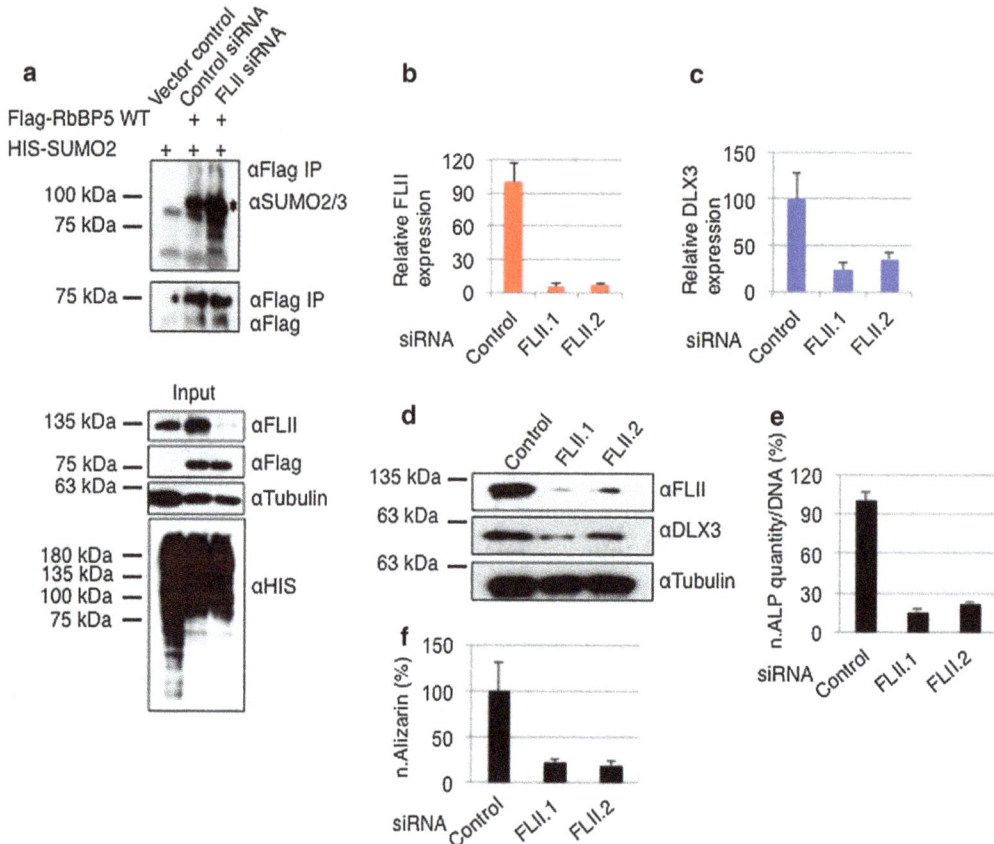

Fig. 5 FLII regulates RbBP5 sumoylation and osteogenic differentiation of DFCs. **a** HeLa cells treated with control siRNA or FLII siRNA were transfected with Flag-RbBP5 and SUMO2 as indicated, and SUMOylation of RbBP5 was monitored by anti-SUMO2 immunoblotting following anti-Flag-IP. The *asterisk* represents the sumoylated RbBP5 species. **b, c** Total RNA was isolated from DFCs, after treating cells with two independent FLII siRNAs or control siRNA. Expression of FLII (**b**) and DLX3 mRNA (**c**) was determined by RT-qPCR. Values (normalized for GAPDH expression) represent the mean of three experiments ± SEM. **d** Cell lysates originating from DFCs (**b, c**) were separated by SDS-PAGE and analyzed by immunoblotting with the indicated antibodies. **e** ALP activity was quantified in differentiating DFCs, 10 days after transfection with FLII siRNAs or control siRNA and incubation with osteogenic differentiation medium (ODM). **f** Mineral deposits in DFC cultures were measured by alizarin red staining after 5 weeks of culture with ODM and transfection with two independent FLII siRNAs or a control siRNA

is characterized by intracellular deposition of calcium, which can be monitored by alizarin staining [54]. Importantly, FLII depletion caused drastically reduced alizarin staining indicating an impaired differentiation process (Fig. 5f and Additional file 1: Fig. S8b). Taken together, these findings establish a critical role of FLII in the control of osteogenic differentiation in human mesenchymal stem cells.

Discussion

Our previous study unraveled a pathway, in which SENP3-mediated desumoylation of the RbBP5 subunit triggers the assembly of a functional MLL1/2 complex by recruitment of the Ash2L and menin subunits. Here, we define FLII as a central regulator of this pathway and provide evidence that FLII determines the chromatin

recruitment of SENP3. FLII association with SENP3 and MLL1/2 complex is crucial for *HOX* gene expression. Mechanistically, we could delineate that FLII action triggers molecular events necessary for MLL1/2 complex association and subsequent epigenetic modifications permissive for *HOX* gene transcription required for human osteogenic differentiation.

FLII harbors an N-terminal leucine-rich repeat domain and two gelsolin-related domains, each of which is comprised of three repeats of around 150 amino acids [55]. Gelsolin domains define a family of actin-remodeling proteins, and accordingly FLII has been characterized as a regulator of cytoskeletal organization and actin dynamics. In addition to its role in actin remodeling, a transcriptional co-activator function in nuclear receptor signaling has been assigned to FLII. FLII interacts with estrogen receptor α (ERα) and the co-activators GRIP1 and CARM1 [50]. Moreover, FLII interacts with BAF53, a component of SWI/SNF chromatin remodeler complexes, and recruits SWI/SNF complex to its target genes [55]. Thus, FLII facilitates chromatin accessibility on ERα target genes, which was proposed to result in a proliferative advantage to breast cancer cells [56]. Our data are consistent with the dual roles for FLII as a transcriptional co-activator in the nucleus and a cytoplasmic modulator of the cytoskeleton. However, our work expands the regulatory function of FLII on transcriptional processes. We provide evidence that FLII is physically and functionally connected to MLL1/2 histone methyltransferase complexes and controls the recruitment of the desumoylating enzyme SENP3 to a subset of MLL1-/2-regulated *HOX* genes. Our interactome studies identified FLII as a very robust direct binding partner of SENP3. Binding is mediated predominantly through the GelA domain. Importantly, in the absence of FLII, SENP3 is not properly recruited to our model homeobox gene *DLX3*. As a consequence of reduced SENP3 occupancy, the MLL complex is not properly assembled, H3K4 trimethylation is impaired and expression of *DLX3* is down-regulated. Intriguingly, there is a striking overlap of *HOX* genes co-regulated by both FLII and SENP3. The expression of *HOXA9, HOXB3, MEIS1, MEOX1* and *DLX3* is all affected by depletion of either FLII or SENP3, while *HOXC8* expression remains unaltered in the absence of either SENP3 or FLII. Together with the fact that co-depletion of FLII and SENP3 is epistatic in controlling DLX3 expression, this strongly supports the idea that FLII acts on *HOX* gene expression through recruitment of SENP3. One possible scenario is that FLII, SENP3 and RbBP5 form a complex, in which FLII acts as a scaffold or adaptor that positions SENP3 in proximity to its substrate RbBP5. This allows desumoylation of RbBP5, which is a prerequisite for proper assembly of the MLL1/2 complexes at chromatin. Altogether, these data indicate that FLII contributes to SENP3 target selection by determining its recruitment to specific chromatin locations. How FLII acquires selectivity for addressing SENP3 to only a subset of MLL1/2 targets remains to be determined, but generally our data further strengthen the concept that substrate specificity of SENPs is dictated by spatial control. SENP3 stands as a paradigm for this concept as its activity is concentrated in the nucleolus, where it functions in ribosome maturation in complex with PELP1–TEX10–WDR18, and at transcriptionally active chromatin sites in conjunction with SET1/MLL methyltransferase complexes. Our data indicate that chromatin association of SENP3 requires FLII, while its nucleolar sequestration is mainly determined by the nucleolar scaffold protein nucleophosmin (NPM1) [11]. Noteworthy,

FLII is largely excluded from the nucleolus making it unlikely that it is involved in the nucleolar functions of SENP3.

Our data also unravel an important and so far unprecedented implication of FLII in cell differentiation. In dental follicle progenitor cells, which serve as a model system of osteogenic differentiation, FLII depletion recapitulates the loss of SENP3 and impairs expression of the osteogenic master regulator DLX3. This in turn affects the expression of osteogenic regulators, such as RUNX2 and alkaline phosphatase, ultimately preventing proper osteogenesis. So far, FLII has been mainly linked to ERα-regulated transcriptional processes. Interestingly, however, at least on some *HOX* genes a cross talk between ERα activation and MLL-dependent H3K4 trimethylation has been reported [57–60]. Based on these observations, it is tempting to speculate that the FLII–SENP3 axis integrates multiple signaling pathways in cellular differentiation processes. Consistent with this idea, it has recently been shown that ERα regulates DLX3-mediated osteoblast differentiation [61]. Generally, our work expands the emerging concept of actin-binding proteins functioning in transcriptional processes. One paradigm for this concept is Wave1, which has a cytoplasmic function in actin reorganization as a downstream target of RAC. Intriguingly, however, reminiscent to our findings on FLII, nuclear Wave1, associates with SET proteins and modulates their activities on *HOX* genes [62]. Understanding how the nuclear and cytosolic functions of actin-binding proteins are coordinated will be one key question for future research.

Conclusions

In summary, our work defines an unrecognized molecular mechanism that connects the actin-remodeling protein FLII to SENP3-sensitive *HOX* gene regulation. By recruiting SENP3 to chromatin at a subset of *HOX* genes, FLII coordinates MLL complex assembly and recruitment of active RNA polymerase II. FLII therefore functions as a critical specificity factor in the control of SENP3-mediated gene expression programs.

Methods
Cell culture/SILAC and transfection
HeLa cells were grown using standard conditions in Dulbecco's modified Eagle's medium (DMEM) supplemented with 10% fetal bovine serum and standard antibiotics. For SILAC labeling, cells were cultured in DMEMs (Thermo Scientific, product no. 89985) supplemented with 10% dialyzed serum (Invitrogen), antibiotics and amino acid isotope (R_0K_0 or R_6K_4, Cambridge Isotope Laboratories). Cells were grown for five passages to ensure the incorporation of labeled amino acids [63]. After five passages,

incorporation of labeled amino acids to more than 95% was checked. siRNA-mediated knockdown experiments in HeLa cells were performed using Oligofectamine (Invitrogen) according to the manufacturer's instructions.

Mass spectrometry and data processing

Overnight in-gel trypsin (12.5 ng/µl) digested peptide mixture was desalted and purified by C-18 StageTip method. An Easy-nLCII liquid chromatography—coupled to a LTQ Orbitrap Elite mass spectrometer (Thermo Scientific)—was used for peptide elution and mass spec analysis. Desalted peptide mixtures were eluted with a 5–33% gradient HPLC solvent B (80% acetonitrile in 0.5% acetic acid). A mass range of m/z 150–2000 with a resolution of 120,000 was set to acquire full MS scan. Twenty most intense ions (Top20 method) were sequentially selected for CID fragmentation process. Data-dependent scanning mode with 90 s of dynamic inclusion was set for MS/MS scan. Data analysis was performed using MaxQuant software [64] version 1.3.0.5 with supported by Mascot as the database search engine for peptide identification. False discovery rate (FDR) was kept at 1% for data filtering, and two missed cleavages were allowed. Initial mass tolerance was set to 7 ppm and 0.5 Da for the fragment ion level. Cysteine carbamidomethylation was set as fixed modification, whereas protein N-terminal acetylation and oxidation of methionine were defined as variable modifications. Information about peptide MS/MS spectra was extracted from MaxQuant viewer. Cytoscape version 3.3.0 was used to generate protein network.

Immunoprecipitation and chromatin immunoprecipitation

Immunoprecipitations were performed as described [18]. For endogenous SENP3 IP-MS, 80 millions SILAC-labeled cells were used. Cells were lysed in buffer containing 50 mM HEPES pH 7.5, 150 mM NaCl, 2 mM EDTA, 0.5% TritonX-100 with freshly added protease and phosphatase inhibitors (Complete/PhosphoSTOP, Roche). SENP3 IP (for 2 h) was performed with the lysate originated from R_6K_4 labeled cells and equal amount of R_0K_0 for control rabbit IgG IP. Immune complexes were captured by proteinG dynabeads (Invitrogen) for 1 h. After washing, immunoprecipitates were eluted by boiling the beads with SDS-PAGE loading buffer. Before loading in a SDS-PAGE gradient (4–20%, Biorad) gel, SENP3 IP and control IP were mixed into 1:1 ratio. ChIP assays were performed according to the published method from [65]. After real-time PCR of ChIPed DNA, data calculation was performed by percentage of input method. Two percentage starting chromatin was used as input. The diluted input was adjusted to 100% by subtracting raw diluted input Cp values with 5.64. If the starting input fraction is 2%, then a dilution factor of 50 is 5.64 cycles

(i.e., log2 of 50 is 5.64). Finally, the percentage of input was derived from the following formula, i.e., $100 \times 2^{\wedge}(Cp$ of adjusted input $-$ Cp (ChIP replicates)). The average of at least three technical replicate was considered for a particular experimental setup. Unperturbed condition (such as control siRNA) was set as 100. Changes in chromatin occupancy were then calculated relative to 100. Data were presented as an average of three biological replicates (unless mentioned otherwise in the corresponding legend) with ± SEM. Details of western blot and ChIP antibodies are described in Supplemental information.

In vitro transcription/translation and GST pull-down assay

One microgram of pCI vector encoding the respective proteins was translated with the TNT T7 quick-coupled transcription/translation system (catalog no. L1170) from Promega. GST-fused FLII domains were purification, and GST pull-downs were performed as described previously [66].

Osteogenic differentiation of DFCs and measurement of alkaline phosphatase activity

Isolation, differentiation of DFC, siRNA-mediated protein knockdown and alkaline phosphatase assays were performed as described [18, 24, 54, 67, 68].

For the preparation of human dental follicle cells (DFCs), impacted human third molars were surgically removed and collected from patients with informed consent. DFCs were isolated as described [18, 54]. Differentiation of DFCs, siRNA-mediated protein knockdown and alkaline phosphatase assays were performed according to Nayak et al. [18].

Alizarin red staining

Long-term cultures (after 5 weeks) were washed in PBS and fixed with 4% formaldehyde/0.1 M PBS. Calcium deposits were detected by treatment with a 2% alizarin red S solution (Morphisto). For quantitative measurement, the alizarin red staining was solved in a 10% cetylpyridinium chloride monohydrate solution (PBS) for 30 min. The optical density of samples was measured in a plate reader at 540 nm.

Quantitative reverse transcription polymerase chain reaction (RT-qPCR)

For total RNA isolation, high-pure RNA isolation kit (Roche) was used. Five hundred nanograms of that was used for cDNA synthesis (Transcriptor First strand cDNA synthesis kit, Roche) by using oligo dT primer. RT-qPCR was performed with Fast Start DNA Master SYBR® Green I kit (Roche) and the Light Cycler PCR System (LightCycler 480 II, Roche). For normalization, GAPDH gene expression was used as a control gene expression marker. Quantification of mRNA expression was performed with the standard curve method.

SENP3 catalytic activity assay

About 10 million HeLa cells (Control siRNA in parallel with FLII siRNA for 72 h) were lysed in 1 ml SEM buffer (0.25 M sucrose, 20 mM MOPS–KOH and 1 mM EDTA–NaOH, pH 8) with or without 20 mM N-ethylmaleimide (NEM). Protease inhibitor cocktail and 1 mM DTT were added freshly. After brief sonication (3 strokes for 5 s at 40% amplitude), protein concentration was measured by Bradford assay. Two hundred micrograms of extract was diluted in activity assay buffer (50 mM Tris–HCl pH 7.5, 0.1 mg/ml BSA and 10 mM DTT) and mixed with 100 ng SUMO-2 vinyl sulfone derivative (HA-SUMO2-VS; Boston Biochem) compound. The reaction was kept at room temperature for 10 min. To stop the reaction, sample buffer was added followed by heating at 95° for 5 min.

Immunofluorescence

HeLa cells were fixed in methanol (MeOH) at −20 °C for 5 min, permeabilized with 0.5% Triton X-100 and processed using standard protocols. Images were acquired by Leica TCS SP8 confocal microscope. An anti-SENP3 (clone D20A10 cat no. 5591, Cell signaling) and anti-FL II (SC-21716, Santa Cruz Biotechnology Inc.) were used as primary antibodies for detecting the respective proteins. The following secondary antibodies were used: Alexa Fluor 488–donkey anti-rabbit (Invitrogen), Cy3–donkey anti-mouse (Invitrogen).

Additional files

Additional file 1: Fig S1. Proteome map of SENP3 derived from SILAC-based mass spectrometry. (a) Schematic representation of SENP3 proteomics. Equal no. of HeLa cells (as mentioned in "Methods") either unlabeled or metabolically labeled with amino acid isotope (R6K4) was used for IP. Control IP and SENP3 IP were mixed in a 1:1 ratio and loaded in a SDS-PAGE. The whole lane was cut into several small pieces and processed for mass spec (as described in Materials and Methods section). (b) One representative western blot shows the enrichment of endogenous SENP3 in IP lane that was used for MS analysis. (c) The cytoscape map of SENP3 interactome was accomplished after filtering the whole protein group file (generated from MaxQuant analysis) through 4 tier of following selection criteria—(i) normalized H/L SILAC ratio cutoff was set as 2; that is proteins with minimum twofold enrichment compare to IgG control were considered. (ii) PEP score cutoff was set as (0.0001). PEP score is like p value that represents statistical significance of an observed peptide as a true one. Therefore, smaller PEP score is significant. (iii) Minimum three peptides were considered for any proteins and (iv) reproducibility in both the independent experiments. (d) Cytoscape network of SENP3 interactome obtained from two independent SILAC-MS assays. Details of the generation of the cytoscape map are described in appendix figure S1c. (e) A representative MS/MS spectrum of FLII peptide that was generated by MaxQuant Viewer program. **Fig S2.** SENP3 protein network. (a) The filtered protein candidates (27) generated from SENP3 proteomics were entered into STRING database as input to check for clustering coefficient. Red

arrow indicates the bait, SENP3. (b) Information about FLII protein–protein interaction was extracted from STRING database and combined together with SENP3 network from our experiment. **Fig S3.** FLII–SENP3 interaction is mostly in the nucleus. (a) Related to Fig. 1. HeLa cells were transfected with Flag-SENP3. Two days after transfection, Flag-agarose bead pull down was performed to check the presence of FLII in the western blot. (b) Subcellular fractionation of HeLa cells was performed according to the user protocol (subcellular protein fractionation kit, ThermoFisher Scientific, catalog no. 78840). Endogenous FLII was immunoprecipitated from cytosolic and nuclear fraction. SDS-PAGE of the immunoprecipitate was performed and probed for indicated antibodies. (c) Subcellular localization of endogenous SENP3 and FLII was studied by immunofluorescence using primary antibodies detecting the respective proteins. **Fig S4.** FLII interaction with RbBP5. (a) Related to Fig. 3. Endogenous FLII was immunoprecipitated from HeLa cells, and blot was probed against indicated antibodies. (b) Related to Fig. 3. Endogenous SENP3 was immunoprecipitated from HeLa cells, and blot was checked for indicated antibodies. (c) Related to Fig. 3. Endogenous WDR5 was immunoprecipitated, and blot was probed against indicated antibodies. (d) Same as Fig. 1d except RbBP5 construct was used for in vitro transcription/translation. (e) Same as additional file 1, Fig S3a, except Flag-tagged FLII constructs were used to check the interaction of various FLII domains (as indicated in the figure) with RbBP5. **Fig S5.** SENP3 is not involved in ERα and FLII does not influence SENP3 catalytic activity. (a) Post-transfection (control, SENP3 and FLII siRNA), MCF7 cells were cultured for 3 days in phenol red-free DMEM medium supplemented with 5% charcoal-dextran-stripped fetal bovine serum. Cells were then treated overnight with 100 nM estradiol (E2) before RNA extraction. Data represent the average of triplicates from two biological experiments ± SEM. (b) 72 h after siRNA treatment directed against FLII, cell lysate was prepared in the presence or absence of NEM in the lysis buffer. Equal amount of protein (200µg) of protein from control and FLII siRNA cell lysate was mixed with SUMO2-VS [69] substrate at room temperature for 10 min. The SDS-PAGE was probed by using SENP3 and a loading control tubulin antibody. The asterisk mark represents slow-migrating catalytic active SENP3 form appeared as a result of conjugation between the substrate and SENP3. Right panel shows FLII knockdown efficiency. (c) Same as in (a), except anti-HA antibody (that detects SUMO2-VS substrate) was used. **Fig S6.** FLII influences MLL1/2 complex assembly on *DLX3* gene. (a) Same as Fig. 4b except 5µg of flag-FLII plasmid was transfected to HeLa cells. 48 h. post-transfection cells were fixed and processed for ChIP using rabbit flag antibody. Data represent average of at least two biological experiments performed in duplicate. (b) Related to Fig. 4B. HeLa cell lysate was probed with indicated antibodies to monitor the depletion of FLII and SENP3. (C) Same as Fig. 4e, but DLX3.2 primer was used in qPCR (n = 3, T test, *p < 0.05; **p < 0.01; ***p < 0.001, n.s. not significant). (d) Same as Fig. 4e, except anti-MLL1 and anti-MLL2 antibodies were used for ChIP. DLX3.1 primer was used for qPCR. (e) Same as in d, except DLX3.2 primer was used for qPCR. **Fig S7.** Expression of MLL1/2 complex subunits is unperturbed upon FLII depletion. (a, b) To check the effect of FLII knockdown on MLL1/2 complex subunit protein stability, cell lysates from control and FLII siRNA were probed in western blot for the indicated antibodies. (c) Same as Fig. 4e except *HOXC8* promoter primer was used (n = 3, n.s. not significant). (d, e) Related to Fig. 4f and g except *HOXC8* promoter primer was used. (f) HeLa cells were transfected with FLII siRNA for 72 h followed by endogenous SENP3 and RbBP5 immunoprecipitation. Blot was probed with indicated antibodies. (g) Same as (f) but SENP3 siRNA was performed and RbBP5 antibody was used for immunoprecipitation. **Fig S8.** FLII is required for osteogenic differentiation. (a) Expression of RUNX2 and ALP mRNA was determined by RT-qPCR in control or DFCs depleted from FLII. Values (normalized for GAPDH expression) represent the mean of three experiments ± SEM. (b) The status of mineral deposition in DFC was measured by alizarin stain. Cells were transfected with two different FLII siRNA and grown in ODM culture for 5 weeks.

Additional file 2: Table S1. Table shows the detail of proteins that passed the filtering criteria of Fig S1.

Additional file 3. Supplementary materials and methods.

Abbreviations

Ash2L: absent, small or homeotic 2-like; MLL: mixed lineage leukemia; RbBP5: retinoblastoma-binding protein 5; SUMO: small ubiquitin-like modifier; SENP3: sentrin-specific protease 3; pol II: RNA polymerase II; DFC: dental follicle stem cells; FLII: flightless-I homolog.

Authors' contributions

AN designed and performed most of the experiments and wrote the manuscript. Experiments using DFCs were carried out by AR and CM. SM supervised the project and wrote the manuscript. All authors read and approved the final manuscript.

Author details

[1] Institute of Biochemistry II, Goethe University Medical School, University Hospital Building 75, Theodor-Stern-Kai 7, 60590 Frankfurt am Main, Germany.
[2] Department of Oral and Maxillofacial Surgery, University of Regensburg, 93042 Regensburg, Germany.

Acknowledgements

We thank all members of our Institute for support and stimulating discussions. We thank M. Stallcup, K. Jeong and X. Tong for providing plasmids used in this work.

Competing interests

The authors declare that they have no competing interests.

Funding

This work was funded by the DFG (SFB 815/1177 and MU-1764/4), LOEWE Ub-Net and the Fritz Thyssen Foundation.

References

1. Hay RT. SUMO: a history of modification. Mol Cell. 2005;18:1–12.
2. Gareau JR, Lima CD. The SUMO pathway: emerging mechanisms that shape specificity, conjugation and recognition. Nat Rev Mol Cell Biol. 2010;11:861–71.
3. Flotho A, Melchior F. Sumoylation: a regulatory protein modification in health and disease. Annu Rev Biochem. 2013;82:357–85.
4. Cubenas-Potts C, Matunis MJ. SUMO: a multifaceted modifier of chromatin structure and function. Dev Cell. 2013;24:1–12.
5. Raman N, Nayak A, Muller S. The SUMO system: a master organizer of nuclear protein assemblies. Chromosoma. 2013;122:475–85.
6. Wang L, Wansleeben C, Zhao S, Miao P, Paschen W, Yang W. SUMO2 is essential while SUMO3 is dispensable for mouse embryonic development. EMBO Rep. 2014;15:878–85.
7. Nacerddine K, Lehembre F, Bhaumik M, Artus J, Cohen-Tannoudji M, Babinet C, Pandolfi PP, Dejean A. The SUMO pathway is essential for nuclear integrity and chromosome segregation in mice. Dev Cell. 2005;9:769–79.
8. Hickey CM, Wilson NR, Hochstrasser M. Function and regulation of SUMO proteases. Nat Rev Mol Cell Biol. 2012;13:755–66.
9. Mukhopadhyay D, Dasso M. Modification in reverse: the SUMO proteases. Trends Biochem Sci. 2007;32:286–95.
10. Nayak A, Muller S. SUMO-specific proteases/isopeptidases: SENPs and beyond. Genome Biol. 2014;15:422.
11. Raman N, Nayak A, Muller S. mTOR signaling regulates nucleolar targeting of the SUMO-specific isopeptidase SENP3. Mol Cell Biol. 2014;34:4474–84.
12. Yun C, Wang Y, Mukhopadhyay D, Backlund P, Kolli N, Yergey A, Wilkinson KD, Dasso M. Nucleolar protein B23/nucleophosmin regulates the

13. vertebrate SUMO pathway through SENP3 and SENP5 proteases. J Cell Biol. 2008;183:589–95.
13. Finkbeiner E, Haindl M, Muller S. The SUMO system controls nucleolar partitioning of a novel mammalian ribosome biogenesis complex. EMBO J. 2011;30:1067–78.
14. Haindl M, Harasim T, Eick D, Muller S. The nucleolar SUMO-specific protease SENP3 reverses SUMO modification of nucleophosmin and is required for rRNA processing. EMBO Rep. 2008;9:273–9.
15. Huang C, Han Y, Wang Y, Sun X, Yan S, Yeh ET, Chen Y, Cang H, Li H, Shi G, et al. SENP3 is responsible for HIF-1 transactivation under mild oxidative stress via p300 de-SUMOylation. EMBO J. 2009;28:2748–62.
16. Fanis P, Gillemans N, Aghajanirefah A, Pourfarzad F, Demmers J, Esteghamat F, Vadlamudi RK, Grosveld F, Philipsen S, van Dijk TB. Five friends of methylated chromatin target of protein-arginine-methyltransferase[prmt]-1 (chtop), a complex linking arginine methylation to desumoylation. Mol Cell Proteomics. 2012;11:1263–73.
17. Huang W, Ghisletti S, Saijo K, Gandhi M, Aouadi M, Tesz GJ, Zhang DX, Yao J, Czech MP, Goode BL, et al. Coronin 2A mediates actin-dependent de-repression of inflammatory response genes. Nature. 2011;470:414–8.
18. Nayak A, Viale-Bouroncle S, Morscheck C, Muller S. The SUMO-specific isopeptidase SENP3 regulates MLL1/MLL2 methyltransferase complexes and controls osteogenic differentiation. Mol Cell. 2014;55:47–58.
19. Duboule D. The rise and fall of Hox gene clusters. Development. 2007;134:2549–60.
20. Montavon T, Soshnikova N. Hox gene regulation and timing in embryogenesis. Semin Cell Dev Biol. 2014;34:76–84.
21. Montavon T, Duboule D. Chromatin organization and global regulation of Hox gene clusters. Philos Trans R Soc Lond B Biol Sci. 2013;368:20120367.
22. Lawrence HJ, Christensen J, Fong S, Hu YL, Weissman I, Sauvageau G, Humphries RK, Largman C. Loss of expression of the Hoxa-9 homeobox gene impairs the proliferation and repopulating ability of hematopoietic stem cells. Blood. 2005;106:3988–94.
23. Takechi M, Adachi N, Hirai T, Kuratani S, Kuraku S. The Dlx genes as clues to vertebrate genomics and craniofacial evolution. Semin Cell Dev Biol. 2013;24:110–8.
24. Viale-Bouroncle S, Felthaus O, Schmalz G, Brockhoff G, Reichert TE, Morscheck C. The transcription factor DLX3 regulates the osteogenic differentiation of human dental follicle precursor cells. Stem Cells Dev. 2012;21:1936–47.
25. Samee N, de Vernejoul MC, Levi G. Role of DLX regulatory proteins in osteogenesis and chondrogenesis. Crit Rev Eukaryot Gene Expr. 2007;17:173–86.
26. Li H, Marijanovic I, Kronenberg MS, Erceg I, Stover ML, Velonis D, Mina M, Heinrich JG, Harris SE, Upholt WB, et al. Expression and function of Dlx genes in the osteoblast lineage. Dev Biol. 2008;316:458–70.
27. Duverger O, Isaac J, Zah A, Hwang J, Berdal A, Lian JB, Morasso MI. In vivo impact of Dlx3 conditional inactivation in neural crest-derived craniofacial bones. J Cell Physiol. 2013;228:654–64.
28. Piunti A, Shilatifard A. Epigenetic balance of gene expression by Polycomb and COMPASS families. Science. 2016;352:aad9780.
29. Ringrose L, Paro R. Epigenetic regulation of cellular memory by the Polycomb and Trithorax group proteins. Annu Rev Genet. 2004;38:413–43.
30. Di Croce L, Helin K. Transcriptional regulation by Polycomb group proteins. Nat Struct Mol Biol. 2013;20:1147–55.
31. Geisler SJ, Paro R. Trithorax and Polycomb group-dependent regulation: a tale of opposing activities. Development. 2015;142:2876–87.
32. Steffen PA, Ringrose L. What are memories made of? How Polycomb and Trithorax proteins mediate epigenetic memory. Nat Rev Mol Cell Biol. 2014;15:340–56.
33. Shilatifard A. Chromatin modifications by methylation and ubiquitination: implications in the regulation of gene expression. Annu Rev Biochem. 2006;75:243–69.
34. Shilatifard A. Molecular implementation and physiological roles for histone H3 lysine 4 (H3K4) methylation. Curr Opin Cell Biol. 2008;20:341–8.
35. Ruthenburg AJ, Allis CD, Wysocka J. Methylation of lysine 4 on histone H3: intricacy of writing and reading a single epigenetic mark. Mol Cell. 2007;25:15–30.
36. Santos-Rosa H, Schneider R, Bannister AJ, Sherriff J, Bernstein BE, Emre NC, Schreiber SL, Mellor J, Kouzarides T. Active genes are tri-methylated at K4 of histone H3. Nature. 2002;419:407–11.

37. Azuara V, Perry P, Sauer S, Spivakov M, Jorgensen HF, John RM, Gouti M, Casanova M, Warnes G, Merkenschlager M, Fisher AG. Chromatin signatures of pluripotent cell lines. Nat Cell Biol. 2006;8:532–8.

38. Bernstein BE, Mikkelsen TS, Xie X, Kamal M, Huebert DJ, Cuff J, Fry B, Meissner A, Wernig M, Plath K, et al. A bivalent chromatin structure marks key developmental genes in embryonic stem cells. Cell. 2006;125:315–26.

39. Hu D, Garruss AS, Gao X, Morgan MA, Cook M, Smith ER, Shilatifard A. The Mll2 branch of the COMPASS family regulates bivalent promoters in mouse embryonic stem cells. Nat Struct Mol Biol. 2013;20:1093–7.

40. Denissov S, Hofemeister H, Marks H, Kranz A, Ciotta G, Singh S, Anastassiadis K, Stunnenberg HG, Stewart AF. Mll2 is required for H3K4 trimethylation on bivalent promoters in embryonic stem cells, whereas Mll1 is redundant. Development. 2014;141:526–37.

41. Dou Y, Milne TA, Ruthenburg AJ, Lee S, Lee JW, Verdine GL, Allis CD, Roeder RG. Regulation of MLL1 H3K4 methyltransferase activity by its core components. Nat Struct Mol Biol. 2006;13:713–9.

42. Ali A, Veeranki SN, Tyagi S. A SET-domain-independent role of WRAD complex in cell-cycle regulatory function of mixed lineage leukemia. Nucleic Acids Res. 2014;42:7611–24.

43. Ernst P, Vakoc CR. WRAD: enabler of the SET1-family of H3K4 methyltransferases. Brief Funct Genomics. 2012;11:217–26.

44. Cao F, Chen Y, Cierpicki T, Liu Y, Basrur V, Lei M, Dou Y. An Ash2L/RbBP5 heterodimer stimulates the MLL1 methyltransferase activity through coordinated substrate interactions with the MLL1 SET domain. PLoS ONE. 2010;5:e14102.

45. Li Y, Han J, Zhang Y, Cao F, Liu Z, Li S, Wu J, Hu C, Wang Y, Shuai J, et al. Structural basis for activity regulation of MLL family methyltransferases. Nature. 2016;530:447–52.

46. Kopecki Z, Cowin AJ. Flightless I: an actin-remodelling protein and an important negative regulator of wound repair. Int J Biochem Cell Biol. 2008;40:1415–9.

47. Lin CH, Waters JM, Powell BC, Arkell RM, Cowin AJ. Decreased expression of Flightless I, a gelsolin family member and developmental regulator, in early-gestation fetal wounds improves healing. Mamm Genome. 2011;22:341–52.

48. Davy DA, Campbell HD, Fountain S, de Jong D, Crouch MF. The flightless I protein colocalizes with actin- and microtubule-based structures in motile Swiss 3T3 fibroblasts: evidence for the involvement of PI 3-kinase and Ras-related small GTPases. J Cell Sci. 2001;114:549–62.

49. Cowin AJ, Adams DH, Strudwick XL, Chan H, Hooper JA, Sander GR, Rayner TE, Matthaei KI, Powell BC, Campbell HD. Flightless I deficiency enhances wound repair by increasing cell migration and proliferation. J Pathol. 2007;211:572–81.

50. Lee YH, Campbell HD, Stallcup MR. Developmentally essential protein flightless I is a nuclear receptor coactivator with actin binding activity. Mol Cell Biol. 2004;24:2103–17.

51. Jensen LJ, Kuhn M, Stark M, Chaffron S, Creevey C, Muller J, Doerks T, Julien P, Roth A, Simonovic M, et al. STRING 8—a global view on proteins and their functional interactions in 630 organisms. Nucleic Acids Res. 2009;37:D412–6.

52. Kopecki Z, O'Neill GM, Arkell RM, Cowin AJ. Regulation of focal adhesions by flightless i involves inhibition of paxillin phosphorylation via a Rac1-dependent pathway. J Invest Dermatol. 2011;131:1450–9.

53. Wang P, Lin C, Smith ER, Guo H, Sanderson BW, Wu M, Gogol M, Alexander T, Seidel C, Wiedemann LM, et al. Global analysis of H3K4 methylation defines MLL family member targets and points to a role for MLL1-mediated H3K4 methylation in the regulation of transcriptional initiation by RNA polymerase II. Mol Cell Biol. 2009;29:6074–85.

54. Saugspier M, Felthaus O, Viale-Bouroncle S, Driemel O, Reichert TE, Schmalz G, Morsczeck C. The differentiation and gene expression profile of human dental follicle cells. Stem Cells Dev. 2010;19:707–17.

55. Jeong KW, Lee YH, Stallcup MR. Recruitment of the SWI/SNF chromatin remodeling complex to steroid hormone-regulated promoters by nuclear receptor coactivator flightless-I. J Biol Chem. 2009;284:29298–309.

56. Jeong KW. Flightless I (Drosophila) homolog facilitates chromatin accessibility of the estrogen receptor alpha target genes in MCF-7 breast cancer cells. Biochem Biophys Res Commun. 2014;446:608–13.

57. Ansari KI, Kasiri S, Hussain I, Bobzean SA, Perrotti LI, Mandal SS. MLL histone methylases regulate expression of HDLR-SR-B1 in presence of estrogen and control plasma cholesterol in vivo. Mol Endocrinol. 2013;27:92–105.

58. Ansari KI, Hussain I, Shrestha B, Kasiri S, Mandal SS. HOXC6 Is transcriptionally regulated via coordination of MLL histone methylase and estrogen receptor in an estrogen environment. J Mol Biol. 2011;411:334–49.

59. Ansari KI, Shrestha B, Hussain I, Kasiri S, Mandal SS. Histone methylases MLL1 and MLL3 coordinate with estrogen receptors in estrogen-mediated HOXB9 expression. Biochemistry. 2011;50:3517–27.

60. Ansari KI, Kasiri S, Hussain I, Mandal SS. Mixed lineage leukemia histone methylases play critical roles in estrogen-mediated regulation of HOXC13. FEBS J. 2009;276:7400–11.

61. Lee SH, Oh KN, Han Y, Choi YH, Lee KY. Estrogen receptor alpha regulates Dlx3-mediated osteoblast differentiation. Mol Cells. 2016;39:156–62.

62. Miyamoto K, Teperek M, Yusa K, Allen GE, Bradshaw CR, Gurdon JB. Nuclear Wave1 is required for reprogramming transcription in oocytes and for normal development. Science. 2013;341:1002–5.

63. Ong SE, Mann M. Stable isotope labeling by amino acids in cell culture for quantitative proteomics. Methods Mol Biol. 2007;359:37–52.

64. Cox J, Mann M. MaxQuant enables high peptide identification rates, individualized p.p.b.-range mass accuracies and proteome-wide protein quantification. Nat Biotechnol. 2008;26:1367–72.

65. Nelson JD, Denisenko O, Bomsztyk K. Protocol for the fast chromatin immunoprecipitation (ChIP) method. Nat Protoc. 2006;1:179–85.

66. Schmidt D, Muller S. Members of the PIAS family act as SUMO ligases for c-Jun and p53 and repress p53 activity. Proc Natl Acad Sci USA. 2002;99:2872–7.

67. Morsczeck C, Gotz W, Schierholz J, Zeilhofer F, Kuhn U, Mohl C, Sippel C, Hoffmann KH. Isolation of precursor cells (PCs) from human dental follicle of wisdom teeth. Matrix Biol. 2005;24:155–65.

68. Viale-Bouroncle S, Klingelhoffer C, Ettl T, Reichert TE, Morsczeck C. A protein kinase A (PKA)/beta-catenin pathway sustains the BMP2/DLX3-induced osteogenic differentiation in dental follicle cells (DFCs). Cell Signal. 2015;27:598–605.

69. Kunz K, Wagner K, Mendler L, Holper S, Dehne N, Muller S. SUMO signaling by hypoxic inactivation of SUMO-specific isopeptidases. Cell Rep. 2016;16:3075–86.

3

Comprehensive evaluation of genome-wide 5-hydroxymethylcytosine profiling approaches in human DNA

Ksenia Skvortsova[1], Elena Zotenko[1], Phuc-Loi Luu[1], Cathryn M. Gould[1], Shalima S. Nair[1,2], Susan J. Clark[1,2*†] and Clare Stirzaker[1,2*†]

Abstract

Background: The discovery that 5-methylcytosine (5mC) can be oxidized to 5-hydroxymethylcytosine (5hmC) by the ten-eleven translocation (TET) proteins has prompted wide interest in the potential role of 5hmC in reshaping the mammalian DNA methylation landscape. The gold-standard bisulphite conversion technologies to study DNA methylation do not distinguish between 5mC and 5hmC. However, new approaches to mapping 5hmC genome-wide have advanced rapidly, although it is unclear how the different methods compare in accurately calling 5hmC. In this study, we provide a comparative analysis on brain DNA using three 5hmC genome-wide approaches, namely whole-genome bisulphite/oxidative bisulphite sequencing (WG Bis/OxBis-seq), Infinium HumanMethylation450 BeadChip arrays coupled with oxidative bisulphite (HM450K Bis/OxBis) and antibody-based immunoprecipitation and sequencing of hydroxymethylated DNA (hMeDIP-seq). We also perform loci-specific TET-assisted bisulphite sequencing (TAB-seq) for validation of candidate regions.

Results: We show that whole-genome single-base resolution approaches are advantaged in providing precise 5hmC values but require high sequencing depth to accurately measure 5hmC, as this modification is commonly in low abundance in mammalian cells. HM450K arrays coupled with oxidative bisulphite provide a cost-effective representation of 5hmC distribution, at CpG sites with 5hmC levels >~10%. However, 5hmC analysis is restricted to the genomic location of the probes, which is an important consideration as 5hmC modification is commonly enriched at enhancer elements. Finally, we show that the widely used hMeDIP-seq method provides an efficient genome-wide profile of 5hmC and shows high correlation with WG Bis/OxBis-seq 5hmC distribution in brain DNA. However, in cell line DNA with low levels of 5hmC, hMeDIP-seq-enriched regions are not detected by WG Bis/OxBis or HM450K, either suggesting misinterpretation of 5hmC calls by hMeDIP or lack of sensitivity of the latter methods.

Conclusions: We highlight both the advantages and caveats of three commonly used genome-wide 5hmC profiling technologies and show that interpretation of 5hmC data can be significantly influenced by the sensitivity of methods used, especially as the levels of 5hmC are low and vary in different cell types and different genomic locations.

Keywords: DNA methylation, Methylome, Epigenetics, 5-Hydroxymethylation, HM450K Bis/OxBis, hMeDIP-seq

*Correspondence: s.clark@garvan.org.au; c.stirzaker@garvan.org.au
†Susan J. Clark and Clare Stirzaker contributed equally to this work
[1] Epigenetics Research Laboratory, Genomics and Epigenetics Division, Garvan Institute of Medical Research, 384 Victoria Street, Darlinghurst, Sydney, NSW 2010, Australia
Full list of author information is available at the end of the article

Background

DNA cytosine methylation is one of the key epigenetic determinants of mammalian gene expression in normal development [1, 2] and disease [3]. DNA methylation is particularly dynamic during early embryonic development followed by "dynamic homeostasis" of the methylation landscape in normal functioning somatic cells [2]. The discovery that 5-methylcytosine (5mC) can be oxidized to 5-hydroxymethylcytosine (5hmC) by the ten-eleven translocation (TET) proteins has prompted wide interest in the potential roles of 5hmC in reshaping the mammalian DNA methylation landscape during early embryonic development [4, 5], during differentiation towards extra-embryonic lineages [6, 7] and in metabolically active normal adult tissues [8, 9] and disease cells [10]. 5hmC levels vary substantially in somatic tissues [11] and the abundance and genomic distribution of 5hmC is dramatically altered during development [8, 12, 13]. Notably, 5hmC has been purported to play a key role as an intermediate in DNA demethylation; this may occur either passively during DNA replication [14], or actively through base excision repair of one or more oxidized intermediates [15, 16]. Other studies, however, regard 5hmC as a distinct epigenetic mark with a characteristic function independent of DNA demethylation [17–20]. The importance of the dynamic interplay between 5mC and 5hmC in maintaining normal DNA methylation patterns and gene expression and the causes and consequences of an imbalance are key questions yet to be answered.

Given the evidence that 5hmC plays a critical role in modulation of the DNA methylation landscape, it is essential to be able to distinguish 5hmC from 5mC and accurately detect and quantitate the levels of 5hmC at single-base resolution. In general, the levels of total 5hmC detected across the genome are approximately 14-fold lower than those of 5mC [21] although these levels vary substantially across tissue types: 5hmC is relatively abundant in brain tissues (~0.15–0.6% of total nucleotides) [22, 23], but is an order of magnitude lower (0.01–0.05%) in other mouse and human tissues [23–25]. In human cell lines, 5hmC abundance is at even lower levels (~0.007–0.009% of total nucleotides) [26]. Such low and variable abundance means that the method of detection has to be highly sensitive and specific for the 5hmC modification.

Importantly, bisulphite sequencing, the "gold standard" for 5mC analysis, does not distinguish 5mC from 5hmC, as both modified bases are resistant to conversion to uracil, in contrast to unmodified cytosine, which is converted to uracil (Fig. 1). This has precluded conventional bisulphite sequencing as a tool for 5hmC detection.

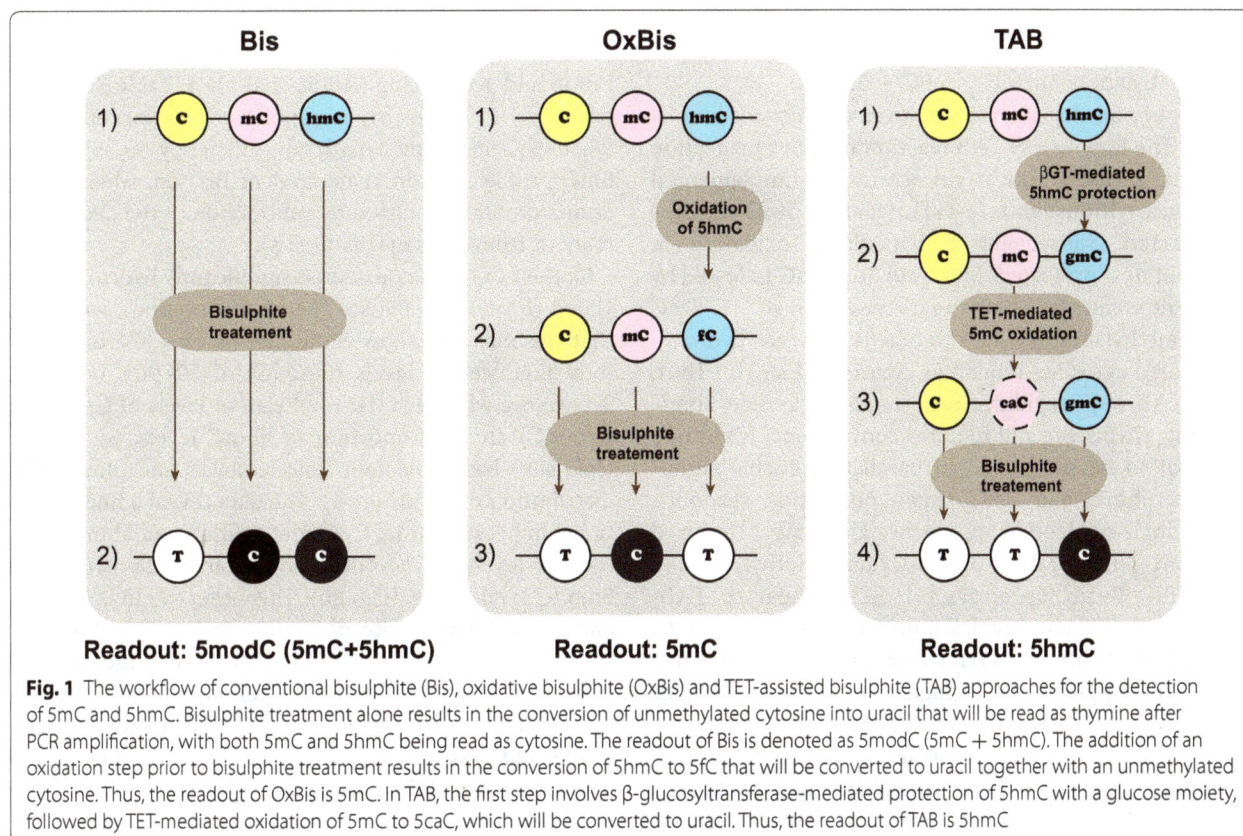

Fig. 1 The workflow of conventional bisulphite (Bis), oxidative bisulphite (OxBis) and TET-assisted bisulphite (TAB) approaches for the detection of 5mC and 5hmC. Bisulphite treatment alone results in the conversion of unmethylated cytosine into uracil that will be read as thymine after PCR amplification, with both 5mC and 5hmC being read as cytosine. The readout of Bis is denoted as 5modC (5mC + 5hmC). The addition of an oxidation step prior to bisulphite treatment results in the conversion of 5hmC to 5fC that will be converted to uracil together with an unmethylated cytosine. Thus, the readout of OxBis is 5mC. In TAB, the first step involves β-glucosyltransferase-mediated protection of 5hmC with a glucose moiety, followed by TET-mediated oxidation of 5mC to 5caC, which will be converted to uracil. Thus, the readout of TAB is 5hmC

Several approaches for hydroxymethylation mapping have been developed over the recent years. These include capture-based techniques, such as antibody-based hydroxymethylated DNA immunoprecipitation followed by sequencing (hMeDIP-seq) [6, 27, 28], and enrichment by hydroxymethyl selective chemical labelling (hMeSeal) [26]. These affinity-based methods have been widely used and provided the first genome-wide 5hmC profiles and biological insights of 5hmC [6, 26–28]. However, such approaches have relatively low resolution and cannot quantitatively determine 5hmC abundance in a single-base resolution manner.

Single nucleotide 5hmC mapping approaches have also been developed. The two best-described approaches are whole-genome oxidative bisulphite in combination with conventional bisulphite sequencing (WG Bis/OxBis-seq) [29] and TET-assisted bisulphite sequencing (TAB-seq) [30]. The principle of WG Bis/OxBis-seq relies on the specific oxidation of 5hmC by potassium perruthenate to form 5-formylcytosine (5fC) and/or 5-carboxylcytosine (5caC). 5fC and 5caC behave as unmethylated cytosines during bisulphite conversion. Therefore, the readout of OxBis-seq is specific for 5mC and does not contain the 5hmC fraction (Fig. 1a). Hence, subtraction of OxBis-seq readout (5mC) from the conventional Bis-seq readout (5mC + 5hmC) evaluates the hydroxymethylated proportion at a single CpG site (Bis-seq (5mC + 5hmC) − OxBis-seq (5mC) = 5hmC). For clarity, we use 5modC to denote Bis-seq readout (i.e. total methylation: 5modC = 5mC + 5hmC).

TAB-seq, on the other hand, gives a *direct* readout of 5hmC. The first step of TAB reaction includes protection of 5hmC residues with a glucose moiety implemented by β-glucosyltransferase (β-GT), whereas 5mC remains unprotected. Subsequent TET-mediated oxidation of 5mC, but not "protected" 5hmC, to 5fC/5caC followed by bisulphite treatment results in a conversion of 5fC/5caC and unmethylated cytosines to uracils, whereas hydroxymethylated cytosines remain as cytosines (Fig. 1b). Thus, while TAB-seq provides direct single nucleotide 5hmC profiling, OxBis-seq still requires conventional bisulphite (Bis-seq) to be performed in parallel to enable simultaneous 5hmC and 5mC single nucleotide readouts. These single nucleotide approaches also have been used for 5hmC interrogation on the HumanMethylation450 (HM450K) BeadChip arrays [31, 32]. Similar to TAB-seq, HM450K-TAB provides a direct readout of 5hmC, while HM450K Bis and OxBis arrays must be performed in parallel. Applying these single nucleotide approaches for whole-genome 5hmC mapping is ideal; however, the need for deep sequencing coverage for accurately resolving the 5hmC levels imposes restrictions and limitations

on the widespread use of WG Bis/OxBis-seq and TAB-seq for mammalian genomes.

Here, we evaluate and compare three commonly used whole-genome 5hmC profiling approaches using WG Bis/OxBis-seq, HM450K arrays coupled with oxidative bisulphite (HM450K Bis/OxBis) and antibody-based immunoprecipitation of hydroxymethylated DNA (hMeDIP-seq). We also validate 5hmC at single nucleotide resolution using TAB-seq of selected candidate genomic regions. The comparisons were made on adult frontal lobe cerebellum DNA, as the highest levels of 5hmC have been reported in adult brain DNA [6, 33]. We highlight the advantages and caveats of 5hmC profiling methods and show that the interpretation of 5hmC data can be significantly influenced by the sensitivity of the method.

Results

Hydroxymethylation profiling by whole-genome Bis-seq/OxBis-seq

To compare the different technologies for analysis of whole-genome 5hmC profiling, we first performed 5hmC profiling using WG Bis/OxBis-seq, which allows interrogation of both 5hmC and 5mC at single nucleotide resolution. Human frontal lobe DNA, spiked with M.SssI CpG-methylated λDNA and fully hydroxymethylated 5hmC APC controls ("Methods" section), was treated with conventional bisulphite and oxidative bisulphite (CEGX TrueMethyl reagents), followed by library preparation and sequencing (Additional file 1: Table S1, Additional file 2: Figure S1; bisulphite conversion efficiency 98.30–99.76%, 5hmC oxidation efficiency 99.33%). The 5mC profile is a direct readout of Bis-seq, whereas the 5hmC profile is deduced by subtraction of the OxBis-seq readout from Bis-seq (Bis-OxBis).

Of the CpG sites considered significantly hydroxymethylated (p value ≤ 0.05: see "Methods" section), we found that the majority show 5hmC levels of ~30% and high total methylation levels (5modC) of 80–90% (Fig. 2a). To explore whether total methylation levels of CpG sites (5modC) are in proportion to 5hmC levels, we binned CpG sites based on their 5modC levels and plotted the distribution of 5hmC levels. The data reveal a linear relationship between total CpG methylation and 5hmC, with 5modC levels <50%, but a nonlinear relationship with 5modC levels >50% (Fig. 2b). This observation is consistent with a dynamic interplay between 5hmC and 5mC at sites that display lower levels (<50%) of total methylation or higher "plasticity" across the genome. In contrast, the more extensively methylated sites (>50%) show a more stable and less "plastic" methylation state, due to the decreasing 5hmC/5mC ratio.

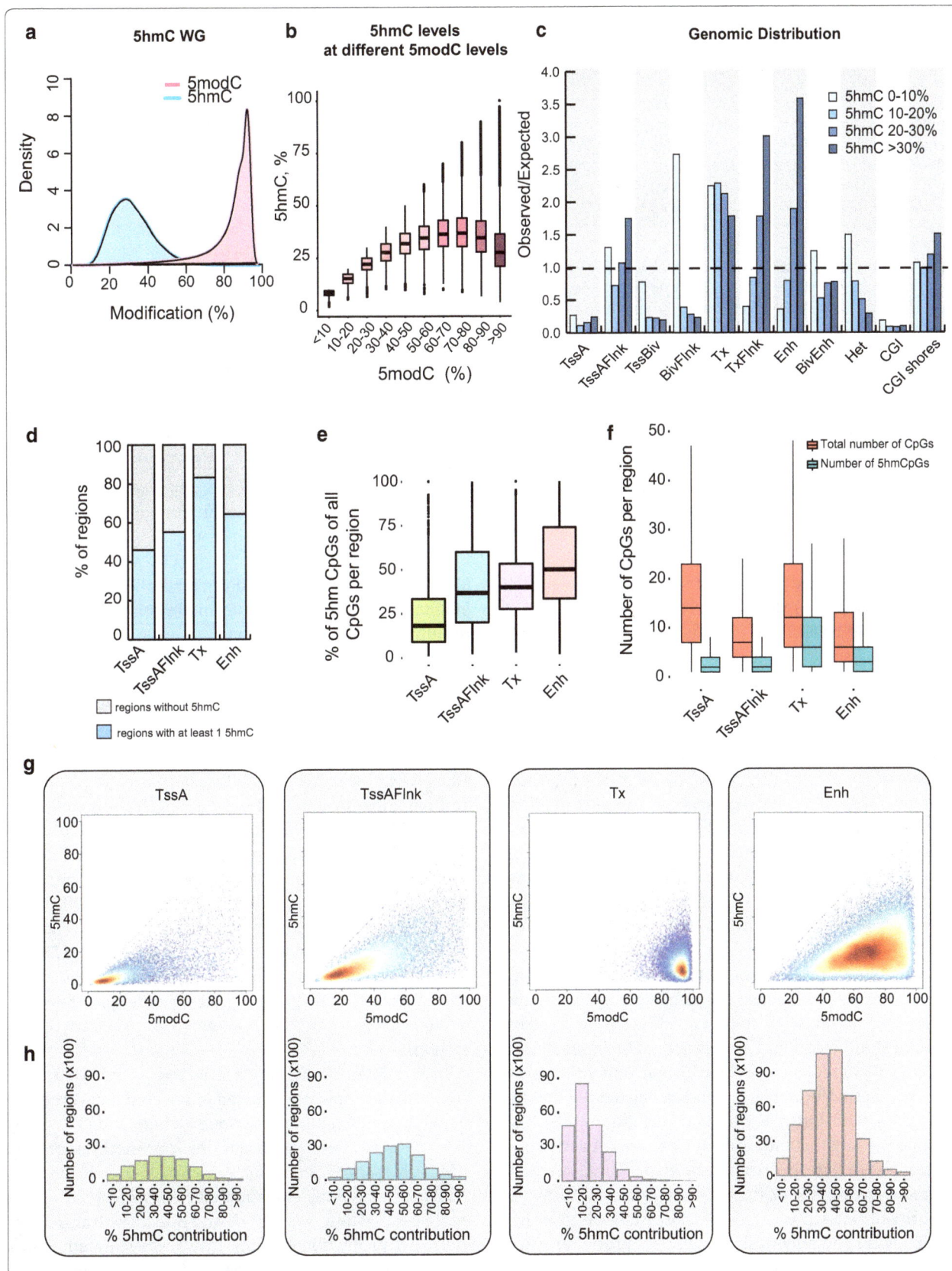

(See figure on previous page.)

Fig. 2 DNA methylation and hydroxymethylation profiling by whole-genome Bis-seq/OxBis-seq. **a** Methylation density plot showing the distribution of 5hmC and 5modC levels. Both 5modC and 5hmC density plots include CpG sites showing significant hydroxymethylation. **b** The relationship between the hydroxymethylated fraction and total methylation levels at each CpG site. CpG sites were binned into groups based on the total methylation levels, and the distribution of hydroxymethylation levels was calculated for each of these groups. **c** *Bar plot* showing observed over expected by chance enrichment of CpGs with different 5hmC levels at multiple genomic locations. Genomic regions comprise of Brain Frontal Lobe ChromHMM features as well as CpG islands and CpG island shores. **d** The percentage of genomic regions harbouring no hydroxymethylated CpG sites (*grey*) and those harbouring at least one hydroxymethylated CpG site (*blue*). **e** The proportion of hydroxymethylated CpGs of the total number of CpGs per genomic region. For each region, the total number of CpGs and number of hydroxymethylated CpGs were calculated (Additional file 2: Figure S1D). From that, for each region, the percentage of hydroxymethylated CpGs of all CpGs was calculated and distribution of those percentages was plotted. **f** The distribution of the total number of CpG sites and the number of hydroxymethylated CpG sites per genomic region. **g** The relationship between average total methylation (5modC) (*x*-axis) and average hydroxymethylation (5hmC) (*y*-axis) at different genomic regions. Each *dot* represents a single region. Hydroxymethylation levels of CpG sites that did not pass the statistical significance criteria were assigned to zero. **h** The hydroxymethylation contribution to the average total methylation at different genomic regions. For each genomic region, the percentage of hydroxymethylation contribution to the total methylation was calculated; the numbers of regions with the corresponding 5hmC contribution were plotted

Next, to investigate the genomic distribution of 5hmC, we annotated CpG sites, according to the 5hmC levels, to Brain Frontal Lobe ChromHMM features [34, 35], as well as CpG islands and CpG island shores. We show that active and bivalent promoters (TssA and TssBiv) and CpG islands (CGI) are depleted of hydroxymethylated CpGs (Fig. 2c), coinciding with their prevalent unmethylated state. However, regions flanking active and bivalent promoters are enriched predominantly for CpGs with low 5hmC levels (<10%) (Fig. 2c), while transcribed regions (Tx), flanking transcribed regions (TxFlnk) and enhancers (Enh) possess the highest levels of 5hmC enrichment (Fig. 2c).

Finally, we asked whether the hydroxymethylated cytosines make a significant contribution to the overall total methylation levels (5modC) at different genomic elements. To address this question, we focused on four genomic regions: active promoters, regions flanking active promoters, transcribed regions and enhancers (TssA, TssAFlnk, Tx, Enh, respectively). For each genomic region harbouring at least one 5hmC site (Fig. 2d), we calculated the percentage of hydroxymethylated CpGs of all CpGs and plotted the distribution (Fig. 2e). The analysis revealed ~50% of CpGs at enhancers are hydroxymethylated (3 CpGs out of 6 on average), whereas only ~15% of CpG sites at active promoters are hydroxymethylated (2 CpGs out of 13 on average) (Fig. 2e, f). To assess the contribution of 5hmC to the total methylation in each region, we plotted average 5hmC per region versus average total methylation (Fig. 2g) and calculated the number of regions with different 5hmC contributions (Fig. 2h). Despite the fact that only 15% of CpG sites at active promoters (TssA) are hydroxymethylated, 5hmC accounts for more than ~50% of the total methylation at the vast majority of active promoters (Fig. 2g). Similarly, over half of enhancers and regions flanking active promoters show that 5hmC contributes significantly to the total average methylation (Fig. 2h). However, most of the transcribed

regions have low 5hmC levels relative to the total average methylation (Fig. 2h). Overall, these data reveal that 5hmC, when present, makes a substantial contribution to the total methylation levels at different genomic regions in brain DNA, highlighting the importance of discerning 5hmC from total methylation (5modC) in analysis of DNA methylomes from brain tissue.

Methylation and hydroxymethylation profiling using Infinium HumanMethylation450 methylation arrays

Infinium HumanMethylation450 BeadChip (HM450K) arrays have been widely used to study genome-wide 5mC DNA methylation profiles [36, 37], but more recently the technology has been adjusted to assess 5hmC [31, 32]. Here, we used the HM450K array in conjunction with OxBis to interrogate both 5mC and 5hmC. Frontal lobe adult DNA was treated with conventional bisulphite and oxidative bisulphite reaction, respectively, followed by hybridization to HM450K arrays (HM450K Bis and HM450K OxBis). Arrays were performed in duplicate with technical replicates showing high correlation (Spearman correlation of 0.987 and 0.986 for HM450K Bis and HM450K OxBis replicates, respectively (Fig. 3a). Of 482,422 cytosines interrogated on the array, 175,000 CpG sites possess significant hydroxymethylation levels (i.e. probes showing a significant difference between Bis and OxBis: adjusted p value ≤ 0.05: see "Methods" section).

We validated 5hmC levels detected by HM450K Bis/OxBis using amplicon TAB-seq of selected loci: 4 regions showing significant hydroxymethylation and 3 regions that showed no significant hydroxymethylation by HM450K Bis/OxBis (Fig. 3c, Additional file 3: Figure S2). TET-mediated 5mC oxidation efficiency of M.SssI CpG-methylated λDNA was 98.74% and β-glucosyltransferase-mediated protection of fully hydroxymethylated 5hmC pUC18 was ~100% ("Methods" section; Additional file 2: Figure S1). In parallel with TAB-seq, we performed

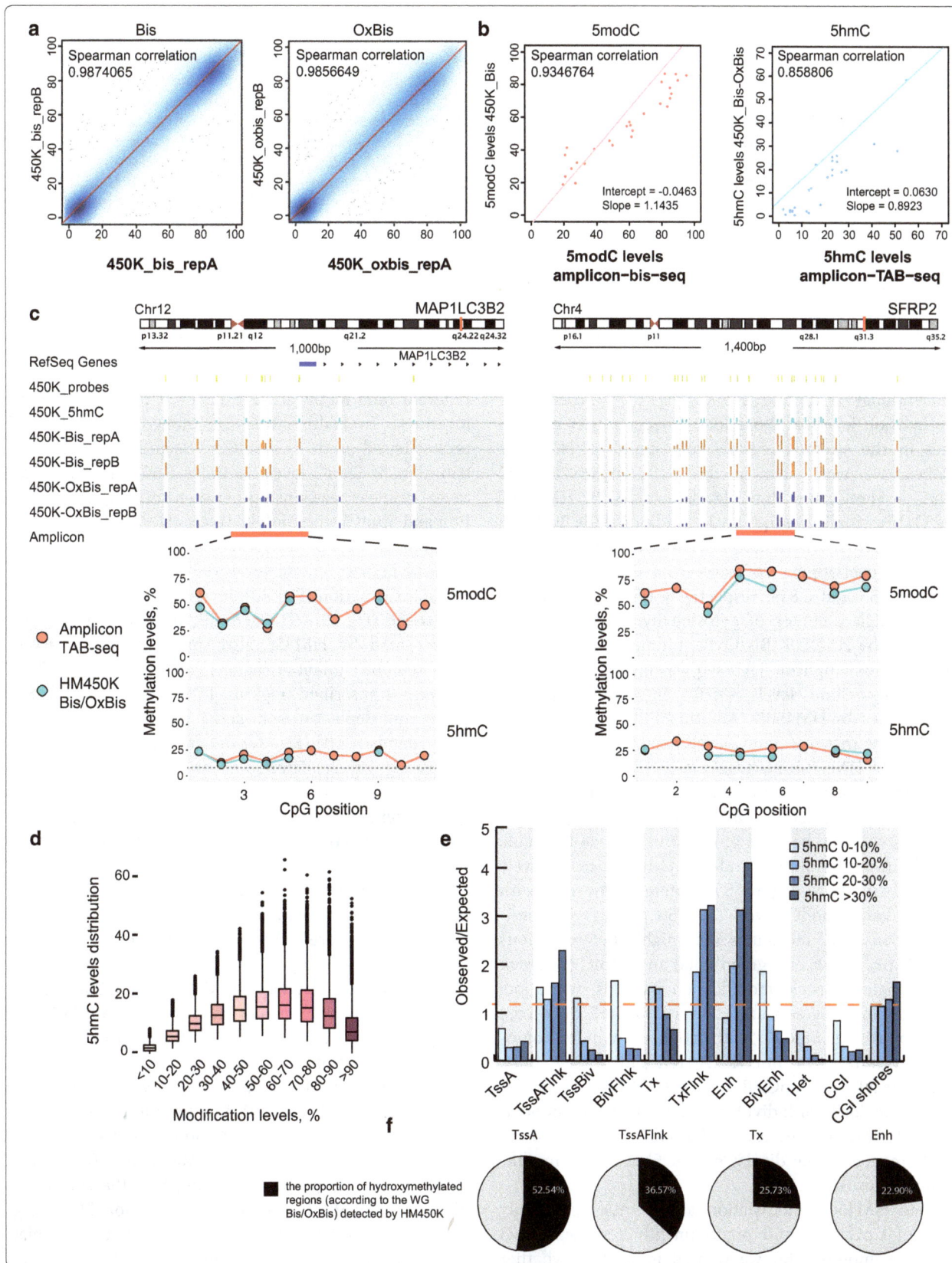

(See figure on previous page.)

Fig. 3 DNA methylation and hydroxymethylation profiling by HM450K Bis/OxBis. **a** *Scatter plots* showing the high correlation between HM450K Bis (*left*) and HM450K OxBis (*right*) replicates. Each *dot* (smoothed) reflects each probe 5modC (*left*) or 5mC (*right*) levels detected by two technical replicates (*x*- and *y*-axes). Spearman's correlation 0.987406 (*left*) and 0.985664 (*right*), respectively. **b, c** 5hmC and 5modC MiSeq amplicon validation of candidate loci using Bis- and TAB-Seq. **b** *Scatter plots* showing the correlation of 5modC (*left*) and 5hmC (*right*) levels between HM450K and amplicon Bis- and TAB-seq. Each *dot* represents single CpG site/probe. The *pink* and *blue* regression *lines* show the intercept and slope of the plots. **c** HM450K screen shots of candidate regions, showing 450K_5hmC, 450K_Bis, 450K_OxBis. Amplicon validation of these genomic regions shows agreement in total methylation (5modC) and hydroxymethylation levels detected by HM450K Bis/OxBis and loci-specific Bis/TAB-seq, respectively. *Red dots* depict 5modC (*top*) and 5hmC (*bottom*) levels of each CpG site detected by loci-specific Bis-/TAB-seq, respectively. *Green dots* depict 5modC (*top*) and 5hmC (*bottom*) levels of HM450K CpG probes on the array. **d** The relationship between hydroxymethylation and different levels of total methylation detected by HM450K. CpG probes were binned into groups based on the total methylation levels, and the distribution of hydroxymethylation levels was calculated for each of these groups. **e** *Bar plot* showing observed over expected by chance enrichment of CpGs with different 5hmC levels (0–10, 10–20, 20–30, >30%) at multiple genomic locations; computationally derived chromatin segmentation (Chrom-HMM) of Brain Frontal Lobe genome, as well as CpG islands and CpG island shores. **f** *Pie charts* showing the proportion of genomic regions defined as hydroxymethylated by HM450K Bis/OxBis compared to the total number defined as hydroxymethylated according to the WG Bis/OxBis-seq. A region is considered hydroxymethylated if it contains >1 significantly hydroxymethylated CpG site

conventional bisulphite sequencing to validate total methylation levels of those loci. Since a subset of CpG sites in the selected regions correspond to HM450K probes, we were able to directly compare 5modC (5mC + 5hmC) and 5hmC levels detected by HM450K Bis/OxBis with TAB-seq (Fig. 3c, Additional file 3: Figure S2A). Spearman's correlation of 5modC and 5hmC levels across overlapping CpG sites from all loci sequenced shows 0.935 and 0.859, respectively (Fig. 3b), confirming the high accuracy of cytosine hydroxymethylation detection by HM450K Bis/OxBis Infinium arrays. Notably, however, amplicon TAB-seq was more sensitive and could detect 5hmC levels (~<10%) in regions that were not detected by HM450K (Additional file 3: Figure S2B), highlighting that that 5hmC levels need to be >10% to be detected by HM450K Bis/OxBis.

Notably, loci-specific Bis- and TAB-seq, with sequencing coverage ~50,000× and average of 13 CpG sites per loci, can also be used to determine single-molecule 5modC (5mC + 5hmC) and 5hmC levels, respectively (Additional file 4: Figure S3). Interestingly, loci with intermediate 5modC levels (50–75%, *x*-axis) predominantly consist of molecules with highly heterogeneous 5modC (A, purple and green dots) and 5hmC (B, green dots) single-molecule modification patterns, while loci with very low (<25%, *x*-axis) or very high (>75%, *x*-axis) levels of average 5modC methylation show more homogeneous patterns of 5modC (A, red and oranges dots, respectively) and 5hmC (B, red dots). This suggests that the vast majority of individual brain cells in the population with intermediate levels of 5modC methylation are not homogenous but display mosaic 5hmC/5mC modification patterns.

Since HM450K methylation arrays interrogate only 1.8% of CpG sites and predominantly represent CpG island promoters [38], we wanted to explore whether the array results in biased 5hmC calling of CpGs with

different total methylation levels or at different genomic locations. To address this, we binned HM450K CpG probes based on their 5modC levels and plotted the distribution of 5hmC levels at those CpG sites. The data reveal a linear relationship between total CpG methylation and 5hmC, with 5modC levels <50%, but a nonlinear relationship with 5modC levels >50%, and notably CpGs with the highest 5modC levels (90–100%) are depleted in the 5hmC fraction (Fig. 3d), in agreement with WG Bis/OxBis-seq (Fig. 2b). The genomic distribution of 5hmC interrogated by HM450 also shows 5hmC observed/expected enrichment at regions flanking Tss (TssAFlnk), flanking transcribed regions (TxFlnk) and enhancers (Enh), and depletion from active promoters (TssA) and CpG islands (CGI) (Fig. 3e) in agreement with WG Bis/OxBis-seq (Fig. 2c). 5hmC was interrogated by HM450 at ~53% of TssA and 37% of TssAFlnk; however, a smaller proportion of hydroxymethylated Enh and Tx regions (23 and 26%, respectively) are detected by HM450K (Fig. 3f).

Comparison of HM450K and whole-genome Bis-seq/OxBis-seq for 5hmC mapping

Next, we compared the performance of 5modC and 5hmC profiling using HM450K Bis/OxBis arrays with WG Bis/OxBis-seq. Of all HM450K probes, 207,125 had sequencing coverage greater than 10× in both WG Bis-seq and OxBis-seq. Bis-seq and OxBis-seq values showed a high correlation between WG and HM450K data (Spearman's correlation 0.907 and 0.914, respectively; Additional file 5: Figure S4). In addition, we determined the CpG sites common to both approaches with significant hydroxymethylation in both WG Bis/OxBis-seq and HM450K (42,537 probes) and compared the correlation of 5modC and 5hmC levels (Fig. 4a, Additional file 5: Figure S4). We found a good correlation between the platforms for 5modC (Bis) and 5mC (OxBis) methylation (Spearman's correlation 0.809 and 0.826, respectively).

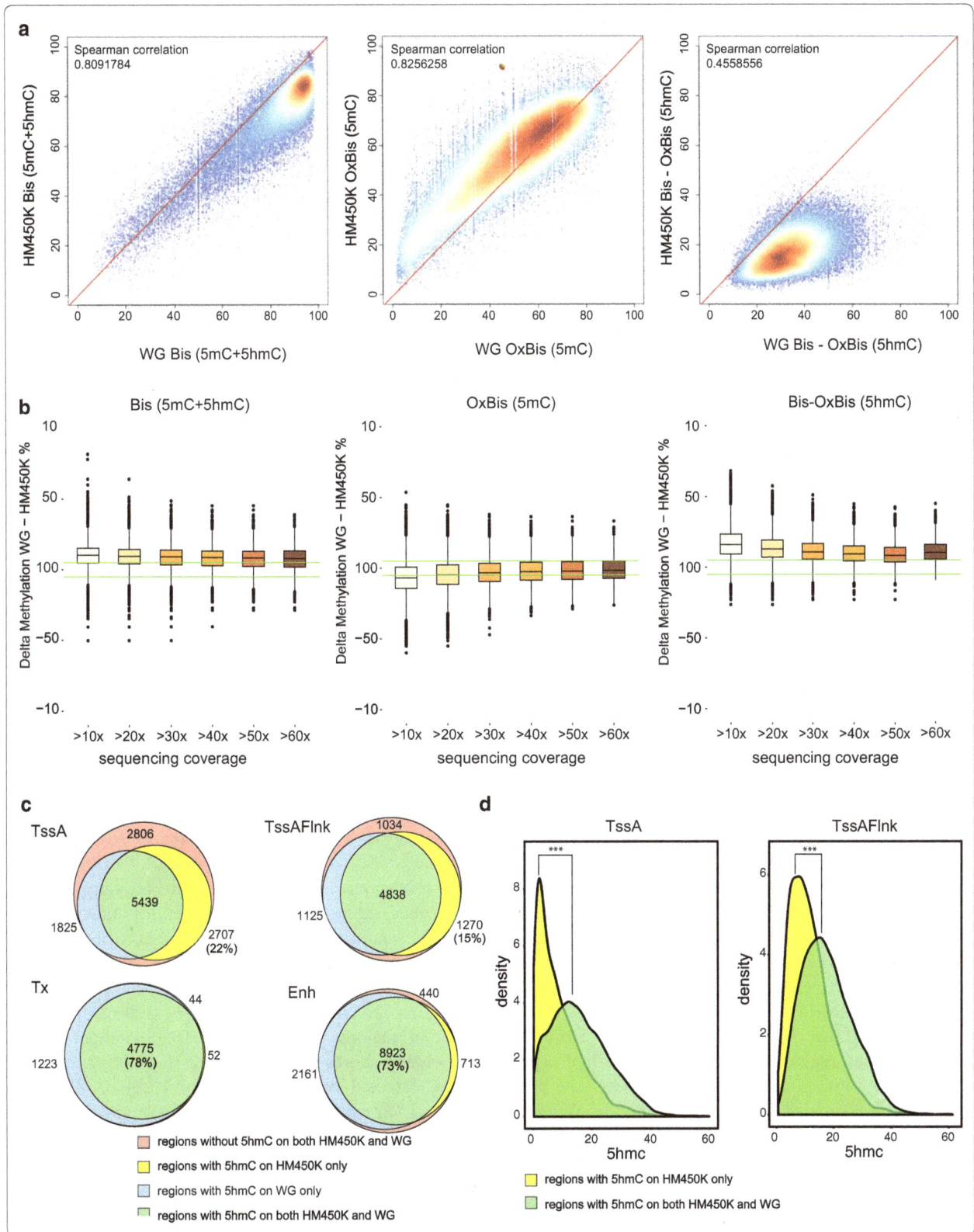

Fig. 4 Comparative evaluation of HM450K Bis/OxBis and whole-genome Bis-/OxBis-seq for 5hmC profiling. **a** *Scatter plots* showing the correlation of 5modC (Bis, *left*), 5mC (OxBis, *middle*) and 5hmC (Bis-OxBis, *right*) between WG and HM450K Bis/OxBis across CpG sites ($n = 42,537$) considered as significantly hydroxymethylated by both approaches. Spearman's correlation is indicated on each *scatter plot*. **b** The agreement in 5modC, 5mC and 5hmC levels detected by WG and HM450K Bis/OxBis as a function of WG sequencing coverage. The difference in methylation calling is plotted along the y-axis for each bin with defined sequencing coverage indicated (>10×, >20×, >30×->60×). *Green lines* indicate ± 5% difference in methylation value detected between approaches (WG-HM450K). **c** *Venn diagrams* show the overlap of hydroxymethylated regions between WG and HM450K Bis/OxBis. Genomic regions with at least one CpG probe and at least 10× WG sequencing coverage were chosen for this analysis (11,625 of 25,235 total TssA; 14,945 of 27,078 total TssA_Flnk; 22,249 of 26,707 total Tx; 47,082 of 73,063 total Enh). Of those, the number of regions with at least one hydroxymethylated CpG according to HM450K only, WG only or both approaches was calculated and overlap was plotted. **d** The distribution of maximal hydroxymethylation values at the regions identified as hydroxymethylated according to the HM450K only (*yellow*) and according to the both WG and HM450K (*green*). For the active promoters (TssA), the median of the max(5hmC) distribution is 0.067 for the *yellow* group and 0.147 for the *green* group. For the regions flanking active promoters (TssA Flank), the median of the max(5hmC) distribution is 0.105 for the *yellow* group and 0.167 for the *green* group. The difference between the distributions of maximal 5hmC values between *yellow* and *green* groups of regions is statistically significant as determined by Kruskal–Wallis nonparametric test ($p < 0.01$)

However, 5modC displays a greater discrepancy at high methylation levels (>80%) with WG Bis-seq exceeding HM450K (Fig. 4a, Additional file 5: Figure S4). Such discordance between WG sequencing and HM450K at high 5modC levels has been reported previously, highlighting the need for improved normalization strategies for HM450K analysis [39].

We next addressed the effect of sequencing coverage and found that increasing sequencing depth (>10×->60×) does not improve the concordance between WG Bis-seq and HM450K Bis (Fig. 4b, left). However, increasing sequencing depth improves agreement between the methods for 5mC (OxBis) methylation comparison (Fig. 4b, middle). In contrast, correlation of 5hmC levels between the two approaches does not improve with greater sequencing coverage (Fig. 4b, right) possibly due to the disagreement between WG and HM450K at high methylation levels.

We next determined whether the same genomic regions are defined as hydroxymethylated by both approaches. To this end, we took genomic regions covered by at least one HM450K probe, and at least 10× coverage on WG Bis/OxBis-seq and only considered hydroxymethylation of those CpGs for the subsequent analysis. Of these regions, we defined hydroxymethylated regions (>1× 5hmC CpG site) according to HM450K only, WG Bis/OxBis only or both. We found that the vast majority of enhancers (73%) and transcribed regions (78%) are defined as hydroxymethylated by both approaches (Fig. 4c, green). The presence of hydroxymethylated regions defined by WG only (Fig. 4c, blue) is potentially due to the 5hmC undercalling of HM450K. Such agreement is lower for the active promoters and flanking regions (Fig. 4c, green), where a substantial proportion of regions (22 and 15%, respectively) are not identified as hydroxymethylated on the WG (Fig. 4c, yellow). More detailed analysis revealed that regions identified as hydroxymethylated on HM450K only (yellow) possess lower levels of 5hmC compared

to regions identified as hydroxymethylated by both approaches (Fig. 4d, green). This highlights that regions with lower 5hmC require higher sequencing coverage to detect 5hmC in the WG Bis/OxBis-seq data and therefore have not been detected as hydroxymethylated in the WG Bis/OxBis-seq data.

Hydroxymethylation profiling by hMeDIP-seq

One of the advantages of Bis/OxBis-seq and TAB-seq for hydroxymethylation mapping is that these methods quantitatively resolve 5hmC at a single nucleotide resolution; however, the depth of sequencing needed to yield meaningful results is generally not cost-effective. An alternative approach for genome-wide 5hmC profiling is antibody-based enrichment of 5hmC, hMeDIP-seq, which is widely used due to its ease of use and cost-effectiveness. We performed hMeDIP on the same adult brain DNA (in duplicate), followed by library preparation and sequencing on HiSeq2500 to compare 5hmC calling with WG Bis/OxBis-seq and HM450K Bis/OxBis arrays.

To first compare signal enrichment between replicates, we binned the genome into 300-bp tiles and calculated logCPM values per bin. The data reveal a high correlation between replicates at high logCPM values with overall Spearman's correlation of 0.756 (Fig. 5a). MACS2 peak calling resulted in 49,292 hMeDIP-seq peaks with enrichment at regions flanking active TSSes, enhancers and CpG island shores (Fig. 5b), similar to the distribution of 5hmC detected by WG (Fig. 2c) and HM450K Bis/OxBis (Fig. 3e). To next assess the specificity of antibody-based 5hmC detection, we compared the data with WG Bis/OxBis-seq single nucleotide 5hmC mapping. We expanded each hMeDIP peak summit 150-bp up- and downstream (summits; Fig. 5c) and calculated average hydroxymethylation from WG Bis/OxBis-seq data and compared this to 5hmC at the regions not captured by hMeDIP-seq (gaps: Fig. 5c). The data revealed that hMeDIP peaks have significantly higher 5hmC levels (>10%) compared to the

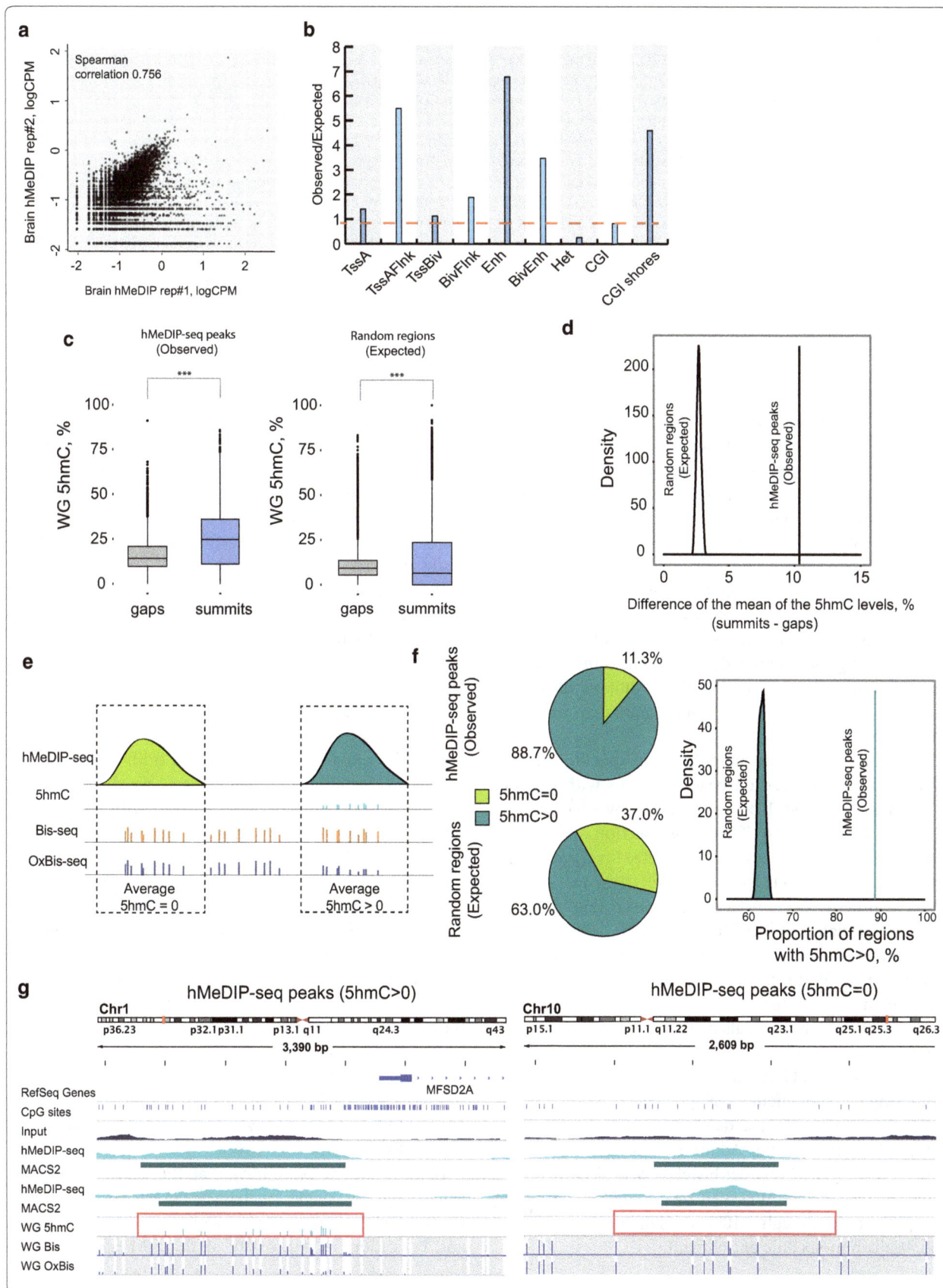

a Spearman correlation 0.756

Brain hMeDIP rep#2, logCPM vs Brain hMeDIP rep#1, logCPM

b Observed/Expected: TssA, TssAFlnk, TssBiv, BivFlnk, Enh, BivEnh, Het, CGI, CGI shores

c hMeDIP-seq peaks (Observed) — WG 5hmC, % — gaps, summits ***

Random regions (Expected) — WG 5hmC, % — gaps, summits ***

d Density — Random regions (Expected), hMeDIP-seq peaks (Observed) — Difference of the mean of the 5hmC levels, % (summits - gaps)

e hMeDIP-seq, 5hmC, Bis-seq, OxBis-seq — Average 5hmC = 0, Average 5hmC > 0

f hMeDIP-seq peaks (Observed) 11.3% / 88.7% — 5hmC=0, 5hmC>0

Random regions (Expected) 37.0% / 63.0%

Density — Random regions (Expected), hMeDIP-seq peaks (Observed) — Proportion of regions with 5hmC>0, %

g hMeDIP-seq peaks (5hmC>0) — Chr1 — p36.23, p32.1, p31.1, p13.1, q11, q24.3, q43 — 3,390 bp — RefSeq Genes MFSD2A, CpG sites, Input, hMeDIP-seq, MACS2, hMeDIP-seq, MACS2, WG 5hmC, WG Bis, WG OxBis

hMeDIP-seq peaks (5hmC=0) — Chr10 — p15.1, p11.1, q11.22, q23.1, q25.1, q25.3, q26.3 — 2,609 bp

(See figure on previous page.)
Fig. 5 hMeDIP-seq hydroxymethylation profiling in the brain. **a** *Scatter plot* showing the correlation between hMeDIP-seq replicates. For 300-bp genomic tiles logCPM (count per million) values were calculated for each replicate. Each *dot* represents one genomic tile. **b** *Bar plot* showing observed over expected by chance enrichment of 5hmC peaks detected by hMeDIP-seq at multiple genomic locations; computationally derived chromatin segmentation (ChromHMM) of Brain Frontal Lobe genome, as well as CpG islands and CpG island shores. **c, d** Correlation of 5hmC profiling between WG Bis/OxBis and hMeDIP-seq. **c** The distribution of WG-derived 5hmC levels at hMeDIP summits expanded $\pm$ 150 bp and gaps between them (observed, *left*). Random permutation of hMeDIP summits and gaps and 5hmC levels distribution (expected, *right*). The differences between gaps and summits are statistically significant as determined by Kruskal–Wallis nonparametric test ($p < 0.01$). **d** The mean of the differences of 5hmC levels between hMeDIP-seq summits and gaps (observed) and the distribution of the mean for the permuted summits and gaps (expected). **e** Schematic representation of hMeDIP-seq peaks with no 5hmC detected by WG Bis/OxBis-seq (average 5hmC = 0) and with 5hmC detected by WG Bis/OxBis-seq (average 5hmC > 0). **f** *Pie charts* showing the proportion of real hMeDIP peaks (*top*) and randomly permuted hMeDIP peaks (*bottom*) with and without 5hmC detected by WG Bis/OxBis-seq. The distribution of the proportion of randomly permuted hMeDIP peaks with 5hmC (expected) and the actual proportion of hMeDIP peaks is shown on the *right*. For the analysis, hMeDIP-seq peaks with all associated CpGs having at least 10× WG coverage have been selected, which accounts for approximately 9000 peaks. **g** Genomic regions showing specific (*left*) and non-specific (*right*) hMeDIP-seq peaks

regions not captured by hMeDIP-seq (gaps; Fig. 5c, left). Random permutation of the hMeDIP peak summits and gaps does not result in the same trend (Fig. 5c, right). The mean of the differences of 5hmC levels between hMeDIP-seq peak summits and gaps (observed) is plotted alongside the distribution of the mean for the permuted summits and gaps (expected), showing the observed mean exceeding the expected mean (Fig. 5d), confirming the specificity of hMeDIP to 5hmC. Interestingly, splitting hMeDIP signal based on the relative enrichment over input signal read counts (logFC, "Methods" section) into low, medium and high categories did not show differences in 5hmC levels according to the WG Bis/OxBis (Additional file 6: Figure S5B). This highlights the semi-quantitative nature of hMeDIP-seq.

Finally, to interrogate potential non-specific binding of hMeDIP, we focused on hMeDIP peaks with all CpG sites having at least 10× sequencing coverage in WG Bis-seq and OxBis-seq ("Methods" section). We separated hMeDIP-seq peaks with detected average 5hmC (5hmC > 0, "Methods" section) and with no detected 5hmC (5hmC = 0) in WG Bis/OxBis-seq (Fig. 5e). The relative abundance of hMeDIP peaks with detected average 5hmC > 0 is 88.7%, revealing high specific hMeDIP binding (Fig. 5f, top). Random permutation of peaks resulted in a maximum of 65% of random peaks with 5hmC > 0, which is smaller than the observed value (88.7%) (Fig. 5f, right). We noted that 11.3% of 5hmC was not detected by WG Bis/OxBis (Fig. 5f), suggesting that either higher sequencing coverage of WG Bis/OxBis-seq is required and/or a degree of non-specific binding of hMeDIP (Fig. 5g, Additional file 6: Figure S5C). Together, the data show that hMeDIP-seq displays a high concordance with the WG Bis/OxBis-seq 5hmC profiling approach, even though ~10% hMeDIP peaks represent potentially non-specific enrichment.

Profiling low abundance 5hmC in cell line DNA

In contrast to adult brain DNA with a known high abundance of 5hmC, we next performed hMeDIP-seq on cancer cell line LNCaP DNA as cell lines are known to have low levels of 5hmC [18]. We observed less correlation (0.625) between replicates (Fig. 6a) than for the brain DNA (Fig. 5a). However, we show that the hMeDIP-seq peaks detected in prostate cancer cell line LNCaP show more correlation (0.384) with those identified in breast cancer cell line MCF7 (publically available: GSM1479831) (Fig. 6b, d) than brain DNA (0.123) (Fig. 6c), as expected between different DNA samples displaying low and high 5hmC content.

Next, we compared the performance of the single nucleotide 5hmC approaches (WG and HM450K Bis/OxBis) for brain and LNCaP DNA. Both single nucleotide 5hmC profiling approaches show the distribution of the 5hmC signal centred around zero with similar proportion of positive and negative values (Fig. 6e, g); however, the 5hmC signal in brain tissue is shifted towards positive values in both approaches (Fig. 6e, g). Since 5hmC is a subtraction of OxBis from the Bis signal, the distribution of 5hmC signal around zero reflects, per se, the error rates of both approaches. Further, the proportion of significantly hydroxymethylated CpGs of all CpGs (see "Methods" section) in the brain is 10 times greater than that of LNCaP cells (27.7 and 2.5% respectively), as detected by WG Bis/OxBis-seq (Fig. 6f). Importantly, HM450K Bis/OxBis did not detect any significant 5hmC in LNCaP cells (Fig. 6h), in contrast to the >35% of CpGs in the brain found to be hydroxymethylated by HM450K Bis/OxBis (Fig. 6h).

To rule out the possibility that the absence of 5hmC calling by the single nucleotide approaches is driven by the lack of sequencing coverage in WG Bis/OxBis and limitations of sensitivity of the HM450K array, we

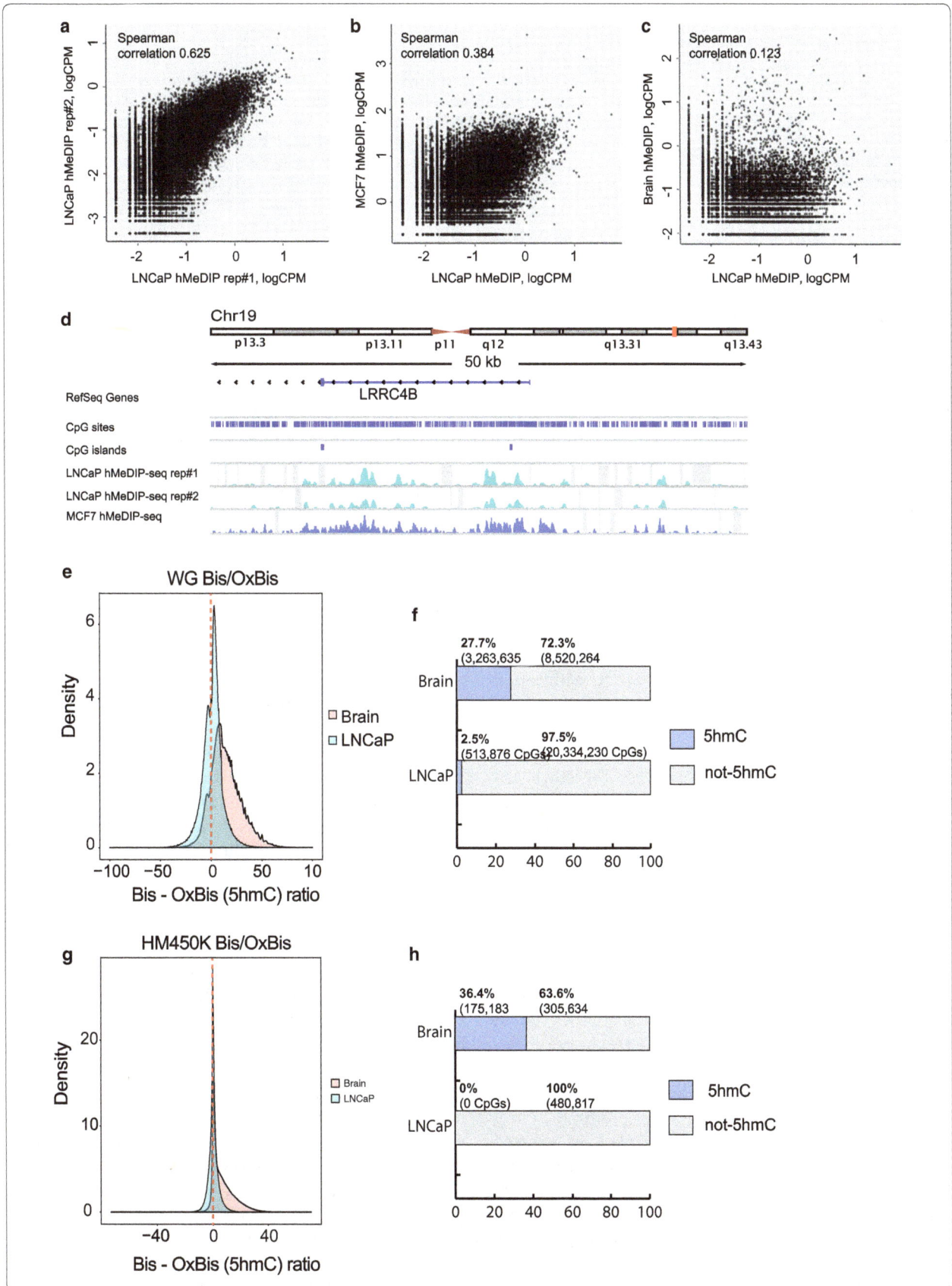

(See figure on previous page.)
Fig. 6 hMeDIP-seq hydroxymethylation profiling in cell line DNA. **a** *Scatter plot* showing the correlation between hMeDIP-seq replicates in LNCaP cells. For each genomic tile from one replicate the average enrichment score for the second replicate was calculated. Each *dot* represents one genomic tile. **b** *Scatter plot* showing the correlation between hMeDIP-seq in LNCaP cells and public hMeDIP-seq in MCF7 cells. **c** *Scatter plot* showing the correlation between hMeDIP-seq in LNCaP cells and brain hMeDIP-seq. **d** Genomic region showing the correspondence of hMeDIP-seq replicates in LNCaP cells as well as MCF7 cells. **e** Density plot showing the distribution of $p_{Bis} - p_{OxBis}$ values in the Brain versus LNCaP WG Bis/OxBis data. **f** The percentages of significantly hydroxymethylated CpGs of all CpGs with at least 10× coverage on the WG Bis/OxBis in the Brain versus LNCaP. **g** Density plot showing the distribution of $p_{Bis} - p_{OxBis}$ values in the Brain versus LNCaP HM450K Bis/OxBis data. **h** The percentages of significantly hydroxymethylated CpGs of all CpGs with at least 10× coverage on the HM450K Bis/OxBis in the Brain versus LNCaP

performed loci-specific TAB-seq with high (~50,000×) sequencing coverage. We chose regions that were detected by hMeDIP-seq (Additional file 7: Figure S6). While loci-specific TAB-seq showed 5hmC levels of <2–4% (Additional file 7: Figure S6), a proportion of the levels are comparable to the technique error rate (~1.8%, "Methods" section), suggesting potential non-specificity of hMeDIP-seq in genomes with low (<2–4%) 5hmC abundance.

These findings highlight that for DNA of high 5hmC abundance (5hmC levels >10%) hMeDIP-seq shows high specificity. However, in the absence or presence of very low levels of 5hmC (<~2–4%), such as in cell line DNA, hMeDIP-seq signals cannot be used as the representative of the actual 5hmC distribution without further validation.

Discussion

For more than two and half decades, DNA methylation studies in higher organisms have relied on the use of bisulphite conversion technologies to study DNA methylation [38, 40]. However, it is now understood that these technologies do not distinguish between 5-methylcytosine (5mC) and 5-hydroxymethylcytosine (5hmC) [41, 42]. Since the discovery of 5-hydroxymethylcytosine and its potential role in modulating the DNA methylation landscape, there has been significant interest in defining the genome-wide distribution of 5hmC. Importantly, there is a critical need to distinguish 5hmC from 5mC and accurately detect and quantitate the levels of 5hmC at single-base resolution. A number of strategies have been developed to map 5hmC, including affinity-based approaches [6, 27, 28] and chemical modification approaches that allow single nucleotide resolution analyses [29, 30, 43]. In general, the abundance of total 5hmC detected across the genome is approximately tenfold to 100-fold lower [18] than that of 5mC, posing technical challenges for many approaches. Here, we have assessed three 5hmC genome-wide profiling approaches, using whole-genome bisulphite/oxidative bisulphite sequencing (WG Bis/OxBis-seq), Infinium HumanMethylation450 BeadChip arrays coupled with OxBis (HM450K Bis/OxBis) and antibody-based immunoprecipitation of hydroxymethylated DNA (hMeDIP-seq).

WG Bis/OxBis-seq approach enables analysis of the relationship between 5mC and 5hmC at the single nucleotide level. We were able to identify hydroxymethylated genomic regulatory regions (Fig. 2c, d), in particular at enhancer regions (Fig. 2e), highlighting the importance of taking 5hmC into consideration, especially when studying DNA methylation dynamics at distal regulatory regions in brain DNA. Interestingly, the vast majority of CpG sites with moderate modification levels possess the highest hydroxymethylation proportion compared to the more densely methylated CpGs that generally possess lower 5hmC signal (Fig. 2b). In addition, genomic regions with moderate methylation levels show mosaic 5hmC/5mC single-molecule methylation patterns. This observation is consistent with the hypothesis that 5hmC is prevalent at dynamic CpG sites showing higher "plasticity" across the genome [44]. Thus, the ability to discriminate 5mC and 5hmC on a genome-wide scale allows the contribution of 5hmC to total methylation to be elucidated, and to identify where regions are more dynamic or stably methylated.

Single nucleotide whole-genome 5hmC profiling is theoretically the most ideal approach to interrogating unbiased distribution of 5hmC across the genome. However, the use of WG approaches is restricted due to its high cost. To accurately assess 5hmC levels at single nucleotide resolution, 5hmC profiling requires higher sequencing depth, compared to 5mC methylation profiling since hydroxymethylation levels are more than a magnitude lower than DNA methylation levels [45]. As an alternative, HM450 BeadChip arrays have been modified to allow the detection of 5hmC separate from 5mC, used in conjunction with both oxidative [31] and TET-assisted bisulphite conversion [32]. Due to the fact that HM450K interrogates only 1.8% of all CpG sites in the human genome, 5hmC detection is restricted to the regions interrogated on the array. Using HM450K Bis/OxBis, we were able to detect hydroxymethylation at ~50% of promoters and ~20% of enhancers, which were identified as hydroxymethylated according to the WG Bis/OxBis-seq (Fig. 3f). However, since enhancers possess the highest enrichment of 5hmC among all genomic elements, a substantial proportion of potentially 5hmC-dependent functional genomic elements are not detected by HM450K.

Importantly, the newly released Illumina EPIC array covers over 850,000 CpG sites, including >90% of the CpGs from the HM450 BeadChip and an additional 413,743 CpGs. The additional probes improve the coverage of regulatory elements, including enhancers, and therefore, this will offer significantly enhanced coverage of 5hmC-enriched genomic elements.

Overall, detection of hydroxymethylation at different genomic regions is highly concordant between approaches (Fig. 4c). However, HM450K Bis/OxBis possess a tendency of 5hmC signal underestimation, driven by CpG sites with extreme levels of total methylation, highlighting the necessity of improved normalization approaches of the array signal. This results in the "loss" of a significant number (~20%) of hydroxymethylated regions at different regulatory elements. On the other hand, a subset of promoters and flanking regions are identified as hydroxymethylated on HM450K Bis/OxBis *only* (Fig. 4d). Those regions possessed lower levels of 5hmC and therefore require higher sequencing coverage highlighting the higher sensitivity of the HM450 compared to the WG at a given sequencing coverage.

In contrast, antibody-based enrichment hMeDIP-seq of 5hmC has been widely used as a relatively easy and cost-effective approach. Overall, we observed a good agreement between hMeDIP-seq and WG Bis/OxBis-seq 5hmC signal as well as high concordance in the patterns of 5hmC genomic distribution. Moreover, we showed that, despite the presence of 5hmC signal in the hMeDIP-seq in cell line DNA, neither HM450K Bis/OxBis nor WG Bis/OxBis-seq and loci-specific TAB-seq were able to detect any significant 5hmC signal. This finding, together with the 5hmC profiling in adult brain DNA, suggests that in the presence of high levels of 5hmC, hMeDIP-seq is reliable, whereas in the absence or presence of very low levels of 5hmC, such as cell line DNA, hMeDIP-seq signal is potentially subject to misinterpretation of 5hmC distribution. An alternative affinity-based approach, hMeSeal, has been used for cost-effective whole-genome 5hmC profiling [24, 26]. While it has been shown to be a reliable and sensitive approach [18, 26] and successfully performed on low input DNA material [46], a detailed comparative analysis would be required to assess its performance compared to single nucleotide 5hmC profiling techniques on a whole-genome scale.

Conclusions

In this study, we provide a detailed comparison of three genomic 5hmC profiling approaches. 5hmC profiling with WG Bis/OxBis-seq provides the most comprehensive quantitative overview of 5mC and 5hmC distribution across the whole genome in cells displaying higher

levels of 5hmC, such as brain tissue. HM450K Bis/OxBis provides a user-friendly, high-throughput and affordable approach for cells that display higher levels of 5hmC, but will miss regions not present on the array and shows undercalling of 5hmC signal driven by CpGs with high methylation. Finally, hMeDIP-seq is a widely used and accepted approach due to its ease of use and cost-effectiveness. However, it is semi-quantitative, does not allow single nucleotide resolution and has the potential for non-specific enrichment in DNA displaying low 5hmC levels. Overall, we find a high correlation of hMeDIP with both WG Bis/OxBis and HM450K Bis/OxBis. Ultimately the method of choice for whole-genome profiling of 5hmC will depend on the abundance of 5hmC, number and quantity of DNA samples, cost consideration, bioinformatics expertise and the question being addressed.

Methods
DNA samples
Adult Brain Frontal Lobe genomic DNA from a single donor was obtained from Banksia Scientific Company (Bulimba, Australia, Cat No D1234035). Genomic DNA from the prostate cancer cell line LNCaP was extracted using QIAamp DNA Mini kit (Qiagen, USA). LNCaP prostate cells were cultured as described previously [47].

Whole-genome bisulphite and oxidative bisulphite sequencing
Adult brain genomic DNA was sheared to an average size of 800 bp; 200 ng was used for the bisulphite (Bis) and oxidative bisulphite (OxBis) reactions. To assess the efficiency of potassium perruthenate-mediated oxidation of 5hmC and the behaviour of 5mC, genomic DNA was spiked with 5-hydroxymethylated 338-bp PCR product of APC genomic locus and M.SssI λDNA, respectively (described in "Spike-in controls for WG Bis/OxBis and TAB-seq" section). The bisulphite and oxidative bisulphite reactions were performed according to the manufacturer's instructions (CEGX TrueMethyl® WG user guide v2). Library preparation and indexing were also carried out as described (CEGX TrueMethyl® WG). Library quality was assessed with the Agilent 2100 Bioanalyzer using the high-sensitivity DNA kit (Agilent, CA, USA). DNA was quantified using the KAPA Library Quantification kit by quantitative PCR (KAPA Biosystems). Paired-end 150-bp sequencing was performed for each library on the Illumina HiSeqX platform using the HiSeq X™ Ten Reagent Kit v2.

Infinium HM450K bisulphite and oxidative bisulphite beadchip arrays
DNA was treated in separate aliquots with CEGX Bis and OxBis reagents according to the manufacturer's

specifications (CEGX TrueMethyl®_UGuide). 2 µg DNA was first sheared to 10 kb and then purified on BioRad® P6 Micro-Bio spin column. Two aliquots were subjected to oxidation and two to mock oxidation prior to the Bis treatment (see CEGX protocol). Following the Bis reaction, the samples were re-quantified using the Qubit ssDNA assay. A minimum sample concentration of 20 ng/µl was required for the next step of the 450K process to ensure 7 µl of bisulphite-converted sample contains 140–160 ng of bisulphite-converted DNA. 7 µl of recovered TrueMethyl template was used in the HM450K protocol with 1 µl of 0.4 N NaOH (see Infinium Methylation assay). All subsequent steps were completed according to Illumina Infinium HM450K beadarray chip instructions.

Hydroxymethylation profiling by hMeDIP-seq

DNA was sonicated with Covaris to produce fragments in size range of 300–500 bp. Prior to hMeDIP procedure, Illumina adaptors were ligated to (5×1 µg) of fragmented DNA as described in the TruSeq LT DNA Sample Preparation kit, Illumina. The hMeDIP assay was performed according to the manufacturer's instructions (Active Motif, hMeDIP, Cat No 55010). Briefly, 3×1 µg of fragmented adapter-ligated DNA was spiked with 50 ng of either unmethylated, 5mC methylated or 5hmC hydroxymethylated 338-bp PCR product of APC genomic locus. The DNA was denatured for 10 min at 95 °C and immunoprecipitated overnight at 4 °C with 4 µl of 5hmC polyclonal antibody (Active Motif Cat No 55010). To allow selective enrichment of immune-captured DNA fragments, the mixture was incubated with 25 µl of Protein G magnetic beads for 2 h at 4 °C prior to washing all unbound DNA fragments. The bound hydroxymethylated DNA was eluted, treated with proteinase K and purified by Phenol/chloroform/isoamyl alcohol extraction and ethanol precipitation. The specificity of the hMeDIP assay was validated by qPCR of the unmethylated, methylated and hydroxymethylated spike-in APC controls and dot blots.

Bisulphite and TET-assisted bisulphite treatment for amplicon sequencing

Bisulphite reaction

Bisulphite reaction was performed using EZ DNA Methylation-Gold Kit (Zymo Research, USA, Cat No D5005) according to the manufacturer's instructions. M.SssI λDNA (5mC) was used as a spike-in control (described in "Spike-in controls for WG Bis/OxBis and TAB-seq" section) to assess the efficiency of bisulphite conversion reaction.

TET-assisted bisulphite treatment

TET-assisted bisulphite treatment was performed using the 5hmC TAB-seq Kit (WiseGene, USA, Cat No K001) according to the manufacturer's instruction. Briefly, 1 µg genomic DNA was sonicated to the size of approximately 2kbp according to the manufacturer's instructions. After sonication, DNA was spiked with 10 ng (1%) M.SssI λDNA (5mC) control and 10 ng (1%) 5hmC pUC18 control DNA. The β-GT-based reaction was performed at 37 °C for 1 h and the DNA purified using QIAquick PCR Purification kit (Qiagen, USA, Cat No 28106) according to the protocol and eluted in 27 µl water. The eluted DNA was split into two separate reactions to ensure no more than 300 ng DNA per TET1-based oxidation reaction. The TET1 oxidation reaction was performed at 37 °C for 1 h, followed by the treatment of 1 µl of proteinase K (20 mg/ml) at 50 °C for 1 h. The oxidized DNA was purified using QIAquick PCR purification kit (Qiagen, USA, Cat No 28106) and eluted in 50 µl water. TET1-oxidized DNA was then bisulphite-treated above using the EZ DNA Methylation-Gold kit (Zymo Research, USA Cat No D5005) as described in the protocol. Post-bisulphite conversion PCR amplification was performed in triplicate (4 ng/PCR); PCRs were pooled and purified using Wizard® SV Gel and PCR Clean-Up System (Promega, USA, Cat No A9282). Library prep was performed following the instructions as per the Illumina TruSeq DNA sample prep kit (Cat No FC-121-2001) described below.

Spike-in controls for WG Bis/OxBis and TAB-seq

Spike-in controls were used to assess the efficiency of the bisulphite, oxidative bisulphite and TET-assisted bisulphite reactions.

M.SssI λDNA (5mC)

We generated an in vitro CpG (5mC)-methylated λDNA control to (1) assess the bisulphite conversion efficiency of unmethylated cytosines (in a non-CpG context) to uracils and (2) to assess the efficiency of TET-mediated oxidation of 5mC in TAB-seq reaction. Unmethylated λDNA (Promega, USA, Cat No D1521) was sonicated to the average size of approximately 2 kbp according to the manufacturer's instructions. 3 µg of sonicated DNA was used in methylation reaction using 4U of CpG methyltransferase M.SssI (New England Biolabs, USA, Cat No M0226S) in the presence of 640 µM SAM. Methylation reaction was allowed to proceed at 37 °C for 2 h and was stopped by heating at 65 °C for 20 min. CpG-methylated λDNA was purified using QIAquick PCR purification kit according to the protocol. After the completion of TAB reaction, 290-bp fragment of λDNA was amplified using following primers: forward 5'-TTTGGGTTATGTAAGTTGATTTTATG-3' and reverse 5′-CACCCTACTTACTAAAATTTACACC-3′ (Additional file 8: Table S2). The PCR product was 3′-adenylated and ligated into the pGEM-T-easy plasmid (Promega, USA, Cat No A1360), followed by MiSeq amplicon sequencing.

For WG Bis/OxBis, the efficiency of bisulphite conversion was 98.30% for Bis and 99.76% for OxBis. For TAB-seq, the efficiency of bisulphite conversion was 99.58% and TET-mediated 5mC-to-T oxidation efficiency was 98.74% (Additional file 2: Figure S1C, D). The efficiency of M.SssI λDNA CpG methylation was assessed by clonal Sanger sequencing after the completion of the conventional bisulphite reaction, which was performed alongside.

5hmC pUC18 control

To assess the efficiency of β-GT-mediated protection of 5hmC, an in vitro hydroxymethylated pUC18 control was generated. 1.64-kbp region of pUC18 plasmid was amplified in the presence of 5-hydroxymethyl-dCTP, 5hmdCTP (Zymo Research, USA, Cat No D1045) using following primers: forward 5′-GCAGATTGTACTGA-GAGTGC-3′ and reverse 5′-TGCTGATAAATCTG-GAGCCG-3′ (Additional file 8: Table S2). After the completion of TAB reaction, 190-bp fragment of 1.64-kbp 5hmC pUC18 control was amplified using the following primers: forward 5'-GTAGATTGTATTGA-GAGTGT-3' and reverse 5'-TACCCAACTTAATCGC-CTTG-3' (Additional file 8: Table S2), followed by the clonal Sanger sequencing as described for the M.SssI λDNA control. Due to the contamination of 5hmdCTP with unmodified 5dCTP, the actual degree of hydroxymethylation of 5hmC pUC18 control had to be assessed by clonal Sanger sequencing after the completion of the conventional bisulphite reaction, which was performed alongside. The β-GT-mediated protection was ~100% as estimated using 5hmC-to-T conversion efficiency of 5hmC pUC18 control in TAB reaction normalized to that in the conventional bisulphite reaction (Additional file 2: Figure S1D).

5hmC APC control

To assess the efficiency of potassium perruthenate ($KRuO_4$)-mediated oxidation of 5mC in the OxBis reaction, Illumina TruSeq DNA adapters were ligated to the commercially available 5hmC APC PCR product of APC genomic locus (Active Motif, USA, Cat No 55008). Briefly, 1 µg of 5hmC APC control was used for the end repair, A-tailing and ligation of Illumina adapter according to the Illumina instructions. After the completion of the OxBis reaction, 5hmC APC PCR product was amplified using PCR primer cocktail supplied by Illumina followed by the clonal Sanger sequencing as described for

the M.SssI λDNA control (Additional file 2: Figure S1B). The $KRuO_4$-mediated 5mC-to-5caC/(T) oxidation efficiency was 99.33% as estimated using 5hmC-to-5caC/(T) conversion efficiency of 5hmC pUC18 control in OxBis reaction normalized to that in the conventional bisulphite reaction.

MiSeq amplicon TAB sequencing

Validation of candidate regions was performed on bisulphite- and TET-assisted bisulphite-treated DNA described above. First, PCRs were performed on a temperature gradient between 50 and 62 °C to achieve optimal amplification temperature. Each PCR was checked on an agarose gel for specific PCR products. To test for the amplification bias, we used the following bisulphite-treated control DNA: (1) human genomic blood DNA (Roche Cat No 11691112001) (unmethylated control DNA); (2) serological DNA from Chemicon (100% methylated control); (3) 50:50 mix of Roche and serological DNA under three different concentrations of $MgCl_2$. For each PCR, optimal temperature and $MgCl_2$ concentrations were determined as described previously [48].

PCRs of Adult Brain and LNCaP DNA (bisulphite or TET-assisted bisulphite treated) were performed in triplicate using optimized conditions (Additional file 8: Table S2). PCRs were pooled and purified using Wizard SV and PCR Clean-Up System (Promega, USA, Cat No A9282) and quantitated by Qubit. Library prep was performed following the instructions as per the Illumina TruSeq DNA sample prep kit (Cat No FC-121-2001). Briefly, 1000 ng of pooled amplicon input DNA was used for each library preparation. End repair, A-tailing and ligation of Illumina adapter to pooled PCR library were performed according to the Illumina instructions. After PCR clean-up, the library was quantified by Qubit, diluted to 10 nM according to Qubit, and accurately quantitated by KAPA SYBR FAST Universal qPCR (KAPA Biosystems, USA, Cat No KK4835) before being sequenced on the Illumina MiSeq™ sequencer (Illumina, CA, USA).

Data analysis

Data processing and alignment was performed using in-house computational pipelines. Statistical analyses were conducted in the R statistical software.

Whole-genome Bis/OxBis sequencing data

Bisulphite reads were aligned to the human genome using version 1.2 of an internally developed pipeline, publicly available for download from http://github.com/astatham/Bisulphite_tools. Briefly, adaptor sequences and poor-quality bases were removed using TrimGalore (version 0.2.8, http://www.bioinformatics.babraham.ac.uk/projects/trim_galore/) in paired-end mode with default parameters. Bismark v0.8.326 was then used to align reads to hg19 using the parameters "-p 4 –bowtie2 –X 1000 –unmapped –ambiguous –gzip –bam". PCR duplicates were removed using Picard v1.91 (http://broadinstitute.github.io/picard). Count tables of the number of methylated and unmethylated bases sequenced at each CpG site in the genome were constructed using bismark_methylation_extractor with the parameters "-p –no_overlap –ignore_r2 4 –comprehensive –merge_non-CpG –bedgraph –counts –report –gzip –buffer_size 20G". The adult brain Bis and OxBis libraries had a total of 517,530,911 and 489,841,771 reads, respectively. Both libraries passed basic quality control checks with 89/90% alignment rate and 18×/16× mean coverage for adult brain Bis/OxBis, respectively.

To assess the level of 5hmC at a CpG locus, we compared the number of retained cytosines in Bis experiment to that in OxBis experiment. More specifically, we assumed that the number of retained cytosines follows a binomial distribution $NC \sim$ Binomial (N, p), where N is the coverage and p is the proportion of modified cytosines. For the Bis experiment $p_{Bis} = p_{5mC} + p_{5hmC}$ is the proportion of both 5mC and 5hmC modifications, whereas for OxBis experiment $p_{OxBis} = p_{5mC}$ is the proportion of 5mC modifications. Thus, to assess the proportion of 5hmC modifications we used *proportion test*, as implemented in R's prop.test() function, to estimate the difference between p_{Bis} and p_{OxBis}, given NC_{Bis}, N_{Bis}, NC_{OxBis} and N_{OxBis}.

Average coverage for Bis-seq and OxBis-seq was 18× and 16×, respectively (Additional file 1: Table S1). Before applying proportion test, we filtered out CpG loci having less than 10x coverage in either Bis or OxBis dataset. This reduced the number of tested CpG loci from 28,269,977 to 11,783,899 for adult brain data. CpGs with statistically significant difference between Bis and OxBis (p value <0.05, $p_{Bis} - p_{OxBis} > 0$) were considered as significantly hydroxymethylated resulting in 3,263,635 CpG sites. Only these CpG sites are considered in the subsequent analyses. For the calculations of the *average hydroxymethylation* per region, we discarded CpGs with p value ≥ 0.05 and $p_{Bis} - p_{OxBis} > 10\%$ as well as CpGs with $p_{Bis} - p_{OxBis} < 0$ and imputed CpGs with p value ≥ 0.05 and $0 < p_{Bis} - p_{OxBis} < 10\%$ to zero.

We performed power calculations using R's power.prop.test() function to determine power as a function of coverage in both Bis and OxBis experiments.

HM450K Bis/OxBis data

Two replicates of adult brain samples per treatment condition, Bis or OxBis, were profiled on Illumina's HumanMethylation450K array [37]. The raw data were preprocessed and background normalized with Biconductor minfi package [49] using preprocessIllumina(..., bg.correct = TRUE, normalize = "controls", reference = 1) normalization function. We used the limma Bioconductor package [49, 50] to perform differential methylation analysis between Bis and OxBis treatments to determine levels of 5hmC. We only considered probes for which there was a reliable methylation readout (detection p value <0.01) in all four samples. We then transformed β-values into M-values using logit transformation: $M = \log\left(\frac{\beta}{1-\beta}\right)$. (To avoid extreme M-values, the β-values were capped at 0.01 and 0.99.) Standard limma workflow with unpaired contrast was then applied to computed M-values to call differentially methylated probes between Bis and OxBis and thus to determine levels of 5hmC. This analysis resulted in 175,183 probes having significant hydroxymethylation (p value <0.05, $p_{Bis} - p_{OxBis} > 0$).

TAB-seq data

Paired-end fastq files were obtained for each library and aligned to hg19 using bwa-meth (http://github.com/brentp/bwa-meth, arXiv:1401.1129). Downstream analysis was performed using the "ampliconAnalysis" function of the R package aaRon (http://github.com/astatham/aaRon). Data quality was checked by assessing the number of reads obtained and the bisulphite conversion efficiency per amplicon and per sample. All samples had high bisulphite conversion efficiency >98% and amplicons had >10,000× coverage. Percent hydroxymethylation/methylation at each CpG site was calculated.

hMeDIP sequencing data

Sequenced reads from hMeDIP immunoprecipitated and input control human brain samples were mapped to the reference human genome (hg19) with bowtie v.1.1.0 [51], allowing up to three mismatches. Reads mapping to multiple locations and/or deemed as PCR duplicates were filtered out. Reads were extended 300 bp and overlapped with the 300-bp tiling of the human genome to create a table of counts to be used for statistical modelling. Bins with low total number of reads (less than 20) across the three samples were removed from the analysis. Standard analysis flow as implemented in the edgeR Bioconductor package [52] was then applied to contrast read counts in hMeDIP samples to the input control. The library normalization step as implemented in calcNormFactors() was omitted and dispersion was set to 0.01. The same analysis was repeated for regions of 300 bp centred on summits of broad peaks identified with MACS2 algorithm [53]. Summit regions from two hMeDIP replicates were merged resulting in 137,598 regions used for the analysis. Of those, 137,373 regions had p value <0.05 and logFC > 0 (with the smallest logFC 1.391). Regions with FDR < 0.1

and logFC > 0 (smallest logFC 1.391) were called as having 5hmC modification resulting in 21,553 marked regions for summit-centred analysis. For the specificity analysis, we selected only hMeDIP-seq regions with all CpGs per region having at least 10× coverage in the WG Bis/OxBis resulting in 7563 regions. Next, we calculated average hydroxymethylation (as described in "Whole-genome Bis/OxBis sequencing data" section) per each region and separated into two categories: average 5hmC > 0 and 5hmC = 0. For LNCaP, hMeDIP replicates were merged resulting in 240,216 regions used for the analysis.

Genome annotation
Genomic locations
Genomic coordinates (hg19) of CpG islands were obtained from UCSC genome browser. Genomic coordinates of CpG island shores were derived by taking ±2 -kb flanking regions around CpG islands.

ChromHMM annotations
A bed-formatted annotation file of chromatin states was downloaded from the Encode Roadmap (http://egg2.wustl.edu/roadmap/data/byFileType/chromhmmSegmentations/ChmmModels/coreMarks/jointModel/final/E073_15_coreMarks_dense.bb) [34, 35]. We used hypergeometric testing to determine statistical significance of overlap between regional hypomethylated probes and the above functional annotations of the genome.

Additional files

Additional file 1. Table S1. Summary of sequencing metrics.

Additional file 2. Figure S1. Spike-in controls showing the efficiency of the WG OxBis and loci-specific TAB reactions. **A**, **B** Each lollipop represents single CpG in M.SssI λDNA 5mC control and single cytosine in 5hmC pUC18 control. **C**, **D** The efficiency of TET-mediated oxidation of 5mC (**C**) and β-glucosyltransferase-mediated protection of 5hmC (**D**) of spike-in controls in loci-specific Bis/TAB-seq. **C** In vitro M.SssI CpG-methylated λDNA was used as a spike-in control to estimate the TET-mediated 5mC oxidation efficiency. Deep Bis-seq shows the degree of CpG methylation of M.SssI λDNA and TAB-seq shows the efficient oxidation of methylated CpGs, and therefore, very low methylation signal (TET-mediated 5mC-to-T oxidation efficiency) was calculated as the ratio of mC signal in TAB and Bis and was 98.74%; non-conversion rate of unmodified cytosine was 0.42%. **D** Region from pUC18 plasmid amplified in the presence of 5hm-dCTPs was used as a spike-in control to estimate the efficiency of β-glucosyltransferase-mediated 5hmC protection from the oxidation by TET enzymes. Deep Bis-seq shows the degree of cytosines hydroxymethylation of 5hmC pUC18 and deep TAB-seq shows the efficiency of protection (β-glucosyltransferase-mediated protection efficiency equals to 100%).

Additional file 3. Figure S2. Loci-specific Bis/TAB-seq for HM450K validation. Genomic regions showing agreement in total methylation and hydroxymethylation levels detected by HM450K Bis/OxBis and loci-specific Bis/TAB-seq, respectively. *Red dots* depict 5modC (*top*) and 5hmC (*bottom*) levels of each CpG site detected by loci-specific Bis/TAB-seq, respectively. *Blue dots* depict 5modC (*top*) and 5hmC (*bottom*) levels of HM450K CpG probes. **A** Regions with significant hydroxymethylation according

to HM450K (SLC12A6_2: chr15: 34,628,635–34,628,921; PEG10: chr7: 94,285,834 94,286,118). **B** Regions with no hydroxymethylation according to HM450K. The selected negative regions had a similar range of total methylation values to positive regions and serve as a control to eliminate the differences in 5hmC detection that could be caused by different levels of total methylation (e.g. efficiency of TET-mediated oxidation).

Additional file 4. Figure S3. Single-molecule total methylation (**A**) and hydroxymethylation (**B**) patterns of each region (represented as one *vertical blue line*) were determined based on the deep loci-specific Bis/TAB-seq, respectively. Methylation patterns were separated into five groups based on the percentage of CpGs per region being methylated and/or hydroxymethylated (**A**) and hydroxymethylated (**B**). For each region (*vertical blue line*), the average total methylation (**A**) or hydroxymethylation (**B**) was calculated (*x-axis*) and the frequency of each pattern was plotted along the y-axis (summing up to 100%). Five different patterns were defined based on the proportion of methylated or hydroxymethylated CpGs of all CpGs per region (0, 0–10, 10–50, 50–80, 80–100%).

Additional file 5. Figure S4. WG Bis/OxBis-seq and HM450K Bis/OxBis correlation analysis. **A** Comparison of 5modC (Bis, *left*) and 5mC (OxBis, *right*) between WG and HM450K Bis/OxBis across CpG sites interrogated by the HM450K and having at least 10× coverage on the WG Bis/OxBis. **B** Boxplots showing the difference in 5modC (Bis, *left*), 5mC (OxBis, *middle*) and 5hmC (Bis-OxBis, *right*) between WG and HM450K Bis/OxBis at different levels of the corresponding modification. Only CpG sites ($n = 42,537$) considered as significantly hydroxymethylated by both approaches are included. **C** Boxplots showing the relationship between the total methylation levels according to the WG Bis/OxBis (*x-axis*) and the difference in 5mC (*left*) and 5hmC (*right*) between approaches. The difference is calculated as HM450K methylation value subtracted from the WG methylation value (*y-axis*).

Additional file 6. Figure S5. Validation of the hMeDIP-seq approach with the WG Bis/OxBis-seq. **A** The histogram showing the distribution of the hMeDIP-seq signal normalized to the input (logFC). The *colours* indicate the cut-offs set to depict low (log FC < 1.8), medium (log FC > 1.8 and <2.4) and high (log FC > 2.4) hMeDIP signal enrichment. **B** The distribution of WG-derived 5hmC levels at hMeDIP summits expanded ±150 bp binned based on their logFC values; and gaps between them. **C** *Screen shots* of genomic regions showing specific (above) and non-specific (below) hMeDIP-seq peaks.

Additional file 7. Figure S6. TAB-Seq amplicon validation of hMeDIP-seq in cell line DNA. *Screen shots* showing hMeDIP-seq peaks. WG Bis/OxBis data are shown. *Grey* denotes regions not covered >10× by WG Bis/OxBis. Locus-specific TAB-Seq validation showing mod C (5mC + 5hmC) and 5hmC across the amplicons with %5mC and %5hmC for each CpG site indicated.

Additional file 8. Table S2. Primers.

Abbreviations
5hmC: 5-hydroxymethylcytosine; 5mC: 5-methylcytosine; 5modC: total methylation (5mC + 5hmC); TAB-seq: TET-assisted bisulphite sequencing; Bis-seq: bisulphite sequencing; OxBis: oxidative bisulphite; WG: whole genome.

Authors' contributions
KS, CS and SJC were involved in conception and design. KS, EZ, CMG and PLL were involved in analysis and interpretation of data (e.g. statistical analysis, computational analysis). SSN and KS performed WG Bis/OxBis and hMeDIP. KS, CS, EZ and SJC wrote and reviewed the manuscript. All authors read and approved the final manuscript.

Author details

[1] Epigenetics Research Laboratory, Genomics and Epigenetics Division, Garvan Institute of Medical Research, 384 Victoria Street, Darlinghurst, Sydney, NSW 2010, Australia. [2] St Vincent's Clinical School, UNSW Australia, Sydney, NSW 2010, Australia.

Acknowledgements

We thank Madhavi Maddugoda for the preparation of figures and reviewing the manuscript. We thank Cambridge Epigenetix (CEGX) for conducting the HM450K Bis and OxBis array experiments.

Competing interests

The authors declare that they have no competing interests.

Funding

This work was supported by National Health and Medical Research Council (NHMRC) project Grant (#1088144); NHMRC Fellowship (S.J.C.) (#1063559); Cancer Australia project Grant (#1044458). The contents of the published material are solely the responsibility of the administering institution and individual authors and do not reflect the views of the NHMRC.

References

1. Bird AP. CpG-rich islands and the function of DNA methylation. Nature. 1986;321:209–13.
2. Li E, Beard C, Jaenisch R. Role for DNA methylation in genomic imprinting. Nature. 1993;366:362–5.
3. Jones PA, Baylin SB. The epigenomics of cancer. Cell. 2007;128:683–92.
4. Shen L, Inoue A, He J, Liu Y, Lu F, Zhang Y. Tet3 and DNA replication mediate demethylation of both the maternal and paternal genomes in mouse zygotes. Cell Stem Cell. 2014;15:459–70.
5. Guo F, Li X, Liang D, Li T, Zhu P, Guo H, Wu X, Wen L, Gu TP, Hu B, et al. Active and passive demethylation of male and female pronuclear DNA in the mammalian zygote. Cell Stem Cell. 2014;15:447–58.
6. Ficz G, Branco MR, Seisenberger S, Santos F, Krueger F, Hore TA, Marques CJ, Andrews S, Reik W. Dynamic regulation of 5-hydroxymethylcytosine in mouse ES cells and during differentiation. Nature. 2011;473:398–402.
7. Koh KP, Yabuuchi A, Rao S, Huang Y, Cunniff K, Nardone J, Laiho A, Tahiliani M, Sommer CA, Mostoslavsky G, et al. Tet1 and Tet2 regulate 5-hydroxymethylcytosine production and cell lineage specification in mouse embryonic stem cells. Cell Stem Cell. 2011;8:200–13.
8. Lister R, Mukamel EA, Nery JR, Urich M, Puddifoot CA, Johnson ND, Lucero J, Huang Y, Dwork AJ, Schultz MD, et al. Global epigenomic reconfiguration during mammalian brain development. Science. 2013;341:1237905.
9. Wang T, Pan Q, Lin L, Szulwach KE, Song CX, He C, Wu H, Warren ST, Jin P, Duan R, Li X. Genome-wide DNA hydroxymethylation changes are associated with neurodevelopmental genes in the developing human cerebellum. Hum Mol Genet. 2012;21:5500–10.
10. Ficz G, Gribben JG. Loss of 5-hydroxymethylcytosine in cancer: cause or consequence? Genomics. 2014;104:352–7.
11. Branco MR, Ficz G, Reik W. Uncovering the role of 5-hydroxymethylcytosine in the epigenome. Nat Rev Genet. 2012;13:7–13.
12. Wossidlo M, Nakamura T, Lepikhov K, Marques CJ, Zakhartchenko V, Boiani M, Arand J, Nakano T, Reik W, Walter J. 5-Hydroxymethylcytosine in the mammalian zygote is linked with epigenetic reprogramming. Nat Commun. 2011;2:241.
13. Hahn MA, Qiu R, Wu X, Li AX, Zhang H, Wang J, Jui J, Jin SG, Jiang Y, Pfeifer GP, Lu Q. Dynamics of 5-hydroxymethylcytosine and chromatin marks in Mammalian neurogenesis. Cell Rep. 2013;3:291–300.
14. Inoue A, Zhang Y. Replication-dependent loss of 5-hydroxymethylcytosine in mouse preimplantation embryos. Science. 2011;334:194.
15. Pastor WA, Aravind L, Rao A. TETonic shift: biological roles of TET proteins in DNA demethylation and transcription. Nat Rev Mol Cell Biol. 2013;14:341–56.
16. He YF, Li BZ, Li Z, Liu P, Wang Y, Tang Q, Ding J, Jia Y, Chen Z, Li L, et al. Tet-mediated formation of 5-carboxylcytosine and its excision by TDG in mammalian DNA. Science. 2011;333:1303–7.
17. Shen L, Zhang Y. 5-Hydroxymethylcytosine: generation, fate, and genomic distribution. Curr Opin Cell Biol. 2013;25:289–96.
18. Song CX, Yi C, He C. Mapping recently identified nucleotide variants in the genome and transcriptome. Nat Biotechnol. 2012;30:1107–16.
19. Thomson JP, Hunter JM, Lempiainen H, Muller A, Terranova R, Moggs JG, Meehan RR. Dynamic changes in 5-hydroxymethylation signatures underpin early and late events in drug exposed liver. Nucleic Acids Res. 2013;41:5639–54.
20. Bachman M, Uribe-Lewis S, Yang X, Williams M, Murrell A, Balasubramanian S. 5-Hydroxymethylcytosine is a predominantly stable DNA modification. Nat Chem. 2014;6:1049–55.
21. Tahiliani M, Koh KP, Shen Y, Pastor WA, Bandukwala H, Brudno Y, Agarwal S, Iyer LM, Liu DR, Aravind L, Rao A. Conversion of 5-methylcytosine to 5-hydroxymethylcytosine in mammalian DNA by MLL partner TET1. Science. 2009;324:930–5.
22. Kriaucionis S, Heintz N. The nuclear DNA base 5-hydroxymethylcytosine is present in Purkinje neurons and the brain. Science. 2009;324:929–30.
23. Ito S, Shen L, Dai Q, Wu SC, Collins LB, Swenberg JA, He C, Zhang Y. Tet proteins can convert 5-methylcytosine to 5-formylcytosine and 5-carboxylcytosine. Science. 2011;333:1300–3.
24. Thomson JP, Hunter JM, Nestor CE, Dunican DS, Terranova R, Moggs JG, Meehan RR. Comparative analysis of affinity-based 5-hydroxymethylation enrichment techniques. Nucleic Acids Res. 2013;41:e206.
25. Globisch D, Munzel M, Muller M, Michalakis S, Wagner M, Koch S, Bruckl T, Biel M, Carell T. Tissue distribution of 5-hydroxymethylcytosine and search for active demethylation intermediates. PLoS ONE. 2010;5(12):e15367.
26. Song CX, Szulwach KE, Fu Y, Dai Q, Yi C, Li X, Li Y, Chen CH, Zhang W, Jian X, et al. Selective chemical labeling reveals the genome-wide distribution of 5-hydroxymethylcytosine. Nat Biotechnol. 2011;29:68–72.
27. Williams K, Christensen J, Pedersen MT, Johansen JV, Cloos PA, Rappsilber J, Helin K. TET1 and hydroxymethylcytosine in transcription and DNA methylation fidelity. Nature. 2011;473:343–8.
28. Wu H, D'Alessio AC, Ito S, Wang Z, Cui K, Zhao K, Sun YE, Zhang Y. Genome-wide analysis of 5-hydroxymethylcytosine distribution reveals its dual function in transcriptional regulation in mouse embryonic stem cells. Genes Dev. 2011;25:679–84.
29. Booth MJ, Branco MR, Ficz G, Oxley D, Krueger F, Reik W, Balasubramanian S. Quantitative sequencing of 5-methylcytosine and 5-hydroxymethylcytosine at single-base resolution. Science. 2012;336:934–7.
30. Yu M, Hon GC, Szulwach KE, Song CX, Zhang L, Kim A, Li X, Dai Q, Shen Y, Park B, et al. Base-resolution analysis of 5-hydroxymethylcytosine in the mammalian genome. Cell. 2012;149:1368–80.
31. Field SF, Beraldi D, Bachman M, Stewart SK, Beck S, Balasubramanian S. Accurate measurement of 5-methylcytosine and 5-hydroxymethylcytosine in human cerebellum DNA by oxidative bisulfite on an array (OxBS-array). PLoS ONE. 2015;10:e0118202.
32. Nazor KL, Boland MJ, Bibikova M, Klotzle B, Yu M, Glenn-Pratola VL, Schell JP, Coleman RL, Cabral-da-Silva MC, Schmidt U, et al. Application of a low cost array-based technique—TAB-Array—for quantifying and mapping both 5mC and 5hmC at single base resolution in human pluripotent stem cells. Genomics. 2014;104:358–67.
33. Ruzov A, Tsenkina Y, Serio A, Dudnakova T, Fletcher J, Bai Y, Chebotareva T, Pells S, Hannoun Z, Sullivan G, et al. Lineage-specific distribution of high levels of genomic 5-hydroxymethylcytosine in mammalian development. Cell Res. 2011;21:1332–42.
34. Ernst J, Kellis M. ChromHMM: automating chromatin-state discovery and characterization. Nat Methods. 2012;9:215–6.
35. Ernst J, Kheradpour P, Mikkelsen TS, Shoresh N, Ward LD, Epstein CB, Zhang X, Wang L, Issner R, Coyne M, et al. Mapping and analysis of chromatin state dynamics in nine human cell types. Nature. 2011;473:43–9.
36. Bibikova M, Le J, Barnes B, Saedinia-Melnyk S, Zhou L, Shen R, Gunderson KL. Genome-wide DNA methylation profiling using Infinium(R) assay. Epigenomics. 2009;1:177–200.
37. Bibikova M, Barnes B, Tsan C, Ho V, Klotzle B, Le JM, Delano D, Zhang L, Schroth GP, Gunderson KL, et al. High density DNA methylation array with single CpG site resolution. Genomics. 2011;98:288–95.
38. Stirzaker C, Taberlay PC, Statham AL, Clark SJ. Mining cancer methylomes: prospects and challenges. Trends Genet. 2014;30:75–84.

39. Pidsley R, Zotenko E, Peters TJ, Lawrence MG, Risbridger GP, Molloy P, Van Djik S, Muhlhausler B, Stirzaker C, Clark SJ. Critical evaluation of the Illumina MethylationEPIC BeadChip microarray for whole-genome DNA methylation profiling. Genome Biol. 2016;17:208.

40. Clark SJ, Harrison J, Paul CL, Frommer M. High sensitivity mapping of methylated cytosines. Nucleic Acids Res. 1994;22:2990–7.

41. Huang Y, Pastor WA, Shen Y, Tahiliani M, Liu DR, Rao A. The behaviour of 5-hydroxymethylcytosine in bisulfite sequencing. PLoS ONE. 2010;5:e8888.

42. Jin SG, Kadam S, Pfeifer GP. Examination of the specificity of DNA methylation profiling techniques towards 5-methylcytosine and 5-hydroxymethylcytosine. Nucleic Acids Res. 2010;38:e125.

43. Yu M, Hon GC, Szulwach KE, Song CX, Jin P, Ren B, He C. Tet-assisted bisulfite sequencing of 5-hydroxymethylcytosine. Nat Protoc. 2012;7:2159–70.

44. Nestor CE, Ottaviano R, Reinhardt D, Cruickshanks HA, Mjoseng HK, McPherson RC, Lentini A, Thomson JP, Dunican DS, Pennings S, et al. Rapid reprogramming of epigenetic and transcriptional profiles in mammalian culture systems. Genome Biol. 2015;16:11.

45. Ziller MJ, Hansen KD, Meissner A, Aryee MJ. Coverage recommendations for methylation analysis by whole-genome bisulfite sequencing. Nat Methods. 2015;12:230–2.

46. Han D, Lu XY, Shih AH, Nie J, You QC, Xu MM, Melnick AM, Levine RL, He C. A highly sensitive and robust method for genome-wide 5hmC profiling of rare cell populations. Mol Cell. 2016;63:711–9.

47. Song JZ, Stirzaker C, Harrison J, Melki JR, Clark SJ. Hypermethylation trigger of the glutathione-S-transferase gene (GSTP1) in prostate cancer cells. Oncogene. 2002;21:1048–61.

48. Consortium B. Quantitative comparison of DNA methylation assays for biomarker development and clinical applications. Nat Biotechnol. 2016;34:726–37.

49. Aryee MJ, Jaffe AE, Corrada-Bravo H, Ladd-Acosta C, Feinberg AP, Hansen KD, Irizarry RA. Minfi: a flexible and comprehensive Bioconductor package for the analysis of Infinium DNA methylation microarrays. Bioinformatics. 2014;30:1363–9.

50. Ritchie ME, Phipson B, Wu D, Hu Y, Law CW, Shi W, Smyth GK. limma powers differential expression analyses for RNA-sequencing and microarray studies. Nucleic Acids Res. 2015;43:e47.

51. Langmead B, Trapnell C, Pop M, Salzberg SL. Ultrafast and memory-efficient alignment of short DNA sequences to the human genome. Genome Biol. 2009;10:R25.

52. Robinson MD, McCarthy DJ, Smyth GK. edgeR: a Bioconductor package for differential expression analysis of digital gene expression data. Bioinformatics. 2010;26:139–40.

53. Zhang Y, Liu T, Meyer CA, Eeckhoute J, Johnson DS, Bernstein BE, Nusbaum C, Myers RM, Brown M, Li W, Liu XS. Model-based analysis of ChIP-Seq (MACS). Genome Biol. 2008;9:R137.

4

The quest for epigenetic regulation underlying unisexual flower development in *Cucumis melo*

David Latrasse[1], Natalia Y. Rodriguez-Granados[1†], Alaguraj Veluchamy[2†], Kiruthiga Gayathri Mariappan[2], Claudia Bevilacqua[3], Nicolas Crapart[3], Celine Camps[1], Vivien Sommard[1], Cécile Raynaud[1], Catherine Dogimont[4], Adnane Boualem[1], Moussa Benhamed[1,2] and Abdelhafid Bendahmane[1*]

Abstract

Background: Melon (*Cucumis melo*) is an important vegetable crop from the *Cucurbitaceae* family and a reference model specie for sex determination, fruit ripening and vascular fluxes studies. Nevertheless, the nature and role of its epigenome in gene expression regulation and more specifically in sex determination remains largely unknown.

Results: We have investigated genome wide H3K27me3 and H3K9ac histone modifications and gene expression dynamics, in five melon organs. H3K9ac and H3K27me3 were mainly distributed along gene-rich regions and constrained to gene bodies. H3K9ac was preferentially located at the TSS, whereas H3K27me3 distributed uniformly from TSS to TES. As observed in other species, H3K9ac and H3K27me3 correlated with high and low gene expression levels, respectively. Comparative analyses of unisexual flowers pointed out sex-specific epigenetic states of TFs involved in ethylene response and flower development. Chip-qPCR analysis of laser dissected carpel and stamina primordia, revealed sex-specific histone modification of MADS-box genes. Using sex transition mutants, we demonstrated that the female promoting gene, *CmACS11*, represses the expression of the male promoting gene *CmWIP1* via deposition of H3K27me3.

Conclusions: Our findings reveal the organ-specific landscapes of H3K9ac and H3K27me3 in melon. Our results also provide evidence that the sex determination genes recruit histone modifiers to orchestrate unisexual flower development in monoecious species.

Background

Melon (*Cucumis melo*) belongs to the *Cucurbitaceae* family that includes about 800 species. Besides melon, the *Cucurbitaceae* family comprises several important vegetable crops, such as cucumber (*C. sativus*), watermelon (*Citrullus lanatus*), squash and pumpkin (*Cucurbita* spp.), and many neglected cultivated species that are major food crops in many developing countries. For decades, melon has been serving as a reference model organism for sex determination studies, and vascular fluxes [1, 2]. Melon also exhibits extreme genetic diversity for fruit traits, including fruit ripening, fruit shape, size, flesh color, texture, sweetness and aroma. Due to its relatively late entry into the genomic era, melon is a species with a barely explored epigenetic landscape. This limits the investigation of the cellular reprogramming of gene regulatory networks that drive development, growth and evolution of species of the *Cucurbitaceae* family.

Genome function, connectivity and structure rely on several epigenetic mechanisms that dynamically regulate the conformation of DNA, and thereby, its exposure to different regulators of gene transcription—e.g., transcription factors, non-coding RNAs, transposons [3–7].

---

*Correspondence: abdelhafid.bendahmane@inra.fr

†Natalia Y. Rodriguez-Granados and Alaguraj Veluchamy contributed equally to this work

[1] Institute of Plant Sciences Paris-Saclay (IPS2), CNRS, INRA, University Paris-Sud, University of Evry, University Paris-Diderot, Sorbonne Paris-Cite, University of Paris-Saclay, Batiment 630, 91405 Orsay, France

Full list of author information is available at the end of the article

At the chromatin level, DNA conformation can be modified by the addition or removal of chemical groups at specific N-terminal residues of histones [8, 9]. Some of these modifications are in charge of genome connectivity and heterochromatin stability, while others regulate gene expression or repression along the euchromatic regions. The distribution and abundance of these marks, at the gene and genome-wide scales, define an epigenetic landscape that directs cell differentiation and identity [10–13]. In the plant kingdom, genome-wide analyses of epigenetic landscapes have been performed in several species including *Arabidopsis* [14–17], maize [18–20], rice [21–23], common bean [24] and barley [25].

Histone methylation and acetylation are epigenetics marks that are usually associated with various biological processes ranging from transcriptional regulation to epigenetic silencing via heterochromatin assembly. While histone acetylation is commonly associated with transcription activation, histone methylation can either promote or inhibit gene transcription [5, 11, 18, 26–29]. Histone H3 lysine 9 acetylation (H3K9ac) is an epigenetic mark commonly found in active promoters [5, 20, 30, 31]. Genome-wide studies in *Arabidopsis* have found around 5200 non-Transposable Element (TE)-related genes that are targeted and regulated by H3K9ac [31]. In contrast, H3 lysine 27 trimethylation (H3K27me3) is commonly associated with gene repression, thereby contributing to the mitotic heritability of PRC2 (Polycomb repressive complex)-mediated silencing of numerous developmental and homeotic genes [29, 32, 33]. In the plant kingdom, the study of these epigenetic marks has elucidated the epigenetic component regulating plant growth and adaptation to their harsh environment [34–44].

The majority of angiosperms are hermaphrodite producing exclusively bisexual flowers. Sex determination is a developmental evolutionary process that leads to unisexual flowers. Monoecious species, such as melon, exhibit male and female flowers on the same plant. Dioecious species have separate male and female individuals [45, 46].

In melon, floral primordia are initially bisexual with sex determination occurring by the selective developmental arrest of either the stamen or the carpel primordia, resulting in unisexual flowers [47, 48]. This sexual organ arrest is genetically governed by the interplay of alleles of the *andromonoecious* (*M*), *androecious* gene (*A*) and *gynoecious* (*G*) genes [47, 49–51]. The cloning and characterization of *M*, *G* and *A* genes have shown that the *gynoecious* (*G*) gene encodes a zinc finger transcription factor, *CmWIP1* [47], the andromonoecious (*M*) gene and the androecious gene (*A*) encode ethylene biosynthesis enzymes: CmACS-7 [49] and CmACS11 [51], respectively. Genetic analysis revealed a mechanistic model in

which expression of the carpel inhibitor, *CmWIP1*, is dependent on non-expression of *CmACS11* and expression of the stamina inhibitor, *CmACS-7*, is dependent on non-expression of *CmWIP1*. In monoecious plants, male flowers result from non-expression of *CmACS11* that permits *CmWIP1* expression. Female flowers develop on branches expressing *CmACS11*, which represses the expression of *CmWIP1*, and thus, releasing the expression of *CmACS7* inhibiting stamina development [51–53].

Now that the identity of the androecy, monoecy and gynoecy sex genes are revealed and genetic pathway controlling sex determination in *Cucurbitaceae* discovered, among the next challenges is to decipher how the sex determination signals are perceived and how the information is translated to cause organ-specific abortion at the flower and the plant level. Several studies have reported a wide-range of epigenetic processes and environmental cues that determine sex in animals [54–57]. In plants, it is well known that sex ratio and determination can be influenced by different environmental factors such as day length and light intensity [58], water restriction [59] and some plant hormones such as ethylene [49–51, 60–63], auxin [64] and gibberellins [65–67]. This multifactorial regulation of sex determination is still poorly characterized from the epigenetic point of view, where limited literature have reported the association for instance of small RNA (sRNAs) transcription and DNA hypomethylation with sex determination [54, 68–71].

In melon, the epigenetic control of sex determination remains poorly characterized and is mainly attributed to a transposon-induced DNA methylation that leads to gynoecy [47]. To gain new information on the epigenetic control of sex determination, we have examined the genome-wide landscape of H3K27me3 and H3K9ac histone modifications and gene expression dynamics, in five melon organs, focusing our analysis on unisexual flowers. The combination of epigenomic and gene ontology analysis pointed out sex-specific epigenetic states of TFs involved in ethylene response and sexual organs development. Furthermore, using sex transition mutants, we demonstrated that the female promoting gene represses the expression of the male promoting genes via deposition of H3K27me3.

Results

Genome-wide landscape of H3K9ac and H3K27me3 in melon

Genome-wide studies in *Arabidopsis* [14–17], and other plant species such as maize [19, 20], rice [21–23], and barley [25], have evidenced the important role of H3K9ac and H3K27me3 in gene activation and repression, respectively. The roles of these histone modifications in melon development remain unknown and represent a limit to fully understand how thousands of bioprocesses are regulated. To determine the genomic landscape and the organ specificity of those marks, we performed ChIP-seq analyses using H3K27me3 and H3K9ac antibodies on different organs, namely fruits, leaves, roots, female and male flowers. Two replicates per tissue were processed and present a correlation coefficient of at least 0.99 (Additional file 1: Figures S1 and S2). A minimum of 14 millions of mapped reads were obtained in the replicates implemented (Additional file 2: Table I). The MACS2 algorithm, which is designed to detect typical histone peaks [72], was used to determine loci that are significantly enriched with H3K9ac or with H3K27me3 (Additional file 1: Figures S3 and S4). We identified 18,424 H3K9ac marked genes in leaves, 18,024 in roots, 13,655 in fruits, 17,660 in male flowers and 16,348 in female flowers. On the other hand, we identified 6486 H3K27me3 marked genes in leaves, 7741 in fruits, 6823 in roots, 6055 in male and 6054 in female flowers (Fig. 1a; Additional file 3: Table II).

Next we analyzed the distribution of those two marks at the chromosome and gene levels. For the sake of simplicity, we will describe mainly results obtained in leaves, and results obtained in the other organs are shown in the additional file. The chromosomal distributions of annotated genes, H3K9ac and H3K27me3 and in each organ on the 12 chromosomes, are illustrated in Additional file 1: Figure S3. At the chromosomal scale we observed an enrichment of both marks in gene-rich regions, which are localized at chromosome arms, compare to centromeric and pericentromeric regions, where the majority of transposable and retro transposable elements are located [73], consistent with the role of these histone marks in the control of gene expression (Fig. 1b; Additional file 1: Figures S5 and S6). In order to detail the H3K27me3 and H3K9ac distributions at the gene level, the peaks obtained in each tissue for both modifications were analyzed and averaged within a 4000-bp region, flanking the TSS and TES of all the annotated genes. We found that

the peak length of H3K9ac ranged from 200 bp to 700 bp (Additional file 1: Figure S7), located preferentially at the TSS regions, which correspond to the first nucleosomes of the genes (Fig. 1c; Additional file 1: Figure S8). In contrast, H3K27me3 peaks presented an averaged length that ranged from 2500 to 2700 bp, which covered the entire gene body (Fig. 1c; Additional file 1: Figure S7). Those patterns were observed in all the organs (Additional file 1: Figures S7 and S8) and were consistent with previous studies on different plant species, highlighting conserved aspects of the epigenetic system in the plant kingdom. Integration of H3K9ac and H3K27me3 datasets showed a clear anti-correlation between those two marks (Fig. 1d). Altogether, these results showed that in melon, as in other plant species, H3K27me3 and H3K9ac are not enriched in intergenic regions but distributed along the gene body, supporting the role of these two marks in gene regulation in an exclusive manner.

Identification of organ-specific H3K27me3 and H3K9ac landscapes

We next aimed to determine the organ-specific profile of both H3K27me3 and H3K9ac marks, and to determine whether these chromatin modifications contribute to the establishment of organ-specific gene expression programs. To this end, first, for each organ, we determined the genes that are marked either by H3K9ac or by H3K27me3 or by both. Secondly, we performed a comparison between all the possible paired combinations of the five different tissues using DiffReps, a highly sensitive program for the detection of differential sites from ChIP-seq data [74]. Third, a Venn diagram was generated to identify genes that were over-represented in the chosen organ regardless of which organ it was compared to and thus to fish out genes that are specifically marked in this organ (Fig. 1e, g; Additional file 1: Figures S9–S12; Additional file 4: Table III). To determine the biological function of these genes, we performed a gene ontology analysis (GO) using Plant MetGenMAP [75]. Organ-specific hyperacetylated and hypomethylated genes,

(See figure on next page.)
Fig. 1 Genome-wide distribution and tissue-specificity of H3K9ac and H3K27me3 in melon. **a** H3K27me3 and H3K9ac target genes. Number of genes presenting H3K9ac (*green bar*) and H3K27me3 (*red bar*) peaks were quantified and plotted for the five different tissues. **b** H3K27me3 and H3K9ac distribution at the chromosome level. Distribution of H3K9ac (*green, upper panel*), H3K27me3 (*red, medium panel*) in leaves and annotated genes (*gray, bottom panel*) are plotted along the chromosome 1. **c** Average tag density profile of H3K27me3 and H3K9ac along the gene body. ChIP-Seq densities of equal bins were plotted along the gene body and 2-kb region flanking the TSS or the TES. **d** Correlation between the genome-wide distribution of H3K27me3 and H3K9ac. Heat map representing the tag density distribution of H3K27me3 and H3K9ac across all genes and a 2 kb flank. **e** Leaf-specific H3K9 hyper-acetylated genes. Paired comparisons of leaves vs. all the organs are shown as a Venn diagram, leaf-specific hyperacetylated genes correspond to the central overlap (highlighted in *green*). **f** GO terms of biological processes enriched for leaf-specific hyperacetylated genes. **g** Leaf-specific H3K27 hypo trimethylated genes. Paired comparisons of leaves versus all the tissues are shown as a Venn diagram, leave-specific hypo methylated genes correspond to the central overlap (highlighted in *red*). **h** GO terms of biological processes enriched for leave-specific hypo trimethylated genes. *p* values for each enriched class are presented in Additional file 9: Table VIII

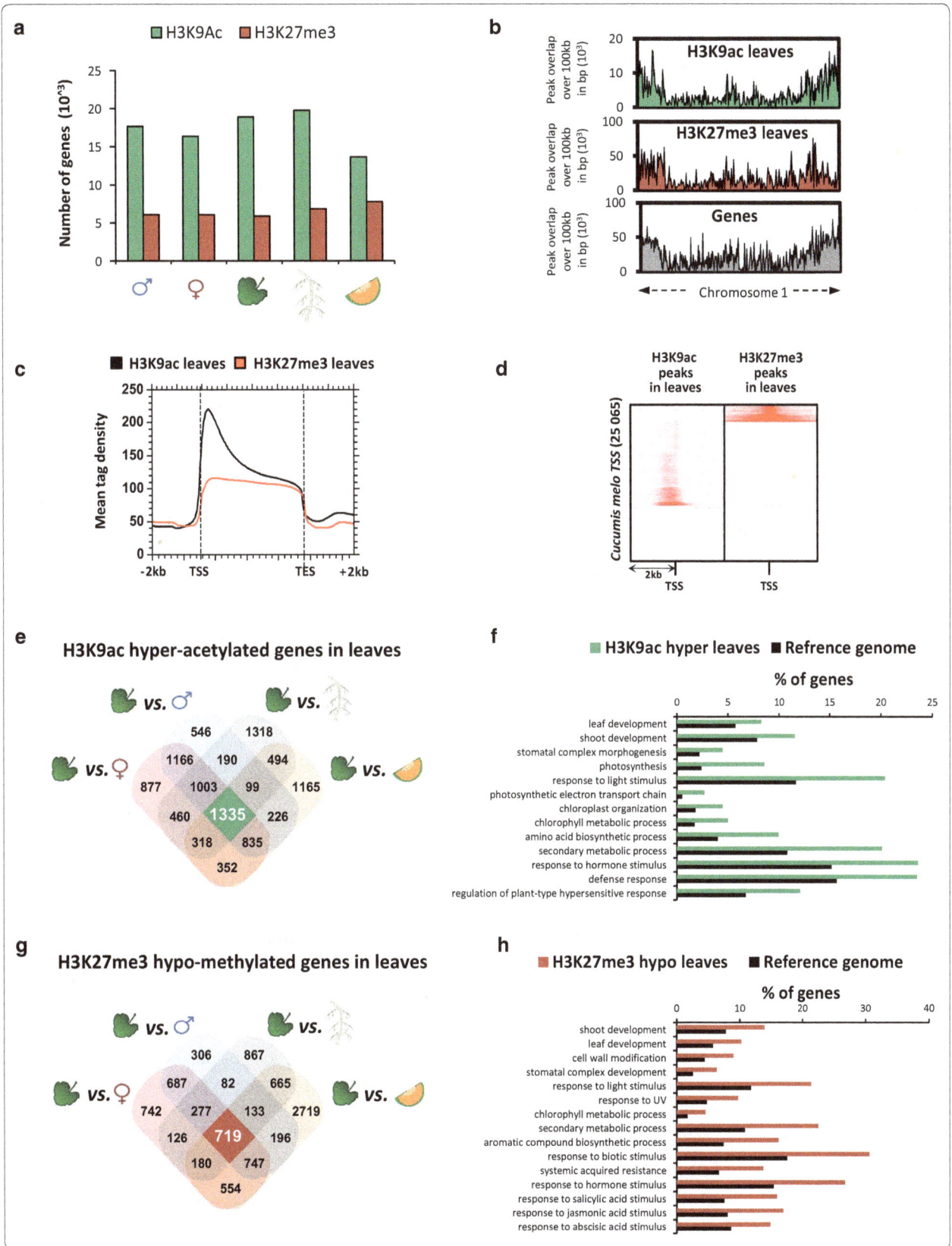

displayed biological processes terms consistent with metabolic needs and physiological aspects of their corresponding organ, providing evidence for the robustness of our data (Fig. 1f, h; Additional file 1: Figures S9–S12). For example in leaves, we found 1335 specifically hyperacetylated genes involved in photosynthesis, leaf development, stomatal morphogenesis (Fig. 1f). Consistent results were also obtained for fruits, roots, female and male flowers (Additional file 1: Figures S9–S12).

These results suggest that both H3K9ac and H3K27me3 display an organ-specific profile crucial for cell identity and physiology.

Relationship between gene expression and histone modifications in melon

To connect H3K9ac and H3K27me3 histone marks with gene expression, we generated and integrated RNA-seq data from the five different organs. Firstly, we confirm that H3K9ac and H3K27me3 in *Cucumis melo* are, respectively, associated with gene expression and gene repression (Fig. 2a). Secondly, in order to determine whether the level of acetylation or methylation could be correlated with gene expression, we divided all the protein coding genes into four quantiles based on their expression levels and plotted their H3K9ac or H3K27me3 profile (Fig. 2b). We observed that H3K9ac occupancy increases with expression level, while H3K27me3 showed the opposite pattern, where it displayed a high enrichment in the lowest-expressed genes but it remained unresponsive to intermediary and high expression levels. This results suggest that in *Cucumis melo* the more a gene is marked by H3K9ac and H3K27me3, the more it will be expressed and repressed, respectively (Fig. 2b).

To confirm the relationship between gene expression and organ-specific landscape of H3K9ac and H3K27me3, we identified genes specifically expressed in each organ (Fig. 2c; Additional file 5: Table IV) and we compared them with organ-specific H3K9ac marked genes. Interestingly, in the case of leaves, we found that a significant part of genes specifically hyperacetylated (43%) and hypomethylated (32%) are specifically expressed in this organ (Fig. 2d, e). Gene ontology analyses of these genes showed a significant enrichment in photosynthesis and response to light. In the case of roots, flowers and fruits we observed the same tendency, suggesting that both H3K9ac and H3K27me3 controlled organ-specific gene expression (Additional file 1: Figure S13).

Female and male flower-specific epigenome landscapes

Beyond elucidating general aspects of the melon epigenome, the analyses we performed so far evidenced how H3K27me3 and H3K9ac organ-specific distributions can be crucial to study gene expression regulation and organ identity. Thus, we directed epigenomic and transcriptomic analyses toward the elucidation of female and male flower-specific epigenome landscapes and their role in sex determination. To this end, we integrated RNA-seq and, H3K9ac and H3K27me3 ChIP-seq data of male and female flowers in order to identify genes that are epigenetically regulated in a sex-specific manner (Fig. 3; Additional file 6: Table V). We observed that 22% of genes that are hypoacetylated in female flowers compared to male flower are downregulated and only 3% of the set of hypoacetylated genes are upregulated (Fig. 3a). Regarding hyperacetylated genes, we also found that a significant proportion of them (28%) are upregulated in female flowers compared to male flower and only 5% are dow-regulated. Furthermore, gene ontology analysis of female-specific epigenetically regulated genes showed enrichment in biological processes such as carpel development, response to ethylene and response to jasmonic acid. For male-specific epigenetically regulated genes, we observed enrichment in biological processes such as, stamen development and lipid metabolic process (Fig. 3a). The same analyses were performed for hypo- and hypermethylated genes, obtaining similar results and confirming that a significant portion of sex-specific genes is epigenetically controlled (Fig. 3b).

After analyzing the biological processes involving the sex-specific epigenetically regulated genes, we classified these genes based on their molecular function. We found enrichment for transporter activity, catalytic activity, oxidoreductase activity and interestingly, for transcription factor (TF) activity (Fig. 3c; Additional file 1: Figures S14–S18; Additional file 7: Table VI). Since transcription factors are the best described and commonly associated components of gene expression regulation, we further focused on them and grouped them according to their TF families (Fig. 3d). We found that the ethylene-responsive factor (ERF) family displayed the highest number of gene members, consistent with previous studies that reported the role of ethylene and ethylene transduction pathways in sex determination in the cucurbitaceae family (22). In addition, we found other TF families such as Zinc Finger (17 genes), basic Helix-Loop-Helix (15 genes), MYB (13 genes) and NAC families (13) (Fig. 3d). Interestingly, we also detected 11 genes belonging to the MADS-box TF family. Several members of this family are conserved master regulators of flower ontogenesis [76–78]. When we assessed the identity of these 11 genes, we observed that all of them are potential homologs of different *Arabidopsis* genes that are part of the ABC model of flower development (Additional file 7: Table VI).

Altogether, these results highlight the organ-specificity of H3K27me3 and H3K9ac as well as its role in the regulation of key genes encoding transcription factors, which

a

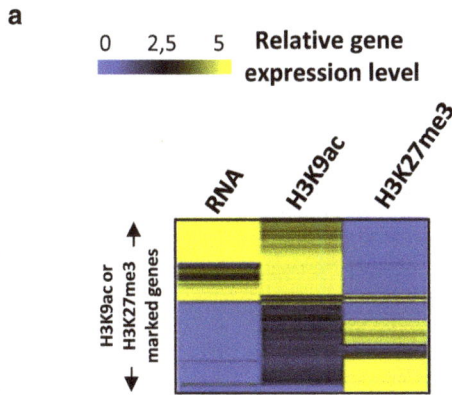

b

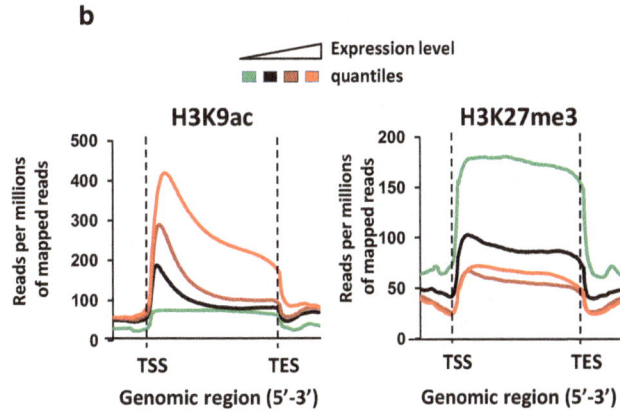

c

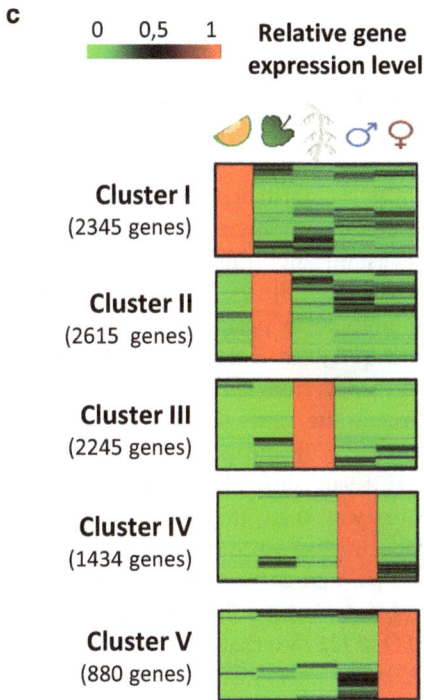

d

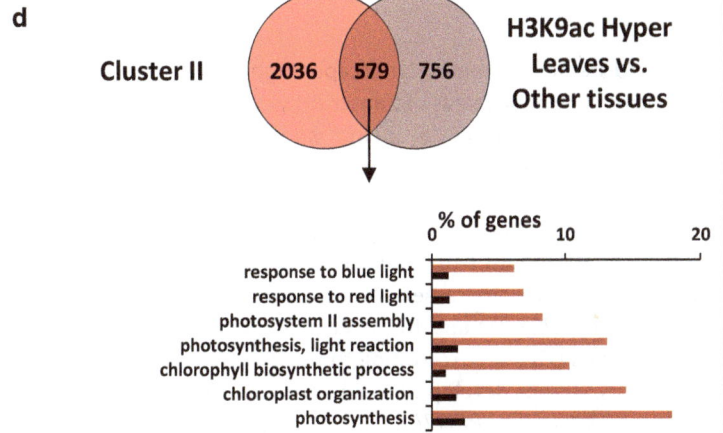

e

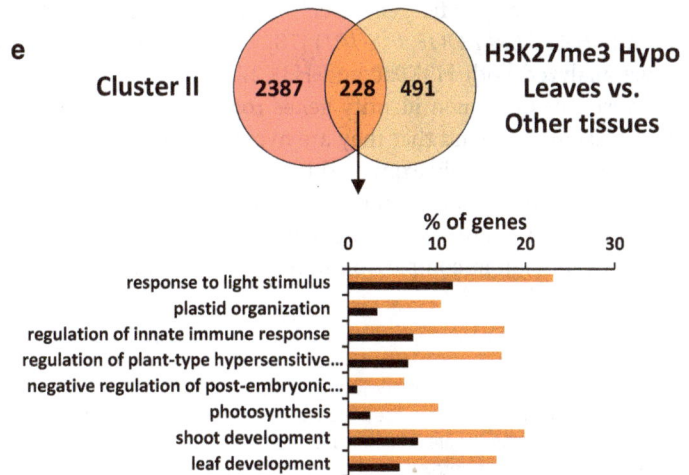

(See figure on previous page.)

Fig. 2 Correlation of H3K27me3 and H3K9ac with gene expression level and organ-specificity in leaves. **a** Correlation between H3K27me3, H3K9ac and gene expression. Genes specifically marked with H3K9ac or H3K27me3 in leaves were clustered based on the log2 value of their normalized tag densities of H3K9ac, H3K27me3 and RNA-seq data. Genes displaying the highest enrichment for H3K27me3 and H3K9ac were used with their corresponding gene expression levels in a heat map. **b** Correlation of between H3K27me3, H3K9ac and gene expression level. All the melon protein-coding genes were divided in 4 quantiles according to their gene expression levels (lowest and highest expression level corresponding to *green* and *red*, respectively). For each quantile the number of H3K27me3 and H3K9ac mapped reads was averaged and plotted along the gene body and 2-kb region flanking the TSS or the TES. **c** Gene clustering analyses based on organ-specific expression. Genes were sorted depending on their expression in each organ. 5 clusters, each one corresponding to an organ, are illustrated and numbered from 1 to 5. Number of genes of each cluster is also indicated. **d** Correlation of hyper-H3K9ac with tissue-specific gene activation. Venn diagram illustrating the genes whose high and tissue-specific expression is correlated with their tissue-specific hyper-acetylation (Overlapping set of genes). A Chi-squared test was done and confirmed the observed correlation. For this set of genes, GO terms of biological processes are shown below. **e** Correlation of hypo-H3K27me3 with tissue-specific gene activation. Venn diagram illustrating the genes whose high and tissue-specific expression is correlated with their tissue-specific hypomethylation (Overlapping set of genes). A Chi-squared test was done and confirmed the observed correlation. For this set of genes, GO terms of biological processes are shown below. *p* values for each enriched class are presented in Additional file 9: Table VIII

regulate processes such as ethylene response and flower development.

Carpel and stamen epigenetic landscapes highlight sex-specific chromatin states of transcription factor genes

Since sex determination in melon and other cucurbits relies on the selective regulation of carpel and stamen development, we analyzed in detail some melon TFs, which are potential homologs of *Arabidopsis* genes involved in the control of sexual organs development. For such genes, we first assessed the genic distribution of H3K27me3, H3K9ac and RNA-seq reads in male and female flowers (Fig. 4a). According to previous studies in *Arabidopsis*, stamen development relies on the expression of *APETELA3* (*AP3*), *PISTILATA* (*PI*) and *AGAMOUS* (*AG*), whereas carpel development depends on the expression of ovule-specific genes such as *SEEDSTICK* (*STK*) and *SHATTERPROOF 1* (*SHP1*) [78]. As expected, when we analyzed both H3K9ac and H3K27me3 epigenetic landscapes of stamen identity genes together with their expression, we found that they are hyperacetylated, hypomethylated and highly expressed in the male flower when compared to the female flower. Carpel-promoting genes, however, presented an enrichment of H3K9ac and high gene expression in comparison to male flowers, but no significant differences were observed for H3K27me3, which may result from the cell heterogeneity of the sample or from the cooperative action of additional histone repressive marks.

Tissue heterogeneity is an important variable that limits the interpretation of ChIP-seq data and the identification of epigenetic events occurring in a specific cell population. This is particularly true in this context of flower development, where it was reported that organ identity relies on the whorl-specific expression of MADS-box genes [76–78]. We therefore decided to overcome such limitation, by performing an H3K27me3 and H3K9ac ChIP-qPCR in isolated carpels and stamens. We isolated carpels from female and stamens from male flowers through the Laser Capture Microdissection (LCM) technique [79] and performed ChIP using H3K27me3 and H3K9ac antibodies. The immunoprecipitated chromatin was then amplified using primers annealing a neighboring sequence to the TSS of *AP3, PI, AG, STK-like* genes; which corresponded to the highest peak of H3K9ac observed along the gene body. A negative control, *MELO3C115188*, that does not display any H3K9ac or H3K27me3 signals was also used (Additional file 1: Figure S19). In all cases, the obtained ChIP-qPCR results were consistent with the ChIP-seq data previously

(See figure on next page.)

Fig. 3 H3K9ac and H3K27me3 differentially marked genes between female and male flowers. **a** Venn diagram showing the relationship between H3K9 hyper-acetylated or H3K9 hypo-acetylated genes and up- or down-regulated genes in females flowers compared to male flowers (*middle panel*). Hyper-acetylated genes are predominantly up-regulated and hypo-acetylated genes are down-regulated, as highlighted in the *pink* and *blue boxes* respectively. Gene ontology analysis (GO) of the genes in the *pink* and *blue boxes* are shown in the *left panel* and the *right panel* respectively. Histograms of the values highlight the enrichment of genes compared to the reference. **b** Venn diagram showing the relationship of H3K27 hyper-methylated or H3K27 hypo-methylated genes and up- or down-regulated genes in females flowers compared to male flowers (*middle panel*). Hyper-methylated genes are predominantly down-regulated and hypo-acetylated genes are up-regulated, as highlighted in the *purple* and *blue boxes* respectively. Gene ontology analysis (GO) of the genes in the *purple* and *blue boxes* are shown in the *left panel* and the *right panel* respectively. Histograms of the values highlight the enrichment of genes compared to the reference genome (*black bars*). *p* values for each enriched class are presented in Additional file 9: Table VIII. **c** Percentage of transcription factors among the H3K9ac and H3K27me3 differentially regulated genes between male and female flowers. **d** Percentage of different known transcription factor families among the H3K9ac and H3K27me3 differentially regulated genes between male and female flowers

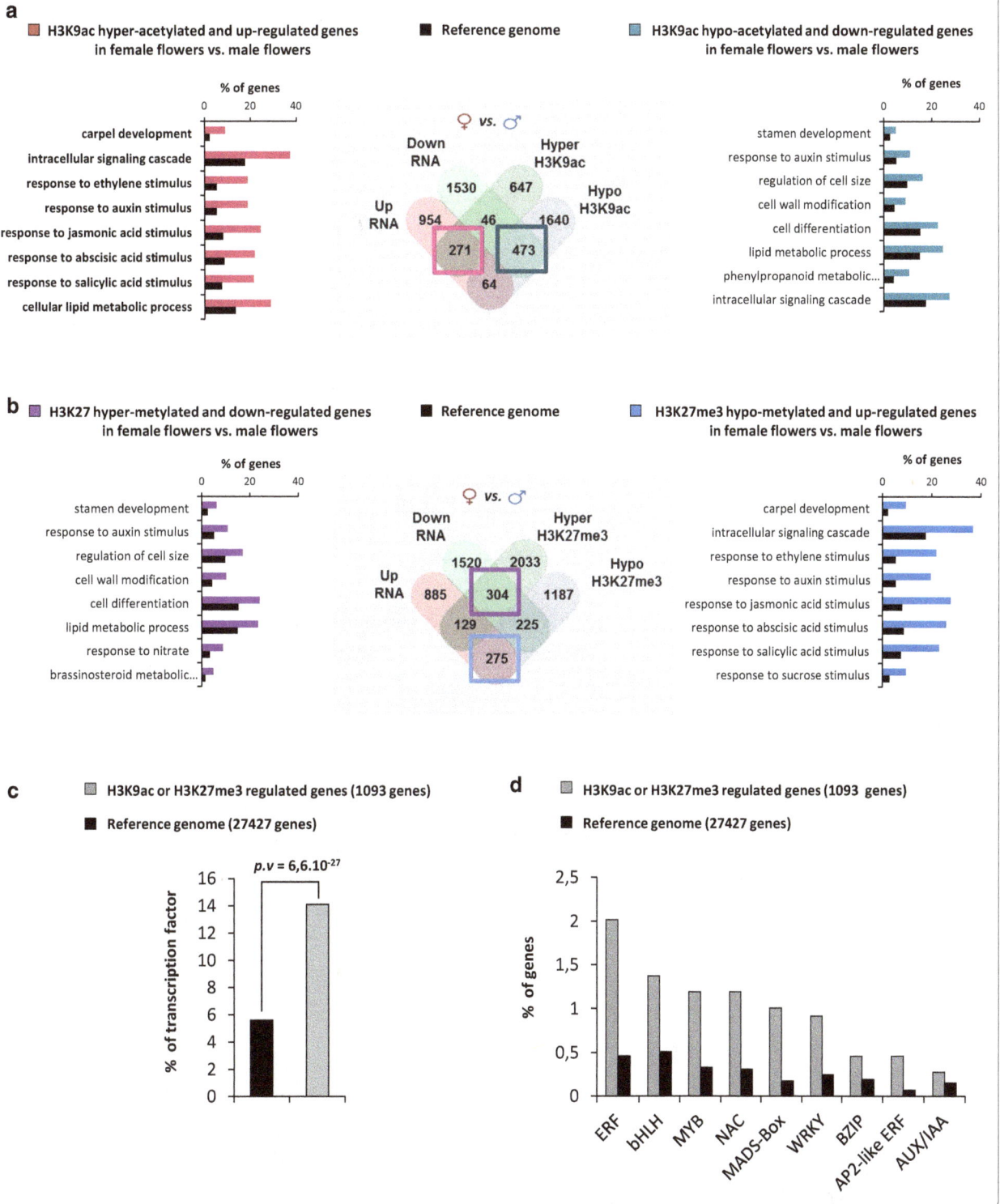

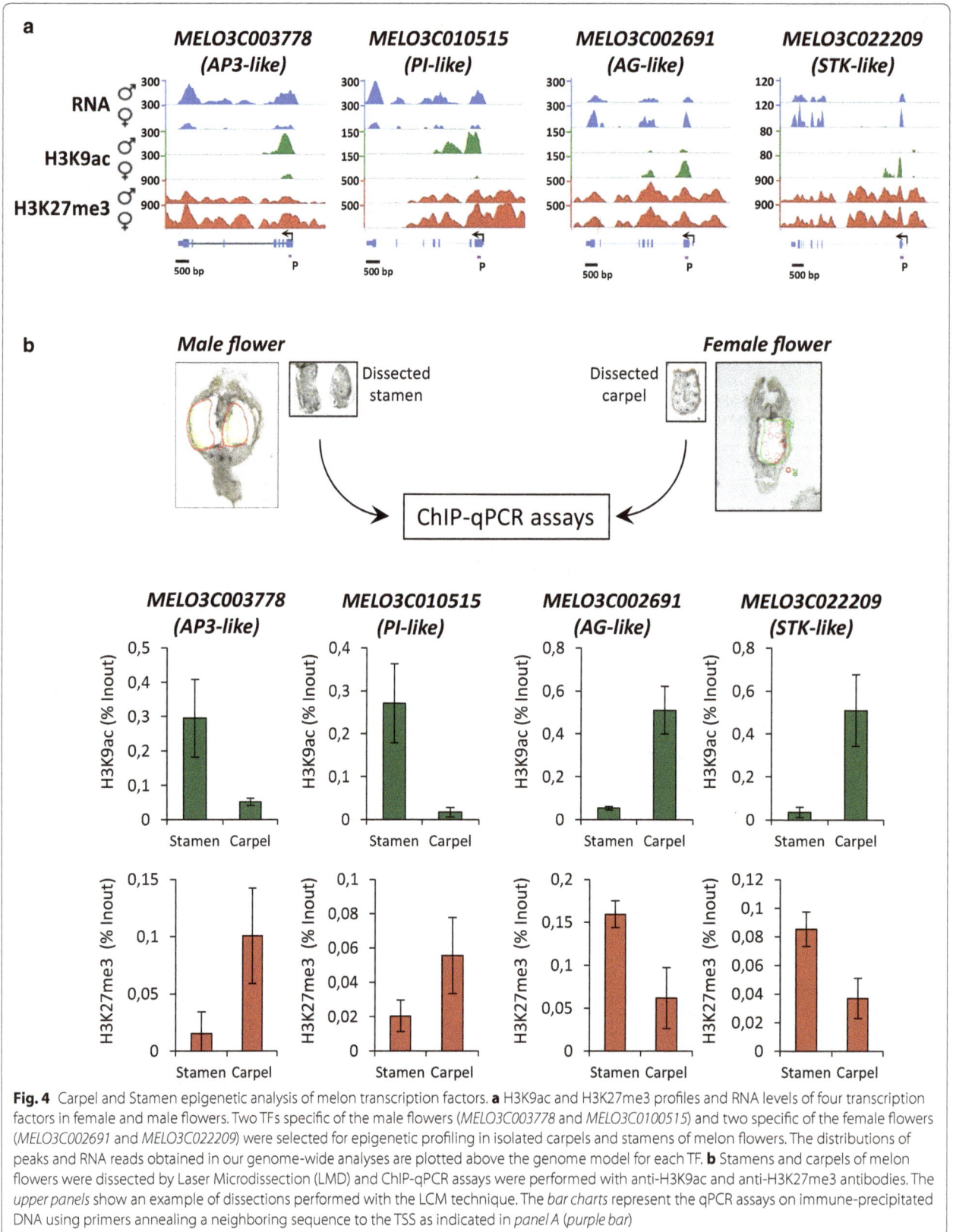

Fig. 4 Carpel and Stamen epigenetic analysis of melon transcription factors. **a** H3K9ac and H3K27me3 profiles and RNA levels of four transcription factors in female and male flowers. Two TFs specific of the male flowers (*MELO3C003778* and *MELO3C0100515*) and two specific of the female flowers (*MELO3C002691* and *MELO3C022209*) were selected for epigenetic profiling in isolated carpels and stamens of melon flowers. The distributions of peaks and RNA reads obtained in our genome-wide analyses are plotted above the genome model for each TF. **b** Stamens and carpels of melon flowers were dissected by Laser Microdissection (LMD) and ChIP-qPCR assays were performed with anti-H3K9ac and anti-H3K27me3 antibodies. The *upper panels* show an example of dissections performed with the LCM technique. The *bar charts* represent the qPCR assays on immune-precipitated DNA using primers annealing a neighboring sequence to the TSS as indicated in *panel A* (*purple bar*)

generated from entire flowers; however, the differences observed between stamen and carpel are more evident, presenting at least a fold change of 2.3. This allowed us to observe clear differences in H3K27me3 occupancy between male and female organs (Fig. 4b).

Altogether, these results indicate that H3K27me3 and H3K9ac target MADS-box TFs regulating carpel and stamen identity and are presented in a sex-specific manner, thereby suggesting the role of these two epigenetic marks in the regulation of their expression. Furthermore, these results showed how tissue homogeneity could represent an advantage for the elucidation of cell-specific epigenetic landscapes in melon and other plant species.

Role of CmACS11 and CmWIP1 genes in sex-specific epigenome acquisition

We previously showed that sex determination in melon relies on the interplay between alleles of three sex determination genes, *M*, *G* and *A* [47, 49–51]. In monoecious melon, most of the flowers are male and female flowers develop at the youngest nodes of the growing vines (Fig. 5a). Male flowers result from the expression of the male promoting gene, *CmWIP1*. Female flowers develop on branches expressing *CmACS11*, which represses the expression of *CmWIP1*. Androecious plants result from a loss-of-function of *CmACS11* leading to expression of *CmWIP1* in all flowers on a plant. Gynoecious plants are obtained by inactivation of *CmWIP1* function [47, 51] (Fig. 5a).

To bring new insight on the role of these sex-determining genes in establishing epigenetic signatures important for sex determination in melon, we investigated the H3K27me3 genome-wide landscape in male and female flowers of wild-type monoecious melon and two androecious and gynoecious isogenic lines, carrying a loss-of-function mutation in *CmACS11* and *CmWIP-1*, respectively.

To determine the proportion of genes whose differential epigenetic state between male and female flowers can depend on *CmACS11* and *CmWIP-1*, we compared male and female flowers from the wild type with flowers occupying the same position on the inflorescence from androecious and gynoecious sex transition mutants, respectively (Fig. 5b; Additional file 1: Figures S20–S22). The gene lists generated by these comparisons are illustrated in a Venn diagram, where the overlapping genes correspond to *CmWIP1*—(Additional file 1: Figure S20B) and *CmACS11*-related (Additional file 1: Figure S20C) H3K27me3 changes linked to sex identity. We observed that 54% (1200/2197) of sex-specific regulated gene was under the control of the *G* locus and that 60% (1324/2197) was under the control of the *A* locus (Fig. 5b). Altogether, these data suggested that *CmACS11*

and *CmWIP-1* are two master regulator controlling sex-specific epigenome acquisitions.

The androecious gene *CmACS11* plays central role in sex determination as it controls the expression the gynoecious gene *CmWIP1* [51]. Knowing from this study that *CmWIP1* is epigenetically regulated through histone methylation (Additional file 1: Figure S23), we predicted that *CmACS11* represses the expression of the *CmWIP1* by modifying its methylation status. To test this hypothesis, we analyzed the H3K27me3 status of *CmWIP1* in flowers that do or do not express *CmACS11* and in flowers impaired in *CmACS11* function. We performed an H3K27me3 ChIP followed by qPCR using 9 different primers covering the entire *CmWIP1 gene*. As predicted, we observed that *CmWIP1* is hypermethylated in female buds that express *CmACS11* but not in male buds that do not express *CmACS11*. The hypermethylation of *CmWIP1* is also lost in *CmACS11* loss of function mutant (Fig. 5c). The hypomethylation correlates with the expression *CmWIP1* and the hypermethylation with the non-expression of *CmWIP1*. Based on this, we concluded that the *A* gene represses the expression of *CmWIP1* via deposition of H3K27me3.

Discussion

Deposition of chromatin modifications is instrumental for the concerted expression or repression of thousands of genes, thereby allowing the establishment of organ or tissue specific transcriptomes [10–13]. H3K9ac and H3K27me3 are among the main chromatin marks involved in the control of gene expression [5, 20, 30–33]. In this study, we generated genome-wide and organ-specific maps of H3K27me3 and H3K9ac in melon, thereby generating a repertoire for the elucidation of mechanisms controlling the expression of the melon genome. This work will provide new information for the analysis of melon development or several agricultural traits, that have until now been mainly studied via classical genetic approaches. The distributions of H3K9ac and H3K27me3 in the melon genome displayed common chromosomal and genic distributions when compared to other plant species such as *Arabidopsis* [14–17], maize [19, 20] and rice [21–23]: H3K27me3 and H3K9ac were mainly enriched at the chromosome arms, resembling the distribution of genes, but were confined to the gene body and not spread along intergenic regions. Within the gene body, H3K27me3 distributes uniformly from the TSS to the TES, whereas H3K9ac is mainly enriched at the TSS.

As described in other plant species, we observed that H3K9ac and H3K27me3 marks display opposite correlation with gene expression. Interestingly, we found a direct proportionality between H3K9ac and gene expression levels, thereby suggesting the crucial role of H3K9ac in

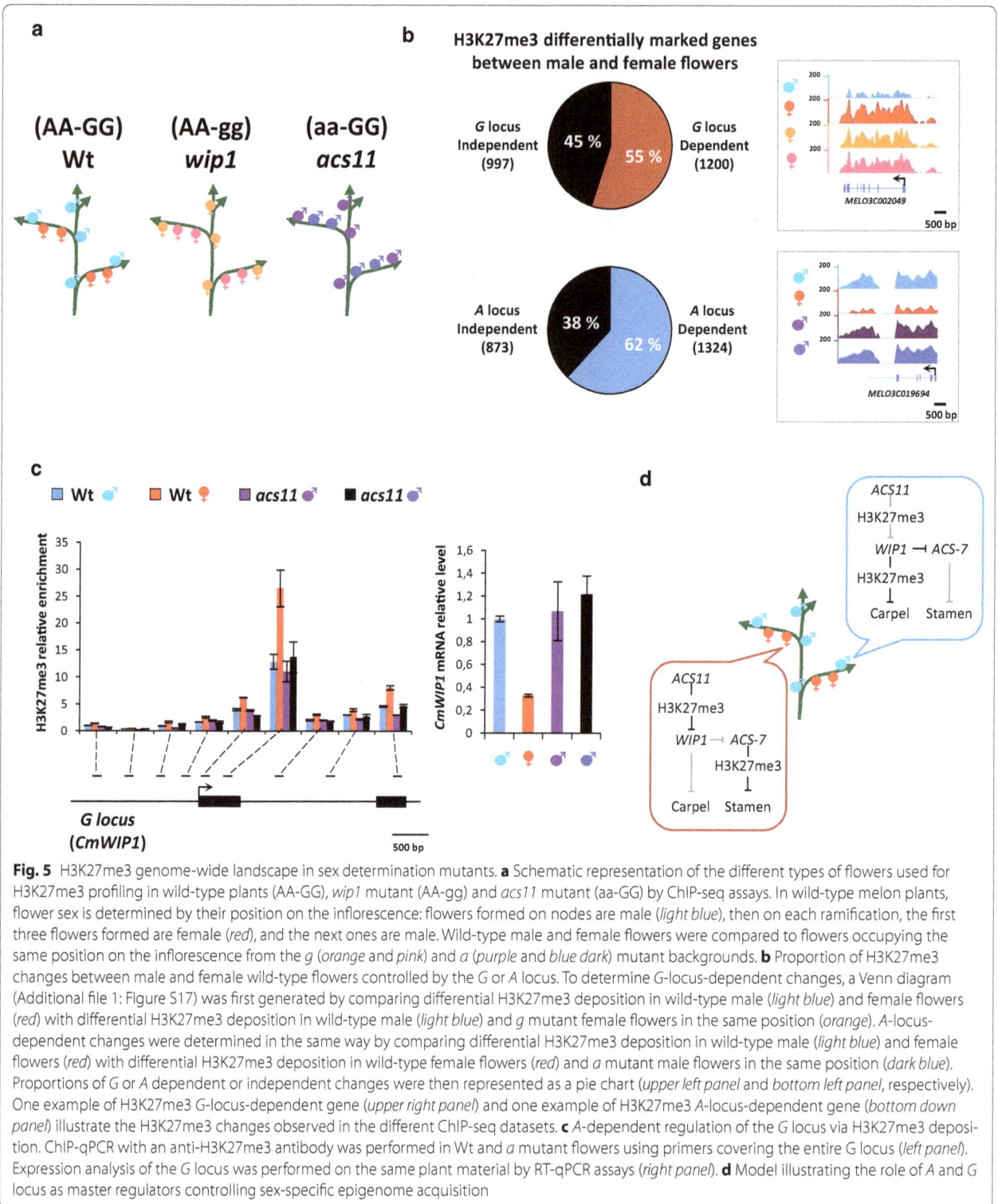

Fig. 5 H3K27me3 genome-wide landscape in sex determination mutants. **a** Schematic representation of the different types of flowers used for H3K27me3 profiling in wild-type plants (AA-GG), *wip1* mutant (AA-gg) and *acs11* mutant (aa-GG) by ChIP-seq assays. In wild-type melon plants, flower sex is determined by their position on the inflorescence: flowers formed on nodes are male (*light blue*), then on each ramification, the first three flowers formed are female (*red*), and the next ones are male. Wild-type male and female flowers were compared to flowers occupying the same position on the inflorescence from the *g* (*orange* and *pink*) and *a* (*purple* and *blue dark*) mutant backgrounds. **b** Proportion of H3K27me3 changes between male and female wild-type flowers controlled by the *G* or *A* locus. To determine *G*-locus-dependent changes, a Venn diagram (Additional file 1: Figure S17) was first generated by comparing differential H3K27me3 deposition in wild-type male (*light blue*) and female flowers (*red*) with differential H3K27me3 deposition in wild-type male (*light blue*) and *g* mutant female flowers in the same position (*orange*). *A*-locus-dependent changes were determined in the same way by comparing differential H3K27me3 deposition in wild-type male (*light blue*) and female flowers (*red*) with differential H3K27me3 deposition in wild-type female flowers (*red*) and *a* mutant male flowers in the same position (*dark blue*). Proportions of *G* or *A* dependent or independent changes were then represented as a pie chart (*upper left panel* and *bottom left panel*, respectively). One example of H3K27me3 *G*-locus-dependent gene (*upper right panel*) and one example of H3K27me3 *A*-locus-dependent gene (*bottom down panel*) illustrate the H3K27me3 changes observed in the different ChIP-seq datasets. **c** *A*-dependent regulation of the *G* locus via H3K27me3 deposition. ChIP-qPCR with an anti-H3K27me3 antibody was performed in Wt and *a* mutant flowers using primers covering the entire G locus (*left panel*). Expression analysis of the *G* locus was performed on the same plant material by RT-qPCR assays (*right panel*). **d** Model illustrating the role of *A* and *G* locus as master regulators controlling sex-specific epigenome acquisition

gene expression regulation and fine-tuning. In contrast, H3K27me3 did not display an evident proportionality with gene expression, being unresponsive to intermediate gene expression levels as it has been observed for barley [25] and maize [19]; however, it was clearly enriched at the lowest-expressed genes, suggesting that this histone mark is involved in the full repression of genes rather in the fine-tuning of their expression.

Beyond their role in gene expression regulation, we observed that both H3K9ac and H3K27me3 are important in the control of organ identity and development, thus mediating the formation of different cell types and functions from a single genome. A considerable portion of H3K9ac and H3K27me3 target genes was expressed in an organ-specific manner, the differential and unique deposition of these two histone marks in each organ, creates an epigenetic landscape on genes that is crucial for its metabolic needs and physiology.

In the case of flowers, female-specific H3K9 hyperacetylated and H3K27 hypo-trimethylated genes are involved in carpel and flower development as well as in several other bioprocesses previously reported to play a role in flower feminization in cucurbits. Such is the case for the hormone ethylene, whose synthesis and perception have important roles in development of female unisexual flowers. The expression of CmACS7 and CmACS11, two enzymes that are part of the ethylene biosynthesis pathway, and CsETR1, one of the reported ethylene receptors, have been already placed in the regulatory network promoting female flower development in cucurbits [47, 49–51, 80–83]. On the other hand, other genes related to jasmonic acid response were specifically modified with active marks in the female flower. Consistent with this, jasmonic acid (JA) was also shown to be associated with sex determination in maize [84]. Specifically, downregulated genes (H3K9 hypo-acetylated and H3K27 hyper-trimethylated) in the female flower pointed out bioprocesses such as lipid metabolism, which has been previously reported to play a role in pollen-tube growth and pollen-stigma interactions [85].

When we assessed the molecular function of sex-specific epigenetically regulated genes, we found enrichment for transcription factor, DNA and nucleotide binding activities. We, therefore, categorized these TFs according to their corresponding protein families and found that the majority of them belong to the ethylene-responsive factors [86–88]. These results suggest that histone modifications regulate ethylene transduction pathway in melon as it has previously reported in Arabidopsis [89].

Interestingly, we also found genes belonging to the MADS-box TF family, which are well recognized for their important role in flower whorl identity and is in concordance with previous reports regarding the regulation of AP1, AG and SEP3 by the histone demethylase REF6 that removes H3K27me3 [90, 91], and polycomb-dependent regulation of AG [92, 93].

The elucidation of epigenetic control of the class C genes for the development of unisexual flowers demands the assessment of the whorl-specific epigenetic state of carpel- and stamen-related identity of MADS-box genes. Tissue heterogeneity has been considered a persistent constraint for the identification of epigenetic events occurring in a specific cell population thus, masking the natural complexity of gene expression regulation. So far in the plant kingdom, few technical improvements to this limitation have been reported, especially due to the complexity of performing ChIP on small samples. In this study, we provide the first protocol to overcome such limitation that can be transferred to any plant species and demonstrate that epigenetic changes that are undetectable with standard methods using whole organs can be identified by using purified cell populations. By this method, we also demonstrated that stamen—(i.e., PI and AP3) and carpel-related (i.e., STK) identity genes present a contrasting H3K27me3 and H3K9ac landscapes. This suggests that the sex determination genes recruit histone modifiers to target organ identity genes in the sexual whorls of the flower, leading to unisexual flower development.

In the cucurbit sex determination pathway, the gynoecious gene, CmWIP1, plays a central role in unisexual flower development. From this analysis, we showed that CmACS11 represses CmWIP1 expression by inducing H3K27me3 at this locus (Fig. 5d). This data suggest that the local production of ethylene via CmACS11 in companion cells is able to activate histone methyltransferase to inhibit CmWIP1 expression in a defined spatio-temporal manner.

Detailed knowledge about spatio-temporal gene expression and epigenetic regulation dynamics is pivotal for a comprehensive understanding of development of unisexual flowers. In species recalcitrant to plant transformation, such as melon, this has been hampered mainly by difficulties in isolating sufficient amounts of tissues from distinct organ primordia for chromatin analysis. In this context, using laser capture micro-dissection, we succeeded in generating high-quality chromatin, from dissected carpel and stamina primordia, from unisexual flowers, and discovered that the sex determination genes recruit histone modifiers to orchestrate unisexual flower development in monoecious species. Genome-wide analyses of chromatin modifications in the dissected reproductive organs from unisexual and hermaphrodite flowers and from key histone modifier mutants will likely to bring new insight on the epigenetic control of unisexual flower development.

Methods

Plant material and growth conditions

Monoecious melon (Cucumis melo) cultivar charentais (Cucumis melo L. subsp. melo var cantalupensis) were used for H3K9ac and H3K27me3 genome-wide descriptive analyses. For each replicate, 7-day-old seedlings were transferred to individual pots and incubated in a growth chamber for 30 days (long day conditions, temperature: 27 °C (day) and 21 °C (night), relative humidity: 60%). Young male and female flowers (3 mm length; Developmental stage 8 [52]) were collected, organized according to their location in the plant (i.e., main stem or branches) and conserved at −80 °C for epigenomic and transcriptomic analyses.

For the determination of *a*- and *g*-related epigenetic changes, we used two families of EMS-treated *C. melo* lines carrying missense mutations in *CmWIP1* and in *CmACS11*, both previously reported to lead to gynoecy [47] and androecy [51], respectively. Germination, growth and flower collecting conditions were the same to the previously described for the monoecious WT plants.

Gene expression analysis

Total RNA were extracted from leaves, root, fruit, male and female flowers using the Nucleospin RNA kit (Macherey–Nagel), according to the manufacturer's instructions. For RT-QPCR analyses, first-strand cDNA was synthesized from 1 μg of total RNA using Improm-II reverse transcriptase (A3802, Promega) according to the manufacturer's instructions. 1/25th of the synthesized cDNA was mixed with 100 nM of each primer and LightCycler® 480 Sybr Green I master mix (Roche Applied Science) for quantitative PCR analysis. Products were amplified and fluorescent signals acquired with a LightCycler® 480 detection system. The specificity of amplification products was determined by melting curves. ACT2 was used as internal control for signals normalization. Exor4 relative quantification software (Roche Applied Science) automatically calculated relative expression level of the selected genes with algorithms based on ∆∆Ct method. Data were from duplicates of at least two biological replicates. The sequences of primers can be found in Additional file 8: Table VII.

For RNA-seq analysis, libraries were synthetized using NEBNext Ultra Directional RNA library Preparation Kit (NEB) according to the manufacturer's instructions. Two biological replicates were run for each tissue. Raw reads from RNA-Seq were first adaptor trimmed and quality filtered using Trimmomatic [94]. Filtered reads were mapped to the *Cucumis melo* genome v3.5.1 using TopHat v2.0.9 [95]. Transcript quantification was derived using Cufflinks v2.2.0 [96]. CuffDiff is used for differential expression analysis (*p* value 0.05; statistical correction: Benjamini Hochberg; FDR: 0.05). A cutoff of 0.5 fold up- or down-regulation has been chosen to define significant differential expression. CummeRbund v2.0.0 was used for visualization of differential analysis [97]. Functional annotation using GO terms was performed using the Plant MetGenMAP tool [75]. All *p* values are shown in Additional file 9: Table VIII.

Chromatin immunoprecipitation experiments

ChIP assays were performed using anti-H3K9ac (Millipore, ref. 07-352) or anti-H3K27me3 (Millipore, ref. 07-449) antibodies, using a procedure adapted from Veluchamy et al. [17]. Briefly, after plant material fixation in 1% (v/v) formaldehyde, tissues were homogenized, nuclei isolated and lysed. Cross-linked chromatin was sonicated using a water bath Bioruptor UCD-200 (Diagenode, Liège, Belgium) (30 s on/30 s off pulses, at high intensity for 60 min). Protein/DNA complexes were immunoprecipitated with antibodies, overnight at 4 °C with gentle shaking, and incubated for 1 h at 4 °C with 50 μL of Dynabeads Protein A (Invitrogen, Ref. 100-02D). The beads were washed for 2 × 5 min in ChIP Wash Buffer 1 (0.1% SDS, 1% Triton X-100, 20mM Tris-HCl pH 8, 2 mM EDTA pH 8, 150 mM NaCl), 2 × 5 min in ChIP Wash Buffer 2 (0.1% SDS, 1% Triton X-100, 20 mM Tris-HCl pH 8, 2 mM EDTA pH 8, 500 mM NaCl), 2 × 5 min in ChIP Wash Buffer 3 (0.25 M LiCl, 1% NP-40, 1% sodium deoxycholate, 10 mM Tris-HCl pH 8,1 mM EDTA pH 8) and twice in TE (10 mM Tris-HCl pH 8, 1 mM EDTA pH 8). ChIPed DNA was eluted by two 15-min incubations at 65 °C with 250 μL Elution Buffer (1% SDS, 0.1 M NaHCO$_3$). Chromatin was reverse-crosslinked by adding 20 μL of NaCl 5 M and incubated over-night at 65 °C. Reverse-cross-linked DNA was submitted to RNase and proteinase K digestion and extracted with phenol–chloroform. DNA was ethanol precipitated in the presence of 20 μg of glycogen and resuspended in 50 μL of nuclease-free water (Ambion) in a DNA low-bind tube.

For ChIP-qPCR experiments, fold enrichment of targets in ChIPed DNA relative to input was calculated from an average of three replicate qPCRs. The sequences of primers can be found in Additional file 8: Table VII. Positions of the amplified regions on the different loci are indicated in Fig. 5.

For ChIP-seq assays, 10 ng of IP or input DNA was used for ChIP-Seq library construction using NEB-Next Ultra II DNA Library Prep Kit for Illumina (New England Biolabs) according to manufacturer's recommendations. For all libraries, twelve cycles of PCR were used. The quality of the libraries was assessed with Agilent 2100 Bioanalyzer (Agilent), and the libraries were subjected to high-throughput sequencing by Illumina Sequencing technology.

Computational analysis of ChIP-seq

Preprocessing of sequencing reads for quality was performed using FASTQC (http://www.bioinformatics. babraham.ac.uk/projects/fastqc/). Filtering and trimming of reads was done using Trimmomatic with the following parameters: Minimum length of 36 bp; mean Phred quality score greater than 30; leading and trailing bases removal with base quality below 3; sliding window of 4:15. Using Bowtie, the remaining high-quality reads were mapped onto the *Cucumis melo* genome v3.5.1 with maximum mismatch of 1 bp. Unique mapping of reads was adopted. To determine the target regions of H3K9Ac ChIP-Seq, the model-based analysis of ChIP-Seq (MACS2) was adopted (Number of duplicate reads at a location: 1; Bandwidth: 300; mfold of 5:30; q-value cutoff: 0.05) [72]. For peak detection in H3K27me3 modified regions, we used SICER with the following criteria: Window size: 200, Gap size: 600 [98]. Alignment and tag-density were inspected with IGB. HOMER was used to associate peaks to nearby genes [99]. To cluster the H3K9Ac and H3K27me3 peaks, linear normalization and clustering of tag density with Density Array method (window size 50 bp; 2 kb flanking region of genes) was performed using SeqMINER [100]. Annotation of corresponding genes was done using melonomics resource (https://melonomics.net). Average profile of coverage along the genic region (between transcriptional start sites (TSS) and TES) along with the 2 kb flanking region was plotted using NGSplot in binning mode [101]. To identify differentially enriched histone modified sites in leaves, root, fruit and flowers, we used DiffReps with the following settings: Window size 1 kb; step size: 100 bp; p value: 0.0001; Statistical testing method: Chi-square method) [74].

Chromatin immunoprecipitation of Laser capture microdissected primordia

Male and female flowers of 3 mm were collected on monoecious melon plants, fixed in 1% (v/v) formaldehyde for 15 min under vacuum, and the crosslink reaction was stopped by addition of glycine (130 mM final) for 5 min under vacuum. Fixed material was then rinsed with water, dried and snap-frozen in liquid nitrogen. In a cryostat, flowers were put and oriented at the bottom of an empty cryomold and covered by cryoprotector (Tissue-Tek® O.C.T. Compound, Sakura® Finetek, VWR, France) until frozen. After 10 min at −20 °C in the cryostat (Shandon Cryotome® FSE, Excilone, France), samples were sectioned at −20 °C. To locate the targeted structures, sections were collected on glass slide and observed through the optical microscope of an XT® Arcturus Technologies microdissection system (Excilone, France). The following 4 sections (30 µm thickness) containing 10 flowers were collected on Arcturus FRAME membrane (Excilone, France). Then, the frame was put to room temperature for the sections to melt on the membrane and a glass slide was added on top of the membrane. The whole system is then placed upside down in the microdissection system. The LCM process was carried out using the XT® Arcturus Technologies

microdissection system and software. Capture was performed using Arcturus CapSure® LCM Macro Caps (Exilone, France). Carpels and stamens were identified using the 20X. FRAME Membrane Slides allowed the use of standard IR and UV lasers. UV laser was used to cut around the selected regions as IR laser was used to capture the selected regions on the macro caps without contaminating the sample with non-target-tissue. Efficiency of microdissection was evaluated by examining the CapSure® Macro caps after capture and the tissue section remaining on the slide before and after lifting off the CapSure® macro caps. To obtain enough material for downstream analysis, around 5 mm² of captured area was collected on 3 or 4 caps per sample and stored at −80 °C. Tissues were then disrupted by grinding with mortar and pestle in liquid nitrogen. Cellular debris was then resuspended in 2 ml of nuclei isolation buffer (20 mM Hepes pH8; 250 mM sucrose; 1 mM MgCl2; 5 mM KCl; 40% Glycerol; 0,25% Triton X-100), filtered through a strainer with a 63 µm pore size. Nuclei integrity was assessed by Dapi coloration and visualization under an epifluorescence microscope. Next, the samples were centrifuged 10 min at 3500g and the pellet was resuspended in 130 µl of nuclei lysis buffer (50 mM Tris–HCl pH8; 10 mM EDTA). Chromatin was sonicated using Covaris (M220 Focused-ultrasonicator; DC 5%; PIP 75Watts; CPB 200; 20 min) in 130 µl microtubes (Covaris AFA tubes) and centrifuged 10 min at 16,000g. Supernatant was then diluted ten times with ChIP dilution buffer (1,1% Triton; 1,2 mM EDTA; 16,7 mM Tris–HCl pH8; 167 mM NaCl) before immunoprecipitation with H3K9ac or H3K27me3 antibodies overnight at 4 °C with gentle shaking, and incubated for 1 h at 4 °C with 50 µL of Dynabeads Protein A (Invitrogen, Ref. 100-02D). After several washes, immunoprecipitated DNA was then recovered using proteinase K digestion, reverse crosslink and phenol–chloroform extraction as previously described. Fold enrichment of targets in ChIPed DNA relative to input was calculated from an average of three replicate qPCRs. The sequences of primers can be found in Additional file 8: Table VII. Positions of the amplified regions on the different loci are indicated in Fig. 4.

Additional files

Additional file 1: Figure S1. Pearson correlation between leaves and roots ChIP-seq replicates. The color scale represents the degree of correlation between replicates. **Figure S2.** Example of melon genomic regions showing ChIP-seq signals of biological replicates in leaves. Genes are represented in blue. Comparisons of peak positions and peak intensities illustrate the high correlation between the two biological replicates. **Figure S3.** Distribution of mapped reads for H3K27me3 (red shades) and H3K9ac (green shades) along the 12 melon chromosomes of leaves, roots, fruit, male and female flowers. Local peak densities of each epigenetic mark were plotted against the genetic distance (gray) and annotation of sense (dark blue) and antisense transcripts (light blue). **Figure S4.** Number of H3K9ac and H3K27me3 peaks identified in the different melon tissues. The computational methods MACS2 and SICER were used to determine H3K9ac and H3K27me3 target regions, respectively. **Figure S5.** H3K27me3 and H3K9ac distribution at the chromosome level in different melon organs. Distribution of H3K9ac (Green), H3K27me3 (*red*) and annotated genes (gray) are plotted along the chromosome 1. **Figure S6.** H3K9ac and H3K27me3 peaks distribution on chromosome 3 illustrating the enrichment of the marks in gene-rich regions located at the distal part of the chromosome. **Figure S7.** Boxplot showing differential peak length between H3K9ac and H3K27me3 in different melon tissues. **Figure S8.** Average tag density profile of H3K27me3 and H3K9ac along the gene body in different melon tissues. ChIP-Seq densities of equal bins were plotted along the gene body and 2-kb region flanking the TSS or the TES. **Figure S9-S12.** Tissue-specific H3K9 or H3K27me3 differentially marked genes. Venn diagrams of paired comparisons of each tissue vs. all the tissues are shown on the left. The central overlap corresponds to H3K9 or H3K27me3 specifically modified genes in each tissue. Gene ontology analysis of these subsets of genes is shown on the right. **Figure S13.** Correlation of hyper- H3K9ac or hypo-H3K27me3 with tissue-specific gene activation. Venn diagram illustrating the genes whose high and tissue-specific expression is correlated with their tissue-specific H3K9 hyper-acetylation or H3K27me3 hypo-methylation (Overlapping set of genes). A Chi-squared test was done for each comparison and confirmed the observed correlation. **Figure S14.** Molecular function analysis of H3K9ac or H3K27me3 differentially regulated genes in male and female flowers. The most significant enrichment is observed for the transcription factor activity category. **Figure S15-S18.** Molecular function and biological process classification of H3K9ac and H3K27me3 differentially regulated genes in male and female flowers. Bar charts represent the number of genes in each category for the molecular function (left) and the biological process (right). **Figure S19.** ChIP-qPCR assays on microdissected stamens and carpels of a genome region that do not display any H3K9ac or H3K27me3 signals in the ChIP-seq data. **Figure S20.** Analysis of H3K27me3 deposition in *a* and *g* mutant flowers. **A.** Schematic representation of the different type of flowers used for H3K27me3 profiling in wild-type plants (AA-GG), *wip1* mutant (AA-gg) and *acs11* mutant (aa-GG) by ChIP-seq assays. **B.** Comparisons of differential H3K27me3 deposition in wild-type male (light blue) and female flowers (*red*) with differential H3K27me3 deposition in wild-type male (light blue) and *g* mutant female flowers in the same position (orange). **C.** Comparisons of differential H3K27me3 deposition in wild-type male (light blue) and female flowers (*red*) with differential H3K27me3 deposition in wild-type female flowers (*red*) and *a* mutant male flowers in the same position (dark blue). **Figure S21.** Proportion of H3K27me3 changes between male and female wild-type flowers controlled by the *G* or *A* locus. Hypo- and hyper-methylated genes were analyzed separately. **Figure S22.** GO analysis of G- and A-H3K27me3 dependent genes. **Figure S23.** Browser view of the H3K27me3 ChIP-seq signals on the *G* locus in Wt flowers and *a* mutant flowers. Only exon 2 of the *G locus* is present in the pseudomolecule of the 12 chromosomes in the 3.5.1 melon genome version. Promoter and exon 1 of the G locus are present on another pseudomolecule that contains non-anchored BAC clones (identified as chromosome 0 in the 3.5.1 melon genome version).

Additional file 2: Table I. Number of aligned reads for each library.

Additional file 3: Table II. List of H3K9ac and H3K27me3 marked genes in leaves, root, fruit, male and female flowers.

Additional file 4: Table III. List of H3K9ac and H3K27me3 specifically enriched or depleted genes in leaves, root, fruit, male and female flowers.

Additional file 5: Table IV. List of specifically over-expressed genes in leaves, root, fruit, male and female flowers.

Additional file 6: Table V. List of H3K9ac and H3K27me3 regulated genes in flowers.

Additional file 7: Table VI. List of transcription factor epigenetically regulated genes in flowers.

Additional file 8: Table VII. List of primers used in this study.

Additional file 9: Table VIII. *p*-values list of GO analyses.

Authors' contributions

DL and NYRG performed the wet laboratory experiments, and AV, KMG and VS performed the bioinformatics analyses. CC, CD, DL, NYRG, and ABo prepared material. CB, NC, DL and NYRG were in charge of laser micro-dissection. DL, CR, NYRG, ABo, AB and MB analyzed the data. AB and MB designed the research. NYRG, DL, MB and AB wrote the manuscript. All authors read and approved the final manuscript.

Author details

[1] Institute of Plant Sciences Paris-Saclay (IPS2), CNRS, INRA, University Paris-Sud, University of Evry, University Paris-Diderot, Sorbonne Paris-Cite, University of Paris-Saclay, Batiment 630, 91405 Orsay, France. [2] Division of Biological and Environmental Sciences and Engineering, King Abdullah University of Science and Technology, Thuwal 23955-6900, Kingdom of Saudi Arabia. [3] UMR 1313 Génétique Animale et Biologie Intégrative, Institut National de la Recherche Agronomique, 78350 Jouy-en-Josas, France. [4] UR 1052, Unité de Génétique et d'Amélioration des Fruits et Légumes, INRA, BP94, 84143 Montfavet, France.

Competing interests

The authors declare that they have no competing interests.

Funding

This work was supported by the Plant Biology and Breeding department in INRA, the grants Program Saclay Plant Sciences (SPS, ANR-10-LABX-40), Inititiative d'Excellence Paris-Saclay (Lidex-3P, ANR-11-IDEX-0003-02), L'Agence Nationale de la Recherche MELODY (ANR-11-BSV7-0024), and the European Research Council (ERC-SEXYPARTH).

References

1. Nerson H. Seed production and germinability of cucurbit crops. Seed Sci Biotechnol. 2007;1:1–10.
2. Dahmani-Mardas F, Troadec C, Boualem A, Lévêque S, Alsadon AA, Aldoss AA, et al. Engineering melon plants with improved fruit shelf life using the TILLING approach. PLoS ONE. 2010. doi:10.1371/journal.pone.0015776.
3. Berger SL. Histone modifications in transcriptional regulation. Curr Opin Genet Dev. 2002;12:142–8.
4. Margueron R, Trojer P, Reinberg D. The key to development: interpreting the histone code? Curr Opin Genet Dev. 2005;15:163–76.
5. Zhou VW, Goren A, Bernstein BE. Charting histone modifications and the functional organization of mammalian genomes. Nat Rev Genet. 2011;12:7–18.
6. Rodriguez-Granados NY, Ramirez-Prado JS, Veluchamy A, Latrasse D, Raynaud C, Crespi M, et al. Put your 3D glasses on: plant chromatin is on show. J Exp Bot. 2016;67:3205–21.
7. Ramirez-Prado JS, Rodriguez-Granados NY, Ariel F, Raynaud C, Benhamed M. Chromatin architecture: a new dimension in the dynamic control of gene expression. Plant Signal Behav. 2016. doi:10.1080/15592324.2016.1232224.
8. Lauria M, Rossi V. Epigenetic control of gene regulation in plants. Biochim Biophys Acta. 2011;1809:369–78.
9. Slotkin RK. Plant epigenetics: from genotype to phenotype and back again. Genome Biol. 2016. doi:10.1186/s13059-016-0920-5.
10. Goldberg AD, Allis CD, Bernstein E. Epigenetics: a landscape takes shape. Cell. 2007;128:635–8.
11. Jaenisch R, Bird A. Epigenetic regulation of gene expression: how the genome integrates intrinsic and environmental signals. Nat Genet. 2003;33:245–54.
12. Schones DE, Cui K, Cuddapah S. Genome-wide approaches to studying yeast chromatin modifications. Methods Mol Biol. 2011;759:61–71.
13. Wang Z, Zang C, Rosenfeld JA, Schones DE, Cuddapah S, Cui K, et al. Combinatorial patterns of histone acetylations and methylations in the human genome. 2009;40:897–903.
14. Roudier F, Ahmed I, Bérard C, Sarazin A, Mary-Huard T, Cortijo S, et al.

Integrative epigenomic mapping defines four main chromatin states in Arabidopsis. EMBO J. 2011;30:1928–38.

15. Malapeira J, Khaitova LC, Mas P. Ordered changes in histone modifications at the core of the Arabidopsis circadian clock. Proc Natl Acad Sci USA. 2012;109:21540–5.

16. Sequeira-Mendes J, Aragüez I, Peiró R, Mendez-Giraldez R, Zhang X, Jacobsen SE, et al. The functional topography of the Arabidopsis genome is organized in a reduced number of linear motifs of chromatin states. Plant Cell. 2014;26:2351–66.

17. Veluchamy A, Jégu T, Ariel F, Latrasse D, Mariappan KG, Kim SK, et al. LHP1 regulates H3K27me3 spreading and shapes the three-dimensional conformation of the Arabidopsis genome. PLoS ONE. 2016;11:1–25.

18. Shi J, Dawe RK. Partitioning of the maize epigenome by the number of methyl groups on histone H3 lysines 9 and 27. Genetics. 2006;173:1571–83.

19. Wang X, Elling AA, Li X, Li N, Peng Z, He G, et al. Genome-wide and organ-specific landscapes of epigenetic modifications and their relationships to mRNA and small RNA transcriptomes in maize. Plant Cell. 2009;21:1053–69.

20. Zhang W, Garcia N, Feng Y, Zhao H, Messing J. Genome-wide histone acetylation correlates with active transcription in maize. Genomics. 2015;106:214–20.

21. Li X, Wang X, He K, Ma Y, Su N, He H, et al. High-resolution mapping of epigenetic modifications of the rice genome uncovers interplay between DNA methylation, histone methylation, and gene expression. Plant Cell. 2008;20:259–76.

22. He G, Zhu X, Elling AA, Chen L, Wang X, Guo L, et al. Global epigenetic and transcriptional trends among two rice subspecies and their reciprocal hybrids. Plant Cell. 2010;22:17–33.

23. Du Z, Li H, Wei Q, Zhao X, Wang C, Zhu Q, et al. Genome-wide analysis of histone modifications: H3K4me2, H3K4me3, H3K9ac, and H3K27ac in Oryza sativa L. J Mol Plant. 2013;6:1463–72.

24. Ayyappan V, Kalavacharla V, Thimmapuram J, Bhide KP, Sripathi VR, Smolinski TG, et al. Genome-wide profiling of histone modifications (H3K9me2 and H4K12ac) and gene expression in rust (Uromyces appendiculatus) inoculated common bean (Phaseolus vulgaris L.). PLoS ONE. 2015. doi:10.1371/journal.pone.0132174.

25. Baker K, Dhillon T, Colas I, Cook N, Milne I, Milne L, et al. Chromatin state analysis of the barley epigenome reveals a higher-order structure defined by H3K27me1 and H3K27me3 abundance. Plant J. 2015;84:111–24.

26. Barski A, Cuddapah S, Cui K, Roh TY, Schones DE, Wang Z, et al. High-resolution profiling of histone methylations in the human genome. Cell. 2007;129:823–37.

27. Mikkelsen TS, Ku M, Jaffe DB, Issac B, Lieberman E, Giannoukos G, et al. Genome-wide maps of chromatin state in pluripotent and lineage-committed cells. Nature. 2007;448:553–60.

28. Bannister AJ, Kouzarides T. Regulation of chromatin by histone modifications. Cell Res. 2011;21:381–95.

29. Zhang X, Clarenz O, Cokus S, Bernatavichute YV, Pellegrini M, Goodrich J, et al. Whole-genome analysis of histone H3 lysine 27 trimethylation in Arabidopsis. PLoS Biol. 2007. doi:10.1371/journal.pbio.0050129.

30. Heintzman ND, Stuart RK, Hon G, Fu Y, Ching CW, Hawkins RD, et al. Distinct and predictive chromatin signatures of transcriptional promoters and enhancers in the human genome. Nat Genet. 2007;39:311–8.

31. Zhou J, Wang X, He K, Charron J-B, Elling A, Deng X. Genome-wide profiling of histone H3 lysine 9 acetylation and dimethylation in Arabidopsis reveals correlation between multiple histone marks and gene expression. Plant Mol Biol. 2010;72:585–95.

32. Schubert D, Primavesi L, Bishopp A, Roberts G, Doonan J, Jenuwein T, et al. Silencing by plant Polycomb-group genes requires dispersed trimethylation of histone H3 at lysine 27. EMBO J. 2006;25:4638–49.

33. Turck F, Roudier F, Farrona S, Martin-Magniette ML, Guillaume E, Buisine N, et al. Arabidopsis TFL2/LHP1 specifically associates with genes marked by trimethylation of histone H3 lysine 27. PLoS Genet. 2007. doi:10.1371/journal.pgen.0030086.

34. Chouard P. Vernalization and its relations to dormancy. Annu Rev Plant Physiol. 1960;11:191–238.

35. Sheldon CC, Rouse DT, Finnegan EJ, Peacock WJ, Dennis ES. The

molecular basis of vernalization: the central role of FLOWERING LOCUS C (FLC). Proc Natl Acad Sci. 2000;97:3753–8.

36. Amasino R. Vernalization, competence, and the epigenetic memory of winter. Plant Cell. 2004;16:2553–9.

37. Dennis ES, Peacock WJ. Epigenetic regulation of flowering. Curr Opin Plant Biol. 2007;10:520–7.

38. De Lucia F, Crevillen P, Jones AME, Greb T, Dean C. A PHD-polycomb repressive complex 2 triggers the epigenetic silencing of FLC during vernalization. Proc Natl Acad Sci. 2008;105:16831–6.

39. Crevillén P, Sonmez C, Wu Z, Dean C. A gene loop containing the floral repressor FLC is disrupted in the early phase of vernalization. EMBO J. 2012;32:140–8.

40. Lee JT. Lessons from X-chromosome inactivation: long ncRNA as guides and tethers to the epigenome. Genes Dev. 2009;23:1831–42.

41. Ietswaart R, Wu Z, Dean C. Flowering time control: another window to the connection between antisense RNA and chromatin. Trends Genet. 2012;28:445–53.

42. Kinoshita T, Seki M. Epigenetic memory for stress response and adaptation in plants. Plant Cell Physiol. 2014;55:1859–63.

43. Yamamuro C, Zhu J-K, Yang Z. Epigenetic modifications and plant hormone action. Mol Plant. 2016;9:57–70.

44. Álvarez-Venegas R, De-la-Peña C. Editorial: recent advances of epigenetics in crop biotechnology. Front Plant Sci. 2016. doi:10.3389/fpls.2016.00413.

45. Renner SS. The relative and absolute frequencies of angiosperm sexual systems: dioecy, monoecy, gynodioecy, and an updated online database. Am J Bot. 2014;101:1588–96.

46. Ainsworth C, Parker J, Buchananwollaston V. Sex determination in plants. Curr Top Dev Biol. 1997;38:167–223.

47. Martin A, Troadec C, Boualem A, Rajab M, Fernandez R, Morin H, et al. A transposon-induced epigenetic change leads to sex determination in melon. Nature. 2009;461:1135–8.

48. Sebastian P, Schaefer H, Telford IRH, Renner SS. Cucumber (Cucumis sativus) and melon (C. melo) have numerous wild relatives in Asia and Australia, and the sister species of melon is from Australia. Proc Natl Acad Sci USA. 2010;107:14269–73.

49. Boualem A, Fergany M, Fernandez R, Troadec C, Martin A, Morin H, et al. A conserved mutation in an ethylene biosynthesis enzyme leads to andromonoecy in melons. Science. 2008;321:836–8.

50. Boualem A, Troadec C, Kovalski I, Sari M-A, Perl-Treves R, Bendahmane A. A conserved ethylene biosynthesis enzyme leads to andromonoecy in two Cucumis species. PLoS ONE. 2009. doi:10.1371/journal.pone.0006144.

51. Boualem A, Troadec C, Camps C, Lemhemdi A, Morin H, Sari M-A, et al. A cucurbit androecy gene reveals how unisexual flowers develop and dioecy emerges. Science. 2015;350:688–91.

52. Bai SL, Ben Peng Y, Cui JX, Gu HT, Xu LY, Li YQ, et al. Developmental analyses reveal early arrests of the spore-bearing parts of reproductive organs in unisexual flowers of cucumber (Cucumis sativus L.). Planta. 2004;220:230–40.

53. Ming R, Bendahmane A, Renner SS. Sex chromosomes in land plants. Annu Rev Plant Biol. 2011;62:485–514.

54. Werren JH, Beukeboom LW. Sex determination, sex ratios, and genetic conflict. Annu Rev Ecol Syst. 1998;29:233–61.

55. Bisoni L, Batlle-Morera L, Bird AP, Suzuki M, McQueen HA. Female-specific hyperacetylation of histone H4 in the chicken Z chromosome. Chromosom Res. 2005;13:205–14.

56. Angelopoulou R, Lavranos G, Manolakou P. Regulatory RNAs and chromatin modification in dosage compensation: a continuous path from flies to humans? Reprod Biol Endocrinol. 2008. doi:10.1186/1477-7827-6-12.

57. Piferrer F. Epigenetics of sex determination and gonadogenesis. Dev Dyn. 2013;242:360–70.

58. Talamali A, Bajji M, Le Thomas A, Kinet J-M, Dutuit P. Flower architecture and sex determination: how does Atriplex halimus play with floral morphogenesis and sex genes? New Phytol. 2003;157:105–13.

59. Freeman DC, Vitale JJ. The influence of environment on the sex ratio and fitness of spinach. Bot Gaz. 1985;146:137–42.

60. Eveland AL, Satoh-Nagasawa N, Goldshmidt A, Meyer S, Beatty M, Sakai H, et al. Digital gene expression signatures for maize development. Plant Physiol. 2010;154:1024–39.

61. Rudich J, Halevy AH, Kedar N. Ethylene evolution from cucumber plants

as related to sex expression. Plant Physiol. 1972;49:998–9.

62. Sun J-J, Li F, Li X, Liu X-C, Rao G-Y, Luo J-C, et al. Why is ethylene involved in selective promotion of female flower development in cucumber? Plant Signal Behav. 2010;5:1052–6.

63. Liu H, Yang X, Liao X, Zuo T, Qin C, Cao S, et al. Genome-wide comparative analysis of digital gene expression tag profiles during maize ear development. Genomics. 2015;106:52–60.

64. Rocheta M, Sobral R, Magalhães J, Amorim MI, Ribeiro T, Pinheiro M, et al. Comparative transcriptomic analysis of male and female flowers of monoecious Quercus suber. Front Plant Sci. 2014. doi:10.3389/fpls.2014.00599.

65. Cheng H, Qin L, Lee S, Fu X, Richards DE, Cao D, et al. Gibberellin regulates Arabidopsis floral development via suppression of DELLA protein function. Development. 2004;131:1055–64.

66. Plackett ARG, Thomas SG, Wilson ZA, Hedden P. Gibberellin control of stamen development: a fertile field. Trends Plant Sci. 2016;16:568–78.

67. Zhang Y, Liu B, Yang S, An J, Chen C, Zhang X, et al. A cucumber DELLA homolog CsGAIP May inhibit staminate development through transcriptional repression of B class floral homeotic genes. PLoS ONE. 2014. doi:10.1371/journal.pone.0091804.

68. Janoušek B, Široký J, Vyskot B. Epigenetic control of sexual phenotype in a dioecious plant, Melandrium album. Mol Gen Genet. 1996;250:483–90.

69. Parkinson SE, Gross SM, Hollick JB. Maize sex determination and abaxial leaf fates are canalized by a factor that maintains repressed epigenetic states. Dev Biol. 2007;308:462–73.

70. Akagi T, Henry IM, Tao R, Comai L. A Y-chromosome-encoded small RNA acts as a sex determinant in persimmons. Science. 2014;346:646–50.

71. Akagi T, Henry IM, Kawai T, Comai L, Tao R. Epigenetic Regulation of the Sex Determination Gene MeGI in Polyploid Persimmon. Plant Cell. 2016. doi:10.1105/tpc.16.00532.

72. Zhang Y, Liu T, Meyer CA, Eeckhoute J, Johnson DS, Bernstein BE, et al. Model-based analysis of ChIP-Seq (MACS). Genome Biol. 2008. doi:10.1186/gb-2008-9-9-r137.

73. Garcia-Mas J, Benjak A, Sanseverino W, Bourgeois M, Mir G, González VM, et al. The genome of melon (Cucumis melo L.). Proc Natl Acad Sci. 2012;109:11872–7.

74. Shen L, Shao NY, Liu X, Maze I, Feng J, Nestler EJ. diffReps: detecting differential chromatin modification sites from ChIP-seq data with biological replicates. PLoS ONE. 2013. doi:10.1371/journal.pone.0065598.

75. Joung J-G, Corbett AM, Fellman SM, Tieman DM, Klee HJ, Giovannoni JJ, et al. Plant MetGenMAP: an integrative analysis system for plant systems biology. Plant Physiol. 2009;151:1758–68.

76. Weigel D, Meyerowitz EM. The ABCs of floral homeotic genes. Cell. 1994;78:203–9.

77. Krizek BA, Fletcher JC. Molecular mechanisms of flower development: an armchair guide. Nat Rev Genet. 2005;6:688–98.

78. Guo S, Sun B, Looi L-S, Xu Y, Gan E-S, Huang J, et al. Co-ordination of flower development through epigenetic regulation in two model species: rice and Arabidopsis. Plant Cell Physiol. 2015;56:830–42.

79. Espina V, Wulfkuhle JD, Calvert VS, VanMeter A, Zhou W, Coukos G, et al. Laser-capture microdissection. Nat Protoc. 2006;1:586–603.

80. Wang DH, Li F, Duan QH, Han T, Xu ZH, Bai SN. Ethylene perception is involved in female cucumber flower development. Plant J. 2010;61:862–72.

81. Manzano S, Martínez C, García JM, Megías Z, Jamilena M. Involvement of ethylene in sex expression and female flower development in watermelon (Citrullus lanatus). Plant Physiol Biochem. 2014;85:96–104.

82. Byers RE, Baker LR, Sell HM, Herner RC, Dilley DR. Ethylene: a natural regulator of sex expression of Cucumis melo L. Proc Natl Acad Sci USA. 1972;69:717–20.

83. Ando S, Sato Y, Kamachi S, Sakai S. Isolation of a MADS-box gene (ERAF17) and correlation of its expression with the induction of formation of female flowers by ethylene in cucumber plants (Cucumis sativus L.). Planta. 2001;213:943–52.

84. Acosta IF, Laparra H, Romero SP, Schmelz E, Hamberg M, Mottinger JP, et al. Tasselseed1 is a lipoxygenase affecting jasmonic acid signaling in sex determination of maize. Science. 2009;323:262–5.

85. Wolters-Arts M, Lush WM, Mariani C. Lipids are required for directional pollen-tube growth. Nature. 1998;392:818–21.

86. Riechmann JL, Heard J, Martin G, Reuber L, Jiang C, Keddie J, et al. Arabidopsis transcription factors: genome-wide comparative analysis among eukaryotes. Science. 2000;290:2105–10.

87. Heim MA, Jakoby M, Werber M, Martin C, Weisshaar B, Bailey PC. The basic helix-loop-helix transcription factor family in plants: a genome-wide study of protein structure and functional diversity. Mol Biol Evol. 2003;20:735–47.

88. Wang D, Guo Y, Wu C, Yang G, Li Y, Zheng C. Genome-wide analysis of CCCH zinc finger family in Arabidopsis and rice. BMC Genom. 2008. doi:10.1186/1471-2164-9-44.

89. Zhang F, Qi B, Wang L, Zhao B, Rode S, Riggan ND, et al. EIN2-dependent regulation of acetylation of histone H3K14 and non-canonical histone H3K23 in ethylene signalling. Nat Commun. 2016. doi:10.1038/ncomms13018.

90. Lu F, Cui X, Zhang S, Jenuwein T, Cao X. Arabidopsis REF6 is a histone H3 lysine 27 demethylase. Nat Genet. 2011;43:715–9.

91. Smaczniak C, Immink RGH, Muiño JM, Blanvillain R, Busscher M, Busscher-Lange J, et al. Characterization of MADS-domain transcription factor complexes in Arabidopsis flower development. Proc Natl Acad Sci. 2012;109:1560–5.

92. Goodrich J, Puangsomlee P, Martin M, Long D, Meyerowitz EM, Coupland G. A Polycomb-group gene regulates homeotic gene expression in Arabidopsis. Nature. 1997;386:44–51.

93. Krizek BA, Lewis MW, Fletcher JC. RABBIT EARS is a second-whorl repressor of AGAMOUS that maintains spatial boundaries in Arabidopsis flowers. Plant J. 2006;45:369–83.

94. Bolger AM, Lohse M, Usadel B. Trimmomatic: a flexible trimmer for Illumina sequence data. Bioinformatics. 2014;30:2114–20.

95. Trapnell C, Pachter L, Salzberg SL. TopHat: discovering splice junctions with RNA-Seq. Bioinformatics. 2009;25:1105–11.

96. Trapnell C, Williams BA, Pertea G, Mortazavi A, Kwan G, van Baren MJ, et al. Transcript assembly and quantification by RNA-Seq reveals unannotated transcripts and isoform switching during cell differentiation. Nat Biotechnol. 2010;28:511–5.

97. Goff LA, Trapnell C, Kelley D. CummeRbund: visualization and exploration of Cufflinks high-throughput sequencing data. R Package. Version 2.2. 2012.

98. Zang C, Schones DE, Zeng C, Cui K, Zhao K, Peng W. A clustering approach for identification of enriched domains from histone modification ChIP-Seq data. Bioinformatics. 2009;25:1952–8.

99. Heinz S, Benner C, Spann N, Bertolino E, Lin YC, Laslo P, et al. Simple combinations of lineage-determining transcription factors prime cis-regulatory elements required for macrophage and B cell identities. Mol Cell. 2010;38:576–89.

100. Ye T, Krebs AR, Choukrallah MA, Keime C, Plewniak F, Davidson I, et al. seqMINER: an integrated ChIP-seq data interpretation platform. Nucleic Acids Res. 2011;39:1–10.

101. Shen L, Shao N, Liu X, Nestler E. ngs.plot: quick mining and visualization of next-generation sequencing data by integrating genomic databases. BMC Genom. 2014. doi:10.1186/1471-2164-15-284.

5

Genomic imprinting does not reduce the dosage of UBE3A in neurons

Paul R. Hillman[1,2†], Sarah G. B. Christian[1†], Ryan Doan[1,3], Noah D. Cohen[4], Kranti Konganti[5], Kory Douglas[4,6], Xu Wang[7], Paul B. Samollow[6] and Scott V. Dindot[1,2,8*] [iD]

Abstract

Background: The ubiquitin protein E3A ligase gene (*UBE3A*) gene is imprinted with maternal-specific expression in neurons and biallelically expressed in all other cell types. Both loss-of-function and gain-of-function mutations affecting the dosage of UBE3A are associated with several neurodevelopmental syndromes and psychological conditions, suggesting that UBE3A is dosage-sensitive in the brain. The observation that loss of imprinting increases the dosage of UBE3A in brain further suggests that inactivation of the paternal *UBE3A* allele evolved as a dosage-regulating mechanism. To test this hypothesis, we examined *UBE3A* transcript and protein levels among cells, tissues, and species with different imprinting states of *UBE3A*.

Results: Overall, we found no correlation between the imprinting status and dosage of UBE3A. Importantly, we found that maternal Ube3a protein levels increase in step with decreasing paternal Ube3a protein levels during neurogenesis in mouse, fully compensating for loss of expression of the paternal *Ube3a* allele in neurons.

Conclusions: Based on our findings, we propose that imprinting of *UBE3A* does not function to reduce the dosage of *UBE3A* in neurons but rather to regulate some other, as yet unknown, aspect of gene expression or protein function.

Keywords: *Ube3a*, Genomic imprinting, Dosage compensation, Angelman syndrome, *Ube3a* antisense, Evolution

Background

Genomic imprinting is a rare epigenetic phenomenon that leads to the differential expression of paternally and maternally derived alleles of a gene in a parent-of-origin dependent manner [50, 54]. It has been documented only in therian mammals and flowering plants and only at a few loci in mammals, of which fewer than half are imprinted in both mouse and human [1, 4, 8, 39, 42, 50, 52]. Several theories have been proposed to explain the evolution and function of genomic imprinting (e.g., dosage regulation, complementation, parental conflict/kinship, host defense, maternal time-bomb, and developmental plasticity models), but there is currently no unifying theory that can explain the conservation of genomic imprinting across taxa [2, 17–19, 35, 50, 51, 58, 59]. Thus, apart from downregulating or silencing the expression of one allele, the functional significance of imprinting is largely unknown. Nevertheless, imprinting is believed to be an important regulatory mechanism, inasmuch as almost all recognized imprinting abnormalities are associated with pathological states [8, 49].

The ubiquitin protein E3A ligase gene (*UBE3A*) is located at the distal end of a cluster of imprinted genes on human chromosome 15q11-q13 and a homologous region on mouse chromosome 7C. In neurons of the central nervous system (CNS), *UBE3A* is imprinted with maternal-allelic expression, whereas in all other cell types, it is expressed from both alleles [11, 43, 62]. Imprinting of *UBE3A* is regulated by expression of the paternally expressed *UBE3A* antisense transcript (*UBE3A-AS*), which comprises the 3′ end of a long polycistronic transcription unit (PTU) containing multiple clusters of C/D box small nucleolar RNAs (snoRNA) and the *SNRPN-SNURF* bicistronic transcript [7, 29, 30,

*Correspondence: sdindot@cvm.tamu.edu
†Paul R. Hillman and Sarah G. B. Christian contributed equally to this work
8 Department of Veterinary Pathobiology, College of Veterinary Medicine and Biomedical Sciences, Texas A&M University, 4467 TAMU, College Station, TX 77843, USA
Full list of author information is available at the end of the article

44]. In both mouse and human, expression of *Ube3a-AS/UBE3A-AS in cis* is both necessary and sufficient to silence expression of the paternal *Ube3a/UBE3A* allele [[31], [33]], which, at least in mouse, appears to occur by inhibiting transcriptional elongation, giving rise to a paternally expressed, 5′-truncated transcript of unknown function [33, 38].

The *UBE3A* gene is highly conserved among vertebrate and invertebrate species [6, 9, 21, 22, 61]; however, imprinting of *UBE3A* is believed to have evolved in the common ancestor of eutherian mammals after divergence from the metatherian (marsupials) lineage, as studies to date show that *UBE3A* is biallelically expressed in tammar wallaby, platypus, chicken, and fruit-fly brain, and that there is no orthologous imprinted region detectable in marsupials or monotreme (prototherian) mammals [9, 21, 41], suggesting that imprinting of *UBE3A* in neurons is somehow advantageous to biallelic expression in eutherian species. The snoRNAs located in 15q11-q13, which serve as the precursors of *Ube3a-AS/UBE3A-AS*, are also eutherian-specific and appear to have rapidly evolved in a lineage-specific manner [64]; however, the relationship, if any, between these transcripts or the evolution of this region and imprinting of *Ube3a/UBE3A* is currently unknown.

Mutations or epimutations affecting the expression or function of *UBE3A* are associated with several neurodevelopmental disorders and psychological conditions. Loss-of-function or loss-of-expression of the maternally inherited *UBE3A* allele causes Angelman syndrome (AS), which presents with intellectual disability, ataxia, epilepsy, sleep disorders, and an atypical 'happy' disposition [26, 32, 60]. Conversely, maternally derived duplications of 15q11-q13 cause dup15q syndrome—a neurodevelopmental disorder distinctly different from AS—involving intellectual disability, ataxia, speech impairment, epilepsy, and autism spectrum disorder (ASD) [16, 20, 36, 40, 47]. Although dup15q syndrome is a contiguous gene disorder, overexpression of *UBE3A* is believed to be the principal pathological mechanism underlying the syndrome, as *UBE3A* is the only maternally expressed gene located within the duplication and as the severity of the condition correlates with the number of copies and expression levels of *UBE3A* [47]. There are also reports of *UBE3A* gain-of-function mutations (e.g., biochemical and genetic) in individuals with ASD and other psychological conditions [37, 63]. Paternally inherited deletions of 15q11-q13, namely those involving the C/D box snoRNA *SNORD116*, result in Prader–Willi syndrome (PWS), which is characterized by dysregulated hunger and satiety patterns, abnormal thermoregulation, sleep disorders, and behavioral issues [45].

The notion that imprinting of *UBE3A* evolved as a dosage-regulating mechanism stems from the role of *UBE3A* in AS and dup15q syndromes and from observations showing that loss of *Ube3a-AS* reactivates paternal *Ube3a* expression, leading to an increase in Ube3a protein levels in the brain [5, 7, 13, 34]. In the present study, we compared *UBE3A* expression levels (RNA transcript and protein) between cells, tissues, and species with different imprinting states of *UBE3A*. We also examined parental Ube3a protein levels during the acquisition of the imprint in neurons. Overall, our findings show that the dosage of Ube3a/UBE3A is relatively constant regardless of its imprinting status, suggesting that imprinting does not function to regulate the dosage of Ube3a/UBE3A in neurons.

Results

Ube3a/UBE3A is highly expressed from the maternal allele in the CNS

To determine whether imprinting of *Ube3a* in neurons of the mouse CNS reduces the dosage of *Ube3a* relative to other tissues where *Ube3a* is biallelically expressed (non-CNS), we compared the steady-state levels of *Ube3a* RNA (hereafter referred to as transcript) and Ube3a protein among tissues in adult wild-type mice. To estimate the relative expression levels of the maternal *Ube3a* allele, we also examined mice with a paternally inherited mutation in the *Ube3a* gene [23]. For both $Ube3a^{m+/p+}$ and $Ube3a^{m+/p-}$ mice, *Ube3a* transcript levels were significantly higher in CNS (cortex and hippocampus) than in non-CNS (heart, kidney, liver, and lung) tissues [$Ube3a^{m+/p+}$: $F(1, 15.8) = 55.6$, $p < 0.0001$; $Ube3a^{m+/p-}$: $F(1, 14) = 338.6$, $p < 0.0001$ (Fig. 1a)]. Likewise, Ube3a protein levels in both $Ube3a^{m+/p+}$ and $Ube3a^{m+/p-}$ mice were significantly higher in CNS (cerebellum, cortex, hippocampus) than in non-CNS (heart, liver, and lung) tissues [$Ube3a^{m+/p+}$: $F(1, 19) = 22.8$, $p < 0.0001$; $Ube3a^{m+/p-}$: $F(1, 14) = 118.5$, $p < 0.001$ (Fig. 1b)]. Neither *Ube3a* RNA or Ube3a protein levels were significantly different in the CNS of $Ube3a^{m+/p+}$ and $Ube3a^{m+/p-}$ mice (RNA: $t = 0.9$, $p = 0.8$; protein: CNS, $t = 2.1$, $p = 0.2$), indicating that the relatively high level of *Ube3a* expression in mouse CNS is primarily attributable to expression of the maternal allele.

To confirm that the maternal *Ube3a* allele is highly expressed in the CNS, we used RNA-seq data to compare *Ube3a* transcript levels expressed from each parental allele within and among CNS (hippocampus) and non-CNS (heart, liver, lung, and thymus) tissues in adult mice [25]. Total *Ube3a* transcript levels (i.e., those expressed by both maternal and paternal alleles) were significantly different among the tissues [$F(4, 41.8) = 8.3$, $p < 0.0001$],

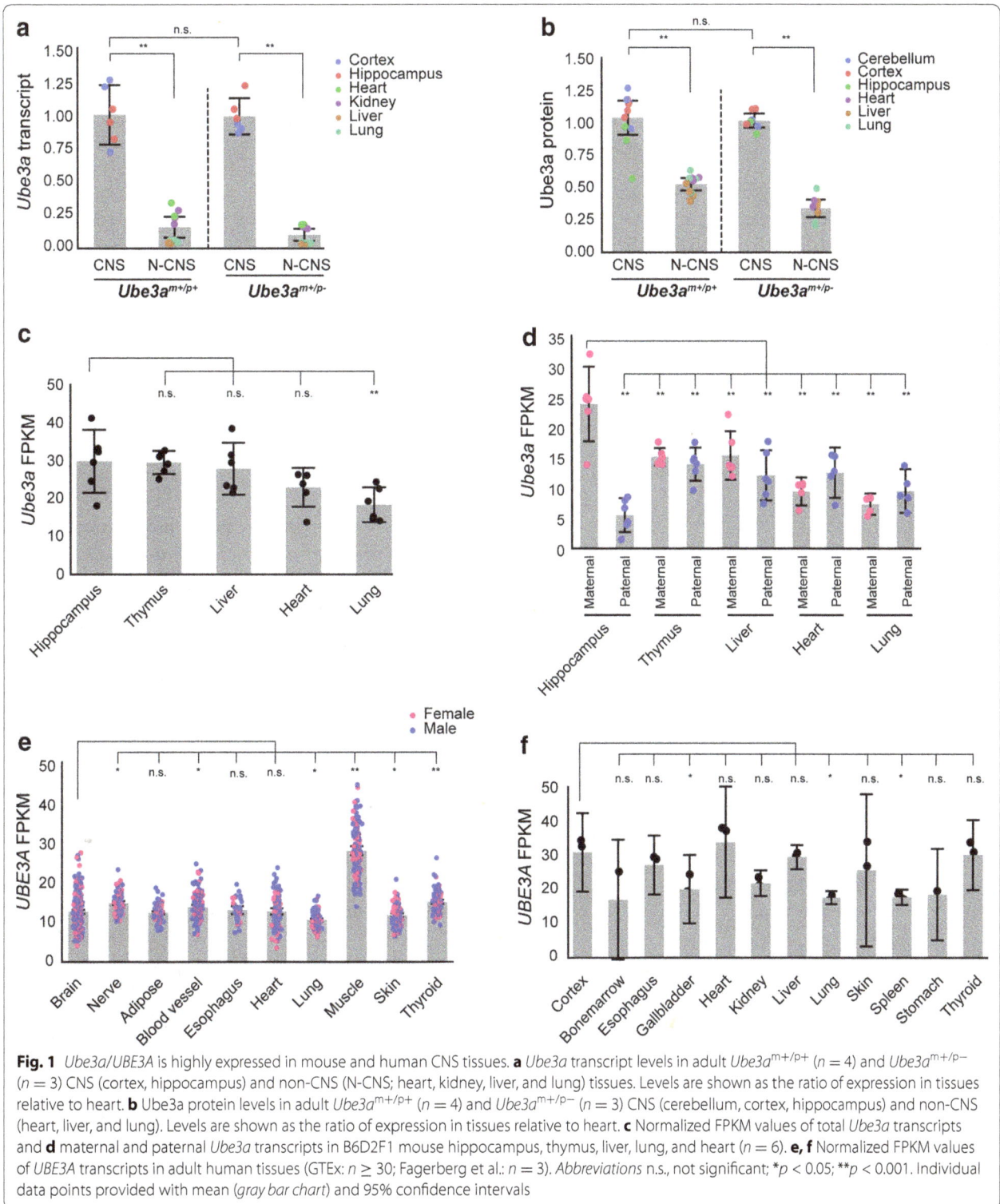

Fig. 1 *Ube3a/UBE3A* is highly expressed in mouse and human CNS tissues. **a** *Ube3a* transcript levels in adult *Ube3a*^m+/p+ (*n* = 4) and *Ube3a*^m+/p− (*n* = 3) CNS (cortex, hippocampus) and non-CNS (N-CNS; heart, kidney, liver, and lung) tissues. Levels are shown as the ratio of expression in tissues relative to heart. **b** Ube3a protein levels in adult *Ube3a*^m+/p+ (*n* = 4) and *Ube3a*^m+/p− (*n* = 3) CNS (cerebellum, cortex, hippocampus) and non-CNS (heart, liver, and lung). Levels are shown as the ratio of expression in tissues relative to heart. **c** Normalized FPKM values of total *Ube3a* transcripts and **d** maternal and paternal *Ube3a* transcripts in B6D2F1 mouse hippocampus, thymus, liver, lung, and heart (*n* = 6). **e, f** Normalized FPKM values of *UBE3A* transcripts in adult human tissues (GTEx: *n* ≥ 30; Fagerberg et al.: *n* = 3). *Abbreviations* n.s., not significant; **p* < 0.05; ***p* < 0.001. Individual data points provided with mean (*gray bar chart*) and 95% confidence intervals

with slightly higher, albeit not significant, *Ube3a* transcript levels in hippocampus compared to those in liver ($t = 0.8$, $p = 0.9$) and thymus ($t = 0.1$, $p = 0.9$) and significantly higher transcript levels than those in heart ($t = 2.9$, $p = 0.02$) and lung [$t = 4.7$, $p < 0.0001$ (Fig. 1c)]. Importantly, we found that maternal *Ube3a* transcript levels in hippocampus were significantly higher than those expressed from either allele in all the other tissues, whereas paternal *Ube3a* transcript levels in hippocampus were lower than those expressed from either allele in all the other tissues, with significantly lower levels relative to hippocampus, thymus, liver, and heart (paternal allele) (Fig. 1d and Additional file 1), confirming that the relatively high levels of *Ube3a* expression in the CNS arise from the maternal allele.

To compare *UBE3A* expression levels between CNS and non-CNS tissues in human, we analyzed *UBE3A* transcript levels in 10 tissues ($n \geq 30$) using RNA-seq data generated by the GTEx consortium [10] and in 12 tissues ($n = 3$) using RNA-seq data generated by Fagerberg et al. [12]. In the GTEx data set, *UBE3A* transcript levels were significantly different among the tissues [$F(9, 1129.8) = 262.5$, $p < 0.0001$], with no significant effect of sex [$F(1, 186.6) = 0.2$, $p = 0.6$] or significant interaction between tissue and sex [$F(9, 1129.8) = 0.8$, $p = 0.6$]. Relative to brain, *UBE3A* transcript levels were similar to those in heart, adipose tissue, and esophagus, significantly higher than those in lung and skin, but significantly lower than those in blood vessel, muscle, thyroid, and nerve (tibial) (Fig. 1e and Additional file 2). In the Fagerberg data set, *UBE3A* transcript levels were also significantly different among the tissues [$F(11, 14) = 7.5$, $p < 0.0004$]. Relative to cortex, *UBE3A* transcript levels were similar to those in bone marrow, esophagus, heart, kidney, liver, skin, and thyroid, and significantly higher than those in gallbladder, lung, spleen, and stomach (Fig. 1f and Additional file 3).

Taken together, these findings show that in both mouse and human cells, expression levels of *Ube3a*/*UBE3A* in CNS are generally equal to or higher than those in non-CNS tissues, despite the different imprinting states.

Maternal *Ube3a* compensates for loss of paternal *Ube3a* expression during neurogenesis

Our findings that the maternal *Ube3a* allele is highly expressed in brain prompted us to examine the expression of each parental *Ube3a* allele during the acquisition of the imprint in neurons. Using the $Ube3a^{YFP}$ reporter mouse model [11], we compared paternal $Ube3a^{YFP}$ ($Ube3a^{m+/pYFP}$) and maternal $Ube3a^{YFP}$ ($Ube3a^{mYFP/p+}$) protein levels in neural stem/progenitor cells (NSC) and in NSC-derived neurons over the course of 16 days of differentiation in vitro. In NSC cultures, maternal and paternal $Ube3a^{YFP}$ protein levels were approximately equal [maternal/paternal ratio = 48.3:51.7; $F(1, 2) = 0.8$, $p = 0.5$ (Fig. 2a, b)], which was not affected by the parent-of-origin of the $Ube3a^{YFP}$ reporter allele [$F(1, 2) = 1.31$, $p = 0.4$] or number of passages in culture (data not shown). In NSC-derived neurons, paternal and maternal $Ube3a^{YFP}$ protein levels were approximately equal at 1 day post-differentiation (DPD; $t = 0.9$, $p = 0.4$); however, at

4, 8, and 16 DPD, maternal $Ube3a^{YFP}$ protein levels were significantly higher than paternal $Ube3a^{YFP}$ protein levels (4 DPD: $t = 2.9$, $p = 0.004$; 8 DPD: $t = 5.7$, $p < 0.0001$; and 16 DPD: $t = 11.8$, $p < 0.0001$), with the protein levels produced by each parental allele diverging at a similar rate (slope: maternal = 0.76, paternal = −0.56; Fig. 2c, d). As a result, total $Ube3a^{YFP}$ protein levels in neurons remained relatively constant during the 16 days of differentiation [$F(3, 3) = 1.3$, $p = 0.3$ (Fig. 2e)], demonstrating that the maternal *Ube3a* allele compensates for loss of expression of the paternal *Ube3a* allele during the acquisition of the imprint in neurons. In contrast, paternal and maternal $Ube3a^{YFP}$ protein levels were approximately equal in NSC-derived astrocytes at 16 DPD [$t = 0.3$, $p = 0.8$; maternal/paternal ratio = 45.4:54.6 (Fig. 2f)]. Taken together, these findings indicate that imprinting of *Ube3a* is initiated upon neuronal differentiation and that the maternal *Ube3a* allele fully compensates for loss of expression of the paternal allele in neurons.

Ube3a/UBE3A protein levels in the mouse and opossum brain do not correlate with their imprinting status

The *UBE3A* gene is highly conserved among vertebrates, but its specific location within the AS-PWS imprinted region in eutherian (placental) mammals suggests that imprinting of *UBE3A* evolved subsequent to divergence of the eutherian and metatherian (marsupial) lineages. Expression data from unspecified tissues of tammar wallaby (*Macropus eugenii*) and platypus (*Ornithorhynchus anatinus*) brain show that *UBE3A* is not imprinted in these species, consistent with this evolutionary scenario [41]. If genomic imprinting of *UBE3A* arose in the common ancestor of modern eutherian mammals as an epigenetic mechanism to reduce the dosage of UBE3A in neurons, we should expect levels of UBE3A in the CNS of eutherian mammals to be reduced relative to that in other (non-CNS) cell types, but have no expectation of a similar pattern of relative reduction in CNS cells of non-eutherian mammals. To test this hypothesis, we contrasted Ube3a/UBE3A protein levels in non-CNS tissues and cortex, within and between mouse and the gray, short-tailed opossum (*Monodelphis domestica*), a metatherian mammal lacking an orthologous AS-PWS region. We first confirmed the non-imprinted imprinting status of *UBE3A* in opossum cortex using RNA-seq. We next used western blot analysis of UBE3A protein levels to obtain measures of relative expression among opossum tissues and normalized, absolute expression between mouse and opossum cortex.

Analysis of RNA-seq data derived from opossum cortex ($n = 4$) revealed 4 informative SNVs in *UBE3A* that were equally represented from each parental allele [SNV-1: $X^2 = 0.2$, $p = 0.7$; SNV-2: $X^2 = 0.02$, $p = 0.9$; SNV-3:

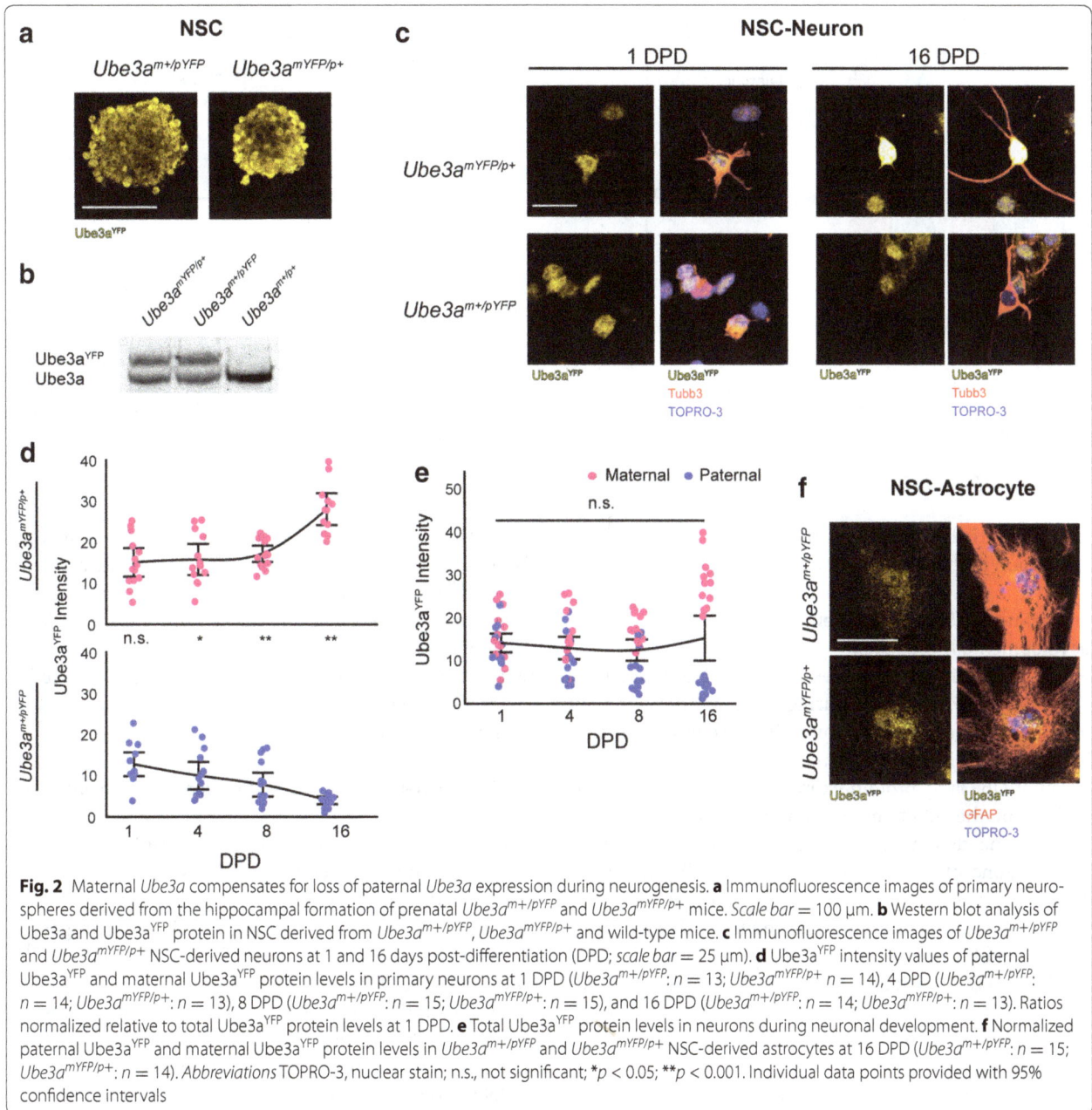

Fig. 2 Maternal *Ube3a* compensates for loss of paternal *Ube3a* expression during neurogenesis. **a** Immunofluorescence images of primary neurospheres derived from the hippocampal formation of prenatal *Ube3a*$^{m+/pYFP}$ and *Ube3a*$^{mYFP/p+}$ mice. *Scale bar* = 100 μm. **b** Western blot analysis of Ube3a and Ube3a^{YFP} protein in NSC derived from *Ube3a*$^{m+/pYFP}$, *Ube3a*$^{mYFP/p+}$ and wild-type mice. **c** Immunofluorescence images of *Ube3a*$^{m+/pYFP}$ and *Ube3a*$^{mYFP/p+}$ NSC-derived neurons at 1 and 16 days post-differentiation (DPD; *scale bar* = 25 μm). **d** Ube3a^{YFP} intensity values of paternal Ube3a^{YFP} and maternal Ube3a^{YFP} protein levels in primary neurons at 1 DPD (*Ube3a*$^{m+/pYFP}$: $n = 13$; *Ube3a*$^{mYFP/p+}$ $n = 14$), 4 DPD (*Ube3a*$^{m+/pYFP}$: $n = 14$; *Ube3a*$^{mYFP/p+}$: $n = 13$), 8 DPD (*Ube3a*$^{m+/pYFP}$: $n = 15$; *Ube3a*$^{mYFP/p+}$: $n = 15$), and 16 DPD (*Ube3a*$^{m+/pYFP}$: $n = 14$; *Ube3a*$^{mYFP/p+}$: $n = 13$). Ratios normalized relative to total Ube3a^{YFP} protein levels at 1 DPD. **e** Total Ube3a^{YFP} protein levels in neurons during neuronal development. **f** Normalized paternal Ube3a^{YFP} and maternal Ube3a^{YFP} protein levels in *Ube3a*$^{m+/pYFP}$ and *Ube3a*$^{mYFP/p+}$ NSC-derived astrocytes at 16 DPD (*Ube3a*$^{m+/pYFP}$: $n = 15$; *Ube3a*$^{mYFP/p+}$: $n = 14$). *Abbreviations* TOPRO-3, nuclear stain; n.s., not significant; $*p < 0.05$; $**p < 0.001$. Individual data points provided with 95% confidence intervals

$X^2 = 0.9$, $p = 0.7$; SNV-4: $X^2 = 2.5$, $p = 0.1$ (Fig. 3a and Additional file 4)]. Combined, the maternal-to-paternal allelic ratio of *UBE3A* transcripts was 49:51 (95% CI maternal: 46–51, paternal: 51–54), confirming that *UBE3A* is biallelically expressed in opossum brain. Alignment of the mouse, human, and opossum Ube3a/UBE3A amino acid sequences revealed a high percent identity (93–95% identical) among the three species, indicating that opossum *UBE3A* is sufficiently similar to the eutherian protein to enable meaningful comparisons by western blot (Additional file 5). Consistent with our findings

in mice, UBE3A protein levels in the opossum CNS (cortex and hippocampus) were significantly higher than those in non-CNS tissues [heart and lung; $F(1, 11) = 38.6$, $p < 0.001$ (Fig. 3b)], suggesting that an increased level of expression in cortex may be a conserved characteristic of all therian mammals, independent of imprinting. Importantly, direct comparisons between mouse and opossum showed significantly higher levels of Ube3a protein in the mouse cortex ($n = 4$) than in the opossum cortex [$t = 2.6$, $p < 0.05$ (Fig. 3c)]. Thus, despite imprinted expression of *Ube3a* in the mouse CNS, Ube3a protein levels in cortex

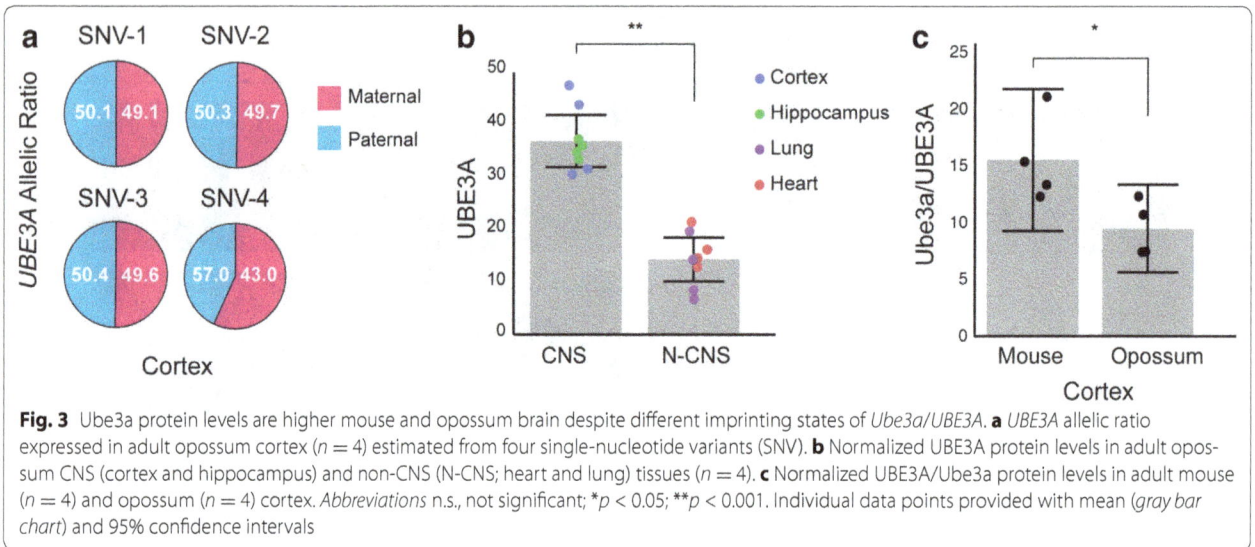

Fig. 3 Ube3a protein levels are higher mouse and opossum brain despite different imprinting states of *Ube3a/UBE3A*. **a** *UBE3A* allelic ratio expressed in adult opossum cortex ($n = 4$) estimated from four single-nucleotide variants (SNV). **b** Normalized UBE3A protein levels in adult opossum CNS (cortex and hippocampus) and non-CNS (N-CNS; heart and lung) tissues ($n = 4$). **c** Normalized UBE3A/Ube3a protein levels in adult mouse ($n = 4$) and opossum ($n = 4$) cortex. *Abbreviations* n.s., not significant; *$p < 0.05$; **$p < 0.001$. Individual data points provided with mean (*gray bar chart*) and 95% confidence intervals

are substantially higher than those produced by biallelic expression in the opossum cortex.

Imprinting of *Ube3a* during neurogenesis is developmentally regulated

Given our findings that expression of the paternal *Ube3a* allele gradually decreases during neuronal development in vitro, we decided to examine the developmental timing of the acquisition of the imprint in vivo by examining parental Ube3a^{YFP} protein expression in two neurogenic regions of the adult mouse brain: (1) the subgranular zone of the dentate gyrus (SGZ) and (2) the subventricular zone of the lateral ventricles (SVZ) and rostral migratory stream (RMS). In the SGZ, neural stem cells differentiate into mature granular neurons that integrate into the dentate gyrus through migration to the granular cell layer (GCL), whereas in the SVZ, neural stem cells differentiate while migrating through the RMS to the olfactory bulb, where they differentiate into mature olfactory neurons [28].

In the SVZ of adult mice, paternal Ube3a^{YFP} protein was detected in NSC/progenitor cells located along the lateral ventricles and expressing the polysialylated-neural cell adhesion molecule (PSA-NCAM) and nestin (NES) (Additional File 6). In the RMS, paternal Ube3a^{YFP} protein was detected in immature neurons expressing doublecortin (DCX); however, in the olfactory bulb, paternal Ube3a^{YFP} protein was barely detectable in mature neurons expressing the RNA-binding protein, Fox-1 homolog 3 [RBFOX3 (Fig. 4a)]. In the dentate gyrus, paternal Ube3a^{YFP} protein was only detected in the SGZ, where it was present in radial glia (i.e., putative neural stem cells) expressing glial fibrillary acidic protein (GFAP), progenitor cells expressing PSA-NCAM, and immature neurons

expressing DCX but not in mature neurons expressing RBFOX3 in the GCL (Fig. 4b, c). In contrast, maternal Ube3a^{YFP} protein was detected in each neurogenic cell type and in mature neurons throughout the CNS (Additional file 6). Taken together, these observations indicate that expression of the paternal *Ube3a* allele is silenced at a specific stage of neurogenesis, preceding the developmental maturation of neurons.

Ube3a is biallelically expressed in myenteric neurons of the PNS

Lastly, we asked whether *Ube3a* is also imprinted in neurons of the peripheral nervous system (PNS). To test this hypothesis, we examined parental Ube3a^{YFP} protein levels in myenteric neurons in Auerbach's ganglia in the colon. In adult *Ube3a$^{mYFP/p+}$* and *Ube3a$^{m+/pYFP}$* mice, maternal Ube3a^{YFP} and paternal Ube3a^{YFP} protein were detected in mature myenteric neurons expressing RBFOX3 (Fig. 5a), in contrast to hippocampal granular neurons in which paternal Ube3a^{YFP} protein was almost undetectable (Fig. 5b). The levels of Ube3a^{YFP} protein expressed from each allele in myenteric neurons, however, was skewed, with higher levels of maternal Ube3a^{YFP} protein albeit not significantly different (maternal/paternal ratio = 60:40; $F(1, 4.2) = 6.3$, $p = 0.06$) (Fig. 5b). Although studies of additional PNS neurons are needed, these findings suggest that imprinting of *Ube3a* is likely specific to neurons of the CNS.

Discussion

In this study, we examined whether imprinting functions to reduce the dosage of Ube3a/UBE3A (i.e., levels of transcripts and/or protein) in neurons. Our results show that imprinting of the mouse *Ube3a* gene in neurons is

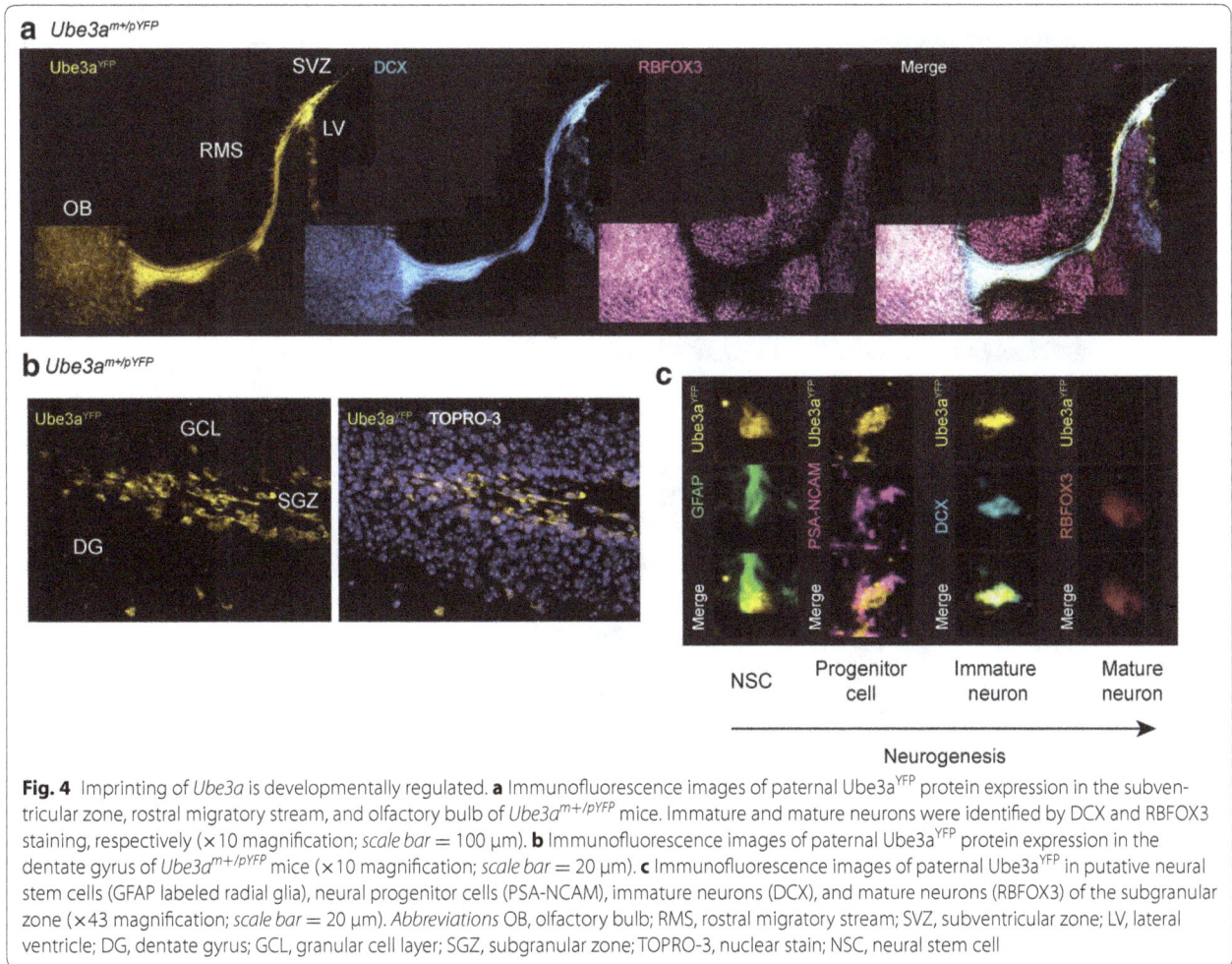

Fig. 4 Imprinting of *Ube3a* is developmentally regulated. **a** Immunofluorescence images of paternal Ube3a^YFP protein expression in the subventricular zone, rostral migratory stream, and olfactory bulb of *Ube3a*^m+/pYFP mice. Immature and mature neurons were identified by DCX and RBFOX3 staining, respectively (×10 magnification; *scale bar* = 100 μm). **b** Immunofluorescence images of paternal Ube3a^YFP protein expression in the dentate gyrus of *Ube3a*^m+/pYFP mice (×10 magnification; *scale bar* = 20 μm). **c** Immunofluorescence images of paternal Ube3a^YFP in putative neural stem cells (GFAP labeled radial glia), neural progenitor cells (PSA-NCAM), immature neurons (DCX), and mature neurons (RBFOX3) of the subgranular zone (×43 magnification; *scale bar* = 20 μm). *Abbreviations* OB, olfactory bulb; RMS, rostral migratory stream; SVZ, subventricular zone; LV, lateral ventricle; DG, dentate gyrus; GCL, granular cell layer; SGZ, subgranular zone; TOPRO-3, nuclear stain; NSC, neural stem cell

achieved through increasing expression of the maternal *Ube3a* allele in step with decreasing expression of the paternal *Ube3a* allele that occurs prior to a specific developmental time point during neurogenesis. These compensating allelic expression states maintain a relatively constant level of Ube3a protein in neurons and relatively high levels of *Ube3a* RNA and protein in the CNS. Overall, we found that the expression levels of *Ube3a/UBE3A* are relatively constant among tissues and between eutherian and metatherian mammals, despite different imprinting states. Taken together, these findings suggest that the acquisition of *UBE3A* imprinting in eutherian mammals coincided with increased expression levels, supporting the conclusion that imprinting of *UBE3A* did not evolve as a dosage-regulating mechanism in neurons.

The dosage model of genomic imprinting suggests that imprinting evolved at some loci to reduce the expression levels of the gene by half [50]. For example, the paternal allele of the *Murr1* gene is imprinted in neurons during brain development by the paternally expressed *U2af1-rs1*

antisense transcript, leading to a substantial reduction in *Murr1* RNA levels in the postnatal brain [57]. Additionally, differential imprinting (i.e., monoallelic and biallelic) of the *Igf2* gene in neurogenic niches of the adult mouse brain has been shown to regulate autocrine and paracrine roles of *Igf2* in a dose-dependent manner [14]. In contrast to these two genes, we show that the levels of *Ube3a/UBE3A* RNA and protein are remarkably stable among tissues and species with different imprinting states, which stems from increased expression of the maternal *Ube3a* allele during the acquisition of the imprint. These findings are consistent with previous studies showing relatively constant levels of *Ube3a/UBE3A* transcripts during mouse brain development and among human tissues [15, 27] and the notion that dosage compensation may in fact be a common feature of imprinted genes, as suggested previously [3].

Given the importance of UBE3A in the brain and the conservation of the imprint in mouse and human neurons, and perhaps all placental mammals, it is likely that

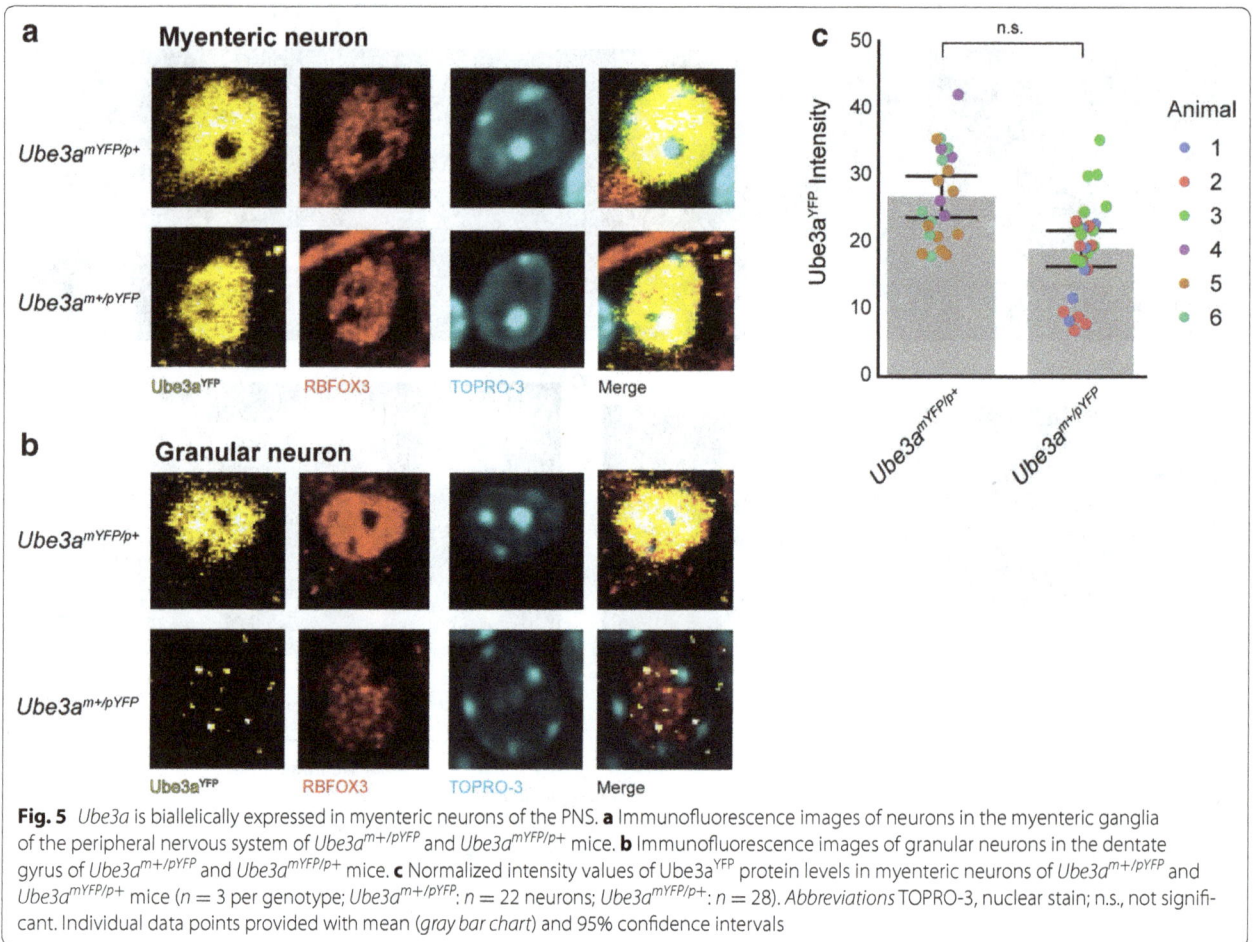

Fig. 5 *Ube3a* is biallelically expressed in myenteric neurons of the PNS. **a** Immunofluorescence images of neurons in the myenteric ganglia of the peripheral nervous system of *Ube3a*$^{m+/pYFP}$ and *Ube3a*$^{mYFP/p+}$ mice. **b** Immunofluorescence images of granular neurons in the dentate gyrus of *Ube3a*$^{m+/pYFP}$ and *Ube3a*$^{mYFP/p+}$ mice. **c** Normalized intensity values of Ube3a^{YFP} protein levels in myenteric neurons of *Ube3a*$^{m+/pYFP}$ and *Ube3a*$^{mYFP/p+}$ mice ($n = 3$ per genotype; *Ube3a*$^{m+/pYFP}$: $n = 22$ neurons; *Ube3a*$^{mYFP/p+}$: $n = 28$). *Abbreviations* TOPRO-3, nuclear stain; n.s., not significant. Individual data points provided with mean (*gray bar chart*) and 95% confidence intervals

imprinting of *UBE3A* is a functionally relevant process in neurons that is somehow linked to the neuron-specific expression of the PWS polycistronic transcription unit (PWS-PTU). For example, imprinting of *UBE3A* might have evolved as a means to permit simultaneous expression of both *UBE3A* and the PWS-PTU in neurons, similar to that proposed in the complementation model of genomic imprinting [24]. Alternatively, transcription of *UBE3A-AS* across *UBE3A* or, perhaps, *UBE3A-AS* itself might expand the functional repertoire of UBE3A isoforms (coding and non-coding) expressed in neurons. Unlike most imprinted genes regulated by an antisense transcript, studies in mouse show that *Ube3a-AS* inhibits transcriptional elongation of the paternal *Ube3a* allele, resulting in a paternally expressed, 5′-truncated transcript expressed exclusively in neurons [33, 38]. Given recent findings that mouse *Ube3a* expresses a non-coding isoform that is important for synapse development [55], it is tempting to speculate that the 5′-truncated *Ube3a* transcript is in fact a functional transcript rather than just a by-product of the imprinting mechanism.

Conclusions

Although future studies are needed to resolve the functional significance of imprinting of *UBE3A* in neurons, the findings presented here provide evidence of an evolutionarily constrained, developmentally regulated process that maintains the dosage of UBE3A despite its monoallelic expression in neurons. Understanding the function of imprinting of *UBE3A* is directly relevant to understanding the function of *UBE3A* in the brain, and understanding the imprinted regulation of *UBE3A* is critically important for approaches aimed at reactivating expression of the paternal *UBE3A* allele as a therapy for individuals with Angelman syndrome.

Methods

Analysis of Ube3a/UBE3A expression levels in tissues

Animals

C57BL/6J mice were obtained from The Jackson Laboratories (http://www.jax.org). The *Ube3a*$^{m-/p+}$ mouse model was obtained from the laboratory of Dr. Arthur Beaudet (Baylor College of Medicine). *Ube3a*$^{m-/p+}$ and animals were generated by crossing wild-type C57BL/6J males with $Ube3a^{m+/p-}$ females, and $Ube3a^{m+/p-}$ animals were generated by crossing $Ube3a^{m+/p-}$ males with wild-type C57BL/6J females. Animals were maintained by the Texas A&M University Comparative Medicine Program. All procedures involving animals were approved by the Texas A&M Institutional Animal Care and Use Committee.

Quantitative RT-PCR

Total RNA was extracted from tissue samples using the PureLink RNA Mini Kit (Life Technologies, Carlsbad, CA). First-strand cDNA synthesis was performed using: (1) the Superscript III First-Strand Synthesis kit and oligo-dT primers (messenger RNA [mRNA]; Life Technologies) and (2) the High Capacity RNA to cDNA kit and random hexamer primers (total RNA [toRNA]; Life Technologies). Real-time PCR was performed using the TaqMan Gene Expression Master Mix and TaqMan Gene Expression Assays per manufacturer's protocol (Life Technologies). *Beta-2 microglobulin* (TaqMan Assay #Mm00437762_m1) was used as an internal control. TaqMan Assay #Mm00839910_m1 was used to assess *Ube3a* toRNA and mRNA levels. The primer and probe set targets an amplicon of 121 base pairs that spans exons 6 and 7 of *Ube3a* isoforms 1 and 3 and exons 8 and 9 of *Ube3a* isoform 2. The reactions were performed using an ABI 7900HT real-time PCR machine.

Statistical analysis

Measurements for inferential statistics were taken using normalized ΔCt values ($2^{-\Delta C_t} = 2^{-C_t[\text{target}]-C_t[\text{internal control}]}$), as outlined previously [46]. A Shapiro–Wilk goodness-of-fit test was used to test normality of sample distribution. A mixed-effect model was used to examine the effect of tissue on *Ube3a* RNA transcript levels, with sample ID as a random effect to account for repeated measurements of individual. A Tukey HSD multiple comparison test was used for pair-wise comparisons of tissues. The tissues were then grouped as CNS or non-CNS, and a mixed-effect model was used to examine the effect of tissue type on *Ube3a* RNA levels, with sample ID as a random effect. Descriptive statistics consist of $\Delta\Delta$Ct values ($2^{-\Delta\Delta C_t} = 2^{-(C_t[\text{target}]-C_t[\text{internal control}])-(C_t[\text{target}]-C_t[\text{internal control}])}$) ; target = *Ube3a*, internal control = *beta-2 microglobulin*).

Western blot analysis

Total protein was isolated by homogenizing tissue samples with a 1% Nonidet P40/0.01% SDS lysis buffer and protease inhibitors (Roche, Indianapolis, IN). The resulting lysates were mixed 1:1 with Laemmli loading buffer (Bio-Rad, Hercules, CA) and heated to 95 °C for 5 min. The samples were then resolved on a SDS-PAGE gel (7.5%) at 25 V for approximately 12 h and then transferred to nitrocellulose membranes at 100 V for 2 h. To normalize samples, the membranes were treated with Ponceau stain (Sigma-Aldrich) and digitally photographed. The membranes were then blocked in 5% milk in Tris-buffered saline plus Tween 20 (T-TBS) for 1 h at room temperature. Ube3a primary antibody (Additional file 7) was diluted in 2.5% milk/T-TBS and incubated on the membrane for 1 h at room temperature. After three 15 min washes in T-TBS, the secondary antibody (Additional file 7) was diluted 1:2000 in 2.5% milk/T-TBS and incubated on the membrane for 1 h at room temperature. Three 15 min washes in T-TBS were performed before developing with Clarity Western ECL Substrate (Bio-Rad), according to the manufacturer's protocol. Membranes were imaged using the FluorChem system.

Statistical analysis

Digital images of western blot membranes (16bit.tif) were imported into ImageJ, and Ube3a protein levels were quantified using the Gel Analysis feature. Protein levels were quantified (as percentage) and then normalized by the amount of total protein per sample using a Ponceau stain. A Shapiro–Wilk goodness-of-fit test was used to test normality of sample distribution. A linear mixed-effects model was used to examine the association of tissue on Ube3a protein levels, with tissue modeled as a fixed, categorical effect and sample ID modeled as a random effect to account for repeated measurements of individual. A Tukey HSD multiple comparison test was used for pair-wise comparisons of tissues. The tissues were then grouped as CNS or non-CNS, and a linear mixed-effects model was used to examine the effect of tissue type on Ube3a protein levels, with sample ID included as a random effect. Descriptive statistics consist of Ube3a protein levels relative to heart.

RNA-sequencing analysis in mouse tissues

Data from Keane et al. [25]: Samples consisted of 76 bp paired-end (PE) libraries generated from total RNA isolated from heart, liver, lung, thymus, and hippocampus of 8-week-old B6D2F1 hybrid mice (C57BL/6 male X DBA female; $n = 6$). FASTQ files were downloaded from the NCBI GEO SRA (accession: ERP000591) and aligned to the mouse reference genome (mm9) using the default settings in TopHat, with the following option: -b2 sensitive.

Normalized RNA steady-state levels of the RefSeq gene annotation were estimated for each sample using Cuffnorm with the following option: -u. The FPKM value of *Ube3a* transcripts for each sample was extracted from the output file and used for descriptive and inferential statistics.

Statistical analysis

The parental allelic ratio of *Ube3a* for each tissue was determined using single-nucleotide polymorphisms (SNPs) located in *Ube3a* exon 5 (SNP-1 = chr7:66,527,581) and 8 (SNP-2 = chr7:66,541,539). Raw counts of the 2 SNPs were determined using the CLC Genomics Workbench quality-based variant detection module with the following settings: neighborhood radius = 5; maximum gap and mismatch count = 2, minimum neighborhood quality = 15, minimum central quality = 20, ignore non-specific matches, minimum coverage = 20, minimum variant % = 1. The raw counts of each SNP for each sample and tissue were then used to estimate the fragments per thousand per million (FPKM) values of *Ube3a* transcripts of the parental alleles using the following equations:

$$\text{Maternal allele} = \left(\left(\left(\sum \text{SNP} - 1^{\text{maternal}}\right)/\left(\sum \text{SNP} - 1^{\text{total}}\right) + \left(\sum SNP - 2^{\text{maternal}}\right)/\left(\sum \text{SNP} - 2^{\text{total}}\right)\right)/2\right) \times \text{FPKM}$$

$$\text{Paternal allele} = \left(\left(\left(\sum \text{SNP} - 1^{\text{paternal}}\right)/\left(\sum \text{SNP} - 1^{\text{total}}\right) + \left(\sum \text{SNP} - 2^{\text{paternal}}\right)/\left(\sum \text{SNP} - 2^{\text{total}}\right)\right)/2\right) \times \text{FPKM}$$

A mixed-effect model was used to examine the effect of tissue and allele on *Ube3a* transcript levels (full factorial), with sample ID as a random effect to account for repeated measurements of individual. A Dunnett's–Hsu multiple comparison test was used to compare total *Ube3a* transcript levels between hippocampus and each tissue, and a Dunnett's–Hsu multiple comparison test was used to compare maternal and paternal *Ube3a* transcript levels between hippocampus and each allele of each tissue.

RNA-sequencing analysis in human tissues

Data from Fagerberg et al. [12]: Samples consisted of 100 bp PE sequencing libraries generated from mRNA isolated from 12 human tissues. Three samples per tissue were used for the analysis. FASTQ files were downloaded (http://www.ebi.ac.uk/arrayexpress/; accession: E-MTAB-1733) and aligned to the human reference genome (hg19) using the default settings in TopHat. Normalized FPKM values of the RefSeq gene annotation were estimated using Cuffnorm using the default settings and the following option: -u. The FPKM value of *UBE3A* for each sample was extracted from the output file and used for descriptive and inferential statistics.

Data from *GTEx*: The FPKM values and accompanying information were downloaded for *UBE3A* from the GTEx Portal (http://www.gtexportal.org/home/GTEx_Analysis_V4_RNA-seq_RNA-SeQCv1.1.8_gene_rpkm.gct) [10, 53]. The initial data set consisted of 25 different histologic samples derived from 175 individuals (1518 tissue samples). Only tissues for which at least 30 samples were available were included in the inferential analyses (1227 tissue samples from 10 tissues): adipose tissue (*n* = 111), brain (*n* = 296), blood vessels (*n* = 138), esophagus (*n* = 34), heart (*n* = 103), lung (*n* = 113), muscle (*n* = 132), nerve (*n* = 86), skin (*n* = 114), and thyroid (*n* = 100). A total of 291 tissue samples from 14 tissues were excluded. Blood samples were also excluded from inferential analysis.

Statistical analysis

A mixed-effect model was used to examine the effect of tissue on *UBE3A* transcript levels, with sample ID as a random effect to account for repeated measurements of individual. For the GTEx data set, tissue and sex were modeled as fixed effects (full factorial), with sample ID as a random effect. A Dunnett's–Hsu multiple comparison test was used to compare *UBE3A* transcript levels among tissues relative to the brain. Goodness-of-fit of models was assessed by AIC and BIC values and visual inspection of diagnostic residual plots.

Analysis of allelic Ube3a^YFP in neural stem cells and differentiated neurons

Animals

The *Ube3a^YFP* mouse model was obtained from the laboratory of Dr. Arthur Beaudet (Baylor College of Medicine). *Ube3a^{mYFP/p+}* animals were generated by crossing wild-type C57BL/6J males with *Ube3a^{m+/pYFP}* females, and *Ube3a^{m+/pYFP}* animals were generated by crossing *Ube3a^{m+/pYFP}* males with wild-type C57BL/6J females. Wild-type mice consisted of siblings lacking the *Ube3a^YFP* allele. PCR to determine the genotypes of mice (i.e., *Ube3a^{m+/p+}* or *Ube3a^YFP*) were performed using methods described previously [11].

Neural stem cell cultures

Neural stem cell cultures were established from *Ube3a^{m+/p+}*, *Ube3a^{mYFP/p+}*, and *Ube3a^{m+/pYFP}* mice using methods described previously [48]. Briefly, the hippocampal formation (HF) was removed from fetuses at embryonic day 17.5 (E17.5), digested using a 10× Trypsin–EDTA solution, triturated into a single cell suspension, and then seeded in neural stem cell (NSC) medium, consisting of DMEM-F12 (Invitrogen, Carlsbad, CA), B-27 supplement (Invitrogen), progesterone, putrescine (Sigma-Aldrich), epidermal growth factor (Sigma-Aldrich), glucose, penicillin/streptomycin (Invitrogen), insulin–transferrin–sodium selenite (Sigma-Aldrich), HEPES, and heparin. The NSC cultures were maintained in humidified incubators at 37 °C and 5% CO_2. Every 3–4 days, the neurospheres were dissociated with TrypLE (Invitrogen) for 20 min and then resuspended in NSC media.

Western blot analysis
Western blot analysis of Ube3a and Ube3a^{YFP} protein levels in NSC cultures was performed using methods described above.

Statistical analysis
A mixed-effect model was used to examine the effect of allele (i.e., *Ube3a* or *Ube3a^{YFP}*) and parent-of-origin (i.e., maternal or paternal) on Ube3a and Ube3a^{YFP} protein levels (full factorial), with sample ID as a random effect to account for repeated measurements.

Neural stem cell differentiation
To differentiate NSC cultures, neurospheres were first dissociated using TrypLE (Invitrogen) following the manufacturer's protocol. The dissociated cells were then resuspended in neuronal growth media (Neurobasal-A [Invitrogen], B-27 supplement [Invitrogen], and Glutamax [Invitrogen]) and plated on glass coverslips coated with poly-ornithine [Sigma-Aldrich] and laminin [Invitrogen] at a density of 380,000 cells per well in a 12-well, cell-culture plate; the plates were maintained in humidified incubators at 37 °C and 5% CO_2.

Immunofluorescence imaging of Ube3a^{YFP}
Immunofluorescence imaging was used to quantify Ube3a^{YFP} protein levels in differentiated neurons as previously described [11]. Briefly, at 1, 4, 8, and 16 days post-differentiation (DPD; day of differentiation = 0 DPD), differentiated cells were washed twice with 1× PBS, fixed in 4% paraformaldehyde in PBS for 15 min, and then washed three times in 1× PBS. The cells were blocked in 0.3% Triton-X100 in PBS (T-PBS) plus 5% goat or donkey serum for 1–2 h at room temperature with gentle agitation. Cells were incubated with the primary antibodies (Additional file 7) for 24 h at 4 °C with gentle agitation. Cells were washed 3 times in 0.1% Tween 20 1× PBS for 15 min each and then incubated with fluorescently labeled secondary antibodies (Additional file 7) for 24 h at 4 °C in the dark. Cells were then washed 4 times in 0.1% Tween 20 1× PBS for 15 min each. Nuclei were labeled using TOPRO-3(Invitrogen) at a dilution of 1:1000 in the third wash. Coverslips were mounted on glass slides with Vectashield (Vector Laboratories, Burlingame, CA) mounting reagent. Confocal images were obtained using a LSM 510 META NLO multiphoton microscope (Zeiss, Oberkochen, Germany). Images of the individual neurons were taken at 43× magnification. Images were imported into ImageJ and converted to the RGB color format. The Plot Profile feature was then used to determine the median gray value of Ube3a^{YFP} in individual neurons. Neurons and astrocytes were identified by immunoreactivity with the Tubb3 and GFAP antibodies, respectively.

Statistical analysis
To examine the effect of day on Ube3a^{YFP} protein levels, a linear regression model was used with normalized intensity values of Ube3a^{YFP} as the dependent variable and days post-differentiation (DPD) and allele (i.e., maternal and paternal) as fixed effects. To compare maternal and paternal Ube3a^{YFP} protein levels at each time point, a least square means contrast linear regression model was used with normalized gray values of Ube3a^{YFP} protein levels as the dependent variable and DPD (dummy variable) and allele as fixed effects.

Analysis of opossum Ube3a/UBE3A expression
Animals
For the analysis of UBE3A protein levels, opossum (*Monodelphis domestica*) tissues (cortex, hippocampus, heart, and lung) samples were obtained from a colony at the Comparative Medicine Program facilities at Texas A&M University. All animals were derived from an outbred stock (LL1) that was founded using wild animals captured in Eastern Brazil. All procedures involving animals were approved by the Texas A&M Institutional Animal Care and Use Committee. For the allelic expression analysis of *UBE3A*, opossum fetal brain (cortex) samples were obtained at 13 days postcopulation (d.p.c.) from F1 animals derived from reciprocal crosses of two semi-inbred stocks (LL1 and LL2) [56].

RNA-seq analysis
The brain tissues were homogenized in TRI Reagent (Invitrogen) and total RNA was extracted using BCP (1-bromo-2 chloropropane), precipitated with isopropanol, and resuspended in RNase-free water. Potential DNA contamination was removed by both DNase I treatment and DNA removal columns in Qiagen RNeasy Plus Mini kit (Qiagen, CA). RNA-seq libraries were made using the Illumina TruSeq RNA Sample Prep Kit (Illumina Inc., CA), and sequenced on an Illumina HiSeq 2000 instrument (Illumina Inc., CA). Details on RNA-seq data analysis could be found in Wang et al. [56].

Statistical analysis
A Pearson's two-sided Chi-square test of the allelic counts of each SNV was used to determine the whether the allelic ratios of *UBE3A* were equal (Ho: maternal/paternal = 50:50; Ha: maternal/paternal = 50:50).

Western blot analysis
Total protein was isolated from adult (21-week-old) opossum tissues ($n = 4$) and adult (8-week-old) mouse

cortex ($n = 4$) as described above. For the analysis of UBE3A protein levels among opossum tissues, a single western blot was performed for each animal for all tissues. UBE3A protein levels were estimated and normalized as described above. For the comparison of Ube3a/UBE3A protein levels between opossum and mouse cortex, samples were run on a single western blot, and Ube3a/UBE3A protein levels were normalized and compared as described above.

Statistical analysis

To compare UBE3A protein levels among opossum tissues, a mixed-effect model was used with normalized UBE3A protein levels as the dependent variable, tissue as a fixed effect, and sample ID as a random effect to account for repeated measurements of individual. To compare Ube3a/UBE3A protein levels between mouse and opossum cortex, a two-tailed Welsh t test was performed.

Analysis of Ube3a^{YFP} in neurogenic niches of the adult mouse brain
Immunofluorescence imaging of Ube3a^{YFP}

$Ube3a^{m+/pYFP}$ and $Ube3a^{mYFP/p+}$ adult mice (6-week-old) were anesthetized with 0.5–1.0 mL of 20 mg/mL Avertin (Sigma-Aldrich, St. Louis, MO) via intraperitoneal injection. Mice were perfused with ice-cold phosphate-buffered saline (PBS) and 4% paraformaldehyde. Dissected brains were post-fixed in 4% paraformaldehyde solution for approximately 12 h and then cryoprotected in 30% sucrose solution. Fifty µm sections (sagital and coronal) were cut on a cryostat and stored in PBS. For immunostaining, sections were blocked in 0.3% Triton-X100 in PBS (T-PBS) plus 5% serum (goat or donkey) for 1–2 h at room temperature in a humidified chamber with gentle agitation. Primary antibodies (Additional file 7) were incubated with sections for 48 h at 4 °C with gentle agitation. Sections were washed 3 times in 0.1% Tween 20 $1\times$ PBS for 15 min each and then incubated with fluorescently labeled secondary antibodies (Additional file 7) for 24 h at 4 °C in the dark with gentle agitation. Sections were washed 4 times in 0.1% Tween 20 plus $1\times$ PBS for 15 min each. Nuclei were labeled by adding TOPRO-3 at 1:1000 dilution in the third wash. Sections were mounted on glass slides with Vectashield (Vector Laboratories, Burlingame, CA) mounting reagent. Confocal images were obtained using a LSM 510 META NLO multiphoton microscope (Zeiss, Oberkochen, Germany). Images were taken using $10\times$ and $43\times$ (oil) objectives. Rostral migratory stream images were taken at $10\times$ magnification, and then images were then stitched together. For resolution of individual cell images in vivo Z-stack images of 8–11 slices were used.

Analysis of Ube3aYFP in myenteric neurons
Immunofluorescence imaging of Ube3a^{YFP}

$Ube3a^{mYFP/p+}$ and $Ube3a^{m+/pYFP}$ adult mice (6-week-old; $n = 3$/genotype) were processed and immunostained as described above. Confocal images of Auerbach's ganglia (8–10 per animal) were obtained using a $43\times$ (oil) objective. Images were imported into ImageJ and converted to the RGB color format. The Plot Profile feature was then used to determine the median gray value of Ube3a^{YFP} in individual neurons (RBFOX3 positive) as described above.

Statistical analysis

A mixed-effect model was used to examine the effect of allele (maternal and paternal) on Ube3a^{YFP} protein levels, with sample ID as a random effect to account for repeated measurements of neurons within an individual animal.

Statistics

Inferential analyses were performed using JMP Pro® (version 12). Mixed-effect models were used for analyses involving repeated measures of cells/tissues from an individual animal. The degrees of freedom were calculated using the Kenward–Roger correction.

Charts

Charts were generated using JMP Pro® (version 12) and formatted in Adobe Illustrator.

Additional files

Additional file 1: Table S1. Pair-wise comparisons of maternal and paternal Ube3a transcript levels.

Additional file 2: Table S2. Pair-wise comparisons of UBE3A transcript levels.

Additional file 3: Table S3. Pair-wise comparisons of UBE3A transcript levels.

Additional file 4: Table S4. RNA-seq analysis of *UBE3A* allelic expression in opossum brain.

Additional file 5: Figure S1. Alignment of mouse, opossum, and human Ube3a/UBE3A amino acid sequences. (A) ClustalW alignment of amino acid sequences of human UBE3A isoform 3, mouse Ube3a isoform 2, and opossum UBE3A. (B) Percent identity values of pair-wise comparisons of UBE3A/Ube3a among human, mouse, and opossum. Values represent conservative, semiconservative, non-conserved, and insertion/deletion percent identity.

Additional file 6: Figure S2. (A-C) Immunofluorescence images of paternal Ube3a^{YFP} protein expression in the lateral ventricle of adult $Ube3a^{+/YFP}$ mice. (D) Immunofluorescence image of maternal Ube3a^{YFP} protein expression in the lateral ventricle of adult $Ube3a^{YFP/+}$ mice. (E) Immunofluorescence images of maternal Ube3a^{YFP} protein expression in the lateral ventricle, subventricular zone, rostral migratory stream, olfactory bulb, and adjacent cortical neurons in $Ube3a^{YFP/+}$ mice. (F) Immunofluorescence images of maternal Ube3a^{YFP} protein expression in the granular cell layer and subgranular zone of $Ube3a^{YFP/+}$ mice.

Additional file 7: Table S5. List of antibodies used for western blot and immunofluorescence analyses.

Abbreviations

UBE3A: ubiquitin protein E3A ligase gene; CNS: central nervous system; PTU: polycistronic transcription unit; UBE3A-AS: UBE3A antisense transcript; SNRPN-SNURF: SNRPN upstream reading frame—small nuclear ribonucleoprotein polypeptide N; snoRNA: small nucleolar RNAs; AS: Angelman syndrome; Dup15q: duplication 15q syndrome; PWS: Prader–Willi syndrome; ASD: autism spectrum disorder; Ube3aYFP: Ube3a yellow fluorescent protein; DPD: days post-differentiation; SVZ: subventricular zone; RMS: rostral migratory stream; SGZ: subgranular zone; PSA-NCAM: polysialylated-neural cell adhesion molecule; NES: nestin; DCX: doublecortin; RBFOX3: NA-binding protein, Fox-1 homolog 3; GFAP: glial fibrillary acidic protein.

Authors' contributions

PRH, SGBC and SVD designed the study; PRH, SGBC, KD, and XW performed experiments; PRH, SGBC, XW, RD, KK, NDC and SVD collected and analyzed data; PBS provided technical support and conceptual advice; and PRH, SGBC, PBS, and SVD wrote the manuscript. All authors read and approved the final manuscript.

Author details

[1] Department of Veterinary Pathobiology, College of Veterinary Medicine and Biomedical Sciences, Texas A&M University, College Station, TX 77845, USA. [2] Department of Molecular and Cellular Medicine, College of Medicine, Texas A&M Health Science Center, College Station, TX 77845, USA. [3] Interdisciplinary Genetics Program, College of Agriculture and Life Sciences, Texas A&M University, College Station, TX 77845, USA. [4] Department of Large Animal Clinical Sciences, College of Veterinary Medicine and Biomedical Sciences, Texas A&M University, College Station, TX, USA. [5] Institute for Genome Science and Society, Texas A&M University, College Station, TX 77845, USA. [6] Department of Veterinary Integrative Biosciences, College of Veterinary Medicine and Biomedical Sciences, Texas A&M University, College Station, TX 77843, USA. [7] Department of Molecular Biology and Genetics, Cornell University, Ithaca, NY 14853, USA. [8] Department of Veterinary Pathobiology, College of Veterinary Medicine and Biomedical Sciences, Texas A&M University, 4467 TAMU, College Station, TX 77843, USA.

Acknowledgements

We thank the Foundation for Angelman Syndrome Therapeutics for funding this project. We also thank Arline Rector for help with the cell-culture studies, Jean Kovar for help with maintaining the *Ube3a^YFP* and *Ube3a^-/+* mouse colonies, and Rola Barhoumi Mouneimne for help with the confocal imaging and quantification of immunofluorescence images. Confocal microscopy was performed in the Texas A&M University College of Veterinary Medicine & Biomedical Sciences Image Analysis Laboratory, supported by NIH-NCRR (1 S10 RR22532-01).

Competing interests

The authors declare that they have no competing interests.

Funding

Foundation for Angelman Syndrome Therapeutics.

References

1. Ager EI, Pask AJ, Gehring HM, Shaw G, Renfree MB. Evolution of the CDKN1C-KCNQ1 imprinted domain. BMC Evol Biol. 2008;8:163.
2. Babak T, DeVeale B, Tsang EK, Zhou Y, Li X, Smith KS, Kukurba KR, Zhang R, Li JB, van der Kooy D, et al. Genetic conflict reflected in tissue-specific maps of genomic imprinting in human and mouse. Nat Genet. 2015;47:544–9.
3. Baran Y, Subramaniam M, Biton A, Tukiainen T, Tsang EK, Rivas MA, Pirinen M, Gutierrez-Arcelus M, Smith KS, Kukurba KR, et al. The landscape of genomic imprinting across diverse adult human tissues. Genome Res. 2015;25:927–36.
4. Baroux C, Spillane C, Grossniklaus U. Genomic imprinting during seed development. Adv Genet. 2002;46:165–214.
5. Bressler J, Tsai TF, Wu MY, Tsai SF, Ramirez MA, Armstrong D, Beaudet AL. The SNRPN promoter is not required for genomic imprinting of the Prader–Willi/Angelman domain in mice. Nat Genet. 2001;28:232–40.
6. Chakraborty M, Paul BK, Nayak T, Das A, Jana NR, Bhutani S. The E3 ligase ube3a is required for learning in *Drosophila melanogaster*. Biochem Biophys Res Commun. 2015;462:71–7.
7. Chamberlain SJ, Brannan CI. The Prader–Willi syndrome imprinting center activates the paternally expressed murine Ube3a antisense transcript but represses paternal Ube3a. Genomics. 2001;73:316–22.
8. Charalambous M, Ferron SR, da Rocha ST, Murray AJ, Rowland T, Ito M, Schuster-Gossler K, Hernandez A, Ferguson-Smith AC. Imprinted gene dosage is critical for the transition to independent life. Cell Metab. 2012;15:209–21.
9. Colosi DC, Martin D, More K, Lalande M. Genomic organization and allelic expression of UBE3A in chicken. Gene. 2006;383:93–8.
10. Consortium GT. Human genomics. The Genotype-Tissue Expression (GTEx) pilot analysis: multitissue gene regulation in humans. Science. 2015;348:648–60.
11. Dindot SV, Antalffy BA, Bhattacharjee MB, Beaudet AL. The Angelman syndrome ubiquitin ligase localizes to the synapse and nucleus, and maternal deficiency results in abnormal dendritic spine morphology. Hum Mol Genet. 2008;17:111–8.
12. Fagerberg L, Hallstrom BM, Oksvold P, Kampf C, Djureinovic D, Odeberg J, Habuka M, Tahmasebpoor S, Danielsson A, Edlund K, et al. Analysis of the human tissue-specific expression by genome-wide integration of transcriptomics and antibody-based proteomics. Mol Cell Proteomics: MCP. 2014;13:397–406.
13. Ferguson-Smith AC, Surani MA. Imprinting and the epigenetic asymmetry between parental genomes. Science. 2001;293:1086–9.
14. Ferron SR, Radford EJ, Domingo-Muelas A, Kleine I, Ramme A, Gray D, Sandovici I, Constancia M, Ward A, Menheniott TR, et al. Differential genomic imprinting regulates paracrine and autocrine roles of IGF2 in mouse adult neurogenesis. Nat Commun. 2015;6:8265.
15. Galiveti CR, Raabe CA, Konthur Z, Rozhdestvensky TS. Differential regulation of non-protein coding RNAs from Prader–Willi syndrome locus. Sci Rep. 2014;4:6445.
16. Glessner JT, Wang K, Cai G, Korvatska O, Kim CE, Wood S, Zhang H, Estes A, Brune CW, Bradfield JP, et al. Autism genome-wide copy number variation reveals ubiquitin and neuronal genes. Nature. 2009;459:569–73.
17. Haig D. Intragenomic conflict and the evolution of eusociality. J Theor Biol. 1992;156:401–3.
18. Haig D. Parental antagonism, relatedness asymmetries, and genomic imprinting. Proc Biol Sci. 1997;264:1657–62.
19. Haig D. Genomic imprinting and kinship: how good is the evidence? Annu Rev Genet. 2004;38:553–85.
20. Hogart A, Wu D, LaSalle JM, Schanen NC. The comorbidity of autism with the genomic disorders of chromosome 15q11.2-q13. Neurobiol Dis. 2010;38:181–91.
21. Hope KA, LeDoux MS, Reiter LT. The *Drosophila melanogaster* homolog of UBE3A is not imprinted in neurons. Epigenetics. 2016;11:637–42.
22. Jana NR. Understanding the pathogenesis of Angelman syndrome through animal models. Neural Plast. 2012;2012:710943.
23. Jiang YH, Armstrong D, Albrecht U, Atkins CM, Noebels JL, Eichele G, Sweatt JD, Beaudet AL. Mutation of the Angelman ubiquitin ligase in mice causes increased cytoplasmic p53 and deficits of contextual learning and long-term potentiation. Neuron. 1998;21:799–811.
24. Kaneko-Ishino T. The regulation and biological significance of genomic imprinting in mammals. J Biochem. 2003;133:699–711.
25. Keane TM, Goodstadt L, Danecek P, White MA, Wong K, Yalcin B, Heger A, Agam A, Slater G, Goodson M, et al. Mouse genomic variation and its effect on phenotypes and gene regulation. Nature. 2011;477:289–94.
26. Kishino T, Lalande M, Wagstaff J. UBE3A/E6-AP mutations cause Angelman syndrome. Nat Genet. 1997;15:70–3.
27. Kohama C, Kato H, Numata K, Hirose M, Takemasa T, Ogura A, Kiyosawa

H. ES cell differentiation system recapitulates the establishment of imprinted gene expression in a cell-type-specific manner. Hum Mol Genet. 2012;21:1391–401.

28. Kriegstein A, Alvarez-Buylla A. The glial nature of embryonic and adult neural stem cells. Annu Rev Neurosci. 2009;32:149–84.

29. Landers M, Bancescu DL, Le Meur E, Rougeulle C, Glatt-Deeley H, Brannan C, Muscatelli F, Lalande M. Regulation of the large (approximately 1000 kb) imprinted murine Ube3a antisense transcript by alternative exons upstream of Snurf/Snrpn. Nucleic Acids Res. 2004;32:3480–92.

30. LaSalle JM, Reiter LT, Chamberlain SJ. Epigenetic regulation of UBE3A and roles in human neurodevelopmental disorders. Epigenomics. 2015;7:1213–28.

31. Martins-Taylor K, Hsiao JS, Chen PF, Glatt-Deeley H, De Smith AJ, Blakemore AI, Lalande M, Chamberlain SJ. Imprinted expression of UBE3A in non-neuronal cells from a Prader-Willi syndrome patient with an atypical deletion. Hum Mol Genet. 2014;23:2364–73.

32. Matsuura T, Sutcliffe JS, Fang P, Galjaard RJ, Jiang YH, Benton CS, Rommens JM, Beaudet AL. De novo truncating mutations in E6-AP ubiquitin-protein ligase gene (UBE3A) in Angelman syndrome. Nat Genet. 1997;15:74–7.

33. Meng L, Person RE, Beaudet AL. Ube3a-ATS is an atypical RNA polymerase II transcript that represses the paternal expression of Ube3a. Hum Mol Genet. 2012;21:3001–12.

34. Meng L, Person RE, Huang W, Zhu PJ, Costa-Mattioli M, Beaudet AL. Truncation of Ube3a-ATS unsilences paternal Ube3a and ameliorates behavioral defects in the Angelman syndrome mouse model. PLoS Genet. 2013;9:e1004039.

35. Moore T, Haig D. Genomic imprinting in mammalian development: a parental tug-of-war. Trends Genet: TIG. 1991;7:45–9.

36. Moreno-De-Luca D, Sanders SJ, Willsey AJ, Mulle JG, Lowe JK, Geschwind DH, State MW, Martin CL, Ledbetter DH. Using large clinical data sets to infer pathogenicity for rare copy number variants in autism cohorts. Mol Psychiatry. 2013;18:1090–5.

37. Noor A, Dupuis L, Mittal K, Lionel AC, Marshall CR, Scherer SW, Stockley T, Vincent JB, Mendoza-Londono R, Stavropoulos DJ. 15q11.2 duplication encompassing only the UBE3A gene is associated with developmental delay and neuropsychiatric phenotypes. Hum Mutat. 2015;36:689-93.

38. Numata K, Kohama C, Abe K, Kiyosawa H. Highly parallel SNP genotyping reveals high-resolution landscape of mono-allelic Ube3a expression associated with locus-wide antisense transcription. Nucleic Acids Res. 2011;39:2649–57.

39. O'Connell MJ, Loughran NB, Walsh TA, Donoghue MT, Schmid KJ, Spillane C. A phylogenetic approach to test for evidence of parental conflict or gene duplications associated with protein-encoding imprinted orthologous genes in placental mammals. Mamm Genome. 2010;21:486–98.

40. Pinto D, Pagnamenta AT, Klei L, Anney R, Merico D, Regan R, Conroy J, Magalhaes TR, Correia C, Abrahams BS, et al. Functional impact of global rare copy number variation in autism spectrum disorders. Nature. 2010;466:368–72.

41. Rapkins RW, Hore T, Smithwick M, Ager E, Pask AJ, Renfree MB, Kohn M, Hameister H, Nicholls RD, Deakin JE, et al. Recent assembly of an imprinted domain from non-imprinted components. PLoS Genet. 2006;2:e182.

42. Renfree MB, Suzuki S, Kaneko-Ishino T. The origin and evolution of genomic imprinting and viviparity in mammals. Philos Trans R Soc Lond B Biol Sci. 2013;368:20120151.

43. Rougeulle C, Glatt H, Lalande M. The Angelman syndrome candidate gene, UBE3A/E6-AP, is imprinted in brain. Nat Genet. 1997;17:14–5.

44. Runte M, Huttenhofer A, Gross S, Kiefmann M, Horsthemke B, Buiting K. The IC-SNURF-SNRPN transcript serves as a host for multiple small nucleolar RNA species and as an antisense RNA for UBE3A. Hum Mol Genet. 2001;10:2687–700.

45. Sahoo T, del Gaudio D, German JR, Shinawi M, Peters SU, Person RE, Garnica A, Cheung SW, Beaudet AL. Prader–Willi phenotype caused by

paternal deficiency for the HBII-85 C/D box small nucleolar RNA cluster. Nat Genet. 2008;40:719–21.

46. Schmittgen TD, Livak KJ. Analyzing real-time PCR data by the comparative CT method. Nat Protoc. 2008;3:1101–8.

47. Scoles HA, Urraca N, Chadwick SW, Reiter LT, Lasalle JM. Increased copy number for methylated maternal 15q duplications leads to changes in gene and protein expression in human cortical samples. Mol Autism. 2011;2:19.

48. Shetty AK. Progenitor cells from the CA3 region of the embryonic day 19 rat hippocampus generate region-specific neuronal phenotypes in vitro. Hippocampus. 2004;14:595–614.

49. Soellner L, Begemann M, Mackay DJ, Gronskov K, Tumer Z, Maher ER, Temple IK, Monk D, Riccio A, Linglart A, et al. Recent advances in imprinting disorders. Clin Genet. 2017;91:3–13.

50. Solter D. Differential imprinting and expression of maternal and paternal genomes. Annu Rev Genet. 1988;22:127–46.

51. Spencer HG, Clark AG. Non-conflict theories for the evolution of genomic imprinting. Heredity. 2014;113:112–8.

52. Stringer JM, Suzuki S, Pask AJ, Shaw G, Renfree MB. GRB10 imprinting is eutherian mammal specific. Mol Biol Evol. 2012;29:3711–9.

53. Sudmant PH, Alexis MS, Burge CB. Meta-analysis of RNA-seq expression data across species, tissues and studies. Genome Biol. 2015;16:287.

54. Surani MA. Genomic imprinting: developmental significance and molecular mechanism. Curr Opin Genet Dev. 1991;1:241–6.

55. Valluy J, Bicker S, Aksoy-Aksel A, Lackinger M, Sumer S, Fiore R, Wust T, Seffer D, Metge F, Dieterich C, et al. A coding-independent function of an alternative Ube3a transcript during neuronal development. Nat Neurosci. 2015;18:666–73.

56. Wang X, Douglas KC, Vandeberg JL, Clark AG, Samollow PB. Chromosome-wide profiling of X-chromosome inactivation and epigenetic states in fetal brain and placenta of the opossum, Monodelphis domestica. Genome Res. 2014;24:70–83.

57. Wang Y, Joh K, Masuko S, Yatsuki H, Soejima H, Nabetani A, Beechey CV, Okinami S, Mukai T. The mouse Murr1 gene is imprinted in the adult brain, presumably due to transcriptional interference by the antisense-oriented U2af1-rs1 gene. Mol Cell Biol. 2004;24:270–9.

58. Wilkins JF, Haig D. What good is genomic imprinting: the function of parent-specific gene expression. Nat Rev Genet. 2003;4:359–68.

59. Wilkins JF, Ubeda F, Van Cleve J. The evolving landscape of imprinted genes in humans and mice: conflict among alleles, genes, tissues, and kin. BioEssays. 2016;38:482–9.

60. Williams CA, Beaudet AL, Clayton-Smith J, Knoll JH, Kyllerman M, Laan LA, Magenis RE, Moncla A, Schinzel AA, Summers JA, et al. Angelman syndrome 2005: updated consensus for diagnostic criteria. Am J Med Genet Part A. 2006;140:413–8.

61. Wu Y, Bolduc FV, Bell K, Tully T, Fang Y, Sehgal A, Fischer JA. A Drosophila model for Angelman syndrome. Proc Natl Acad Sci USA. 2008;105:12399–404.

62. Yamasaki K. Neurons but not glial cells show reciprocal imprinting of sense and antisense transcripts of Ube3a. Hum Mol Genet. 2003;12:837–47.

63. Yi JJ, Berrios J, Newbern JM, Snider WD, Philpot BD, Hahn KM, Zylka MJ. An autism-linked mutation disables phosphorylation control of UBE3A. Cell. 2015;162:795–807.

64. Zhang YJ, Yang JH, Shi QS, Zheng LL, Liu J, Zhou H, Zhang H, Qu LH. Rapid birth-and-death evolution of imprinted snoRNAs in the Prader–Willi syndrome locus: implications for neural development in euarchontoglires. PLoS ONE. 2014;9:e100329.

6

Activation of the alpha-globin gene expression correlates with dramatic upregulation of nearby non-globin genes and changes in local and large-scale chromatin spatial structure

Sergey V. Ulianov[1,2†], Aleksandra A. Galitsyna[1,3,5†], Ilya M. Flyamer[1,2,9], Arkadiy K. Golov[1], Ekaterina E. Khrameeva[4,5], Maxim V. Imakaev[6], Nezar A. Abdennur[7], Mikhail S. Gelfand[3,4,5,8], Alexey A. Gavrilov[1] and Sergey V. Razin[1,2*]

Abstract

Background: In homeotherms, the alpha-globin gene clusters are located within permanently open genome regions enriched in housekeeping genes. Terminal erythroid differentiation results in dramatic upregulation of alpha-globin genes making their expression comparable to the rRNA transcriptional output. Little is known about the influence of the erythroid-specific alpha-globin gene transcription outburst on adjacent, widely expressed genes and large-scale chromatin organization. Here, we have analyzed the total transcription output, the overall chromatin contact profile, and CTCF binding within the 2.7 Mb segment of chicken chromosome 14 harboring the alpha-globin gene cluster in cultured lymphoid cells and cultured erythroid cells before and after induction of terminal erythroid differentiation.

Results: We found that, similarly to mammalian genome, the chicken genomes is organized in TADs and compartments. Full activation of the alpha-globin gene transcription in differentiated erythroid cells is correlated with upregulation of several adjacent housekeeping genes and the emergence of abundant intergenic transcription. An extended chromosome region encompassing the alpha-globin cluster becomes significantly decompacted in differentiated erythroid cells, and depleted in CTCF binding and CTCF-anchored chromatin loops, while the sub-TAD harboring alpha-globin gene cluster and the upstream major regulatory element (MRE) becomes highly enriched with chromatin interactions as compared to lymphoid and proliferating erythroid cells. The alpha-globin gene domain and the neighboring loci reside within the A-like chromatin compartment in both lymphoid and erythroid cells and become further segregated from the upstream gene desert upon terminal erythroid differentiation.

Conclusions: Our findings demonstrate that the effects of tissue-specific transcription activation are not restricted to the host genomic locus but affect the overall chromatin structure and transcriptional output of the encompassing topologically associating domain.

Keywords: Alpha-globin genes, Transcription, CTCF, Chromatin spatial structure, TAD, Chromatin compartment

*Correspondence: sergey.v.razin@usa.net
†Sergey V. Ulianov and Aleksandra A. Galitsyna have contributed equally to this work
2 Faculty of Biology, M.V. Lomonosov Moscow State University, Moscow, Russia 119992
Full list of author information is available at the end of the article

Background

In mammals, tissue-specific genes are often located within permanently active chromosome regions, which are enriched with genes ubiquitously transcribed in various cell types. The current paradigm suggests that tissue-specific transcriptional regulation within complex genome environment depends on a specific spatial organization of the interphase chromatin [1–3]. Non-random long-range contacts between remote regulatory sequences and promoters in mammalian genomes are often anchored by CTCF/cohesin-mediated chromatin loops [4] and preferentially occur within the same topologically associating domain (TAD) [5, 6]. TAD boundaries were found to be important for preventing abnormal enhancer–promoter communication [7, 8] and, consequently, delimiting zones of "licensed" enhancer influence, or regulatory domains [9].

The alpha-globin gene domain (AgGD) is defined here as a cluster of α-globin genes along with the remote regulatory elements. The AgGDs of warm-blooded vertebrates represent a canonical and arguably the most comprehensively studied example of a genomic locus where an array of tissue-specific (erythroid) genes is located within an extended cluster of housekeeping genes [10, 11]. The structure of AgGD is highly conserved among homeotherms. The domain comprises several alpha-globin genes transcribed in erythroid cells in a developmental stage-specific manner, and several enhancers whose exact number may vary slightly in different taxa [11]. A relatively long genomic region upstream of AgGD is syntenic in vertebrates [12]. The major regulatory element (MRE) of the AgGD, a strong and evolutionary conserved erythroid-specific enhancer [13, 14], is obligatorily located within an intron of the gene NPRL3 residing immediately upstream of the alpha-globin cluster. Although the AgGD is located in a permanently open (DNAse I-sensitive) highly acetylated chromatin in all cell types [15], full activation of the alpha-globin transcription in differentiated erythroid cells is accompanied by further histone hyperacetylation [16, 17] and reconfiguration of the local chromatin spatial structure within the domain. In these cells, MRE interacts with the adult alpha-globin gene promoters via direct looping, and this interaction is a prerequisite for proper development stage-specific expression of the adult alpha-globin genes [18, 19].

Previous studies have revealed that full activation of the AgGD in erythroid cells of different origin also has certain distant effects on surrounding transcription and chromatin structure. In primary human erythroid cells, the active status of the alpha-globin genes correlates with upregulation of the nearby NPRL3 gene and the distant non-related gene NME4, located 220 Kb away, which is activated via spatial interaction with the MRE [20]. In the human malignant erythroid cell line K562, as compared to lymphoid cells, transcription of the alpha-globin genes correlates with the decompaction of a 500-Kb chromosomal segment harboring the alpha-globin gene cluster, based on 5C data and polymer simulations [21]. In mouse primary erythroid cells, active alpha-globin genes may be recruited to a transcription factory formed by promoters of housekeeping genes located up to 70 Kb away [22]. Thus, it appears that the activation of globin genes and the assembly of the alpha-globin active chromatin hub occur simultaneously with the modification of the spatial structure of a genomic region at least several hundred kilobases in size.

Here, to get a systematic view on the effects of high-output AgGD activation on neighboring gene loci and chromatin structure, we studied a 2.7 Mb-region of chicken chromosome 14 harboring AgGD, 35 non-globin (predominantly housekeeping) genes, and a gene desert. We performed total rRNA-depleted transcriptome profiling to capture both genic and intergenic transcription, and applied high-resolution 3C, a new variant of Capture Hi-C (Chromatin TArget Ligation Enrichment, C-TALE), and large-scale 5C approaches coupled with a high-throughput analysis of CTCF binding to probe the spatial organization of the genome at the levels of CTCF-anchored loops, TADs, and compartments. We have found that transcription outburst of the alpha-globin genes in terminally differentiated erythroblasts is accompanied by (i) a substantial increase in the level of transcription of several adjacent housekeeping genes and intergenic regions, (ii) dramatic compaction of the encompassing sub-TAD amid the chromatin decompaction and massive attenuation of CTCF binding within the extended chromosome vicinity, and (iii) large-scale changes in the chromatin interaction profile at the level of chromatin compartments including segregation of AgGD from the transcriptionally silent gene desert. Our results suggest that activation of tissue-specific transcription may considerably affect adjacent non-related genes and chromatin folding of a large chromosomal segment. The opposite scenario where reconfiguration of an extended genomic domain enables remodeling of the local chromatin structure within AgGD also cannot be ruled out.

Results

Full activation of the alpha-globin gene expression is accompanied by the emergence of abundant intergenic transcription and upregulation of nearby non-globin genes

We used three cell types—cultured lymphoid DT40 cells, which are not expressing globins, and cultured erythroid HD3 cells that were either proliferating (committed to globin expression) or differentiated (actively expressing globins). To precisely track changes in the transcriptional profile during terminal erythroid differentiation, we performed sequencing of total rRNA-depleted transcriptomes of these cell types. Of note, although chicken erythroblasts are extensively used for studies of the genome biology and function [23–26], a whole-transcriptome analysis and comparison with lymphoblasts and between the differentiation stages, to our knowledge, has not been reported previously.

We performed cluster analysis, differential expression analysis and gene ontology annotation, and expectedly found that genome-wide differences in the transcriptional profile are more pronounced between HD3 and DT40 cells than between proliferating and differentiated HD3 cells (Fig. 1a; Additional file 1: Figure S1, panels a, b and c). The differential expression analysis reveals the upregulation (logFC > −0.6, FDR < 10^{-7}) of numerous erythroid genes in differentiated HD3 cells (Fig. 1a), including genes encoding transcription factors (*KLF1*, *GATA2*, *SCL*, *LMO2*, *NF-E2* and *FOG1*), enzymes involved in heme synthesis (mitochondrial ferrochelatase [*FECH*] and coproporphyrinogen oxidase [*CPOX*]) and cell surface markers (transferrin receptor [*TFRC*]). Gene ontology annotation reveals that the differentiation of HD3 cells is accompanied by a significant decrease in the expression of metabolism-related genes ($P < 10^{-18}$, Additional file 1: Figure S1d), indicating total repression of cellular biosynthetic processes, which recapitulates the normal erythrocyte maturation [27].

The studied genomic region harbors two gene-rich, highly transcribed zones (5′-terminal zone I and 3′-terminal zone III containing AgGD) separated by a 1.6-Mb gene desert (zone II) containing a predicted homologue of the human gene *RBFOX1* (Fig. 1b). In differentiated HD3 cells, about 50% of genes in the entire studied region display higher expression as compared to DT40 and proliferating HD3 cells (logFC > 0.6, FDR < 10^{-7}; Fig. 1; see normalized RNA-seq profiles in panel [b]). Importantly, non-globin genes, *NPRL3* and *TMEM8A*, located within the AgGD and involved in the formation of the alpha-globin active chromatin microcompartment (or active chromatin hub, ACH) [26, 28], show the highest degree

of upregulation among non-globin genes in differentiated HD3 cells (5.4-fold and 10.8-fold, respectively; Fig. 1c; Additional file 2: Table S1).

A remarkable feature of the transcriptome profile in the vicinity of AgGD is the presence of abundant intergenic transcription, especially in differentiated HD3 cells (Fig. 1d). The highest level of intergenic transcription is observed for regions between the *RHBDF1* promoter and the *MPG* promoter, and between *MRPL28* and *AXIN1* genes, i.e., within immediate neighborhoods of the AgGD. Within the alpha-globin gene cluster, intergenic transcription is detected exclusively in differentiated HD3 cells. In these cells, previous studies using RT-PCR and Northern blotting have demonstrated the presence of the so-called alpha-globin full-domain transcript, a long noncoding RNA presumably covering the region from its promoter within the *NPRL3* gene to the 3′-enhancer of AgGD [29–31]. Indeed, our RNA-seq data reveal the presence of low-level intergenic transcription from the *NPRL3* promoter to the *HBZ* (π) gene, and between *HBAD* (α^D) and *HBAA* (α^A) genes (Fig. 1d) in differentiated HD3 cells. Furthermore, in these cells, intergenic transcription is readily detectable between the 3′-enhancer of AgGD and the *TMEM8A* promoter. Systematic analysis reveals that transcription level within intergenic regions is substantially increased in differentiated HD3 cells genome-wide (Additional file 1: Figure S1e), the six intergenic regions from an immediate vicinity of AgGD being among the top 5% most upregulated.

Taken together, our data reveal general transcriptional upregulation and the presence of abundant intergenic transcription within and around AgGD in differentiated HD3 cells.

AgGD is located within A-like chromatin compartment in both lymphoid and erythroid cells, and is further spatially segregated from the gene desert upon erythroid differentiation

Transcription regulation and changes in chromatin epigenetic state upon erythroid differentiation were extensively studied in warm-blooded vertebrate alpha-globin

(See figure on next page.)
Fig. 1 Analysis of transcriptomes of the studied cell types. **a** Genome-wide variation of gene expression in the studied cell types (proliferating and differentiated HD3 cells are designated by HD3pr and HD3dif, respectively). Numbers of upregulated, downregulated, and ubiquitously transcribed genes for each pair of the cell types are shown at the *upper left corner* of the plots. The following genes are highlighted in the plots: (1) erythroid transcription factors GATA2, SCL (Tal1), FOG1, LMO2, NF-E2, Ldb1 and KLF1 (EKLF); (2) enzymes involved in heme synthesis including FECH and CPOX; (3) transferrin receptor (TFRC); and (4) the alpha-globin gene π (HBZ) and beta-globin gene (HBG2). BACH2 and EBF1 are lymphoid transcription factors. **b** Normalized profiles of total rRNA-depleted RNA-seq within the studied region. The functional AgGD from the 3′-end of the NPRL3 gene to the 3′-end of the TMEM8A gene is highlighted in *pink*. Alpha-globin genes are highlighted in *red*, and non-globin genes located within the AgGD are highlighted in *brown*. **c** Transcription level changes between cell types for all genes within the studied region. **d** The RNA-seq profile of the AgGD and closest neighbors. Intergenic transcription profiles are highlighted in *black*, and genic transcription profiles are highlighted in *gray*. Positions of genes from the Ensembl database are highlighted in *red*

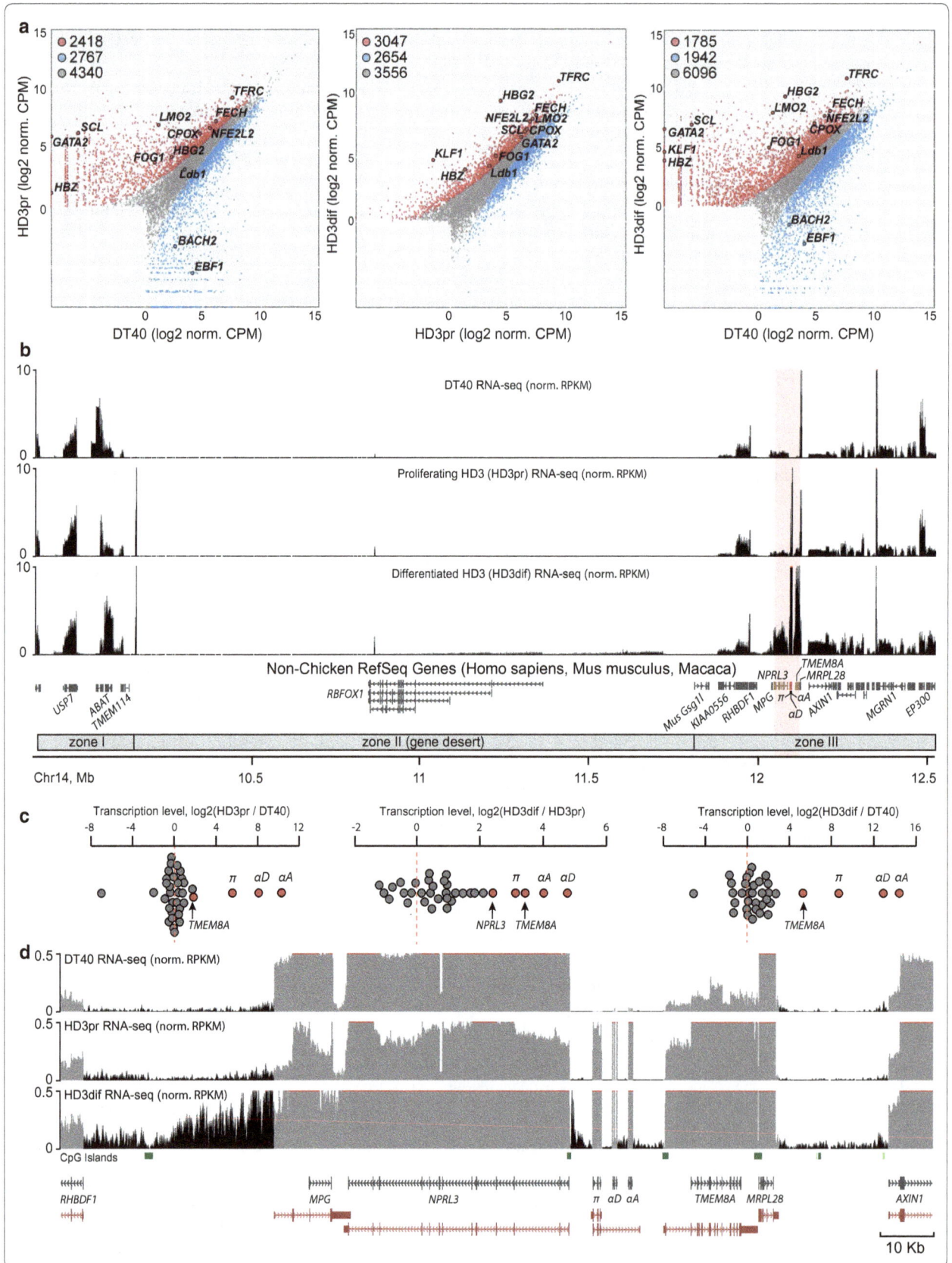

a

Left panel axes: HD3pr (log2 norm. CPM) vs DT40 (log2 norm. CPM)
Legend: 2418, 2767, 4340
Labels: SCL, GATA2, HBZ, FOG1, LMO2, CPOX, HBG2, Ldb1, NFE2L2, FECH, TFRC, BACH2, EBF1

Middle panel axes: HD3dif (log2 norm. CPM) vs HD3pr (log2 norm. CPM)
Legend: 3047, 2654, 3556
Labels: KLF1, HBZ, NFE2L2, LMO2, SCL, CPOX, FOG1, GATA2, Ldb1, HBG2, FECH, TFRC

Right panel axes: HD3dif (log2 norm. CPM) vs DT40 (log2 norm. CPM)
Legend: 1785, 1942, 6096
Labels: SCL, GATA2, KLF1, HBZ, FOG1, LMO2, HBG2, CPOX, Ldb1, NFE2L2, FECH, TFRC, BACH2, EBF1

b

DT40 RNA-seq (norm. RPKM)

Proliferating HD3 (HD3pr) RNA-seq (norm. RPKM)

Differentiated HD3 (HD3dif) RNA-seq (norm. RPKM)

Non-Chicken RefSeq Genes (Homo sapiens, Mus musculus, Macaca)

USP7 ABAT TMEM114   RBFOX1   Mus Gsg1l KIAA0556 RHBDF1 MpG NPRL3 π αA αD AXIN1 TMEM8A MRPL28 MGRN1 EP300

zone I   zone II (gene desert)   zone III

Chr14, Mb   10.5   11   11.5   12   12.5

c

Transcription level, log2(HD3pr / DT40)   −8 −4 0 4 8 12
π αD αA
TMEM8A

Transcription level, log2(HD3dif / HD3pr)   −2 0 2 4 6
π αA αD
NPRL3 TMEM8A

Transcription level, log2(HD3dif / DT40)   −8 −4 0 4 8 12 16
π αD αA
TMEM8A

d

DT40 RNA-seq (norm. RPKM)

HD3pr RNA-seq (norm. RPKM)

HD3dif RNA-seq (norm. RPKM)

CpG Islands

RHBDF1   MPG   NPRL3   π αD αA   TMEM8A MRPL28   AXIN1

10 Kb

gene domains [10, 11, 32–35]. However, large-scale spatial chromatin structure was precisely examined only in human AgGD [21]. Here, we performed 5C analysis of an extended region of the chicken chromosome 14 that harbors the AgGD. The experiments were conducted in two biological replicates for each cell type as described previously [36] with minor modifications (see "Methods" and Additional file 3: Figure S2). A total of 93 forward primers and 92 reverse primers were used (on average, one primer per 14.5 Kb, Additional file 4: Table S2), and 8556 unique pairwise chromatin interactions were detected within the selected area. The raw 5C data were binned at a 30 Kb resolution, iteratively corrected, and smoothed (see "Methods").

The 5C heatmaps show non-uniform chromatin compaction along the studied genomic region. A preliminary visual inspection of the heatmaps reveals that the major part of zone III harboring the AgGD interacts more intensively with zone I than with the gene desert (Fig. 2a; Additional file 5: Figure S3a). In mammals, active and repressed genomic regions are spatially segregated into two compartments, namely compartment A containing active genes, and compartment B harboring silent genes and gene deserts [37, 38]. Using principal component analysis [37], we annotated two compartment borders within the studied region (Fig. 2a, see "Methods"). Remarkably, the first compartment border is located exactly at the border between the highly transcribed gene-reach zone I and the gene desert, and coincides with the right boundary of the 5′-terminal TAD (genomic bin #11; see the next section for the description of TAD annotation). The second compartment border occurs at the genomic bin #73 approximately 60 Kb upstream of AgGD. To validate this observation, we constructed a plot featuring the average interaction profile of bins #74-91 located in zone III (a virtual "anchor") with all bins located upstream of this region. Indeed, the anchor region interacts with zone I more frequently than with the gene desert in all three cell types (Fig. 2b). Notably, a high interaction frequency between the gene-dense

zones of the studied region and their spatial separation from the gene desert is supported by our previously published results of 4C experiments with an anchor primer placed at the *NPRL3* promoter [39] (Additional file 5: Figure S3b). Taken together, these observations indicate that the studied region is partitioned into two chromatin compartments in both lymphoid and erythroid cells. The first compartment (negative values of the first principal component, Fig. 2a) consists of the zone I and the larger part of the zone III, which are gene-dense, contain many actively transcribed genes and intergenic regions, and are enriched with CpG-islands. We classified this compartment as A-like. The second compartment (positive values of the first principal component, Fig. 2a) is depleted in CpG-islands, and includes the gene desert and the poorly transcribed 5′-segment of the zone III, and thus may be considered as B-like.

Being located within the zone III, the AgGD belongs to the A-like chromatin compartment in both lymphoid and erythroid cells. To find out whether full activation of the AgGD affects its interaction frequency with the transcriptionally silent gene desert, we compared interaction frequency of the gene desert with the AgGD and with two control 60 Kb regions flanking the AgGD. Despite control regions being transcriptionally upregulated in differentiated HD3 cells along with globin genes (2.6- and 1.7-fold, respectively, as compared to proliferating cells), we observed a significant decrease in the interaction frequency with the gene desert exclusively for the AgGD ($P < 0.01$, one-tailed Wilcoxon's signed-rank test, Fig. 2c). Thus, transcription activation leads to further segregation of the AgGD from the gene desert.

Upregulation of transcription in zones I and III in differentiated HD3 cells is accompanied by a significant increase in interaction frequency between these two zones belonging to A-compartment ($P < 0.01$, one-tailed Wilcoxon's signed-rank test, Fig. 2d). Importantly, this fact cannot be explained by the general chromatin compaction during erythroid cell maturation, as the

(See figure on next page.)

Fig. 2 Analysis of A/B-like chromatin compartment structure of the studied genomic region. **a** Heatmap of the differentiated HD3 cells demonstrating an increased interaction frequency between gene-rich zones I and III. A-like and B-like chromatin compartments are outlined using *black rectangle* and *black stipple triangle*, respectively. The first principal component value is shown *below the heatmap*. Structural zones I (5′-terminal gene-rich area), II (gene desert), and III (3′-terminal gene-rich area) were selected manually based on the density of CpG-islands and annotated genes. **b** The averaged interaction profiles of 18 genomic bins #74-91 from zone III (virtual "anchor") with all bins from the remaining part of the studied region. *Circles* show average values of 5C counts between genomic bins #74-91 with each of the other bins in the studied region; standard deviation is shown. **c** 5C counts corresponding to interactions of the AgGD and flanking regions from the TAD T3 with the gene desert. *Thick black lines* represent median values. *Two asterisks* represent a significant difference with a *P*-value <0.01 calculated using a one-tailed Wilcoxon's signed-rank test. *Three asterisks* represent a significant difference with a *P*-value <0.001. The same notations are in panels (**d**) and (**e**). **d** The distributions of 5C counts within the A-like chromatin compartment. **e** 5C counts between genomic bins #74-91 and the remaining part of zone III and the gene desert. **f** 5C counts between four groups of bins from zones I and III (differentiated HD3 cells); bins were divided into groups according to their transcription level. **g** 5C counts between four groups of bins from zones I and III; bins were divided into groups according to their degree of the transcription level increase in the HD3dif cells as compared to the HD3pr cells

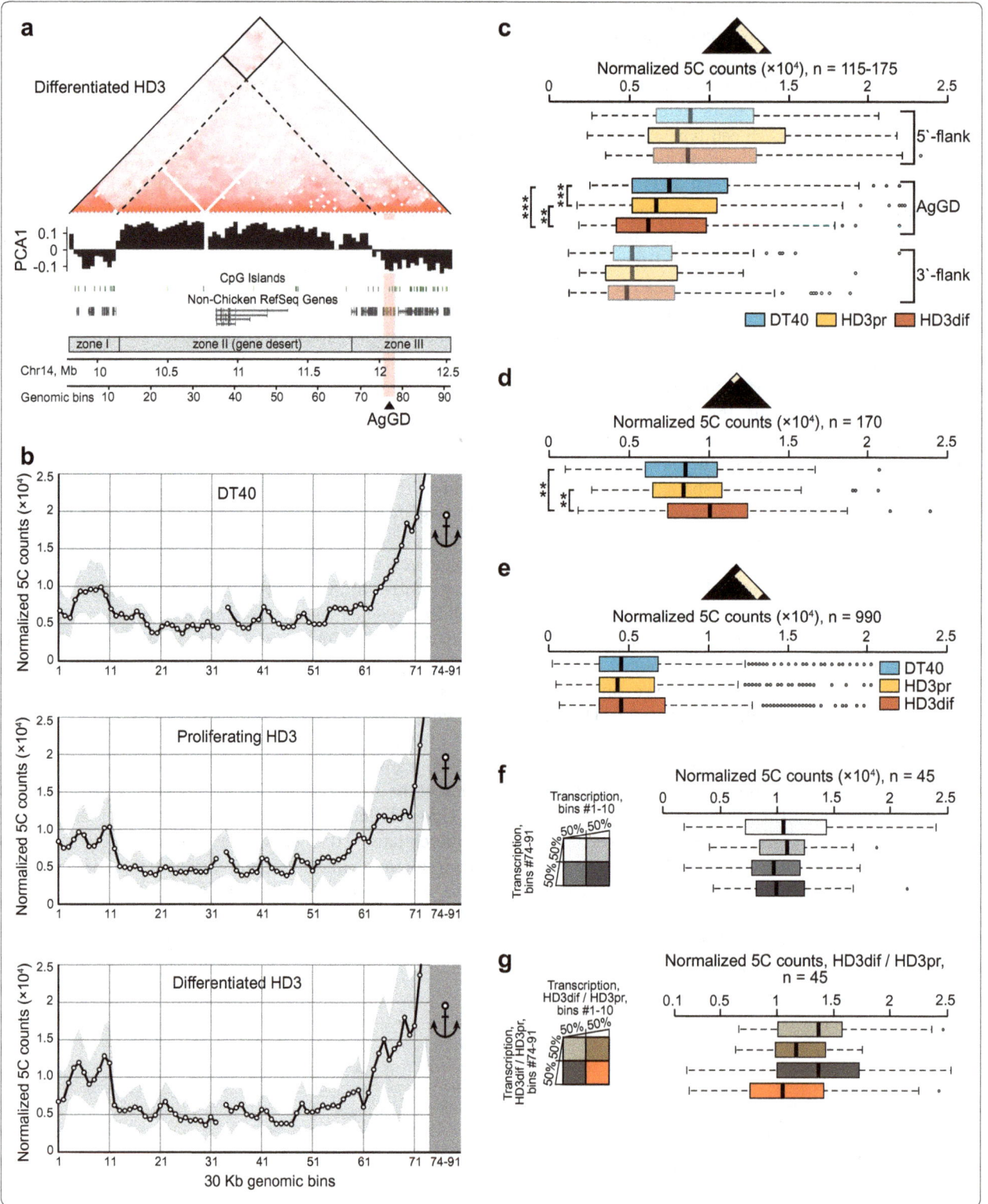

interaction frequency between the zone III and the gene desert is not increased (Fig. 2e). Based on these observations, one may propose that the internal structure of the A-like compartment is affected by the transcription level of the encompassed genes. However, analyzing pairwise interactions of the genomic bins within A-like compartment, we did not find any dependency of the interaction frequency on the level of transcription within the bins (see data for the differentiated HD3 cells in Fig. 2f). Moreover, we observe that the bins characterized by the highest increase of transcription upon HD3 differentiation and show the slightest (if any) increase in contact frequency with each other (orange boxplot in Fig. 2g). Consequently, augmentation of the contact frequency within the A-like compartment and increase in the transcription within zones I and III appear to be concomitant but not causatively related events during HD3 cell differentiation.

Finally, to visualize the overall 3D configuration of the studied region, we performed IMP polymer simulations [40]. The obtained 3D structures (Additional file 6: Figure S4) demonstrate that, in both lymphoid and erythroid cells, the studied region adopts a loop configuration allowing for juxtaposition of gene-rich zones I and III. Terminal differentiation of HD3 cells correlates with compression of the loop, resulting in an increase in interactions between these zones.

Full activation of alpha-globin gene transcription triggers compaction of the encompassing sub-TAD

In all three cell types, self-interacting chromatin domains are clearly visible in the studied region (Fig. 3). To accurately annotate positions of TADs, we used a dynamic programming segmentation algorithm [41] implemented in the Lavaburst package (see "Methods"). Partitioning of the studied region into TADs appears to be robust within a wide range of the parameter settings and largely conserved across the cell types (Additional file 7: Table S3; gamma-value 0.15 was used for the TAD boundary annotation in all cell types). Of note, our data provide the first evidence on the existence of TADs in the chicken interphase chromatin.

The comparison of the TAD positions with the previously reported CTCF ChIP-seq profiles [39] demonstrates high occupancy of CTCF binding sites (CBSs) within TADs located in gene-rich zones I and III in DT40 and proliferating HD3 cells. Notably, most of annotated CBSs in the 5′-terminal TAD contain reverse CTCF binding motifs (reverse CBSs; blue tailless arrows in Fig. 3), the majority of CBSs in the 3′-terminal TAD contain forward motifs (forward CBSs; red tailless arrows in Fig. 3), whereas the area containing AgGD and flanks predominantly contains both forward and reverse CBSs

which are always divergently oriented. Terminal differentiation of the HD3 cells is accompanied by a proliferation arrest and genome-wide depletion of CTCF binding [39]. Within the region of interest, there is an almost complete loss of CTCF ChIP-seq peaks within zone III and the gene desert. A partial loss of CTCF binding was observed inside the 3′- and 5′-terminal TADs (Fig. 3). Surprisingly, the decrease in CTCF binding sites occupancy does not lead to dramatic changes in the TAD profile: twelve out of fourteen TAD boundaries identified in proliferating HD3 cells are present in the differentiated cells (Fig. 3; Additional file 7: Table S3). Thus, there should be a mechanism preserving TAD structure in post-mitotic cells even when the CTCF binding is abrogated or substantially depleted.

In lymphoid DT40 cells, the area containing the AgGD and several neighboring genes (from the *RHBDF1* to the *RGS9*) is partitioned into two TADs that are arbitrary designated as T1 and T2 (Fig. 3). In proliferating erythroid HD3 cells, the boundary between these TADs is not annotated. The self-interacting domains T1 and T2 appeared fused into a large loose TAD T3 (Fig. 3). To examine the TAD profile around AgGD more in details, we have developed and applied C-TALE (Chromatin TArget Ligation Enrichment)—a novel cost-effective derivative of the Capture Hi-C technique [42, 43] (Fig. 4a). C-TALE procedure is based on the hybridization of a standard Hi-C library with biotin-labeled fragments of bacterial artificial chromosomes covering the locus of interest with subsequent biotin pull-down and deep sequencing of the trapped Hi-C junctions (see "Methods", Additional file 8: Supplementary Methods and Additional file 9: Table S5). C-TALE maps for a 735-Kb region centered at the AgGD show that this region is partitioned into eight clearly defined contact domains (CD), and their positions are completely conserved among the three cell types examined (Fig. 4b). Importantly, the boundaries of T1 and T2 TADs perfectly align with the boundaries of C-TALE-identified CDs (Figs. 4b, 5c). At the same time, in contrast to relatively low-resolution 5C data, high-resolution C-TALE analysis does not reveal measurable fusion of T1 and T2 TADs in HD3 cells. C-TALE analysis shows that AgGD-containing T2 TAD has a hierarchical structure and is comprised of four smaller CDs that can be considered as sub-TADs. In the three cell types studied, all genomic elements involved in the formation of the AgGD active chromatin microcompartment (the major enhancer MRE, the *NPRL3*-promoter, the alpha-globin genes, the 3′-enhancer of AgGD and the proximal part of the *TMEM8A* gene containing an erythroid-specific enhancer [28]) are located within a distinct sub-TAD (Fig. 5a). Activation of the alpha-globin gene transcription and the formation of loops between AgGD

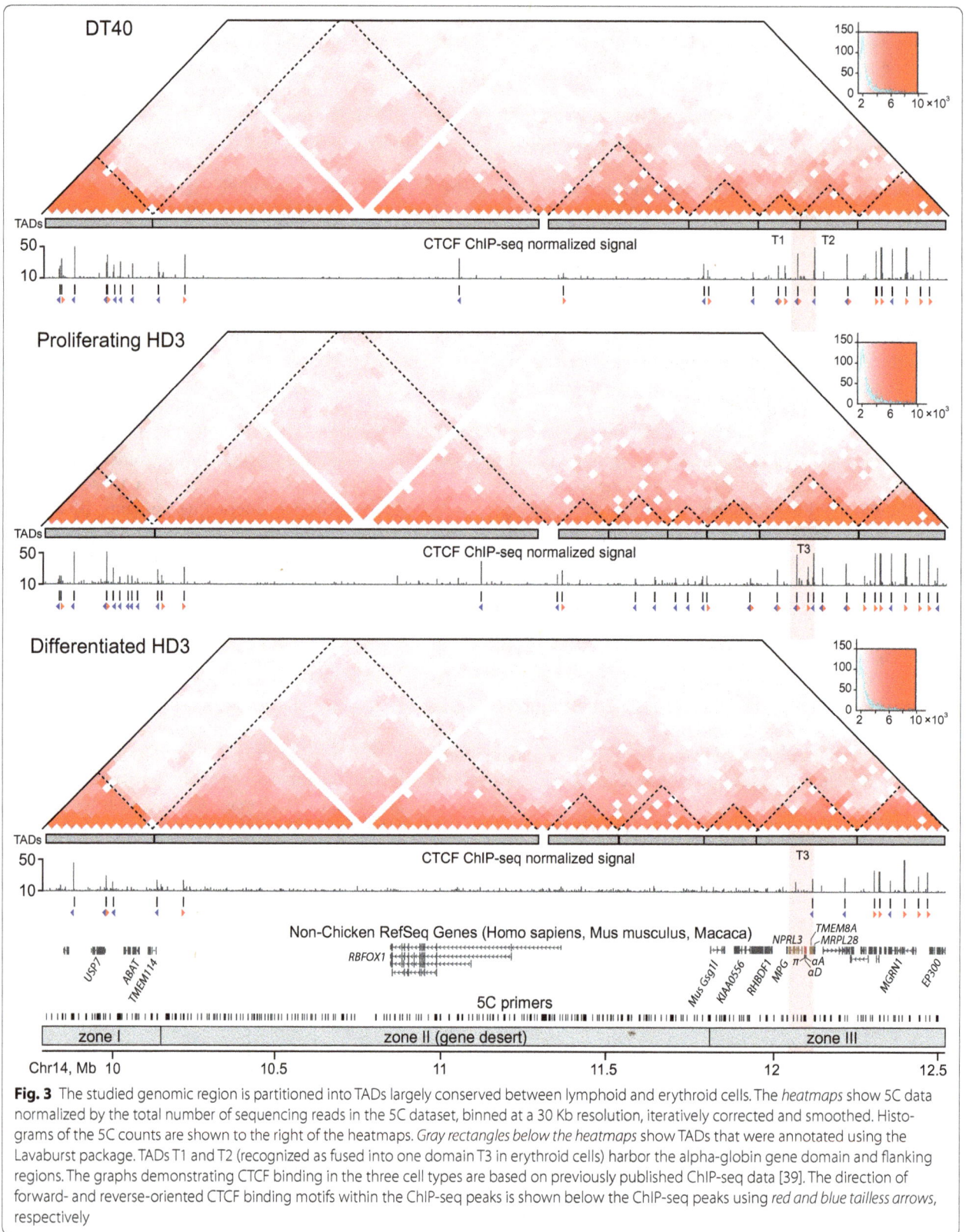

Fig. 3 The studied genomic region is partitioned into TADs largely conserved between lymphoid and erythroid cells. The *heatmaps* show 5C data normalized by the total number of sequencing reads in the 5C dataset, binned at a 30 Kb resolution, iteratively corrected and smoothed. Histograms of the 5C counts are shown to the right of the heatmaps. *Gray rectangles below the heatmaps* show TADs that were annotated using the Lavaburst package. TADs T1 and T2 (recognized as fused into one domain T3 in erythroid cells) harbor the alpha-globin gene domain and flanking regions. The graphs demonstrating CTCF binding in the three cell types are based on previously published ChIP-seq data [39]. The direction of forward- and reverse-oriented CTCF binding motifs within the ChIP-seq peaks is shown below the ChIP-seq peaks using *red and blue tailless arrows*, respectively

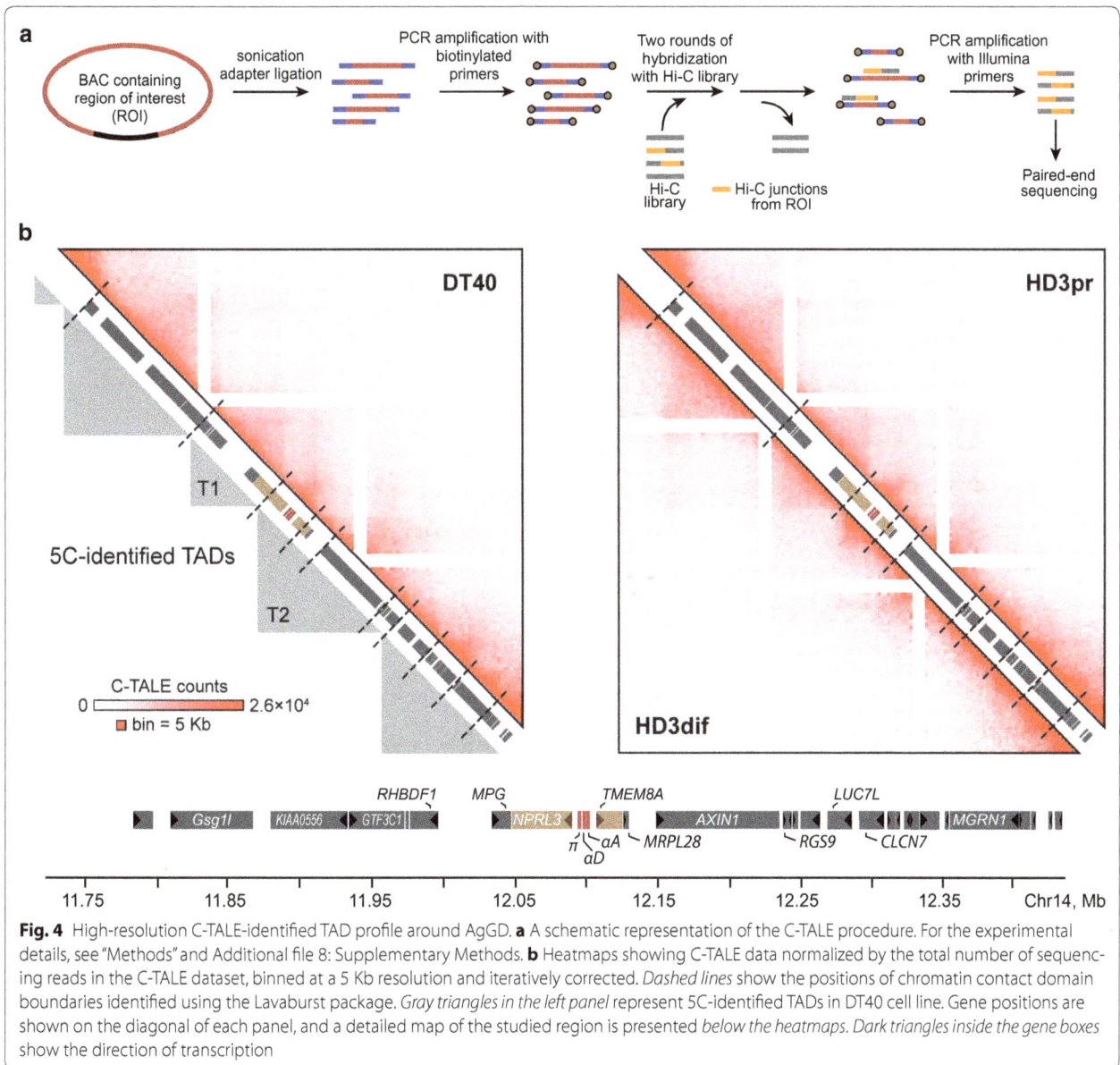

Fig. 4 High-resolution C-TALE-identified TAD profile around AgGD. **a** A schematic representation of the C-TALE procedure. For the experimental details, see "Methods" and Additional file 8: Supplementary Methods. **b** Heatmaps showing C-TALE data normalized by the total number of sequencing reads in the C-TALE dataset, binned at a 5 Kb resolution and iteratively corrected. *Dashed lines* show the positions of chromatin contact domain boundaries identified using the Lavaburst package. *Gray triangles in the left panel* represent 5C-identified TADs in DT40 cell line. Gene positions are shown on the diagonal of each panel, and a detailed map of the studied region is presented *below the heatmaps*. *Dark triangles inside the gene boxes* show the direction of transcription

functional elements [26] are accompanied by a significant compaction of this sub-TAD in differentiated HD3 cells ($P < 10^{-4}$, one-tailed Wilcoxon's signed-rank test, Fig. 5b, d). Along with it, upregulation of six out of seven nonglobin genes located within T1 and T2 TADs and almost complete loss of CTCF binding within these domains are accompanied by a remarkable depletion of middle- and long-range chromatin contacts within T1 and T2 (Fig. 5b, e, f), indicating a partial decompaction of an extended 300-Kb region around the alpha-globin gene cluster. To validate this observation, we performed 3D FISH with probes placed within T1 and T2 TADs and separated by 165 Kb (Fig. 5g). In agreement with the results of C-TALE analysis, we found that in differentiated HD3

cells the probes are separated from each other by a significantly larger distance than in DT40 ($P = 4.3 \times 10^{-4}$, Mann–Whitney rank test) and in proliferating HD3 cells ($P = 8.7 \times 10^{-3}$, Mann–Whitney rank test).

Alpha-globin MRE is involved in long-range looping interactions with CBSs outside of the alpha-globin locus

In the three cell types studied, the strong erythroid-specific enhancer of AgGD known as the major regulatory element (MRE) [44] is located in an immediate vicinity of the T1–T2 boundary (Fig. 5a). In DT40 cells, this boundary contains two closely positioned divergent CBSs located 3.5 Kb upstream of the MRE (designated as "−3.5 CBSs"; see Fig. 6a). In proliferating HD3 cells,

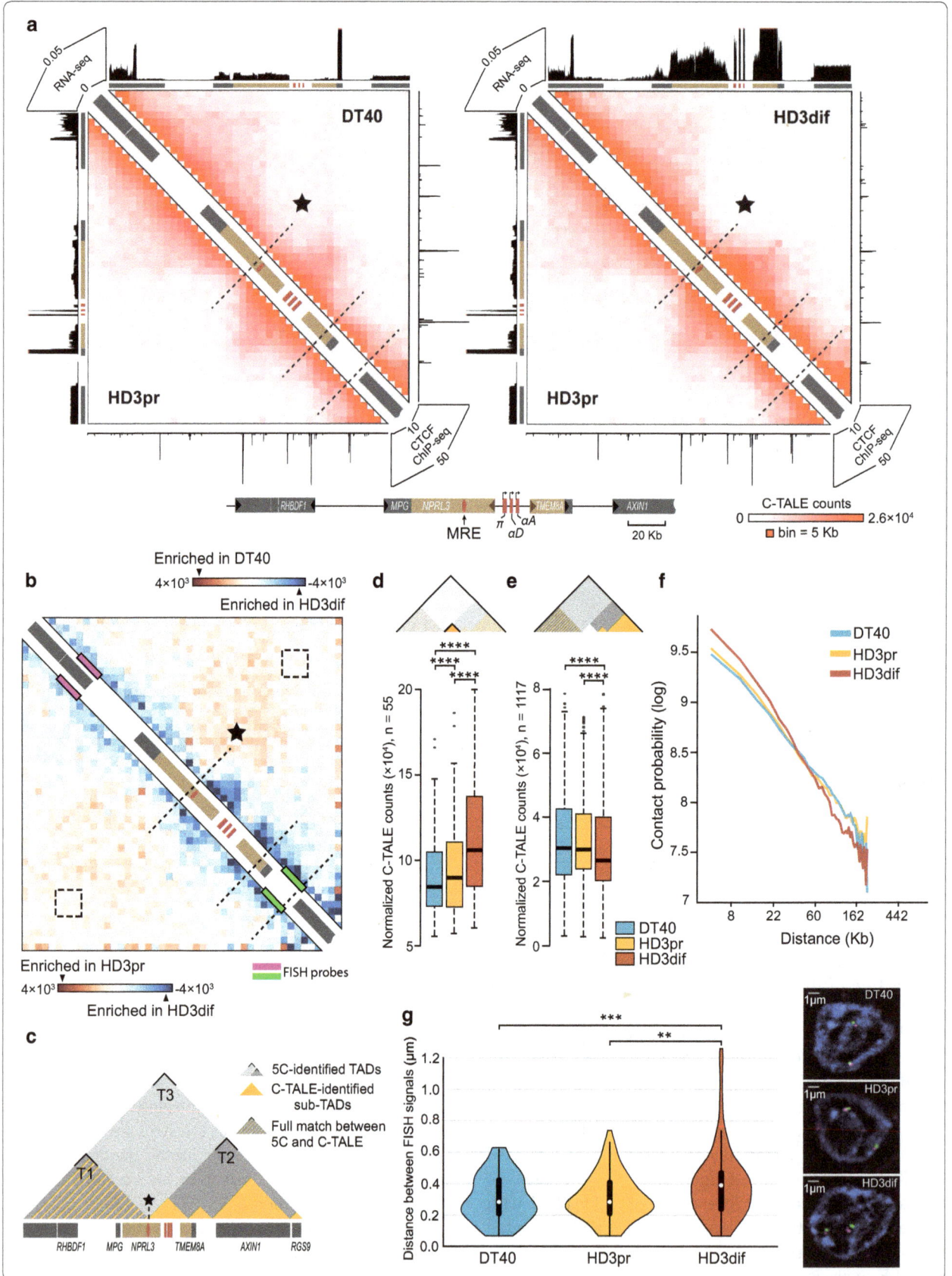

(See figure on previous page.)
Fig. 5 Activation of alpha-globin gene transcription is accompanied by local changes in chromatin interaction profile. **a** C-TALE heatmaps showing local chromatin interaction profile in a close vicinity to AgGD (region chr14:11957600-12236390). *Black star* denotes the boundary between T1 and T2 TADs. All other notations are as in Fig. 4b. **b** C-TALE heatmap showing interactions enriched in differentiated HD3 cells as compared to DT40 cells (right part of the map) and to proliferating HD3 cells (*left part of the map*). The positions of FISH probes are shown with *magenta and green rectangles*. Regions of the map corresponding to interactions between FISH probes are encircled with *dashed squares*. **c** Schematic representation C-TALE-identified sub-TADs within T1 and T2 TADs. **d** The distributions of C-TALE counts corresponding to chromatin interactions inside the sub-TAD harboring alpha-globin gene cluster. *Thick black lines* represent median values. *Asterisks* represent a significant difference with a *P*-value $<10^{-4}$ (one-tailed Wilcoxon's signed-rank test). **e** The distributions of C-TALE counts corresponding to chromatin interactions between genomic bins, separated by the distance more than 60 Kb throughout the shown region without including the sub-TAD harboring alpha-globin gene cluster. All notations are as in panel (**d**). **f** Scale-plots showing the dependency of contact probability on genomic distance within the shown region. **g** *Violin plots* showing the distributions of spatial distance between FISH probes. *White dots* represent the medians. *Asterisks* represent a significant difference with a *P*-value <0.01 (*two asterisks*), and with a *P*-value <0.001 (*three asterisks*) (Mann–Whitney rank-sum test). Representative examples of FISH images are shown on the *right*. *Scale bar* 1 μm

we observed an approximately twofold decrease in CTCF binding at the forward motif within the −3.5 CBSs, and both peaks are strongly reduced in differentiated HD3 cells (Fig. 6b). Considering that in DT40 and proliferating HD3 cells T1 and T2 TADs harbor multiple CBSs containing binding motifs in both orientations and that the majority of them are lost in differentiated HD3 cells, we assumed that the T1–T2 boundary could be involved in multiple cell type-specific looping interactions in DT40 and proliferating HD3 cells, but not in differentiated HD3 cells. To test this hypothesis, we performed a high-resolution 3C analysis with the anchor primers placed in DpnII restriction fragments harboring MRE and the −3.5 CBSs.

As anchor primers at −3.5 CBSs and at MRE are located close to each other, we expected that their 3C profiles would be similar. Indeed, in DT40 cells, both anchors interact strongly with the forward-orientated −42 CBS located 2 Kb upstream of the *MPG* gene (Fig. 6b, c). In both proliferating and differentiated HD3 cells (consistently with the depletion of CTCF binding at the −42 CBS), we observed a two- to threefold depletion of this interaction with both anchors. In a downstream direction relative to the anchor positions, in DT40 and proliferating HD3 cells both anchors exhibit strong looping with the reverse-oriented +46 CBS located within the *TMEM8A* gene. Unexpectedly, in lymphoid cells, the MRE interacts with the +46 CBS much stronger than the −3.5 CBSs, and that was confirmed in a reciprocal experiment with the anchor primer placed at the +46

CBS (Fig. 6d). Moreover, in proliferating HD3 cells, the interaction of the +46 CBS with the MRE considerably decreases, whereas the interaction with the −3.5 CBSs does not change. Taken together, these observations suggest that the MRE potentially forms a direct loop with the +46 CBS in both lymphoid and erythroid cells. Notably, the MRE/+ 46 CBS looping appears to be an alternative for the formation of the alpha-globin active chromatin hub, because our 3C analysis does not reveal reliable spatial interaction between the +46 CBS and the α^D promoter in any cell type.

Additionally, our 3C analysis does not show spatial contacts between the MRE and the promoters of non-globin genes *RHBDF1*, *MPG*, *TMEM8A*, and *MRPL28* flanking the alpha-globin gene locus (Fig. 6c). Thus, upregulation of the three former genes and downregulation of the *MRPL28* gene in differentiated HD3 cells is not related to looping interactions with MRE.

Discussion

Alpha-globin gene domains of warm-blooded animals are located within areas enriched in widely expressed genes [15, 45]. Here, we find that the majority of such genes surrounding the chicken alpha-globin cluster are substantially upregulated in erythroid cells induced to terminal differentiation (Fig. 1b, c; Additional file 2: Table S1). At first glance, such upregulation could be considered generally erythroid-specific, particularly for the *TMEM8A* that is slightly upregulated even in proliferating HD3 cells as compared to DT40 cells (Fig. 1c, left plot). However,

(See figure on next page.)
Fig. 6 DpnII 3C-analysis of chromatin contacts of the alpha-globin MRE and nearby CTCF-occupied region −3.5 CBSs. **a** Map showing the positions of genes and CTCF ChIP-seq peaks. Test and anchor regions are outlined with *gray and pink vertical rectangles*, respectively. CBSs are designated according to their distance (in kilobases) from the MRE. **b** 3C interaction profile of the −3.5 CBSs. Ligation frequencies averaged between biological replicates are shown. *Error bars* represent SEM. Anchor position relative to positions of test regions is outlined with *vertical dashed line*. *Horizontal gray line* represent a relative noise level ±SEM (see "Methods"). A closer view of the CTCF ChIP-seq profile of −3.5 CBS is shown above the diagram. **c** 3C interaction profile of the MRE. All notations are as in panel (**b**). **d** 3C interaction profile of the +46 CBS. All notations are as in panel (**b**). **e** Schematic representation of the CTCF-anchored loops observed around AgGD

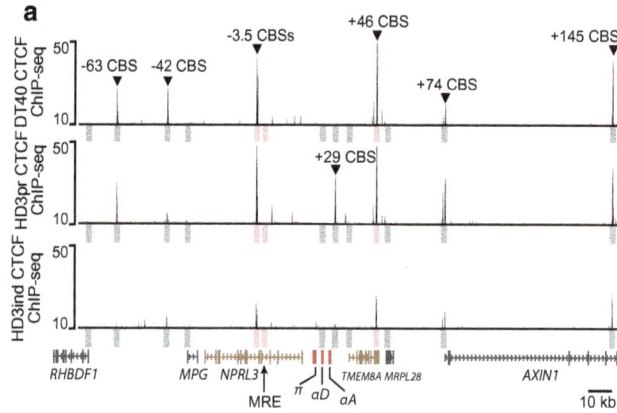

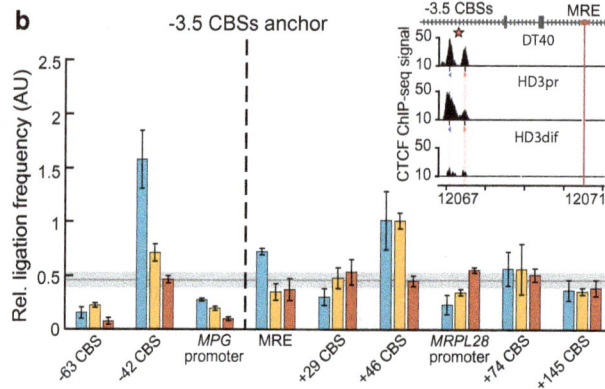

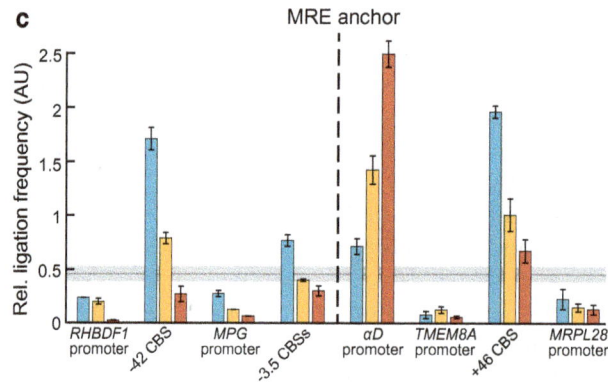

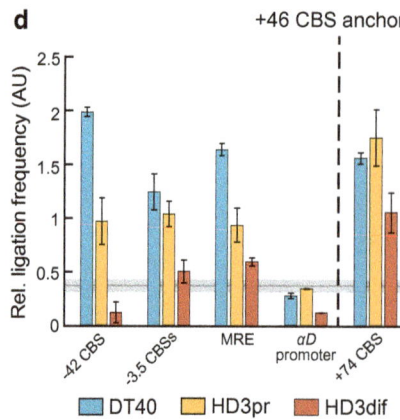

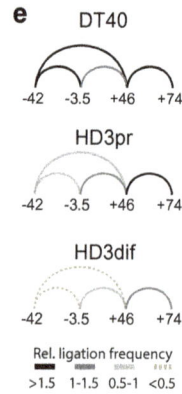

in the cell line K562, a human counterpart of the chicken HD3 cells, induction of differentiation leads to upregulation of the *NPRL3* gene but not *TMEM8A* [28, 46] which in the human genome is located in the neighboring TAD 200 Kb downstream of the AgGD [4]. Consistently, it has been previously suggested that upregulation of the chicken *TMEM8A* gene in differentiated erythroid cells is likely to be a consequence of it being in close proximity to AgGD enabling looping interactions between *TMEM8A* and alpha-globin-specific enhancers (−9DHS and 3′-enhancer) but not with the MRE [28] (the latter is supported by our data, Fig. 6c).

In contrast, the genomic position of *NPRL3* relative to the alpha-globin gene cluster and its involvement in the regulatory and spatial network of AgGD is conserved in chicken and mammals [18, 26]. The *NPRL3* promoter interacts with the MRE, and that could potentially cause its upregulation in differentiated erythroid cells, which, in turn, could have some function in the erythropoiesis. Although there are no studies of the role for the *NPRL3* protein product in the maturation of red blood cells, it has been shown that early defects of hematopoiesis in mice may be caused by the knock-out of the NPRL2, a paralog [47] and interacting partner of the NPRL3 within the GATOR1 complex regulating the mTOR pathway [48]. Moreover, *NPRL3* is transcribed at high level during normal in vitro differentiation of the primary human erythroid cells [20, 49]. Thus, transcriptional upregulation of the widely expressed gene *NPRL3* in differentiated erythroid cells is evolutionarily conserved across mammals and birds, and we propose that it might be functionally related to the definitive erythropoiesis in homeotherms.

Similarly to the *NPRL3*, genomic location of two more housekeeping genes *RHBDF1* and *MPG* upstream of the *NPRL3* is highly conserved in evolution of vertebrates [50]. Upregulation of these genes in differentiated HD3 cells (Additional file 2: Table S1) could be explained by some indirect mechanisms of stimulation rather than by immediate impact of MRE, as our 3C analysis shows that MRE does not spatially interact with their promoters (Fig. 6c). In this regard, abundant intergenic transcription covering the entire AgGD in differentiated HD3 cells might be of particular interest. It has been proposed that intergenic transcription promotes activation of genomic domains via facilitated spreading of histone acetyltransferases and chromatin remodeling complexes bound to the C-terminal tail of the elongating RNA polymerase II [51]. Possibly, this intergenic transcription contributes to the chromatin decompaction around AgGD (Fig. 5b) in differentiated HD3 cells that induces upregulation of adjacent genes via the increased chromatin accessibility to transcription factors.

At the same time, activation of alpha-globin gene transcription is accompanied by an extensive increase in chromatin interactions within the encompassing sub-TAD (Fig. 5b, d). In chicken erythroid cells, the alpha-globin locus is assembled into an active chromatin hub (ACH) [26]. Formation of this ACH comprised of several looping interactions between functional elements of the locus may explain the overall chromatin compaction within the entire sub-TAD. One of the first steps in the assembly of the alpha-globin ACH is the establishment of spatial contacts between the MRE, the *NPRL3* promoter, and the promoter of the α^D gene [26]. The results of our experiments suggest that, to participate in the assembly of ACH, the MRE should be first released from its interactions with the upstream (−42) and the downstream (+46) CBSs. It is tempting to speculate that the removal of CTCF from the T1–T2 boundary and weakening of the corresponding loops (Fig. 6a, e) triggers a spatial reconfiguration of the domain, allowing for the assembly of the alpha-globin ACH and reconfiguration of the encompassing sub-TAD. The spectrum of spatial contacts of the MRE in the lymphoid cells (and partially in proliferating HD3 cells) could be constrained by strong interactions of the upstream −3.5 CBS (Fig. 6b). In other words, when the −3.5 CBS located close to MRE is occupied by CTCF, MRE could preferentially contact the interaction partners of this CBS outside of AgGD. Such looping would restrict the possibility of establishing spatial contacts between the MRE and the downstream promoters of alpha-globin genes.

In differentiated HD3 cells, an almost complete loss of the CTCF binding is observed throughout the AgGD neighborhoods (Figs. 3, 5a). On the one hand, it reflects a general decrease in CTCF binding genome-wide—8768 and 4820 CBSs are detected in proliferating and differentiated HD3 cells, respectively (GSE51939 [39]). On the other hand, attenuation of CTCF binding within the T1 and T2 TADs is much more drastic than within two other gene-rich regions of the studied chromosome segment (5′- and 3′-terminal TADs) (Fig. 3). Numerous studies performed in human and murine cultured and primary cell lines have identified CTCF as a key regulator of TAD formation in mammals [4, 52–54]. However, we do not see the loss of TADs or dramatic changes in the TAD profile in differentiated HD3 cells in spite of dramatic loss of CTCF occupancy at the binding sites within the studied area (Fig. 3; Additional file 7: Table S3). Importantly, using the 3C analysis we demonstrate that the strength of all CTCF-anchored loops detected within the T1 and T2 TADs is considerably decreased in differentiated HD3 cells (Fig. 6e). Surprisingly, this does not lead to obvious changes in the sub-TAD profile in this region. These results are in agreement with recently reported observations, suggesting that induced degradation of

CTCF in mouse embryonic stem cells does not largely affect the positions of TAD boundaries [55]. Thus, some other additional mechanisms may be responsible for the maintenance of preformed TADs when the CTCF binding at CBSs is decreased. One possibility is that the tertiary structure of the chromatin fiber (self-interacting globule) that is initially established by the CTCF/cohesin-dependent looping is further stabilized and maintained by interactions between nucleosomes based on the mechanism of TAD formation in the *Drosophila* genome that we recently proposed [56]. To this end, it may be of importance that terminally differentiated HD3 cells do not divide. Consequently, the higher-order chromatin structure is not perturbed by mitosis.

In this study, we demonstrated for the first time the presence of TADs and A- and B-like compartments in chicken genome. Although the profiles of A- and B-like chromatin compartments within the studied genomic segment are similar if not identical in the three cell types studied, the contact frequency within the A-like compartment increases in differentiated HD3 cells (Fig. 2d). Surprisingly, we did not find any relationship between the increase in 5C counts and the increase in of transcription within the respective genomic bins (Fig. 2f, g). This observation argues that the nature of long-range interactions within chromatin compartments could be predominantly stochastic and determined by co-occurrence of chromatin domains within the same volume of nuclear space rather than by specific associations of highly transcribed genes. Consistently with the previously observed escape of upregulated genes from the B-compartment during human ESC differentiation [57], full activation of the alpha-globin transcription in differentiated HD3 cells is accompanied by a further segregation of the AgGD from the gene desert (Fig. 2c). Interestingly, other regions of the T1 and T2 TADs do not demonstrate changes in contact frequency with the gene desert, showing that different parts of the same TAD may possess a certain degree of autonomy in establishing long-range spatial contacts.

Conclusions

Taken together, our findings suggest that full activation of the chicken alpha-globin gene domain in differentiated erythroid cells correlates with local as well as large-scale changes in the chromatin folding and gene expression within an extended chromosome region. It contradicts simple models of the genome partitioning into functionally isolated gene domains sequentially located along the chromosome and argues that the effects of tissue-specific transcription activation are not restricted to the host genomic locus but affect the overall chromatin structure and transcriptional output of an extended genomic

neighbourhood. At the same time, our observations demonstrate that local reconfiguration of a tissue-specific gene domain may be guided by changes in the 3D packaging of an extended genomic region.

Methods
Cell culture

The avian erythroblastosis virus-transformed chicken erythroblast cell line HD3 (clone A6 of line LSCC) and the chicken lymphoid cell line DT40 (CRL-2111, ATCC, London, UK) were grown in Dulbecco's modified Eagle's medium (DMEM) supplemented with 2% chicken serum and 8% fetal bovine serum at 37 °C in 5% CO_2 atmosphere. To induce terminal erythroid differentiation, HD3 cells (8×10^5 cells/ml) were incubated in the full growth media supplemented with 10 mM HEPES (pH 8.0) and 20 μM iso-H-7 (1-(5-isoquinolinylsulfonyl)-3-methyl-piperazine dihydrochloride, Fluka, Seelze, Germany) at 42 °C in 100% air atmosphere.

3C library preparation

3C libraries were prepared as described previously [58] with minor modifications. Briefly, after fixation with 2% formaldehyde, the cells were lysed for 15 min in 1.5 ml of ice-cold isotonic lysis buffer (50 mM Tris–HCl (pH8.0), 150 mM NaCl, 0.5% (v/v) NP-40, 1% (v/v) Triton X-100 and 1× Protease Inhibitor Cocktail (Thermo Scientific #78430)). The nuclei were harvested, washed twice with 1.12× restriction enzyme buffer (NEB), and then treated with 0.3% SDS which was subsequently diluted to a concentration of 0.1% with 1.12× restriction enzyme buffer and quenched with 1.8% Triton X-100. DNA was digested overnight with 600 U of 100 U/μl of HindIII-HF restriction endonuclease (NEB) or 600 U of 50 U/μl of DpnII restriction endonuclease (NEB) at 37 °C with shaking. The next morning, 200 U of the corresponding restriction enzyme was added, and the samples were incubated for 2 h under the same conditions. Then, the samples were heated to 65 °C for 20 min to inactivate the restriction enzyme. The nuclei were washed twice with 100 μl of 1× T4 DNA ligase reaction buffer (Thermo Scientific) and resuspended in 300 μl of 1× T4 DNA ligase reaction buffer, 75 U of T4 DNA ligase (Thermo Scientific) was added, and the DNA was ligated for 6 h at 16 °C with constant shaking (1400 rpm). After cross-link reversal, RNAse A treatment, and single phenol–chloroform extraction followed by ethanol precipitation, the DNA was additionally purified using AMICON Ultra Centrifugal Filter Units (0.5 ml, 30 K, Millipore #UFC5030BK) to remove residual salts and DTT.

For the detection of the ligation products in DpnII-3C libraries, real-time PCR with TaqMan probes was employed. Primers and TaqMan probes for PCR analysis were designed using Primer Premier 5 software (PRIMER Biosoft International). The sequences of the primers and TaqMan probes are present in Additional file 10: Table S4. A total of 400 ng of the DpnII-3C library was used as a template in PCR. Calibration curves were constructed using four tenfold dilutions of the BAC standard template (200, 20, 2, and 0.2 pg of the BAC template supplemented with 50 ng of the chicken genomic DNA per PCR reaction) prepared from an equimolar mixture of the bacterial artificial chromosomes CH261-75C12 and CH261-85E12 (CHORI BACPAC). All 3C experiments were carried out in at least two independent biological replicates. Relative noise level in 3C plots was calculated as described in [59].

5C primer design

5C primers covering region 9.8–12.5 Mb of chicken chromosome 14 were designed for HindIII restriction fragments using the alternating scheme in my5C.primers [60] with the following parameters: (1) an optimal primer length of 30 bp; (2) an optimal Tm of 65 °C; and (3) a default primer quality parameters (mer: 800, U-blast: 3 and S-blast: 50). Primers were not designed for large (>50 Kb) and small (<100 bp) restriction fragments, low-complexity and repetitive sequences. Primers with sequence matches to more than one genomic target were discarded.

The universal T7-tail (5′-TAATACGACTCACTATAG CC-specific part-3′) and T3-tail (5′-TCCCTTTAGTGAG GGTTAATA-specific part-3′) were added to the forward and reverse 5C primers, respectively. 5′-ends of reverse 5C primers were phosphorylated. In total, 93 forward and 92 reverse primers were designed (Additional file 4: Table S2).

5C library preparation

5C libraries were prepared in two biological replicates as described previously [36] with minor modifications. Briefly, 5C primer stocks (17 mM) were mixed and diluted in water to a final concentration of 1.7 nM of each primer. 5C primers were mixed with 3C libraries in 10 µl of annealing mixtures of the following composition: 1 µl of 10× NEBuffer 4 (NEB), 1.7 µl of 5C primer mix, 600 ng of 3C library, 900 ng of salmon testis DNA (Sigma) sonicated to a size of 100–500 bp, and mQ water. Samples were denatured for 5 min at 95 °C and annealed for 16 h at 57 °C. Samples were supplemented with pre-heated 1× Taq DNA ligase buffer, and ligation with 10 U of Taq DNA ligase (NEB) was performed for 1 h at 57 °C. After that, the samples were quickly frozen at −80 °C. A total of 2 µl of each 5C library was then PCR-amplified individually with primers to T7- and T3-tails using KAPA High Fidelity DNA Polymerase (KAPA). The temperature

profile was 5 min at 98 °C, followed by 22, 24, 26, 28, and 30 cycles of 20 s at 98 °C, 15 s at 65 °C, and 20 s at 72 °C. The PCR reactions were separated on a 2% agarose gel supplied with ethidium bromide, and the number of PCR cycles necessary to obtain a sufficient amount of DNA was determined based on visual inspection of gels (typically 28–30 cycles). Four preparative PCR reactions were performed for each sample. The PCR mixtures were combined, and the products were purified with a QIAGEN PCR Purification Kit. The DNA was eluted with 50 µl of 10-mM Tris–HCl (pH 8.0) and separated on a 1.8% agarose gel supplied with ethidium bromide; 100-bp fragments were excised from the gel, purified using a QIAGEN Gel Extraction Kit and sequenced on Illumina HiSeq 2000 using single-end 101-nt reads.

C-TALE library preparation

For the experimental details, see Additional file 8: Supplementary Methods. Briefly, biotinylated hybridization baits were produced from BACs covering the region of chromosome 14 with coordinates 11721574-12436767 (galGal4 genome assembly) by ultrasonic fragmentation of BAC DNA to a size of 200–600 bp with subsequent adapter ligation and PCR amplification with biotinylated primers. Hi-C libraries were hybridized with the baits for 40 h at 65 °C, the trapped portion of the Hi-C libraries was PCR-amplified with Illumina PE1.0 and PE2.0 primers, and the products of amplification were subjected to the second round of hybridization with the same conditions. After that, trapped fragments of the Hi-C library were PCR-amplified with Illumina PE1.0 and PE2.0 primers and sequenced on Illumina HiSeq 2000 using paired-end 101-nt reads.

5C raw data processing and analysis, and CTCF binding motif search

Each 5C experiment was performed in two biological replicates (Additional file 3: Figure S2). 5C reads were checked for the presence of the HindIII restriction site and divided into two groups corresponding to forward and reverse primers in an interacting pair of primers. These read groups were mapped independently with bowtie2 [61] to the studied region of chicken chromosome 14 (genome assembly galGal4). To filter out low-coverage data, we excluded all primers with zero interactions in at least one cell type. The filtered data were binned in 30 Kb windows and the mean 5C count per bin was calculated, resulting in a matrix of interactions between forward and reverse primers. Bins corresponding to interactions within the same fragment were removed. We then applied the iterative correction procedure [38] to reduce intrinsic bin-specific biases. Because biological replicates demonstrated high correlation for all three cell types (Pearson's correlation coefficient >0.9, Additional file 3: Figure S2), we merged them by summing their matrices element by element. To reduce noise and achieve better visualization, we smoothed the data by assigning each matrix element the median 5C count between this element and its adjacent elements. Zero matrix elements were not smoothed to avoid inflating low-confidence interaction values.

TADs were identified using the Lavaburst package (https://github.com/mirnylab/lavaburst), which provides a set of dynamic programming algorithms to assess an ensemble of TAD segmentations derived from a TAD scoring function. We used its optimal segmentation finder, which is based on the Armatus algorithm [41] using the TAD scoring function from that study. The algorithm finds the global TAD segmentation of a contact map having the highest aggregate score.

Chromatin compartments were identified using the principal component analysis, as described previously [37].

A CTCF motif search was performed using the Biopython motif search method [62] and the human CTCF position-frequency matrix from the JASPAR database [63].

C-TALE raw data processing and analysis

Illumina HiSeq 2500 bcl2fastq v1.8.4 Conversion Software was used for base-calling. The reads were mapped independently to the region 11700000-12450000 of chromosome 14 using hiclib [38] (https://bitbucket.org/mirnylab/hiclib/) iterative mapping procedure (minimal mapping length 25, iteration step 5) and Bowtie 2 version 2.2.1.

Reads mapped in close proximity to DpnII restriction sites (5 bp) and possible PCR duplicates were eliminated. We also excluded reads mapped to extremely short (<100 bp) and long (>100 Kb) restriction fragments. The reads in pairs that mapped to the same restriction fragment or closer than 500 bp to each other in genomic coordinates were excluded as potential self-ligations and dangling end products. Uniquely mapped read pairs that passed filtering procedure were retained. Resulting pairs were binned into 5 Kb genomic windows of the region of interest. Bins corresponding to interactions within the same fragment were removed. For each bin, the mean C-TALE read count was calculated. To merge C-TALE replicates, we summed their matrices element by element. Then, the matrices were iteratively corrected [38]. TAD annotation was performed with the Lavaburst package as described for 5C. To minimize the effect of missing data on TADs annotation procedure, we removed bins with no interactions from the matrices prior to TAD annotation.

3D DNA fluorescence in situ hybridization (3D FISH)

The cells were harvested overnight on poly-L-lysine coated coverslips placed in culture flasks. The cells were fixed in 4% paraformaldehyde for 10 min, permeabilized in 0.5% Triton X-100, washed in PBS, dehydrated in ethanol series, air-dried, stored at room temperature for two days, and then frozen at −80 °C. Probes were PCR-amplified from the CH261-75C12 BAC, primer sequences are presented in Table S4 (Additional file 10). Probes were labeled with biotin or digoxigenin using nick-translation. Approximately 150 ng of each probe was used in hybridization. Denaturation was performed at 80 °C for 30 min in 70% formamide (pH7.5), 2xSSC. Hybridization of probes was done for 24 h in 50% formamide, 2x SSC, 10% dextran sulfate, 1% Tween 20. Washing steps were performed in 2x SSC at 45 °C followed by 0.1x SSC at 60 °C and 4x SSC, 0.1% Triton X-100. For imaging, cells were counterstained with DAPI and epifluorescent images were acquired using a Photometrics Coolsnap HQ2 CCD camera and a Zeiss Imager A1 fluorescence microscope with a Plan Apochromat 100x 1.4NA objective, a Lumen 200-W metal halide light source (Prior Scientific Instruments, Cambridge, UK) and Chroma #89014ET single excitation and emission filters (Chroma Technology Corp., Rockingham, VT) with the excitation and emission filters installed in Prior motorized filter wheels. A piezoelectrically driven objective mount (PIFOC model P-721, Physik Instrumente GmbH & Co, Karlsruhe) was used to control movement in the z dimension. Hardware control, image capture, deconvolution (using the fast algorithm) and analysis were performed using Volocity (PerkinElmer Inc, Waltham, MA).

Polymer simulations

We used the integrative modeling platform (IMP) [40] to model the 3D-chromatin structure from the 5C data as described previously (24). We assigned a particle to each 30 Kb bin in smoothed 5C matrices. To use all of the information from the 5C experiment, we set parameters uZ and lZ to zeros; the mP parameter was set to 400 nm. A total of 5000 IMP runs were performed, and one thousand best models were used for further analysis. As a measure of the model quality, we used the Spearman's correlation coefficient between the modeled matrix of interactions and the experimental 5C matrix (Additional file 6: Figure S4). The TADbit program (https://github.com/3DGenomes/tadbit) was used for all calculations. Chimera [64] and PyMOL (The PyMOL Molecular Graphics System, version 1.7.4 Schrodinger, LLC), tools were used to visualize the models. We used the centroid model for the wireframe representation, and the Gaussian approximation for the surface representation.

Total rRNA-depleted RNA-seq: experimental procedures and analysis

RNA-seq experiments were performed in two independent biological replicates, according to ENCODE recommendations (https://genome.ucsc.edu/ENCODE/protocols/dataStandards/ENCODE_RNAseq_Standards_V1.0.pdf). The two biological replicates demonstrated a high correlation in all three cell types (Pearson's $r > 0.95$).

Total RNA was extracted from the cells using TRIzol Reagent (Thermo Scientific) according to the manufacturer's protocol. Total RNA was depleted with ribosomal RNA using the RiboZero Human/Mouse Kit (Illumina). For library preparation, a TruSeq Stranded RNA Sample Prep Kit (Illumina) was used following the manufacturer's instructions. After preparation, the quality of the libraries was evaluated using quantitative PCR and Bioanalyzer.

Illumina Casava 1.8.2 was used for base-calling. Reads were mapped to the *Gallus gallus* reference genome (version galGal4) using the TopHat program [65] taking the ENSEMBL gene annotation into account (Ensembl Release 82) with the -no-novel-juncs option. The Htseq-counts program [66] was used to calculate per gene counts. The BedTools program [67] was used to calculate read coverage.

We used R package edgeR for gene expression analysis. Only genes with CPM > 1 in at least two out of six RNA-seq experiments were retained. We averaged CPM over replicates and calculated fold change of expression between experiments. *P*-values were calculated per gene using quasi-likelihood negative binomial generalized log-linear model (glmQLFit function in edgeR). The *P*-values were adjusted with Benjamini–Hochberg correction for multiple testing. The threshold for false discovery rate was set to 1e-7 for detection of differentially regulated genes. For gene ontology analysis, we used DAVID as implemented in RDAVIDWebService R package. We assessed the enrichment of gene ontology terms of "Biological Process" category for the genes that changed expression level more than twofold with adjusted *P*-value <1e-04. The minimal number of genes present in the gene ontology category was set to 5. Resulting DAVID *P*-values of enriched terms were adjusted with Bonferroni correction.

To assess the intergenic transcription levels, we retrieved the regions between ENSEMBL genes and calculated read coverage with BedTools software [67]. To compare the transcription level between experiments for each intergenic loci, log2-fold change of coverage was calculated. We obtained the distribution of fold change between experiments for all intergenic regions and assessed the percentile rank of loci within the studied region.

Additional files

Additional file 1: Figure S1. Analysis of the total rRNA-depleted RNA-seq data in the three studied cell types. **(A)** The Venn diagram showing the numbers of active genes shared between the studied cell types. **(B)** Cluster analysis of biological replicates based on total rRNA-depleted RNA-seq data. **(C)** Principal component analysis of the RNA-seq data. **(D)** Top-30 gene ontology terms for the genes downregulated (logFC < −0.6, FDR < 10^{-7}) in differentiated HD3 cells compared to proliferating HD3 cells. The number of genes in each term is shown. **(E)** Scatter plots showing normalized level of transcription within intergenic regions genome-wide in the studied cell types.

Additional file 2: Table S1. Transcription level of the genes from the studied region (CPM). Only genes with CPM > 1 in at least 2 replicates are presented. CPM values were averaged over replicates. Log2-fold change and FDR of difference between experiments are presented (see "Methods"). For genes that have not passed filtering procedure, only CPM are present.

Additional file 3: Figure S2. Biological replicates of the 5C experiment. Color intensity in each cell of the heatmaps represents the interaction frequency of two corresponding forward and reverse 5C primers. Histograms of the interaction counts are shown to the right of the heatmaps.

Additional file 4: Table S2. Sequences and characteristics of the 5C primers used in this study. 5C primers were designed using the alternating scheme in my5C.primers and chicken reference genome assembly galGal4.

Additional file 5: Figure S3. The chromatin compartment profile along the studied genomic region. **(A)** Heatmaps demonstrating an increased interaction frequency between gene-rich zones I and III in both lymphoid and erythroid cells. A-like and B-like chromatin compartments are outlined using black opened rectangle and black dotted triangle, respectively. The first principal component is shown below the heatmaps. **(B)** The 4C profiles from [39] revealing an increased interaction frequency of the alpha-globin gene domain (zone III of the studied region) with the USP7-ABAT locus (zone I), and the spatial separation of the gene desert from flanking gene-dense areas in proliferating HD3 and DT40 cells. Positions of the anchor primers are shown using red vertical lines; the alpha-globin gene domain is highlighted with a vertical pink rectangle. The first principal component (PCA1) of the 5C data (DT40 cells) is shown below the 4C profiles. The blue rectangle below the genomic coordinates track represents the genomic region analyzed in this work. The scale is changed at the bottom of the panel to emphasize the borders of 4C counts peak around the anchor. Positions of TADs that were identified using the 5C analysis are highlighted using gray rectangles.

Additional file 6: Figure S4. (A) Interaction heatmaps of the simulated polymer calculated with TADbit. Color bars in the heatmaps represent arbitrary partitioning of the studied region into several distinct fragments. **(B)** IMP-derived 3D models of the studied genome region. The central wireframe colored as the color bar in the map of the studied region represents the centroid model for simulations. The surface represents the Gaussian approximation of 1000 simulated models.

Additional file 7: Table S3. Genomic coordinates of topologically associating domains within the studied region. TADs were identified using the optimal segmentation algorithm from the Lavaburst package with the with the Armatus scoring function.

Additional file 8: Supplementary Methods. A detailed description of the C-TALE library preparation.

Additional file 9: Table S5. C-TALE sequence statistics.

Additional file 10: Table S4. Sequences of primers and TaqMan probes used for the 3C analysis. Sequences of primers used for FISH probe amplification.

Abbreviations
3C: chromosome conformation capture; 3D: three-dimensional; 5C: chromatin conformation capture carbon copy; ACH: active chromatin hub; AgGD: alpha-globin gene domain; bp: base pair; CBS: CTCF-binding site; ChIP-seq: chromatin immunoprecipitation followed with deep sequencing; C-TALE: Chromosome TArget Ligation Enrichment; ESC: embryonic stem cells; IMP: integrative modeling platform; Kbp: kilobases, thousands of base pairs; Mb: megabases, millions of base pairs; MRE: major regulatory element; PCA: principal component analysis; TAD: topologically associating domain.

Authors' contributions
SVU, IMF, and AKG performed the experiments; AAGal, EEK, IMF, and MSG performed processing of 5C, C-TALE, RNA-seq, and ChIP-seq data; AAGal and EEK performed polymer simulations; AAGal, EEK, and SVU performed statistical analysis; NAA and MVI developed the Lavaburst package; SVR and AAGav conceived the study; SVU, SVR, and AAGav wrote the manuscript with the input from all other authors. All authors read and approved the final manuscript.

Author details
[1] Institute of Gene Biology of the Russian Academy of Sciences, Moscow, Russia 119334. [2] Faculty of Biology, M.V. Lomonosov Moscow State University, Moscow, Russia 119992. [3] Faculty of Bioengineering and Bioinformatics, M.V. Lomonosov Moscow State University, Moscow, Russia 119992. [4] Skolkovo Institute of Science and Technology, Skolkovo, Russia 143026. [5] Institute for Information Transmission Problems (the Kharkevich Institute) of the Russian Academy of Sciences, Moscow, Russia 127051. [6] Department of Physics, Massachusetts Institute of Technology, Cambridge, MA 02139, USA. [7] Computational and Systems Biology Graduate Program, Massachusetts Institute of Technology, Cambridge, MA, USA. [8] Faculty of Computer Science, Higher School of Economics, Moscow, Russia 125319. [9] Present Address: MRC Human Genetics Unit, Institute of Genetics and Molecular Medicine, University of Edinburgh, Edinburgh, UK.

Acknowledgements
We thank Sergei Isaev and Alena Potapenko who performed preliminary CTCF motif and ChIP-seq analysis during the Summer Molecular and Theoretical Biology School in Pushchino, 2015. We thank Shelagh Boyle and the staff of the Institute of Genetics and Molecular Medicine (The University of Edinburgh) imaging facility for assistance with FISH imaging and analysis.

Competing interests
The authors declare that they have no competing interests.

Funding
This work was supported by the Russian Science Foundation (RSF) (Grant Numbers 14-24-00022, 14-24-00155).

References
1. Ulianov SV, Gavrilov AA, Razin SV. Nuclear compartments, genome folding, and enhancer–promoter communication. Int Rev Cell Mol Biol. 2015;315:183–244.
2. Vernimmen D, Bickmore WA. The hierarchy of transcriptional activation: from enhancer to promoter. Trends Genet. 2015;31:696–708.
3. Dekker J, Mirny L. The 3D genome as moderator of chromosomal communication. Cell. 2016;164:1110–21.
4. Rao SS, Huntley MH, Durand NC, Stamenova EK, Bochkov ID, Robinson JT, et al. A 3D map of the human genome at kilobase resolution reveals principles of chromatin looping. Cell. 2014;159:1665–80.
5. Mifsud B, Tavares-Cadete F, Young AN, Sugar R, Schoenfelder S, Ferreira L, et al. Mapping long-range promoter contacts in human cells with high-resolution capture Hi-C. Nat Genet. 2015;47:598–606.
6. Remeseiro S, Hornblad A, Spitz F. Gene regulation during development in the light of topologically associating domains. Wiley Interdiscip Rev Dev Biol. 2015;5:169–85.
7. Lupianez DG, Kraft K, Heinrich V, Krawitz P, Brancati F, Klopocki E, et al. Disruptions of topological chromatin domains cause pathogenic rewiring of gene–enhancer interactions. Cell. 2015;161:1012–25.
8. Valton AL, Dekker J. TAD disruption as oncogenic driver. Curr Opin Genet Dev. 2016;36:34–40.
9. Symmons O, Uslu VV, Tsujimura T, Ruf S, Nassari S, Schwarzer W, et al. Functional and topological characteristics of mammalian regulatory domains. Genome Res. 2014;24:390–400.
10. Razin SV, Ulianov SV, Ioudinkova ES, Gushchanskaya ES, Gavrilov AA, Iarovaia OV. Domains of alpha- and beta-globin genes in the context of the structural-functional organization of the eukaryotic genome. Biochemistry (Mosc). 2012;77:1409–23.
11. Vernimmen D. Uncovering enhancer functions using the alpha-globin locus. PLoS Genet. 2014;10:e1004668.
12. Hughes JR, Cheng JF, Ventress N, Prabhakar S, Clark K, Anguita E, et al. Annotation of cis-regulatory elements by identification, subclassification, and functional assessment of multispecies conserved sequences. Proc Natl Acad Sci USA. 2005;102:9830–5.
13. Higgs DR, Wood WG, Jarman AP, Sharpe J, Lida J, Pretorius I-M, et al. A major positive regulatory region located far upstream of the human α-globin gene locus. Genes Dev. 1990;4:1588–601.
14. Higgs DR, Vernimmen D, Wood B. Long-range regulation of alpha-globin gene expression. Adv Genet. 2008;61:143–73.
15. Craddock CF, Vyas P, Sharpe JA, Ayyub H, Wood WG, Higgs DR. Contrasting effects of alpha and beta globin regulatory elements on chromatin structure may be related to their different chromosomal environments. EMBO J. 1995;14:1718–26.
16. Mahajan MC, Karmakar S, Newburger PE, Krause DS, Weissman SM. Dynamics of alpha-globin locus chromatin structure and gene expression during erythroid differentiation of human CD34(+) cells in culture. Exp Hematol. 2009;37:1143–56.
17. Anguita E, Johnson CA, Wood WG, Turner BM, Higgs DR. Identification of a conserved erythroid specific domain of histone acetylation across the alpha-globin gene cluster. Proc Natl Acad Sci USA. 2001;98:12114–9.
18. Vernimmen D, Marques-Kranc F, Sharpe JA, Sloane-Stanley JA, Wood WG, Wallace HA, et al. Chromosome looping at the human alpha-globin locus is mediated via the major upstream regulatory element (HS -40). Blood. 2009;114:4253–60.
19. Vernimmen D, De Gobbi M, Sloane-Stanley JA, Wood WG, Higgs DR. Long-range chromosomal interactions regulate the timing of the transition between poised and active gene expression. EMBO J. 2007;26:2041–51.
20. Lower KM, Hughes JR, De Gobbi M, Henderson S, Viprakasit V, Fisher C, et al. Adventitious changes in long-range gene expression caused by polymorphic structural variation and promoter competition. Proc Natl Acad Sci USA. 2009;106:21771–6.
21. Bau D, Sanyal A, Lajoie BR, Capriotti E, Byron M, Lawrence JB, et al. The three-dimensional folding of the alpha-globin gene domain reveals formation of chromatin globules. Nat Struct Mol Biol. 2011;18:107–14.
22. Zhou GL, Xin L, Song W, Di LJ, Liu G, Wu XS, et al. Active chromatin hub of the mouse alpha-globin locus forms in a transcription factory of clustered housekeeping genes. Mol Cell Biol. 2006;26:5096–105.
23. Therwath A, Mengod G, Scherrer K. Altered globin gene transcription pattern and the presence of a 7–8 kb alpha A globin gene transcript in avian erythroblastosis virus-transformed cells. EMBO J. 1984;3:491–5.
24. Razin SV, Petrov P, Hancock R. Precise localization of the alpha-globin gene cluster within one of the 20- to 300-kilobase DNA fragment released by cleavage of chicken chromosomal DNA at topoisomerase II

sites in vivo: evidence that the fragments are DNA loops or domains. Proc Natl Acad Sci USA. 1991;88:8515–9.

25. Valdes-Quezada C, Arriaga-Canon C, Fonseca-Guzman Y, Guerrero G, Recillas-Targa F. CTCF demarcates chicken embryonic alpha-globin gene autonomous silencing and contributes to adult stage-specific gene expression. Epigenetics. 2013;8:827–38.

26. Gavrilov AA, Razin SV. Spatial configuration of the chicken α-globin gene domain: immature and active chromatin hubs. Nucleic Acids Res. 2008;36:4629–40.

27. Gasaryan KG. Genome activity and gene expression in avian erythroid cells. Int Rev Cytol. 1982;74:95–126.

28. Philonenko ES, Klochkov DB, Borunova VV, Gavrilov AA, Razin SV, Iarovaia OV. TMEM8—a non-globin gene entrapped in the globin web. Nucleic Acids Res. 2009;37:7394–406.

29. Broders F, Zahraoui A, Scherrer K. The chicken alpha-globin gene domain is transcribed into a 17-kilobase polycistronic RNA. Proc Natl Acad Sci USA. 1990;87:503–7.

30. Razin SV, Rynditch A, Borunova V, Ioudinkova E, Smalko V, Scherrer K. The 33 kb transcript of the chicken alpha-globin gene domain is part of the nuclear matrix. J Cell Biochem. 2004;92:445–57.

31. Arriaga-Canon C, Fonseca-Guzman Y, Valdes-Quezada C, Arzate-Mejia R, Guerrero G, Recillas-Targa F. A long non-coding RNA promotes full activation of adult gene expression in the chicken alpha-globin domain. Epigenetics. 2014;9:173–81.

32. Razin SV, Vassetzky JYS, Kvartskhava AI, Grinenko NF, Georgiev GP. Transcriptional enhancer in the vicinity of replication origin within the 5′ region of the chicken alpha-globin gene domain. J Mol Biol. 1991;217:595–8.

33. Razin SV, De Moura Gallo CV, Scherrer K. Characterization of the chromatin structure in the upstream area of the chicken alpha-globin gene domain. Mol Gen Genet. 1994;242:649–52.

34. Knezetic J, Felsenfeld G. Identification and characterization of a chicken alpha-globin enhancer. Mol Cell Biol. 1989;9:893–901.

35. Higgs DR, Vernimmen D, De Gobbi M, Anguita E, Hughes J, Buckle V, et al. How transcriptional and epigenetic programmes are played out on an individual mammalian gene cluster during lineage commitment and differentiation. Biochemical Society Symposia. 2006. pp. 11–22

36. Dostie J, Dekker J. Mapping networks of physical interactions between genomic elements using 5C technology. Nat Protoc. 2007;2:988–1002.

37. Lieberman-Aiden E, van Berkum NL, Williams L, Imakaev M, Ragoczy T, Telling A, et al. Comprehensive mapping of long-range interactions reveals folding principles of the human genome. Science. 2009;326:289–93.

38. Imakaev M, Fudenberg G, McCord RP, Naumova N, Goloborodko A, Lajoie BR, et al. Iterative correction of Hi-C data reveals hallmarks of chromosome organization. Nat Methods. 2012;9:999–1003.

39. Gushchanskaya ES, Artemov AV, Ulyanov SV, Logacheva MD, Penin AA, Kotova ES, et al. The clustering of CpG islands may constitute an important determinant of the 3D organization of interphase chromosomes. Epigenetics. 2014;9:951–63.

40. Alber F, Dokudovskaya S, Veenhoff LM, Zhang W, Kipper J, Devos D, et al. Determining the architectures of macromolecular assemblies. Nature. 2007;450:683–94.

41. Filippova D, Patro R, Duggal G, Kingsford C. Identification of alternative topological domains in chromatin. Algorithms Mol Biol. 2014;9:14.

42. Sanborn AL, Rao SS, Huang SC, Durand NC, Huntley MH, Jewett A. Chromatin extrusion explains key features of loop and domain formation in wild-type and engineered genomes. PNAS. 2015;112(47):E6456–65.

43. Dryden NH, Broome LR, Dudbridge F, Johnson N, Orr N, Schoenfelder S. Unbiased analysis of potential targets of breast cancer susceptibility loci by Capture Hi-C. Genome Res. 2014;24(11):1854–68.

44. Flint J, Tufarelli C, Peden J, Clark K, Daniels RJ, Hardison R, et al. Comparative genome analysis delimits a chromosomal domain and identifies key regulatory elements in the alpha globin cluster. Hum Mol Genet. 2001;10:371–82.

45. Vyas P, Vickers MA, Simmons DL, Ayyub H, Craddock CF, Higgs DR. Cis-acting sequences regulating expression of the human alpha-globin cluster lie within constitutively open chromatin. Cell. 1992;69:781–93.

46. Addya S, Keller MA, Delgrosso K, Ponte CM, Vadigepalli R, Gonye GE, et al. Erythroid-induced commitment of K562 cells results in clusters of

differentially expressed genes enriched for specific transcription regulatory elements. Physiol Genomics. 2004;19:117–30.

47. Kowalczyk MS, Hughes JR, Babbs C, Sanchez-Pulido L, Szumska D, Sharpe JA, et al. Nprl3 is required for normal development of the cardiovascular system. Mamm Genome. 2012;23:404–15.

48. Dutchak PA, Laxman S, Estill SJ, Wang C, Wang Y, Wang Y, et al. Regulation of hematopoiesis and methionine homeostasis by mTORC1 inhibitor NPRL2. Cell Rep. 2015;12:371–9.

49. Kowalczyk MS, Hughes JR, Garrick D, Lynch MD, Sharpe JA, Sloane-Stanley JA, et al. Intragenic enhancers act as alternative promoters. Mol Cell. 2012;45:447–58.

50. Hardison RC. Evolution of hemoglobin and its genes. Cold Spring Harb Perspect Med. 2012;2:a011627.

51. Travers A. Chromatin modification by DNA tracking. Proc Natl Acad Sci USA. 1999;96:13634–7.

52. Ea V, Baudement MO, Lesne A, Forne T. Contribution of topological domains and loop formation to 3D chromatin organization. Genes (Basel). 2015;6:734–50.

53. de Wit E, Vos ES, Holwerda SJ, Valdes-Quezada C, Verstegen MJ, Teunissen H, et al. CTCF binding polarity determines chromatin looping. Mol Cell. 2015;60:676–84.

54. Vietri Rudan M, Barrington C, Henderson S, Ernst C, Odom DT, Tanay A, et al. Comparative Hi-C reveals that CTCF underlies evolution of chromosomal domain architecture. Cell Rep. 2015;10:1297–309.

55. Kubo N, Ishii H, Gorkin D, Meitinger F, Xiong X, Fang R, et al. Preservation of chromatin organization after acute loss of CTCF in mouse embryonic stem cells. bioRxiv. 2016. doi:10.1101/118737.

56. Ulianov SV, Khrameeva EE, Gavrilov AA, Flyamer IM, Kos P, Mikhaleva EA, et al. Active chromatin and transcription play a key role in chromosome partitioning into topologically associating domains. Genome Res. 2016;26:70–84.

57. Dixon JR, Jung I, Selvaraj S, Shen Y, Antosiewicz-Bourget JE, Lee AY, et al. Chromatin architecture reorganization during stem cell differentiation. Nature. 2015;518:331–6.

58. Hagege H, Klous P, Braem C, Splinter E, Dekker J, Cathala G, et al. Quantitative analysis of chromosome conformation capture assays (3C-qPCR). Nat Protoc. 2007;2:1722–33.

59. Braem C, Recolin B, Rancourt RC, Angiolini C, Barthes P, Branchu P, et al. Genomic matrix attachment region and chromosome conformation capture quantitative real time PCR assays identify novel putative regulatory elements at the imprinted Dlk1/Gtl2 locus. J Biol Chem. 2008;283:18612–20.

60. Lajoie BR, van Berkum NL, Sanyal A, Dekker J. My5C: web tools for chromosome conformation capture studies. Nat Methods. 2009;6:690–1.

61. Langmead B, Salzberg SL. Fast gapped-read alignment with Bowtie 2. Nat Methods. 2012;9:357–9.

62. Cock PJ, Antao T, Chang JT, Chapman BA, Cox CJ, Dalke A, et al. Biopython: freely available Python tools for computational molecular biology and bioinformatics. Bioinformatics. 2009;25:1422–3.

63. Sandelin A, Alkema W, Engstrom P, Wasserman WW, Lenhard B. JASPAR: an open-access database for eukaryotic transcription factor binding profiles. Nucleic Acids Res. 2004;32:D91–4.

64. Pettersen EF, Goddard TD, Huang CC, Couch GS, Greenblatt DM, Meng EC, et al. UCSF Chimera–a visualization system for exploratory research and analysis. J Comput Chem. 2004;25:1605–12.

65. Kim D, Pertea G, Trapnell C, Pimentel H, Kelley R, Salzberg SL. TopHat2: accurate alignment of transcriptomes in the presence of insertions, deletions and gene fusions. Genome Biol. 2013;14:R36.

66. Anders S, Pyl PT, Huber W. HTSeq—a Python framework to work with high-throughput sequencing data. Bioinformatics. 2015;31:166–9.

67. Quinlan AR, Hall IM. BEDTools: a flexible suite of utilities for comparing genomic features. Bioinformatics. 2010;26:841–2.

7

Initial high-resolution microscopic mapping of active and inactive regulatory sequences proves non-random 3D arrangements in chromatin domain clusters

Marion Cremer[1*], Volker J. Schmid[2], Felix Kraus[1,3], Yolanda Markaki[1,4], Ines Hellmann[1], Andreas Maiser[1], Heinrich Leonhardt[1], Sam John[5,6], John Stamatoyannopoulos[5] and Thomas Cremer[1*]

Abstract

Background: The association of active transcription regulatory elements (TREs) with DNAse I hypersensitivity (DHS[+]) and an 'open' local chromatin configuration has long been known. However, the 3D topography of TREs within the nuclear landscape of individual cells in relation to their active or inactive status has remained elusive. Here, we explored the 3D nuclear topography of active and inactive TREs in the context of a recently proposed model for a functionally defined nuclear architecture, where an active and an inactive nuclear compartment (ANC–INC) form two spatially co-aligned and functionally interacting networks.

Results: Using 3D structured illumination microscopy, we performed 3D FISH with differently labeled DNA probe sets targeting either sites with DHS[+], apparently active TREs, or DHS[−] sites harboring inactive TREs. Using an in-house image analysis tool, DNA targets were quantitatively mapped on chromatin compaction shaped 3D nuclear landscapes. Our analyses present evidence for a radial 3D organization of chromatin domain clusters (CDCs) with layers of increasing chromatin compaction from the periphery to the CDC core. Segments harboring active TREs are significantly enriched at the decondensed periphery of CDCs with loops penetrating into interchromatin compartment channels, constituting the ANC. In contrast, segments lacking active TREs (DHS[−]) are enriched toward the compacted interior of CDCs (INC).

Conclusions: Our results add further evidence in support of the ANC–INC network model. The different 3D topographies of DHS[+] and DHS[−] sites suggest positional changes of TREs between the ANC and INC depending on their functional state, which might provide additional protection against an inappropriate activation. Our finding of a structural organization of CDCs based on radially arranged layers of different chromatin compaction levels indicates a complex higher-order chromatin organization beyond a dichotomic classification of chromatin into an 'open', active and 'closed', inactive state.

Keywords: Transcription regulatory sequences, DNAse I hypersensitive sites, Super-resolution microscopy, Chromatin domain, Nuclear architecture, Active and inactive nuclear compartment, Chromatin compaction

*Correspondence: Marion.cremer@lrz.uni-muenchen.de;
Thomas.Cremer@lrz.uni-muenchen.de
[1] LMU Biocenter, Department Biology II, Ludwig Maximilians-Universität
(LMU Munich), Grosshadernerstr. 2, 82152 Martinsried, Germany
Full list of author information is available at the end of the article

Background

Transcription regulatory elements (TREs) such as promoters, enhancers or insulators comprise non-coding sequences located within or in close vicinity to a gene but are also found up to ~1 Mb distally from their genes [1, 2]. Active TREs are characterized by a local 'open' chromatin conformation [3–5] associated with nucleosome displacement and specific histone signatures [6, 7] and were shown to have an increased sensitivity to DNAse I digestion, constituting DNAse I hypersensitive sites (DHS[+] of ~100–1000 bp [8–11]. Genome-wide profiling in various human cell types averaged over large cell populations identified >10,000 DHS[+] clusters. Approximately 10% were found to be cell type specific [2, 10, 12]. These sites encompass all experimentally validated cis-regulatory sequences; thus, DHS[+] clusters typically signify the location of active TREs in a genome.

The spatial organization of TREs has been addressed in terms of their contact frequencies with defined chromatin segments by chromosome conformation capture (Hi-C) or ChIP-Seq analyses [13–17], where genome-wide detection of DHS[+] sites in single cells was recently achieved by an ultrasensitive DNase sequencing strategy [18]. Still, little is known about the global 3D and 4D organization of regulatory sequences within the nuclear landscape of individual cells. Accordingly, it is unknown to which extent 'open' and 'closed' chromatin configurations may guide and constrain the accessibility of transcription factors (TFs) or chromatin modifiers to regulatory sequences. This issue has gained interest in the context of strong evidence for a distinct, but also dynamic nuclear architecture [19–24].

Microscopic investigations have demonstrated a structural organization of chromosome territories (CTs) built up from ~1-Mb chromatin domains (CDs) [25–27]. In particular, fluorescence labeling of specific genomic regions combined with 3D super-resolved microscopy has provided unprecedented opportunities to study nuclear arrangements of specific chromatin structures at the single-cell level and their cell-to-cell variability [28–31]. These studies have indicated that ~1-Mb CDs are composed of smaller subdomains (subCDs) and also form larger chromatin domain clusters (CDCs) [21, 26]. Genome-wide chromosome conformation capture methods have confirmed the territorial organization of chromosomes in mammalian cell nuclei and led to the discovery of ~1-Mb-sized topologically associating domains (TADs) [32–35]. TADs are built from smaller subdomains [32, 36] but also form larger units, called 'metaTADs' [37]. CDs and TADs reflect higher-order chromatin entities [38, 39], which provide the structural backbone for tissue-specific regulatory interactions [23, 40, 41].

In our present study, we explored the feasibility of 3D-FISH, 3D structured illumination microscopy (3D-SIM) and new 3D image analysis tools to determine the 3D nuclear topography of active and inactive TREs defined by their DHS[+] or DHS[−] status in human fibroblasts (BJ1 cells) and an adenocarcinoma cell line (A549). Below we use the terms DHS[+] interchangeably for sites with active TREs and DHS[−] for sites lacking (active) TREs. 3D FISH experiments were performed with an appropriately adapted protocol, which was previously shown to preserve key characteristics of the nuclear ultrastructure discernible at the resolution level of 3D-SIM [30, 42, 43], which is set at ~120 nm lateral and 250–300 nm axial [44, 45]. Highly resolved, quantitative measurements of DAPI intensities after DNA staining of nuclei were used as a proxy for local differences in chromatin compaction [46]. We describe the 3D topography of regulatory sequences in the context of a recently proposed model for a functionally defined nuclear landscape, where an active and an inactive nuclear compartment (termed ANC and INC) form two spatially co-aligned and functionally interacting 3D networks (for review, see [21] and Additional file 1 for illustration). The INC is formed by compact chromatin domains with low transcriptional activity for coding genes, which forms the interior core of a chromatin domain cluster (CDC). This compact core of CDCs is lined by a peripheral layer of decondensed, transcriptionally active chromatin domains, enriched in marks for transcriptionally competent chromatin termed perichromatin region (PR) [43, 47, 48]. The PR lines a contiguous channel system, the interchromatin compartment (IC), which starts at nuclear pores and permeates between CDs/CDCs [43, 48]. In addition to its potential role in nuclear import/export functions, the IC harbors nuclear bodies required for functions occurring within the PR. Accordingly, IC and PR together are considered as the active nuclear compartment (ANC).

According to the ANC–INC network model, we expected an enrichment of active TREs within the ANC, whereas a location of inactive TREs seemed possible within either the ANC or the INC. In case of inactive TREs embedded within the INC, their activation would correlate with a relocation toward the ANC. Our study argues for a non-random distinct distribution of targeted sites: Segments harboring active TREs are typically exposed at the outer periphery of CDCs constituting the active nuclear compartment (ANC), whereas inactive TREs are enriched toward the more compacted interior of CDCs, constituting the INC.

Results

Semi-automated quantitative 3D mapping of DNA sequences on chromatin compaction-defined nuclear landscapes

For our investigation, we developed a semi-automated approach for 3D mapping of FISH signals on chromatin compaction-defined nuclear landscapes [46] based on optical serial sections of DAPI-stained nuclei recorded with 3D-SIM. 3D mapping of specific hybridization signals is exemplified in Fig. 1a with two contiguous, differentially labeled 6-kb probes (see below for details and application of these probes). For an unbiased 3D assessment of probe signals, defined preset values such as minimal target size, relative signal intensity and a maximal distance between the centroid position of a given signal to its nearest differently labeled signal were applied for signal segmentation (Fig. 1b). An algorithm for chromatin compaction classification of DAPI-stained nuclei was employed to generate seven DAPI intensity classes with equal intensity variance as a measure for chromatin compaction [46]. Classes of chromatin compaction can be visualized as a color-coded heat map (Fig. 1c). Voxels assigned to class 1 (blue) depict regions with or close to background DAPI intensities, representing the largely DNA-free interchromatin compartment (IC), classes 2–3 (purple and deep red) comprise chromatin with low DAPI intensity, representing decondensed ('open') chromatin. Class 4 (dark orange) is considered as an intermediate zone, classes 5–7 (orange, yellow, white) comprise 'closed' chromatin. 3D coordinates of segmented FISH signals are spatially mapped on chromatin compaction classes (Fig. 1d) and plotted with their relative distributions on the respective classes. Figure 1e exemplifies a highly non-random distribution of 3D FISH targets within the seven chromatin compaction classes. Classes 1–3 represent the active nuclear compartment, class 4 an intermediate zone, classes 5–7 the INC (Fig. 1f, for review, see [21]).

Multilayered shell-like organization of chromatin domain clusters

Color heat maps of DAPI-stained nuclear SIM sections (as exemplified in Fig. 1c, d) suggest a multilayered shell-like chromatin organization of CDCs with compact chromatin (classes 5–7) typically located in their interior, surrounded by a decondensed peripheral layer (classes 2–3) lined by the IC (class 1). The visual impression of a radially arranged compartmentalization of chromatin layers is supported by quantitation of nearest-neighbor voxels of a given chromatin compaction class in 3D SIM serial sections of BJ1 ($N = 45$) and A549 ($N = 30$) cell nuclei (Fig. 2a). Most nearest neighbors belong to the same intensity class, a smaller fraction to the next higher or lower class and only rare voxels to remote classes. For a rough estimate of minimal distances required for potential movements of any target DNA from the most compact (interior) to the most decondensed (peripheral) part of CDCs, the distance for each voxel assigned to class 2 to the nearest voxel assigned to all other classes

was measured. Minimal average distances indicate that movements of ~100 nm may suffice for a relocation of a sequence between the most inner and the most outer layer of CDCs (Fig. 2b).

3D topography of active and inactive TREs in chromatin domain clusters

For the coverage of DHS[+] and DHS[−] sites located on different chromosomes in BJ1 cells, we used two differently labeled DNA probe sets of pooled fosmid clones with each fosmid carrying a human sequence of ~40 kb (Fig. 3). DHS profiles of the respective sites were identified on availability of NIH Roadmap Epigenomics Mapping Consortium data (www.roadmapepigenomics.org hg19). Fosmid pool 1 comprises six genomic targets located on chromosome bands 1p33.1, 2p13.3, 2q37.3, 3p13, 5q35.3 and 12q24.21 peppered with numerous DHS[+] clusters expanding over several kb (Fig. 3a). These targets contain different types of active TREs (for type and 'open' chromatin marks, see Additional file 2). Fosmid pool 2 (Fig. 3b) comprises two DHS[−] segments on chromosome bands 3p22 and 13q21.31. Several TREs were identified in these segments providing evidence for regulatory potential in these regions [49, 50]. However, in line with their DHS[−] status none of these regions shows 'open' chromatin marks in fibroblasts (Additional file 2).

BJ1 is a diploid cell strain (46, XY, Additional file 3). Accordingly, up to twelve distinct segments carrying DHS[+] sites can be targeted in a BJ1 nucleus by fosmid pool 1 and up to four distinct DHS[−] segments by fosmid pool 2 (Additional file 4: movie_1). Representative SIM sections of a BJ1 nucleus demonstrate the preferential localization of DHS[+] targets in low chromatin compaction classes at the periphery of CDCs, and of DHS[−] targets in more interior and compacted regions (Fig. 4a). Quantitative image analysis confirms the highly significant enrichment of DHS[+] segments indicating active TREs in low chromatin compaction classes 2 and 3 and of DHS[−] segments (inactive TREs) in high compaction classes 5 and 6 ($p < 0.001$ for classes 2/3 and 5/6) (Fig. 4b, c). This difference is consistently seen in all single-cell profiles considered as a series of 'snapshots' (Additional file 5).

Both fosmid pools were also hybridized on the chromosomally rearranged adenocarcinoma cell line A549. This cell line was used for comparison with BJ1 cells for two reasons: First, its DHS profile is clearly distinct from BJ1 cells in that all genomic targets delineated by fosmid pool 1 are DHS[−] in A549 (Additional file 6) indicating the lack of active TREs as supported by a lack of 'open' chromatin marks (Additional file 2). Second, their flat nuclear shape with z-diameters of 4–5 μm (data not shown) facilitates 3D-SIM acquisition [51]. Numerical and structural

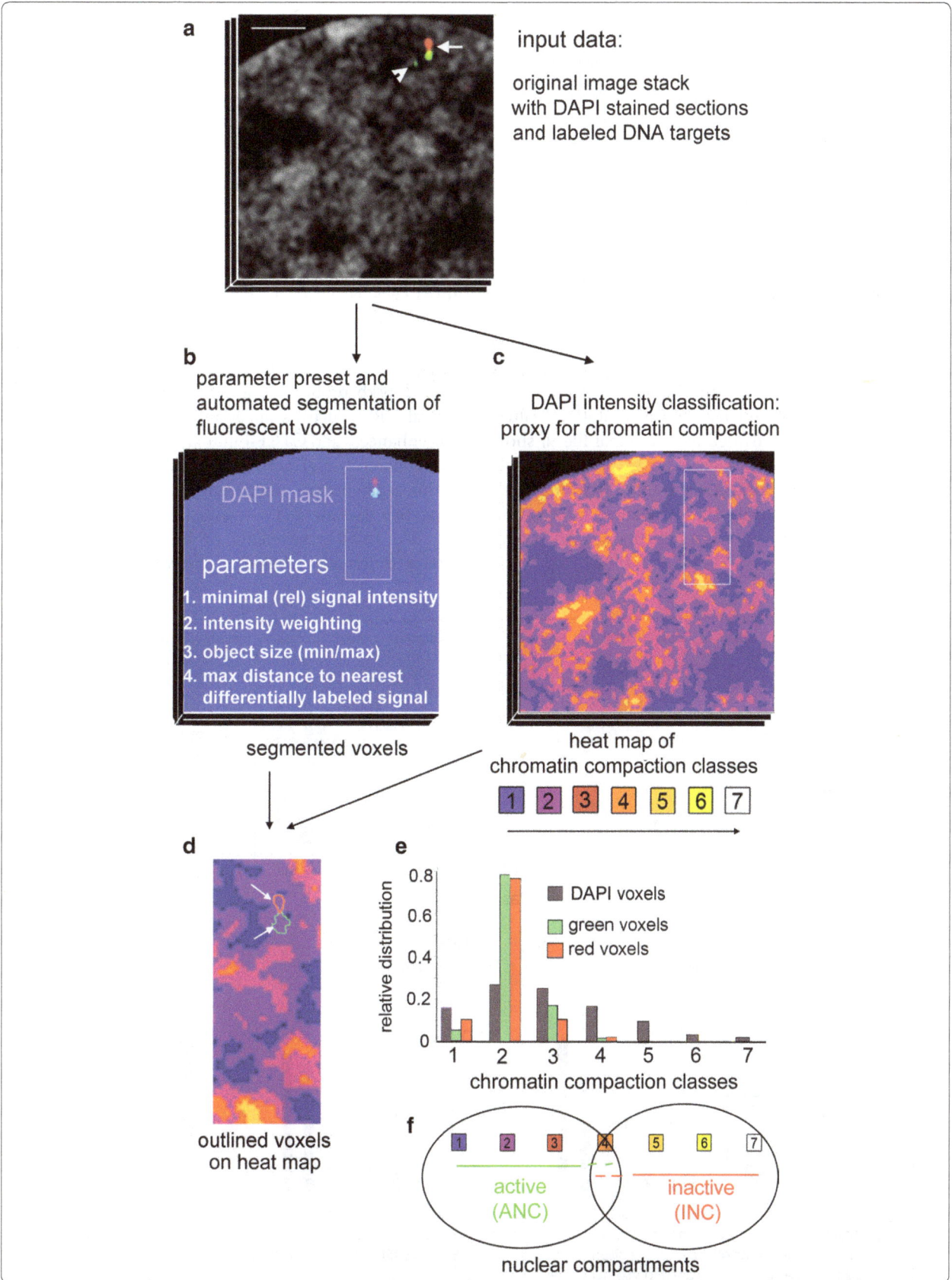

a input data:

original image stack
with DAPI stained sections
and labeled DNA targets

b parameter preset and
automated segmentation of
fluorescent voxels

DAPI mask

parameters

1. minimal (rel) signal intensity
2. intensity weighting
3. object size (min/max)
4. max distance to nearest
   differentially labeled signal

segmented voxels

c DAPI intensity classification:
proxy for chromatin compaction

heat map of
chromatin compaction classes

1 2 3 4 5 6 7

d outlined voxels
on heat map

e
relative distribution

■ DAPI voxels
■ green voxels
■ red voxels

chromatin compaction classes

f
1 2 3 4 5 6 7

active
(ANC)

inactive
(INC)

nuclear compartments

(See figure on previous page.)
Fig. 1 Workflow for quantitative mapping of specific FISH signals on 3D chromatin compaction maps. **a** Representative part of a section from an original 3D-SIM image stack of a whole nucleus acquisition shown by the example of a BJ1 nucleus. Chromatin counterstained with DAPI (*gray*), two adjacent DNA targets visualized by *green* and *red* fluorescent signals (*arrow*). The *arrowhead points* to an additional small *green* fluorescent (background) signal. *Scale bar* 2 μm. **b** Segmented fluorescent voxels of FISH signals within DAPI mask after defined parameter settings. The small isolated *green signal* seen in (*arrowhead* in **a**) is discarded due to a distance >0.5 μm from the nearest *red signal* centroid, set as limit between two differently labeled targets. **c** Same section after classification of DAPI signals into seven intensity classes as proxy for chromatin compaction visualized as *color* heat map. **d** *Inset* magnification from framed area in **b** and **c** with *outlined green* and *red* segmented signals. **e** Relative distribution of *green* and *red* fluorescent voxels in this nucleus mapped on DAPI intensity-defined chromatin compaction classes (*gray*). **f** Assignment of the active and inactive nuclear compartment (ANC/INC) linked to chromatin compaction classes, for explanation see text

rearrangements in A549 cells (see Additional file 7: for karyotype) permit up to 19 distinct hybridization sites for fosmid pool 1 and up to 5 sites for fosmid pool 2 (Additional file 8: movie_2). In contrast to BJ1, A549 nuclei show a fairly similar 3D nuclear topography of targets delineated by fosmid pools 1 and 2 that are both DHS[−]. A representative A549 nucleus and quantitative evaluation of 10 nuclei (as provided in Additional file 9) show the highest enrichment of signals for both fosmid pools in chromatin compaction classes 4 and 5 (for single-cell profiles, see Additional file 10; for detailed information on statistical values for all measurements, see Additional file 11).

To dissect the topography of a specific smaller segment in BJ1 cells harboring a cluster of DHS[+] peaks over its entire length, we performed 3D-FISH with a probe delineating a ~6-kb segment on chromosome 2q37 located within the 17-kb-long first intron of the *COL6A3* gene, termed '6-kb probe 1' (Fig. 5A). This segment contains several TREs such as an annotated transcription start site, enhancers and CTCF-binding sites (Additional file 2) with apparently high activity in BJ1 cells. The topography of this segment was compared in A549 cells where the respective sites are DHS[−] over the entire probe length (Fig. 5A). In all experiments, probe 1 was co-hybridized with a differently labeled probe which delineates an

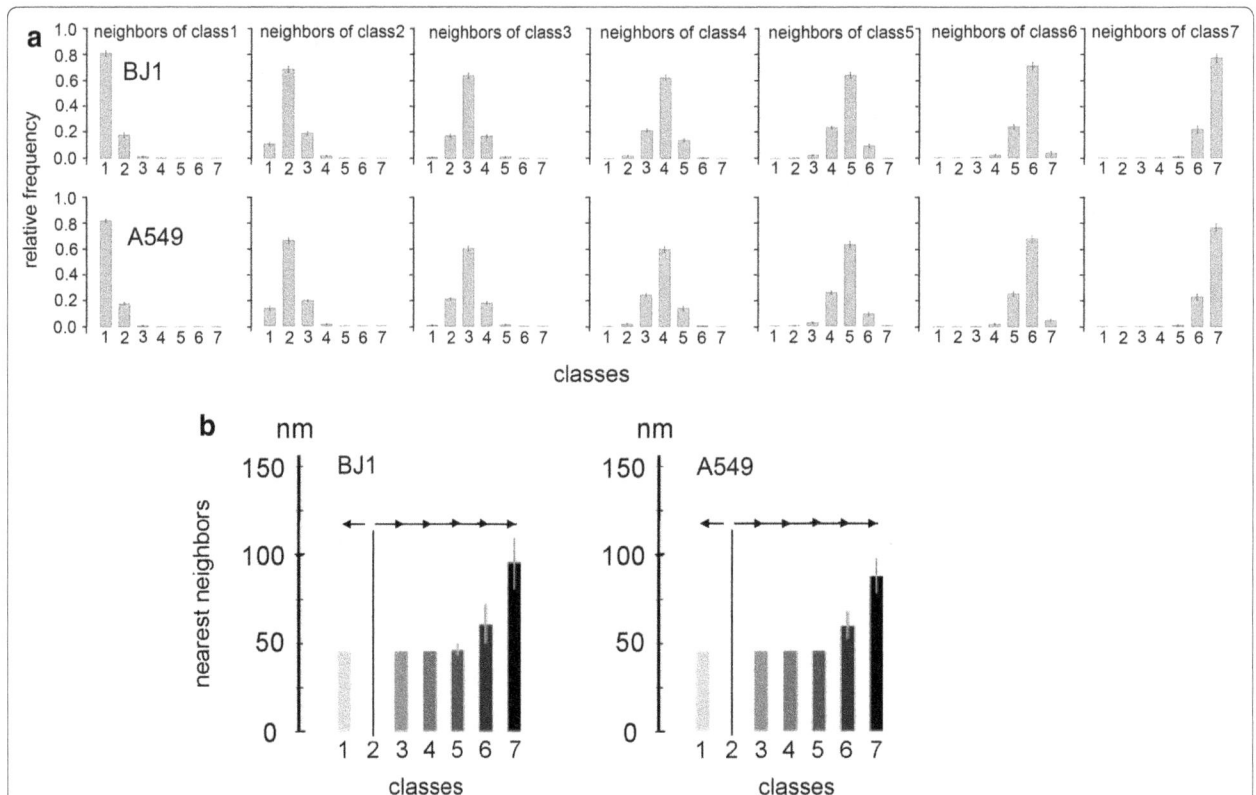

Fig. 2 Quantitative assessment for a compartmentalized organization of distinct chromatin compaction classes. **a** Frequency of chromatin compaction class for each nearest-neighbor voxel of a given compaction class in BJ1 ($N = 45$) and A549 ($N = 30$). **b** Minimal average distances of each voxel of class 2 to the nearest voxel assigned to all other classes

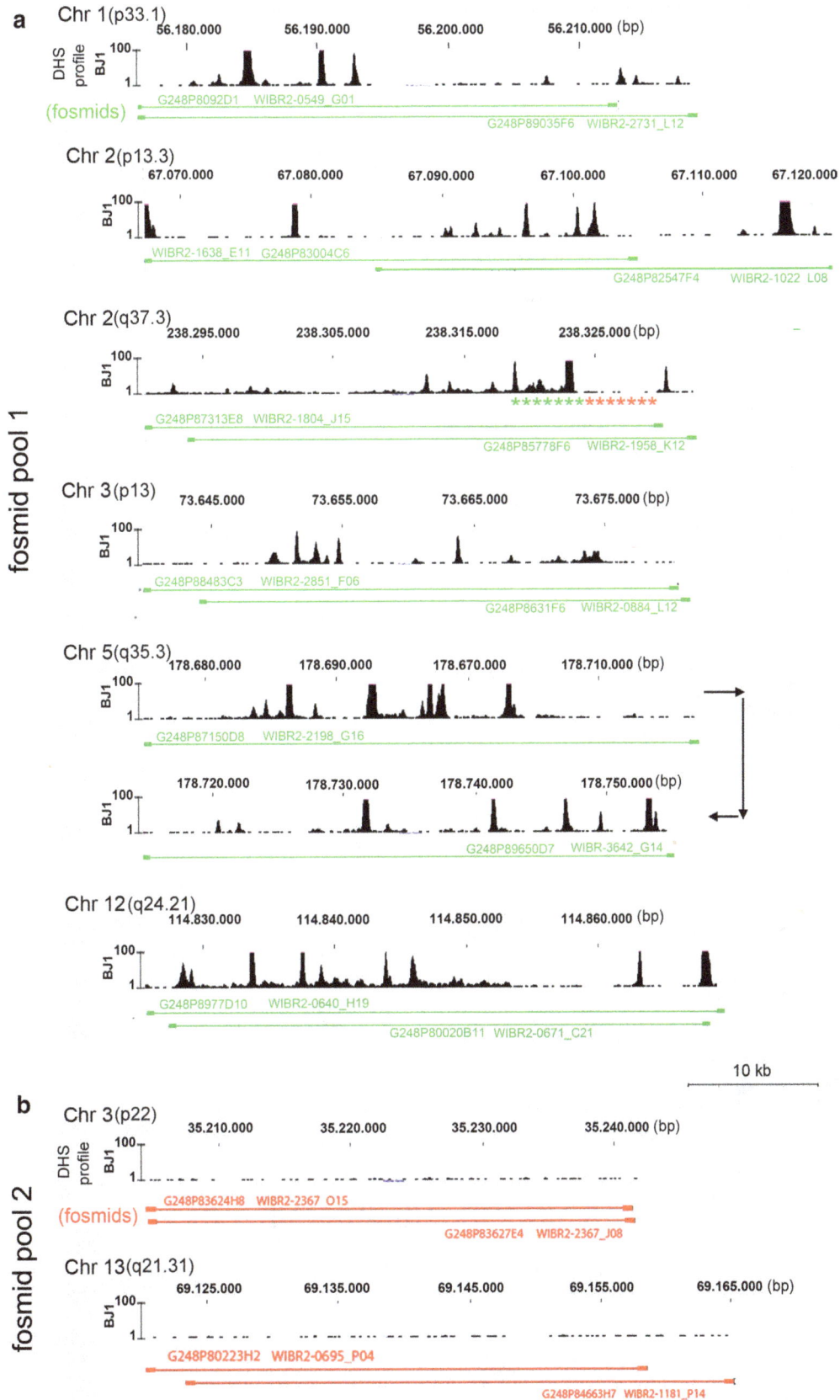

(See figure on previous page.)

Fig. 3 DHS profiles of targeted genomic regions representing DHS[+] or DHS[−] sites on different chromosomes in BJ1 nuclei and fosmid clones used for their delineation by 3D-FISH. **a** Selected regions with interspersed clusters of DHS[+] sites, DHS profiles shown in *black* (*browser shots* adopted from http://encodeproject.org/). Genomic position and assignment to DHS profiles of fosmid clones used in fosmid pool 1 are indicated by *green lines*. **b** Selected regions representing DHS[−] sites, DHS profiles shown in *black* (*browser shots* adopted from http://encodeproject.org/). Genomic position and assignment to DHS profiles of fosmid clones used in fosmid pool 2 are indicated by *red lines*. Using pairs of (partially) overlapping clones for both fosmid pools ensures optimal hybridization efficiency. *Asterisks* mark the approximate location of probes described in Fig. 5

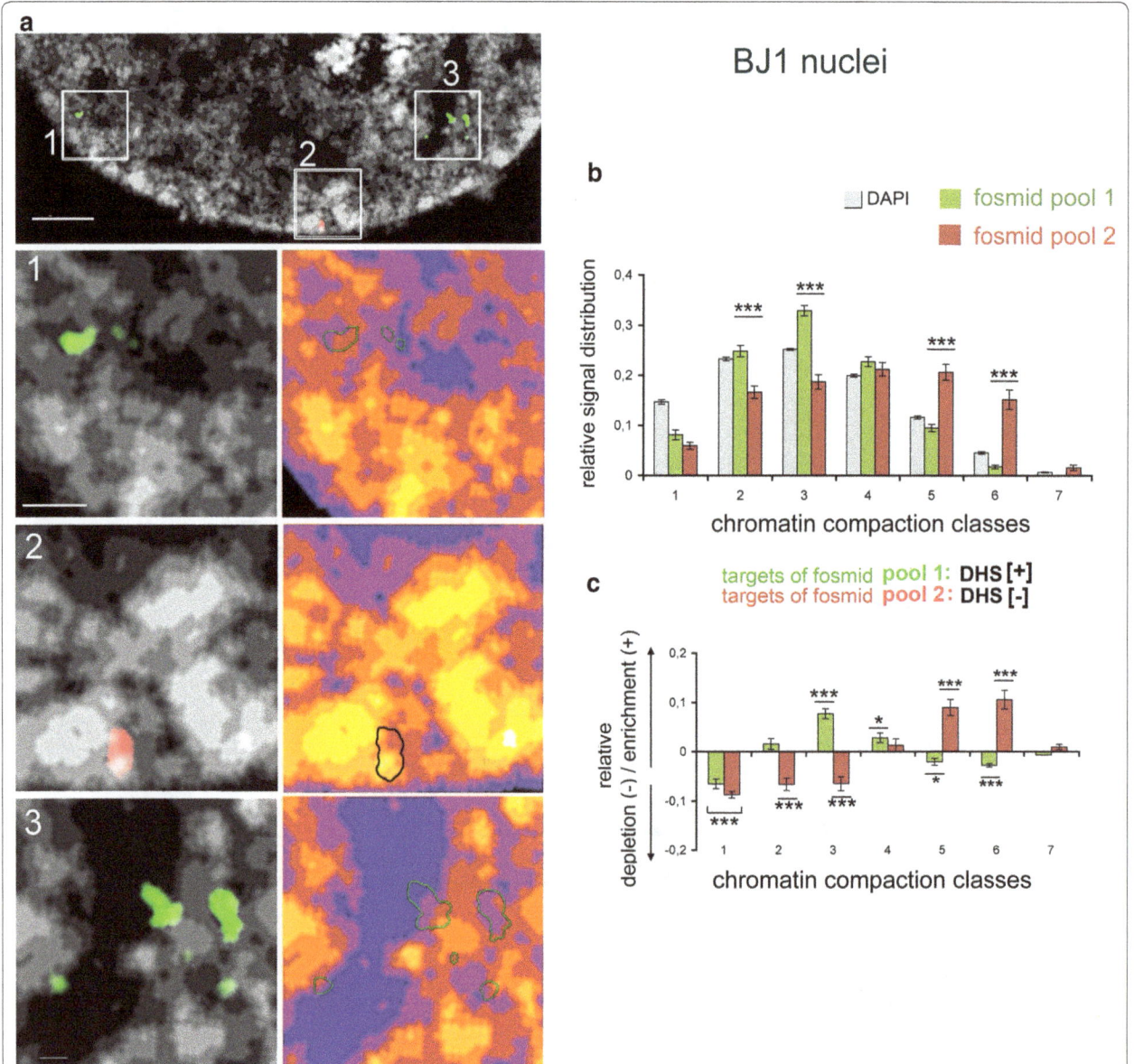

Fig. 4 3D nuclear topography and quantitative mapping of ~40-kb targets of DHS[+] and DHS[−] regions in BJ1 nuclei. **a** Part of a SIM light-optical section from a whole nucleus acquisition with framed areas indicating representative *inset* magnifications 1–3. DAPI-stained DNA after intensity classification shown as *gray* gradations and *color* heat maps, respectively. Segmented signals delineating targets of fosmid pool 1 (DHS[+], *green*) and fosmid pool 2 (DHS[−], *red*) show a preferential localization of pool 1 signals DHS[+] in zones of low DAPI intensity and of pool 2 (DHS[−]) within the more compacted core of CDCs as shown by outlined signals in *color* heat maps: pool 1 = *green*, pool 2 = *black*. *Scale bar* 2 µm, *insets* 0.5 µm. **b** Quantified distributions (N = 25 nuclei) of fosmid pool 1 (DHS[+], *green*) and pool 2 (DHS[−], *red*) within respective chromatin compaction classes (all classes shown in *gray*) confirm the significantly distinct topography for DHS[+] and DHS[−] associated signals with a shift of DHS[−] sites toward higher compaction classes. **c** Quantified levels of relative enrichment (positive values) or depletion (negative values) of fosmid pool 1 and pool 2 signals within chromatin compaction classes. *Error bars* = standard deviation of the mean.*$p \leq 0.05$; **$p \leq 0.01$; ***$p \leq 0.001$

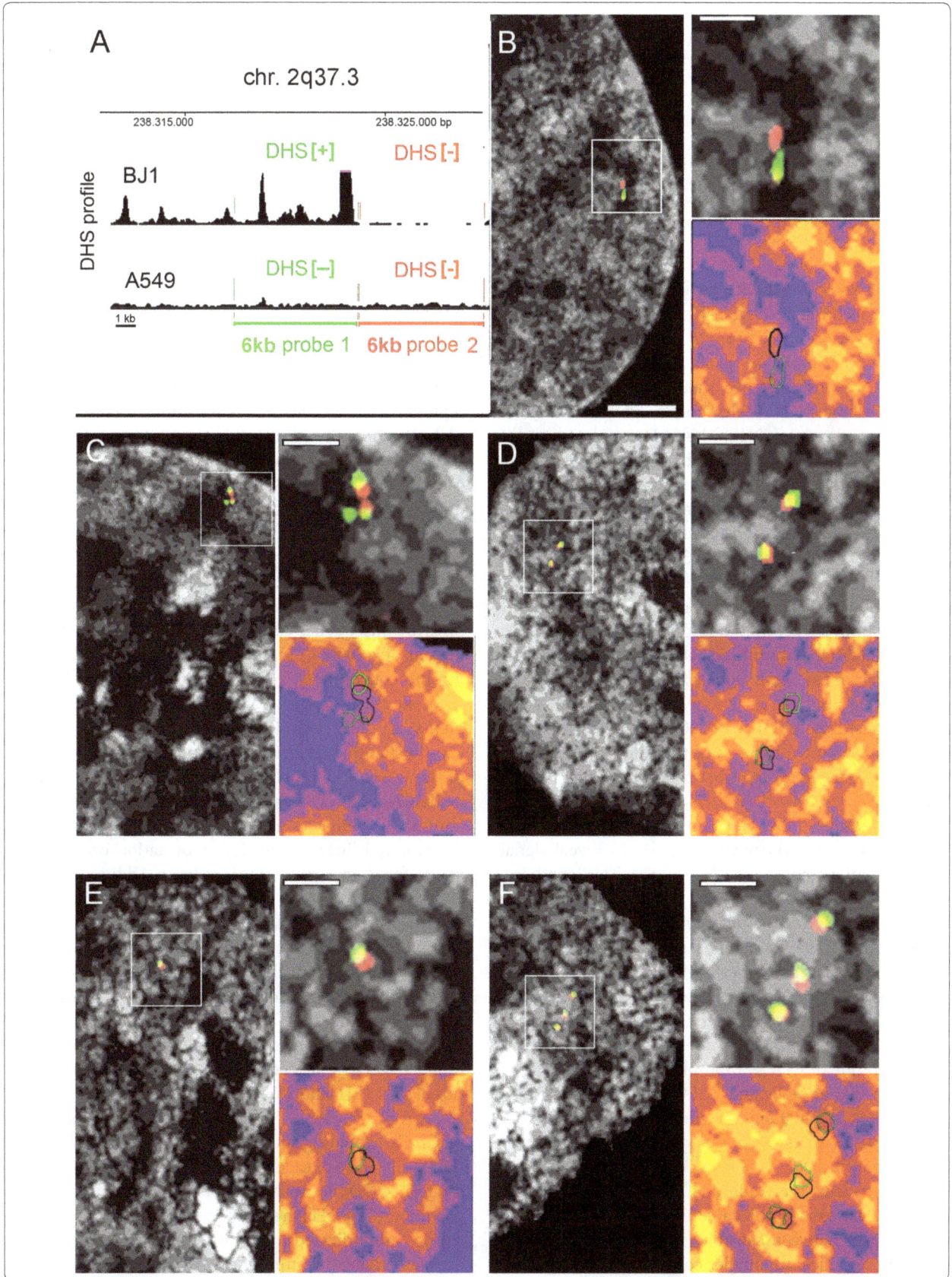

A chr. 2q37.3

(See figure on previous page.)
Fig. 5 Scheme and topography of two contiguous 6-kb segments targeting a DHS[+] and adjacent DHS[−] site. **A** Scheme of probe 1 (*green*; DHS[+] in BJ1, DHS[−] in A549) and probe 2 (*red*; DHS[−] both in BJ1 and in A549 cells) in relation to their DHS profiles (*black*; *browser shots* adopted from http://encodeproject.org/). **B–D** SIM light-optical sections from 3D acquisitions of different BJ1 nuclei and representative *inset* magnifications delineating DAPI-stained DNA after intensity classification (shown as *gray* gradations and as *color* heat maps) and segmented probe signals (probe 1 *green*; probe 2 *red*). Signal positions are shown as outlines in the *color* heat maps (*red signals* outlined in *black* for better visibility). Magnifications reveal the localization of both *red* and *green signals* in compartments of low chromatin compaction. Note variability of signal conformation: **B** elongated signal pair; **C** signal pair with extended *red signal* lined by two separate *green signals* suggestive for ongoing replication; **D** two compact separate signal pairs <0.5 µm apart suggest two chromatids after replication. A second signal pair (see Additional files 12, 13, 14) is seen in a different section of each nucleus. **E, F** Same probe setup in A549 cells shows the preferential topology of targeted sites (both DHS[−]) toward the compacted core of a chromatin domain cluster. *Scale bar* 2 µm, *inset* 0.5 µm

adjacent 6-kb segment with no known content of TREs [52] and accordingly DHS[−] status in both BJ1 and A549 cells, termed '6-kb probe 2' (Fig. 5A). Co-hybridization of the two contiguous probes was performed for the following reasons: First, we wanted to test whether we could detect a distinct localization in different chromatin compaction classes of DHS[+] and DHS[−] sites at this length scale. Second, using a probe set composed of adjacent differently labeled probes for quantitative assessment helps to exclude in an unbiased way any dotted unspecific background signals, which can become a challenge for very small single-copy 3D FISH signals, reflecting with a signal volume of ~0.005 µm^3 the volumetric resolution limit of 3D-SIM. Signals with a similar volume can arise from background fluorophores below this resolution limit. Such background cannot be entirely avoided even under most meticulous experimental conditions. Signals were accepted as true hybridization events only in case of centroid distances ≤500 nm between green and red signals, taking into account that the length of 10 kb as a fully extended 10-nm fiber is approximately 500 nm [53].

Representative SIM sections of hybridized BJ1 (Fig. 5B–D) and A549 nuclei (Fig. 5E, F) reveal signals with a lateral diameter of ~120–150 nm. This length is at the diffraction limit of 3D-SIM, so their true diameters may be smaller. In BJ1 nuclei, both signal pairs of 6-kb probes 1 and 2 are consistently noted in low chromatin compaction zones at the periphery of chromatin domain clusters (CDCs). In A549 nuclei where both probes delineate DHS[−] sites, signals are closer toward the compact core of CDCs, yet excluded from the most compact zones in the interior of CDCs. Examples shown in Fig. 5C, D are suggestive of replicated DNA in close proximity (C) or separated by ~400 nm (D), since a second site of (replicated) signal pairs is seen elsewhere in these nuclei (Additional files 12, 13, 14: movies_3, 5; for a presumable G1 nucleus, see Additional file 15: movie_6). Replicated signal pairs hint to a consistent orientation since 3D distances between centroids of 6-kb probe 1 signals are found significantly smaller compared to 6-kb probe 2

centroid distances in both cell types (Additional file 16). Quantitation in BJ1 cells confirms enrichment of the 6-kb DHS[+] segment delineated by probe 1 in the low chromatin compaction class 2 and a depletion in classes 4–6 (Fig. 6a, b left panels), while the corresponding (DHS[−]) segment in A549 cells shows a significant enrichment in higher compaction classes (Fig. 6a, b right panels, Fig. 6c shows the direct comparison between BJ1 and A549 cells). Notably—except for few nuclei (see Additional file 17: for single-cell profiles)—probe 2 largely mirrors the 3D topography of probe 1 in both cell types. This was expected for A549 cells where both probes delineate DHS[−] segments. In BJ1 cells, the similar topography of both segments within the ANC may be explained as a passive consequence of the functionally important looping out of a DHS[+] segment with an active TRE enforcing the concomitant movement of an adjacent DHS[−] segment at this length scale. Minimal distance measurements between centroids of 6-kb probe 1 and 2 signals show a slightly higher, although not significant extension of the entire ~12-kb segment in BJ1 compared to A549 nuclei (Fig. 6d). This finding may suggest a difference in chromatin compaction of the respective target sites in the two cell lines. A detailed information on statistical values for all measurements is given in Additional file 18.

Discussion

In this study, we further explored the structural and functional organization of chromatin domain clusters (CDCs) at the level of super-resolution microscopy and compared for the first time the 3D nuclear topography of selected DHS[+] and DHS[−] sites which typically reflect active or inactive TREs, respectively. In two human types of cultured cells (BJ1 and A549), we found an enrichment of active TREs (DHS[+]) in the ANC, i.e., at the periphery of CDCs extending into the IC, whereas inactive TREs (DHS[−]) were enriched toward the more compacted interior of CDCs, constituting the 'inactive' nuclear compartment (INC). Further studies with other cell types and species are indicated to test whether our current results present a general feature of the ANC–INC model, which was based on studies of a variety of normal and cancer cell types from different mammalian species [43, 47, 48]. In case that the ANC–INC model stands further experimental scrutiny, the spatial separation of inactive TREs in the INC could be explained as additional protection against their inappropriate activation, but experimental evidence for this is lacking. The distinctly different topography of active and inactive TREs in the ANC and INC, respectively, suggests a dynamic organization of CDCs, which allows positional changes of TREs between the two compartments depending on their functional state.

Structural and functional organization of CDCs

Our study indicates a radially arranged structural organization of CDCs based on layers of different chromatin compaction with most decondensed, transcriptionally competent chromatin at the CDC periphery to the most compact chromatin within the CDC interior. This layered organization suggests transition zones of chromatin compaction between CDs with a fully 'closed' and a fully 'open' configuration [6, 54, 55]. Hi-C experiments revealed two higher-order compartments A and B, respectively, of ~1-Mb CDs corresponding to open (transcriptionally competent) and closed (transcriptionally silent) chromatin [33], but their relationship to CDCs has remained elusive. Current microscopic evidence demonstrates a significant enrichment of transcriptionally competent chromatin at the periphery and of repressed chromatin in interior of CDCs [43, 47, 48]. The term INC suggests a heterochromatic nature of compact CDs. According to the classical view of heterochromatin, facultative heterochromatic contains repressed genes, while constitutive heterochromatin is built up from repetitive blocks without interspersed coding genes. In both cases, the paucity of transcription was formerly considered as a hallmark of heterochromatin. Recent evidence, however, has shown that non-coding RNAs, called hetRNA, are transcribed from heterochromatin at a previously unexpected level, including pericentric and intergenic major satellite repeats [56].

Our initial assessment of the extent of minimal movements required for shifts of regulatory sequences between the decondensed periphery of CDCs and the compact interior shows that positional changes <100 nm may suffice. However, covered distances may be substantially larger since movements of genomic elements within the nuclear landscape typically occur along individual trajectories resembling an anomalous diffusion rather than the shortest possible path [19]. A considerable local dynamic of nucleosomes in living cells that drive chromatin accessibility was reported [57]. An in-depth exploration of the dynamic nature of CDCs, including movements of CDs and chromatin loops, requires live

cell imaging with resolution <100 nm. The necessity to study large numbers of sites of specific types of TREs with special scrutiny of housekeeping, developmental and cell-type-specific genes adds another methodological challenge for future studies.

Perspectives of a functional interplay between CDCs and IC channels

3D/4D super-resolution microscopic studies are limited by the fact that only a few targets can be studied in each experiment. For our initial study, we selected eight targets on seven chromosomes (compare Fig. 3) and altogether mapped the 3D positions of about 300 DHS[+] and 400 DHS[−] sites within a total of 45 BJ1 and 30 A549 cell nuclei. Hi-C of cell populations with millions of fixed cells allows the genome-wide detection of billions (10^9) of pairwise 3D DNA–DNA contacts yielding a 1-kb resolution of topologically associating domains (TADs) [32, 33, 36]. In line with microscopic evidence, Hi-C led to the discovery of ~1-Mb-sized TADs carrying smaller sub-TADs and larger meta-TADs [32, 36, 37]. In contrast to microscopic studies, Hi-C provides the major advantage of direct comparisons with other genome-wide data sets, such as gene expression profiles, histone signatures or DHS sites (see, e.g., Roadmap Epigenomics Mapping Consortium data). Recent advancements of Hi-C have made it possible to explore the 3D organization of the entire genome in individual cell nuclei as well, although at the cost of a strongly reduced number of 3D contact sites detected in individual cell nuclei compared with Hi-C of cell populations [24, 58].

Considering these powerful possibilities, one may argue that Hi-C may replace microscopic studies, which is, however, not the case as outlined below. Hi-C experiments support a CT organization based on a 3D multiloop aggregate/rosette chromatin architecture [59]. At face value such an organization may leave sufficient inter-chromatin space to allow diffusion of macromolecules directly through open loops. Maeshima and colleagues have challenged this view with their proposal of a liquid drop model of chromatin domain organization [60, 61]. According to this model, chromatin domains are composed of irregularly folded, highly compacted 10-nm nucleosome fibers. The diffusion of non-coding RNAs and of single transcription factors into the interior of these compact CDs is possible, yet highly constrained [62]. Based on Monte Carlo simulations, the authors have proposed that small gene-specific transcription factors with a size of ~50 kD can penetrate into compact chromatin domains and search their target sequences, whereas large transcription complexes are excluded.

Taking into account a high level of compaction, the ANC–INC model [21] predicts that transcription factors

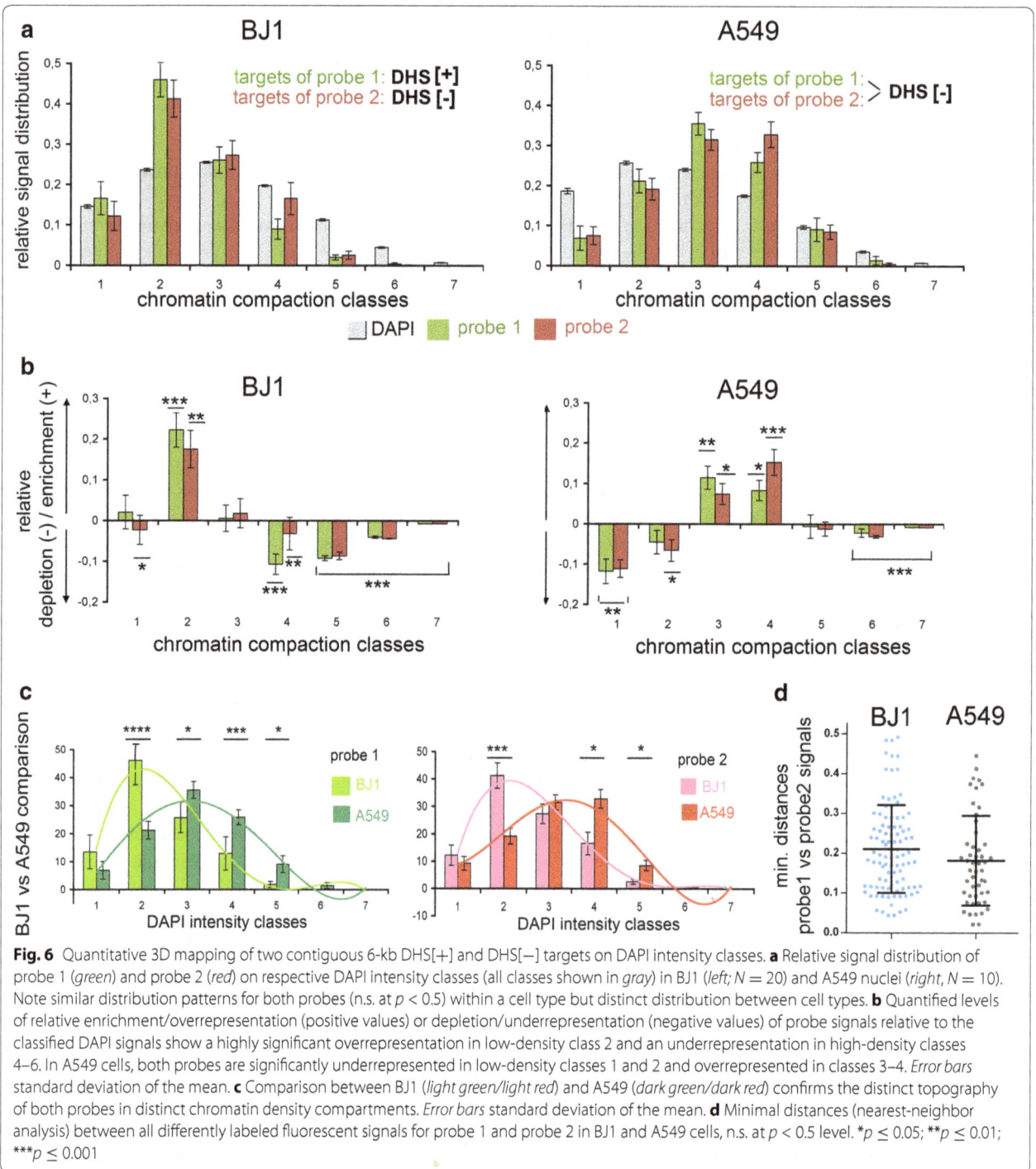

Fig. 6 Quantitative 3D mapping of two contiguous 6-kb DHS[+] and DHS[−] targets on DAPI intensity classes. **a** Relative signal distribution of probe 1 (*green*) and probe 2 (*red*) on respective DAPI intensity classes (all classes shown in *gray*) in BJ1 (*left; N = 20*) and A549 nuclei (*right, N = 10*). Note similar distribution patterns for both probes (n.s. at $p < 0.5$) within a cell type but distinct distribution between cell types. **b** Quantified levels of relative enrichment/overrepresentation (positive values) or depletion/underrepresentation (negative values) of probe signals relative to the classified DAPI signals show a highly significant overrepresentation in low-density class 2 and an underrepresentation in high-density classes 4–6. In A549 cells, both probes are significantly underrepresented in low-density classes 1 and 2 and overrepresented in classes 3–4. *Error bars* standard deviation of the mean. **c** Comparison between BJ1 (*light green/light red*) and A549 (*dark green/dark red*) confirms the distinct topography of both probes in distinct chromatin density compartments. *Error bars* standard deviation of the mean. **d** Minimal distances (nearest-neighbor analysis) between all differently labeled fluorescent signals for probe 1 and probe 2 in BJ1 and A549 cells, n.s. at $p < 0.5$ level. *$p \le 0.05$; **$p \le 0.01$; ***$p \le 0.001$

and other functional proteins, which enter the nucleus via nuclear pores, reach their sites of action preferentially by constrained diffusion along routes provided by IC channels, which start at nuclear pores and pervade the nuclear interior between CDCs with finest branches extending into their interior. Similarly, the IC-channel network may serve for the rapid intranuclear distribution of mobile, non-coding RNAs involved in gene expression and for the export of mRNAs. Our observation of a shell-like organization of CDCs argues for the possibility that CDs with different chromatin compaction levels coexist in individual CDCs. Further studies are necessary to explore the diffusional constraints inflicted by the true physical compaction of individual CDs and TADs, respectively, as well as by structural entities, which may provide 'stumbling blocks' within open loops [21]. We expect that electron and super-resolved fluorescence microscopy will remain the methods of choice to analyze the true 3D geometry of CDCs, including the size, shape and extent of compaction or decondensation of individual CDs. Ongoing efforts to achieve multicolor visualization of specific DNA targets in live cells will help to address dynamic aspects of CDC organization [28, 63].

Conclusions

This study demonstrates the non-random distribution of active and inactive transcription regulatory elements (TREs) within the higher-order chromatin landscape of cell nuclei studied in a diploid human fibroblast line (BJ1) and an aneuploid human lung cancer cell line (A549). Data were obtained by 3D FISH with differently labeled DNA probe sets targeting sites with apparently active or inactive TREs and 3D quantitative image analyses of DAPI-stained nuclei recorded with 3D structured illumination microscopy (3D-SIM). Our results indicate a 3D organization of chromatin domain clusters (CDCs) with radially arranged layers of increasing chromatin compaction from the periphery toward the CDC core. Segments harboring active TREs are significantly enriched at the decondensed periphery of CDCs, while segments with inactive TREs are embedded within the more compacted interior layers. This difference suggests positional changes of TREs within CDCs depending on their functional state. Live cell studies with high resolution are required to directly observe a relocation of TREs within CDCs in line with their state of activity. A further improvement of resolution beyond the reach of SIM, achieved, for example, by single-molecule localization microscopy [64] is essential both with respect to the precision of target localization and with respect to the quantitative measurements of 3D chromatin domain compaction within CDCs, which depends critically on the resolution limit. Comparisons of individual cells in their living state and after 3D-FISH are further necessary to quantify at a given resolution the extent of potential changes of the 3D chromatin landscape due to fixation and DNA denaturation required for 3D-FISH [30, 65].

Materials and methods
Cells and culture conditions

BJ1-tert skin fibroblasts (ATCC # CRL-2522) and A549 cells Human Lung Carcinoma Epithelial Cells (ATCC #: CCL-185) were grown under conditions used for DNase-seq analysis. BJ1 cells: MEM, supplemented with 1.5 g/L sodium bicarbonate, 1 mM sodium pyruvate, 2 mM L-glutamine, 1× non-essential amino acids, 10% FBS, Pen-Strep (1×). A549 cells: F-12 K Medium, supplemented with 10% FBS, Pen-Strep (1×).

Generation of complex DNA probe sets

DHS profiles in BJ1 and A549 cells were identified on availability of NIH Roadmap Epigenomics Mapping Consortium data (www.roadmapepigenomics.org hg19). Genomic segments with or void of DHS[+] sites in BJ1 and A549 cells, respectively, were selected as DNA targets for the generation of two probe sets. Fosmid clones for fosmid pools 1 and 2 are described in Fig. 3. For optimal target representation, a pair of overlapping fosmid clones was selected for each region from the WIBR-2 human fosmid library (http://www.ncbi.nlm.nih.gov/clone/library/genomic/) and purchased from the BACPAC Resources Center. For clone ID and genomic positions, see Additional file 19. In total, 6-kb probes 1 and 2, located within fosmid clone G248P85778F6 (as detailed description in Fig. 5A), were generated by PCR of 1-kb subfragments and subsequent pooling of amplification products (for position and primer_seq, see Additional file 20). PCR was performed in 10 mM Tris–HCl with pH 8.3, 50 mM KCl, 2 mM $MgCl_2$, 200 μM each dNTP, 1 μM each primer, 10 ng template DNA using 25 cycles of 94 °C for 30 s/56 °C for 30 s/72 °C for 1 min. Equal amounts of fosmid DNA- or PCR-amplified DNA assigned for labeling with either biotin or digoxigenin in the respective probe set were pooled and the pooled samples labeled by either biotin or digoxigenin by standard nick translation. Forty nanograms of labeled probe together with 20-fold excess of human COT-1 DNA was dissolved per 1 μl hybridization solution (2 × SSC/10% dextran sulfate/50% formamide). Hybridization efficiency and specificity of probes were verified on human metaphase chromosomes (Additional file 21).

Regulatory element annotation

The following data sets were converted from hg19 into hg38 using the liftover chain downloaded from the UCSC web-page and the liftover-tool implemented in the R package rtracklayer.

To identify regulatory elements, we accessed the Ensembl Regulation Database (v 88) [52] via the R package biomaRt. We used the data sets of Human Other Regulatory Regions, which only identified FANTOM annotations [49] as well as Human Regulatory Features to find annotated regulatory elements and Human Regulatory Evidence to get a better idea, what kind of chromatin marks or binding sites were found [66]. Type and sequence location of TREs targeted by targeted in this study are summarized in Additional file 2.

Pretreatment of cells for 3D-FISH and hybridization/detection setup

Unsynchronized cells grown up to ~60–70% confluency on high-precision borosilicate glass coverslips (170 ± 5 μm thickness) were fixed in 4% *para*-formaldehyde/PBS (10 min) followed by a stepwise replacement with PBS/0.05% Tween 20 and subsequent quenching of free aldehydes by 20 mM glycine (10 min). Cell and nuclear membranes were permeabilized by 0.5% Triton X-100/PBS (10 min), repeated freezing/thawing of cells in liquid N2 and subsequent gradual incubation in 0.1 N HCl (5 min). Cells were equilibrated in 2× SSC; at this step, RNA was removed by RNAse I treatment (100 μg/ml, 1 h at 37°). Cells were incubated in 50% formamide/2× SSC (pH = 7.0) at 4 °C until hybridization, at least overnight.

After simultaneous denaturation of cells and probes (76°/2 min), hybridization was performed at 37 °C for at least 48 h. Stringent washings in 0.1× SSC at 60 °C were followed by extensive blocking in 2% bovine serum albumin/0.5% Fish gelatin/4× SSCT for at least 2 h at RT. Probe detection was performed with avidin-Alexa488 (Molecular probes) and mouse-antidigoxigenin (Sigma) followed by an Alexa594-conjugated anti-mouse IgG (Molecular Probes). Cells were postfixed for 10 min in 4% formaldehyde/PBS, and DNA was counterstained with 1 μg/ml 4′,6-diamidino-2-phenylindole (DAPI) in 2× SSC. Samples were mounted in Vectashield antifade mounting medium (Vector Lab). A detailed protocol for 3D-FISH meeting the requirements for 3D-SIM is provided in [42].

3D-SIM

Super-resolution structured illumination imaging was performed on a DeltaVision OMX V3 system (Applied Precision Imaging/GE Healthcare) equipped with a 100×/1.4 UPlan S Apo oil immersion objective (Olympus), Cascade II:512 EMCCD cameras (Photometrics) and 405, 488 and 593 nm lasers (for detailed description, see [67]). Raw data image stacks were acquired with 15 raw images per plane (5 phases, 3 angles) and an axial distance of 125 nm and then computationally reconstructed with a Wiener filter setting of 0.002 and channel-specific optical transfer functions (OTFs) using SoftWoRx (Applied Precision). The reconstruction process generates 32-bit data sets with the pixel number doubled in the lateral axes, resulting in the pixel size being halved from 79–39.5 nm in order to meet the Nyquist sampling criterion. The level of spherical aberration was minimized and matched to the respective OTFs using immersion oil of different refractive indices (RI). Best results were

typically obtained with OTFs measured on red, green (both 110 nm diameter) and blue (170 nm diameter) fluorescent FluoSpheres (Invitrogen) using RI 1.512, and sample acquisition with RI 1.512 for depth adjustment in the region of optimal reconstruction a few μm into the sample. Images from the different color channels were corrected for chromatic aberration in SoftWoRx with alignment parameters obtained from calibration measurements with 0.2-μm-diameter TetraSpeck beads (Invitrogen). To normalize all image stacks for subsequent image processing and data analysis, the original 32-bit images were shifted to positive values and transformed to 16-bit. All further image processing was performed in ImageJ (http://rsb.info.nih.gov/ij/). For a detailed description of methodological image quality assessment survey, see [68].

Chromatin compaction classification by 3D assessment of DAPI intensity classes

Nuclei voxels were identified automatically from the DAPI channel intensities using Gaussian filtering and automatic threshold determination. For chromatin quantification, a 3D mask was generated in ImageJ to define the nuclear space considered for the segmentation of DAPI signals into seven classes with equal intensity variance by an in-house algorithm described previously [43, 48], available on request. Briefly, a hidden Markov random field model classification was used, combining a finite Gaussian mixture model with a spatial model (Potts model), implemented in the statistics software R [69, 70]. This approach allows threshold-independent signal intensity classification at the voxel level, based on the intensity of an individual voxel. Color or gray value heat maps of the seven intensity classes in individual nuclei were performed in ImageJ. For a detailed description, see [46].

Semi-automatic segmentation of hybridization signals and their allocation on 3D chromatin compaction classes

Individual voxels of FISH signals of the respective marker channels were segmented using a semi-automatic thresholding algorithm (using custom-built scripts for the open-source statistical software R http://www.r-project.org, available on request). Xyz-coordinates of segmented voxels were mapped to the seven DNA intensity classes. The relative frequency of intensity weighted signals mapped on each DAPI intensity class was used to calculate the relative distribution of signals over chromatin classes. For 3D mapping of two contiguous differentially labeled 6-kb DNA probes (6-kb probes 1 and 2) relative to chromatin compaction classes, any fluorescent dot with a distance >0.5 μm from the nearest signal centroid of a differentially labeled target was attributed to background and eliminated from further consideration after signal segmentation with appropriate parameter settings. For each studied nucleus, the total number of voxels counted for each intensity class and the total number of voxels identified for the respective FISH signals were set to 1.

As an estimate of over/under representations (relative depletion/enrichment) of marker signals in the respective intensity classes, we calculated the difference between the percentage points obtained for the fraction of voxels for a given DAPI intensity class and the corresponding fraction of voxels calculated for the FISH signals. Calculations were performed on single-cell level and average values over all nuclei used for evaluation and plotting. For a detailed description, see [46].

Nearest-neighbor/minimal distance measurements

Nearest-neighbor/minimal distance measurements between centroid xyz coordinates of differently labeled segmented FISH 'objects' were taken using the TANGO Plugin for ImageJ/Fiji [43, 71, 72]. Mode-subtracted, 16-bit transformed 3D-SIM image stacks were imported into TANGO. Xyz centroid coordinates from segmented objects were extracted based on the geometrical gravity center of the segmented 3D foci and subsequently used for centroid mapping and nearest-neighbor analysis. Nearest-neighbor distances of different experiments were analyzed by pairwise t test comparison with Bonferroni correction of level of significance. For the measurements of minimal absolute distances between DAPI intensity classes, distances between voxels were calculated from their centroid. For each class, the distance from a voxel of this class to the nearest voxel of each other class was calculated.

Statistical evaluation

GraphPad Prism 6 was used for plots and statistical evaluations. Statistical differences were tested using the Wilcoxon rank sum test with continuity correlation as well as Student's t test (two-tailed, $p < 0.05$). For interpolation models of DHS[+] and DHS[−] distributions, a second-order polynomial fit was used. SD and SEM were used for error bars, as indicated.

Additional files

Additional file 1. ANC–INC network model of nuclear organization based on spatially co-aligned active and inactive nuclear compartments (for detailed information, see [21]). Nuclear organization according to co-aligned 3D networks of an active (ANC) and an inactive nuclear compartment (INC). The ANC is a composite structural and functional entity of a 3D-channel network, the 'Inter-chromatin-compartment' (IC) together with the decondensed periphery of a higher-order chromatin network, which is built up from ~1-Mb chromatin domains (CDs), representing basic units of larger chromatin domain clusters (CDCs). The decondensed periphery of CDCs is known as the perichromatin region (PR). According to this model, the PR harbors regulatory and coding sequences of active genes and represents the preferential nuclear subcompartment for transcription, RNA-splicing, and possibly also for DNA replication and repair. Small chromatin loops expand from the perichromatin region into the interior of IC channels which start/end at nuclear pore complexes. Nuclear bodies are located within the IC, which serves as a transport system for macromolecule complexes. The INC is represented by the compacted core of CDCs enriched in markers for silent chromatin (Fig. modified from [21]).

Additional file 2. Overview of type, sequence location and open chromatin marks of TREs targeted by probe sets used in this study (fosmid pools 1 and 2; 6-kb probes 1 and 2). Data are based on [52]. For the segment covered by 6-kb probe 2, no TREs were identified in the used databases. Sequence coordinates represent the coordinates for hg38 after conversion of hg19 into hg38 by liftover. The used databases do not provide data on 'open chromatin marks' H3K9ac and H3K4me3 as additional indirect information for their state of activity at the respective loci in BJ1 cells. In the sheet 'open chromatin marks,' available data for IMR90 fibroblasts are shown instead. IMR90 cells show an almost identical DHS[+] profile to BJ1 cells (www.roadmapepigenomics.org); accordingly, similar epigenetic signatures between both cell lines can be assumed.

Additional file 3. Metaphase spreads of BJ1 cells: Five out of six randomly selected Giemsa stained metaphase spreads reveal an inconspicuous diploid chromosome set of n = 46, XY (n = 45 in metaphase 2 is likely due to loss of one chromosome during preparation).

Additional file 4. Movie_1 entire image stack of a BJ1 nucleus hybridized with fosmid pools 1 and 2. DAPI-stained DNA after intensity classification shown in gray. Fosmid pool 1 (green), fosmid pool 2 (red).

Additional file 5. Single-cell profiles for target sites of fosmid pools 1 and 2 mapped to chromatin compaction classes in BJ1 cells for illustration of intercellular variability. (**A**) Mapping profiles from ten randomly chosen individual nuclei illustrate consistent distinct distribution profiles of fosmid pool 1 (green) toward low chromatin compaction classes and of pool 2 (red) toward higher chromatin compaction classes. (**B**) Standard deviations of relative probe signal distributions of all evaluated nuclei (compare Fig. 4 for standard errors of the mean (SEM). (**C**) Standard deviations of DAPI signal distribution on classes (compare Fig. 4 for SEM).

Additional file 6. DHS profiles and fosmid clones used for target regions delineating DHS[−] sites on different chromosomes in A549 nuclei. (**A**) Selected regions with clones of fosmid pool 1 (green) and (**B**) with clones of fosmid pool 2 (red). DHS profile in black (browser shots adopted from http://encodeproject.org/). Note: probe sets are identical to probe sets shown in Fig. 3.

Additional file 7. M-FISH karyotype analysis of A549 cells. (**A**) Representative karyotype obtained by M-FISH after combinatorial labeling of chromosome-specific paint probes with seven fluorochromes. (**B**) Quantitation of 20 metaphases reveals a karyotype with 62–66 chromosomes with consistent structural rearrangements involving chromosomes 1,2,3,6,8,11,15,19,20. This constellation allows for fosmid pool 1 up to 19, for fosmid pool 2 up to five distinct hybridization sites in a nucleus (compare Additional file 6).

Additional file 8. Entire image stack of A549 nucleus after hybridization with fosmid pools 1 and 2. DAPI-stained DNA after intensity classification shown in gray. Fosmid pool 1 (*green*), fosmid pool 2 (*red*).

Additional file 9. 3D nuclear topography and quantitative mapping of ~40 kb targets of DHS[−] regions in A549 nuclei. **(A)** Part of a SIM light-optical section from a whole nucleus acquisition with representative inset magnifications. DAPI-stained DNA after intensity classification shown as gray gradations and color heat map, respectively. Segmented signals delineating targets both of fosmid pool 1 (*green*) and fosmid pool 2 (*red*) show a similar location with regard to chromatin compaction classes (asterisks in color heat maps, pool 1 (*green*), pool 2 (*black*). Scale bar 2 μm, insets 0.5 μm. **(B)** Quantified distributions (N = 10 nuclei) of fosmid pools 1 (*green*) and 2 (*red*) within respective chromatin compaction classes (all classes shown in gray). **(C)** Quantified levels of relative enrichment (positive values) or depletion (negative values) of fosmid pool 1 and pool 2 signals show an enrichment of signals in higher compaction classes. Error bars = standard deviation of the mean *$p \leq 0.05$, **$p \leq 0.01$, ***$p \leq 0.001$.

Additional file 10. Single-cell profiles for target sites of fosmid pools 1 and 2 mapped to chromatin compaction classes in A549 cells for illustration of intercellular variability. **(A)** Mapping profiles of A549 nuclei (N = 10) illustrate for most nuclei fairly similar distribution profiles of fosmid pool 1 (*green*) and fosmid pool 2 (*red*). **(B)** Standard deviations of relative probe signal distributions of all evaluated nuclei (compare Additional file 8 for standard errors of the mean (SEM). **(C)** Standard deviations of DAPI signal distribution on classes (compare Additional file 8 for SEM).

Additional file 11. Significance values for relative signal distributions of fosmid pools 1 and 2 in BJ1 and A549 cells.

Additional file 12. Entire image stack of the nucleus shown in Fig. 5B. DAPI-stained DNA after intensity classification (blue). Segmented probe signals (green/red, seen in section z19 and 23) reveal two sites of double signal pairs, suggestive of ongoing/post-replication. Note that the respective section shown in Fig. 5B is horizontally flipped for arrangement.

Additional file 13. Entire image stack of the nucleus shown in Fig. 5C. DAPI-stained DNA after intensity classification (blue). Segmented probe signals (green/red, seen in sections z11 and 15) reveal two sites of double signal pairs, suggestive of ongoing/post-replication.

Additional file 14. Entire image stack of the nucleus shown in Fig. 5D. DAPI-stained DNA after intensity classification (blue). Segmented probe signals (green/red seen in sections z27 and 34) reveal two sites of double signal pairs, suggestive of ongoing/post-replication.

Additional file 15. Entire image stack of a presumable G1 nucleus. DAPI-stained DNA after intensity classification (blue). Segmented probe signals (green/red seen in sections z24 and 35) reveal two sites of a single signal pair, suggestive of G1.

Additional file 16. 3D distances between centroids of 6-kb probe 1 (green) and centroids of 6-kb probe 2 in BJ1 (left) and in A549 cells (right). Distance measurements are restricted to distances <500 nm presumably comprising only sister chromatids of S/G2 nuclei. The smaller distances between centroids of green signals (probe 1; DHS[+] in BJ1 cells, DHS[−] in A549 cells) compared to distances between centroids of red signals (probe 2; DHS[−] both in BJ1 and A549 cells) hint to a consistent orientation of these segments irrespective of DNAse I sensitivity.

Additional file 17. Single-cell profiles of 6-kb probes 1 and 2 targets mapped to chromatin compaction classes in BJ1 **(A–C)** and A549 cells **(D–F)** for illustration of intercellular variability. **(A)** Mapping profiles from ten randomly chosen individual BJ1 nuclei for illustration of intercellular variabilities and similarities of relative signal distribution of probe 1 (green) and probe 2 (red) within DAPI intensity classes. Note an only marginal signal representation in classes 5–7. **(B)** Standard deviations of relative probe signal distributions of all evaluated nuclei (compare Fig. 6 for standard errors of the mean (SEM). **(C)** Standard deviations of DAPI signal distribution on classes (compare Fig. 6 for standard errors of the mean (SEM). **(D)** Respective mapping profiles from 10 individual A549 nuclei of relative signal distribution of probe 1 (green) and probe 2 (red) within DAPI intensity classes. Profiles show an overall broader distribution range compared to

BJ1 nuclei. **(E)** Standard deviations of relative probe signal distributions of all evaluated nuclei (compare Fig. 6 for standard errors of the mean (SEM). **(F)** Standard deviations of DAPI signal distribution on classes (compare Fig. 6 for standard errors of the mean (SEM).

Additional file 18. Significance values for relative signal distributions of 6-kb probes 1 and 2 in BJ1 and A549 nuclei.

Additional file 19. fosmid ID (G248 library #) and sequence alignment of fosmids used. Data are based on hg19.

Additional file 20. Position and primer sequences used for amplification of ~1-kb subfragments for assembling of 6-kb probes 1 and 2. Data are based on hg19.

Additional file 21. FISH of fosmid pairs on normal human metaphases for verification of specificity. **(A)** Human metaphase ideogram with marked positions of tested fosmids. **(B–E)** Metaphase spreads after FISH with **(B)** fosmid pairs G248P8092D1/G248P89035F6 mapped on 1p and G248P80020B1/G248P8977D10 mapped on 12q, **(C)** G248P83004C6/G248P82547F4 mapped on 2p and G248P87313E8/G248P85778F6 mapped on 2q, **(D)** G248P8631F6/G248P88483C3 mapped on 3p and G248P87150D8/G248P89650D7 mapped on 5q, **(E)** G248P83624H8/G248P83627E4 mapped on 3p and G248P80223H2/G248P84663H7 mapped on 13q. All tested probes show a specific hybridization signal at the expected chromosomal position.

Abbreviations

3D FISH: 3D fluorescence in situ hybridization; 3D SIM: 3D structured illumination microscopy; ANC/INC: active/inactive nuclear compartment; CT: chromosome territory; CD(C): chromatin domain (cluster); CTCF: CCCTC-binding factor; DAPI: 4′,6-diamidino-2-phenylindole; DHS[+]: DNAse I hypersensitive site indicates the presence of active regulatory sequences; DHS[−]: DNAse I non-sensitive site in a given cell, although the site is DHS[+] in at least one other cell type or cell line, indicates the presence of inactive regulatory sequences (exception: probe 2 delineating a DHS[−] segment lacks the content of TREs); DNase-seq: DNAse I digestion and sequencing; Hi-C: chromosome conformation capturing combined with deep sequencing; IC: interchromatin compartment; OTF: optical transfer function; PR: perichromatin region; TAD: topologically associating domain; TF: transcription factor; TRE: transcription regulatory element.

Authors' contributions

TC, JS and MC initiated and conceived the study; SJ provided sequence data for PCR-generated probes and fosmid probes; MC planned and performed all 3D-FISH experiments shown in this study; YM initiated and performed initial experiments; VJS developed and performed quantitative 3D image analyses; FK performed quantitative 3D image analyses; IH analyzed regulatory elements in relevant databases; HL provided input for the 3D imaging part; AM carried out 3D SIM microscopy and image processing; MC and TC wrote the manuscript. All authors read and approved the final manuscript.

Author details

[1] LMU Biocenter, Department Biology II, Ludwig Maximilians-Universität (LMU Munich), Grosshadernerstr. 2, 82152 Martinsried, Germany. [2] Biolmaging Group, Department of Statistics, Ludwig Maximilians-Universität (LMU Munich), Munich, Germany. [3] Present Address: Department of Biochemistry and Molecular Biology, Monash Biomedicine Discovery Institute, Monash University, Melbourne 3800, Australia. [4] Present Address: Department of Biological Chemistry, David Geffen School of Medicine at UCLA, Los Angeles, CA, USA. [5] Department of Genome Sciences, University of Washington, Seattle, WA, USA. [6] Present Address: Center for Cancer Research, National Cancer Institute, Bethesda, MD, USA.

Acknowledgements

We thank Anna Jauch (University of Heidelberg, Germany) for kindly performing M-FISH analysis on A549 cells. We are grateful to Shreeram Akilesh (Dept of Pathology, University of Washington Seattle, USA) and to Tobias Ragoczy (Altius Institute for Biomedical Sciences, Seattle, USA) for critical reading the manuscript. The Center of Advanced Light Microcopy (CALM) of the LMU Biocenter (headed by H. Leonhardt/H. Hartz) was essential for the 3D-SIM recording.

Competing interests

The authors declare to have no competing interests.

Funding

This work was supported by funding of the Ludwig Maximilians-Universität (LMU Munich) to TC.

References

1. Maston GA, Evans SK, Green MR. Transcriptional regulatory elements in the human genome. Annu Rev Genomics Hum Genet. 2006;7:29–59.
2. Sakabe NJ, Nobrega MA. Genome-wide maps of transcription regulatory elements. Wiley Interdiscip Rev Syst Biol Med. 2010;2:422–37.
3. Bell O, Tiwari VK, Thoma NH, Schubeler D. Determinants and dynamics of genome accessibility. Nat Rev Genet. 2011;12:554–64.
4. Boyle AP, Davis S, Shulha HP, Meltzer P, Margulies EH, Weng Z, Furey TS, Crawford GE. High-resolution mapping and characterization of open chromatin across the genome. Cell. 2008;132:311–22.
5. Cockerill PN. Structure and function of active chromatin and DNase I hypersensitive sites. FEBS J. 2011;278:2182–210.
6. Boettiger AN, Bintu B, Moffitt JR, Wang S, Beliveau BJ, Fudenberg G, Imakaev M, Mirny LA, Wu CT, Zhuang X. Super-resolution imaging reveals distinct chromatin folding for different epigenetic states. Nature. 2016;529:418–22.
7. Pundhir S, Bagger FO, Lauridsen FB, Rapin N, Porse BT. Peak-valley-peak pattern of histone modifications delineates active regulatory elements and their directionality. Nucleic Acids Res. 2016;44:4037.
8. Crawford GE, Davis S, Scacheri PC, Renaud G, Halawi MJ, Erdos MR, Green R, Meltzer PS, Wolfsberg TG, Collins FS. DNase-chip: a high-resolution method to identify DNase I hypersensitive sites using tiled microarrays. Nat Methods. 2006;3:503–9.
9. Crawford GE, Holt IE, Whittle J, Webb BD, Tai D, Davis S, Margulies EH, Chen Y, Bernat JA, Ginsburg D, et al. Genome-wide mapping of DNase hypersensitive sites using massively parallel signature sequencing (MPSS). Genome Res. 2006;16:123–31.
10. Mercer TR, Edwards SL, Clark MB, Neph SJ, Wang H, Stergachis AB, John S, Sandstrom R, Li G, Sandhu KS, et al. DNase I-hypersensitive exons colocalize with promoters and distal regulatory elements. Nat Genet. 2013;45:852–9.
11. Sabo PJ, Humbert R, Hawrylycz M, Wallace JC, Dorschner MO, McArthur M, Stamatoyannopoulos JA. Genome-wide identification of DNaseI hypersensitive sites using active chromatin sequence libraries. Proc Natl Acad Sci USA. 2004;101:4537–42.
12. Thurman RE, Rynes E, Humbert R, Vierstra J, Maurano MT, Haugen E, Sheffield NC, Stergachis AB, Wang H, Vernot B, et al. The accessible chromatin landscape of the human genome. Nature. 2012;489:75–82.
13. Ay F, Bailey TL, Noble WS. Statistical confidence estimation for Hi-C data reveals regulatory chromatin contacts. Genome Res. 2014;24:999–1011.
14. Kieffer-Kwon KR, Tang Z, Mathe E, Qian J, Sung MH, Li G, Resch W, Baek S, Pruett N, Grontved L, et al. Interactome maps of mouse gene regulatory domains reveal basic principles of transcriptional regulation. Cell. 2013;155:1507–20.
15. Ma W, Ay F, Lee C, Gulsoy G, Deng X, Cook S, Hesson J, Cavanaugh C, Ware CB, Krumm A, et al. Fine-scale chromatin interaction maps reveal the cis-regulatory landscape of human lincRNA genes. Nat Methods. 2015;12:71–8.
16. Ramani V, Cusanovich DA, Hause RJ, Ma W, Qiu R, Deng X, Blau CA, Disteche CM, Noble WS, Shendure J, Duan Z. Mapping 3D genome architecture through in situ DNase Hi-C. Nat Protoc. 2016;11:2104–21.
17. van de Werken HJ, Landan G, Holwerda SJ, Hoichman M, Klous P, Chachik R, Splinter E, Valdes-Quezada C, Oz Y, Bouwman BA, et al. Robust 4C-seq data analysis to screen for regulatory DNA interactions. Nat Methods. 2012;9:969–72.
18. Jin W, Tang Q, Wan M, Cui K, Zhang Y, Ren G, Ni B, Sklar J, Przytycka TM, Childs R, et al. Genome-wide detection of DNase I hypersensitive sites in single cells and FFPE tissue samples. Nature. 2015;528:142–6.
19. Bronshtein I, Kanter I, Kepten E, Lindner M, Berezin S, Shav-Tal Y, Garini Y. Exploring chromatin organization mechanisms through its dynamic properties. Nucleus. 2016;7:27–33.
20. Bronshtein I, Kepten E, Kanter I, Berezin S, Lindner M, Redwood AB, Mai S, Gonzalo S, Foisner R, Shav-Tal Y, Garini Y. Loss of lamin A function increases chromatin dynamics in the nuclear interior. Nat Commun. 2015;6:8044.
21. Cremer T, Cremer M, Hubner B, Strickfaden H, Smeets D, Popken J, Sterr M, Markaki Y, Rippe K, Cremer C. The 4D nucleome: evidence for a dynamic nuclear landscape based on co-aligned active and inactive nuclear compartments. FEBS Lett. 2015;589:2931–43.
22. Gaspar-Maia A, Alajem A, Meshorer E, Ramalho-Santos M. Open chromatin in pluripotency and reprogramming. Nat Rev Mol Cell Biol. 2011;12:36–47.
23. Sexton T, Cavalli G. The role of chromosome domains in shaping the functional genome. Cell. 2015;160:1049–59.
24. Stevens TJ, Lando D, Basu S, Atkinson LP, Cao Y, Lee SF, Leeb M, Wohlfahrt KJ, Boucher W, O'Shaughnessy-Kirwan A, et al. 3D structures of individual mammalian genomes studied by single-cell Hi-C. Nature. 2017;544:59–64
25. Cremer T, Cremer C. Chromosome territories, nuclear architecture and gene regulation in mammalian cells. Nat Rev Genet. 2001;2:292–301.
26. Cremer T, Cremer M. Chromosome territories. Cold Spring Harb Perspect Biol. 2010;2:a003889.
27. Lanctot C, Cheutin T, Cremer M, Cavalli G, Cremer T. Dynamic genome architecture in the nuclear space: regulation of gene expression in three dimensions. Nat Rev Genet. 2007;8:104–15.
28. Anton T, Leonhardt H, Markaki Y. Visualization of genomic loci in living cells with a fluorescent CRISPR/Cas9 system. Methods Mol Biol. 2016;1411:407–17.
29. Ma H, Naseri A, Reyes-Gutierrez P, Wolfe SA, Zhang S, Pederson T. Multi-color CRISPR labeling of chromosomal loci in human cells. Proc Natl Acad Sci USA. 2015;112:3002–7.
30. Markaki Y, Smeets D, Fiedler S, Schmid VJ, Schermelleh L, Cremer T, Cremer M. The potential of 3D-FISH and super-resolution structured illumination microscopy for studies of 3D nuclear architecture: 3D structured illumination microscopy of defined chromosomal structures visualized by 3D (immuno)-FISH opens new perspectives for studies of nuclear architecture. BioEssays. 2012;34:412–26.
31. Beliveau BJ, Boettiger AN, Avendano MS, Jungmann R, McCole RB, Joyce EF, Kim-Kiselak C, Bantignies F, Fonseka CY, Erceg J, et al. Single-molecule super-resolution imaging of chromosomes and in situ haplotype visualization using Oligopaint FISH probes. Nat Commun. 2015;6:7147.
32. Dixon JR, Selvaraj S, Yue F, Kim A, Li Y, Shen Y, Hu M, Liu JS, Ren B. Topological domains in mammalian genomes identified by analysis of chromatin interactions. Nature. 2012;485:376–80.
33. Lieberman-Aiden E, van Berkum NL, Williams L, Imakaev M, Ragoczy T, Telling A, Amit I, Lajoie BR, Sabo PJ, Dorschner MO, et al. Comprehensive mapping of long-range interactions reveals folding principles of the human genome. Science. 2009;326:289–93.
34. Nora EP, Lajoie BR, Schulz EG, Giorgetti L, Okamoto I, Servant N, Piolot T, van Berkum NL, Meisig J, Sedat J, et al. Spatial partitioning of the regulatory landscape of the X-inactivation centre. Nature. 2012;485:381–5.
35. Sexton T, Yaffe E, Kenigsberg E, Bantignies F, Leblanc B, Hoichman M, Parrinello H, Tanay A, Cavalli G. Three-dimensional folding and functional organization principles of the Drosophila genome. Cell. 2012;148:458–72.
36. Rao SS, Huntley MH, Durand NC, Stamenova EK, Bochkov ID, Robinson JT, Sanborn AL, Machol I, Omer AD, Lander ES, Aiden EL. A 3D map of the human genome at kilobase resolution reveals principles of chromatin looping. Cell. 2014;159:1665–80.
37. Fraser J, Ferrai C, Chiariello AM, Schueler M, Rito T, Laudanno G, Barbieri M, Moore BL, Kraemer DC, Aitken S, et al. Hierarchical folding and reorganization of chromosomes are linked to transcriptional changes in cellular differentiation. Mol Syst Biol. 2015;11:852.

38. Boulos RE, Drillon G, Argoul F, Arneodo A, Audit B. Structural organization of human replication timing domains. FEBS Lett. 2015;589:2944–57.

39. Pope BD, Ryba T, Dileep V, Yue F, Wu W, Denas O, Vera DL, Wang Y, Hansen RS, Canfield TK, et al. Topologically associating domains are stable units of replication-timing regulation. Nature. 2014;515:402–5.

40. Neems DS, Garza-Gongora AG, Smith ED, Kosak ST. Topologically associated domains enriched for lineage-specific genes reveal expression-dependent nuclear topologies during myogenesis. Proc Natl Acad Sci USA. 2016;113:E1691–700.

41. Yang R, Kerschner JL, Gosalia N, Neems D, Gorsic LK, Safi A, Crawford GE, Kosak ST, Leir SH, Harris A. Differential contribution of cis-regulatory elements to higher order chromatin structure and expression of the CFTR locus. Nucleic Acids Res. 2016;44:3082–94.

42. Markaki Y, Smeets D, Cremer M, Schermelleh L. Fluorescence in situ hybridization applications for super-resolution 3D structured illumination microscopy. Methods Mol Biol. 2013;950:43–64.

43. Smeets D, Markaki Y, Schmid VJ, Kraus F, Tattermusch A, Cerase A, Sterr M, Fiedler S, Demmerle J, Popken J, et al. Three-dimensional super-resolution microscopy of the inactive X chromosome territory reveals a collapse of its active nuclear compartment harboring distinct Xist RNA foci. Epigenetics Chromatin. 2014;7:8.

44. Schermelleh L, Heintzmann R, Leonhardt H. A guide to super-resolution fluorescence microscopy. J Cell Biol. 2010;190:165–75.

45. Demmerle J, Wegel E, Schermelleh L, Dobbie IM. Assessing resolution in super-resolution imaging. Methods. 2015;88:3–10.

46. Schmid VJ, Cremer M, Cremer T. Quantitative analyses of the 3D nuclear landscape recorded with super-resolved fluorescence microscopy. Methods. 2017;123:33–46

47. Hubner B, Lomiento M, Mammoli F, Illner D, Markaki Y, Ferrari S, Cremer M, Cremer T. Remodeling of nuclear landscapes during human myelopoietic cell differentiation maintains co-aligned active and inactive nuclear compartments. Epigenetics Chromatin. 2015;8:47.

48. Popken J, Brero A, Koehler D, Schmid VJ, Strauss A, Wuensch A, Guengoer T, Graf A, Krebs S, Blum H, et al. Reprogramming of fibroblast nuclei in cloned bovine embryos involves major structural remodeling with both striking similarities and differences to nuclear phenotypes of in vitro fertilized embryos. Nucleus. 2014;5:555–89.

49. Arner E, Daub CO, Vitting-Seerup K, Andersson R, Lilje B, Drablos F, Lennartsson A, Ronnerblad M, Hrydziuszko O, Vitezic M, et al. Transcribed enhancers lead waves of coordinated transcription in transitioning mammalian cells. Science. 2015;347:1010–4.

50. Lawrence M, Huber W, Pages H, Aboyoun P, Carlson M, Gentleman R, Morgan MT, Carey VJ. Software for computing and annotating genomic ranges. PLoS Comput Biol. 2013;9:e1003118.

51. Popken J, Dahlhoff M, Guengoer T, Wolf E, Zakhartchenko V. 3D structured illumination microscopy of mammalian embryos and spermatozoa. BMC Dev Biol. 2015;15:46.

52. Zerbino DR, Wilder SP, Johnson N, Juettemann T, Flicek PR. The ensembl regulatory build. Genome Biol. 2015;16:56.

53. Belmont AS, Bruce K. Visualization of G1 chromosomes: a folded, twisted, supercoiled chromonema model of interphase chromatid structure. J Cell Biol. 1994;127:287–302.

54. Boeger H, Shelansky R, Patel H, Brown CR. From structural variation of gene molecules to chromatin dynamics and transcriptional bursting. Genes (Basel). 2015;6:469–83.

55. Over RS, Michaels SD. Open and closed: the roles of linker histones in plants and animals. Mol Plant. 2014;7:481–91.

56. Saksouk N, Simboeck E, Dejardin J. Constitutive heterochromatin formation and transcription in mammals. Epigenetics Chromatin. 2015;8:3.

57. Hihara S, Pack CG, Kaizu K, Tani T, Hanafusa T, Nozaki T, Takemoto S, Yoshimi T, Yokota H, Imamoto N, et al. Local nucleosome dynamics facilitate chromatin accessibility in living mammalian cells. Cell Rep. 2012;2:1645–56.

58. Nagano T, Lubling Y, Stevens TJ, Schoenfelder S, Yaffe E, Dean W, Laue ED, Tanay A, Fraser P. Single-cell Hi-C reveals cell-to-cell variability in chromosome structure. Nature. 2013;502:59–64.

59. Knoch TA, Wachsmuth M, Kepper N, Lesnussa M, Abuseiris A, Ali Imam AM, Kolovos P, Zuin J, Kockx CE, Brouwer RW, et al. The detailed 3D multiloop aggregate/rosette chromatin architecture and functional dynamic organization of the human and mouse genomes. Epigenetics Chromatin. 2016;9:58.

60. Maeshima K, Ide S, Hibino K, Sasai M. Liquid-like behavior of chromatin. Curr Opin Genet Dev. 2016;37:36–45.

61. Maeshima K, Hihara S, Eltsov M. Chromatin structure: does the 30-nm fibre exist in vivo? Curr Opin Cell Biol. 2010;22:291–7.

62. Maeshima K, Kaizu K, Tamura S, Nozaki T, Kokubo T, Takahashi K. The physical size of transcription factors is key to transcriptional regulation in chromatin domains. J Phys: Condens Matter. 2015;27:064116.

63. Shao S, Zhang W, Hu H, Xue B, Qin J, Sun C, Sun Y, Wei W. Long-term dual-color tracking of genomic loci by modified sgRNAs of the CRISPR/Cas9 system. Nucleic Acids Res. 2016;44:e86.

64. Kirmes I, Szczurek A, Prakash K, Charapitsa I, Heiser C, Musheev M, Schock F, Fornalczyk K, Ma D, Birk U, et al. A transient ischemic environment induces reversible compaction of chromatin. Genome Biol. 2015;16:246.

65. Hubner B, Cremer T, Neumann J. Correlative microscopy of individual cells: sequential application of microscopic systems with increasing resolution to study the nuclear landscape. Methods Mol Biol. 2013;1042:299–336.

66. Andersson R, Gebhard C, Miguel-Escalada I, Hoof I, Bornholdt J, Boyd M, Chen Y, Zhao X, Schmidl C, Suzuki T, et al. An atlas of active enhancers across human cell types and tissues. Nature. 2014;507:455–61.

67. Dobbie IM, King E, Parton RM, Carlton PM, Sedat JW, Swedlow JR, Davis I. OMX: a new platform for multimodal, multichannel wide-field imaging. Cold Spring Harb Protoc. 2011;2011:899–909.

68. Ball G, Demmerle J, Kaufmann R, Davis I, Dobbie IM, Schermelleh L. SIMcheck: a toolbox for successful super-resolution structured illumination microscopy. Sci Rep. 2015;5:15915.

69. Pau G, Fuchs F, Sklyar O, Boutros M, Huber W. EBImage: an R package for image processing with applications to cellular phenotypes. Bioinformatics. 2010;26:979–81.

70. R Core Team. A language and environment for statistical computing. R Foundation for Statistical Computing, Vienna, Austria; 2013. http://www.R-project.org/.

71. Ollion J, Cochennec J, Loll F, Escude C, Boudier T. TANGO: a generic tool for high-throughput 3D image analysis for studying nuclear organization. Bioinformatics. 2013;29:1840–1.

72. Cerase A, Smeets D, Tang YA, Gdula M, Kraus F, Spivakov M, Moindrot B, Leleu M, Tattermusch A, Demmerle J, et al. Spatial separation of Xist RNA and polycomb proteins revealed by superresolution microscopy. Proc Natl Acad Sci USA. 2014;111:2235–40.

8

Histone 4 lysine 8 acetylation regulates proliferation and host–pathogen interaction in *Plasmodium falciparum*

Archana P. Gupta, Lei Zhu, Jaishree Tripathi, Michal Kucharski, Alok Patra and Zbynek Bozdech*

Abstract

Background: The dynamics of histone modifications in *Plasmodium falciparum* indicates the existence of unique mechanisms that link epigenetic factors with transcription. Here, we studied the impact of acetylated histone code on transcriptional regulation during the intraerythrocytic developmental cycle (IDC) of *P. falciparum*.

Results: Using a dominant-negative transgenic approach, we showed that acetylations of histone H4 play a direct role in transcription. Specifically, these histone modifications mediate an inverse transcriptional relationship between the factors of cell proliferation and host–parasite interaction. Out of the four H4 acetylations, H4K8ac is likely the rate-limiting, regulatory step, which modulates the overall dynamics of H4 posttranslational modifications. H4K8ac exhibits maximum responsiveness to HDAC inhibitors and has a highly dynamic distribution pattern along the genome of *P. falciparum* during the IDC. Moreover, H4K8ac functions mainly in the euchromatin where its occupancy shifts from intergenic regions located upstream of 5′ end of open reading frame into the protein coding regions. This shift is directly or indirectly associated with transcriptional activities at the corresponding genes. H4K8ac is also active in the heterochromatin where it stimulates expression of the main antigenic gene family (*var*) by its presence in the promoter region.

Conclusions: Overall, we demonstrate that H4K8ac is a potential major regulator of chromatin-linked transcriptional changes during *P. falciparum* life cycle which is associated not only with euchromatin but also with heterochromatin environment. This is potentially a highly significant finding that suggests a regulatory connection between growth and parasite–host interaction both of which play a major role in malaria parasite virulence.

Keywords: *P. falciparum*, H4K8ac, Chromatin, Transcriptional dynamics, HDAC inhibitors, ChIP-Seq

Background

Plasmodium falciparum, one of the protozoan parasites responsible for malaria in humans, exhibits coordinated mechanisms of transcriptional regulation during the development through its life cycle. There is mounting evidence that epigenetic mechanisms contribute to this unique gene regulation and thus are vital for the parasite's growth and development [1–10]. Besides, absence of the linker histone H1 [11], scarcity of transcription factors [12] and lack of a functional RNA interference system [13] has led to a working model in which histone posttranslational modifications (PTMs) play a pivotal role in *P. falciparum* gene regulation. Based on this model, *Plasmodium* epigenome is considered mainly euchromatic [14–16], marked by unique combinations of histone variants and their posttranslational modifications (PTMs) [11, 17–19]. The abundance of parasite specific histone modifications is suggestive of the role of epigenetic mechanisms regulating parasite virulence [18, 19]. Intriguingly, the genome-wide distribution of histone variants and their PTMs is highly dynamic across the *P. falciparum* intraerythrocytic developmental cycle (IDC) [20–23]. Like mRNA abundance, occupancy of many histone marks such as H4K8ac and H3K9ac shows single peak profiles across the IDC in a large proportion of

*Correspondence: zbozdech@ntu.edu.sg
School of Biological Sciences, Nanyang Technological University, 60 Nanyang Drive, Singapore 637551, Singapore

the *P. falciparum* genes [21]. On the other hand, there are histone PTMs whose occupancy is constant throughout the IDC. These include canonical heterochromatin markers such as H3K9me3 and H3K36me3, implicated in gene silencing [14, 24, 25] and few euchromatin histone marks such as H4K5ac and H3K14ac, either abundantly spread through the majority of the genome or confined to a small number of genomic loci [21]. Occupancy of histone variants also contributes to the dynamic chromatin remodeling throughout the *P. falciparum* IDC with some variants which associate with actively transcribed genes, while others play roles in chromatin structure [20, 26–28]. Chromatin remodeling is also affected by dynamic nucleosome structures [29–33] and chromatin binding proteins [34–37] in *P. falciparum*. Likewise, the dynamics is reflected in occupancy of RNA polymerase II exhibiting distinct patterns for early and late expressed genes [38]. Altogether these findings suggest that the time component of the occupancy profiles across the IDC is one of the variables of the overall "histone code" playing a key fundamental role in gene expression during the *Plasmodium* life cycle.

Studying 13 canonical PTMs of H4 and H3, we have previously shown that acetylation of H4 at lysine residue 8 (H4K8ac) is among the most dynamic modifications occupying predominantly the 5′ intergenic regions (5′IGRs) and 5′ termini of the open reading frames (ORFs) of more than half of the *P. falciparum* genes [21]. The single peak occupancy profiles of the 5′IGR/ORF-bound H4K8ac showed good correlation with the respective mRNA profiles across the IDC. This was highly surprising given that the "neighboring" PTMs at the H4 "tail" (H4K5ac and H4K12ac) exhibited a highly abundant but constant occupancy throughout the vast majority of the genome. Treatment of *P. falciparum* parasites with the Class I and II histone deacetylase (HDAC) inhibitor apicidin resulted in induction of the overall protein levels of H4K8ac as well as its occupancy across the genome [39]. H4K8ac (together with H3K9ac) was also found to be an effector of the DNA damage stress response, being induced by treatment of *P. falciparum* parasites with methyl methanesulfonate (MMS) at multiple stages of the IDC [40]. The MMS-induced levels of H4K8ac in *P. falciparum* coincide with transcriptional induction of stress responses. Intriguingly, artemisinin, the main chemotherapeutics for malaria treatment, solicited a similar effect characterized by increased levels of H4K8ac and upregulation of the stress response genes [40]. This strongly suggests that H4K8ac plays a role in transcription regulation associated with both, progression of the *Plasmodium* life cycle, and, responses to external perturbations/stresses. Surprisingly, in other eukaryotic organisms, H4K8ac is yet to be implicated

in any major processes of epigenetic regulation of gene expression. In yeast and humans, H4K8ac seems to play an auxiliary role in transcription, being a part of an overall euchromatin-linked histone PTM complex that occupies active promoters [41, 42]. This may suggest that in contrast to most eukaryotes, during the *Plasmodium* evolution, H4K8ac acquired new functions in epigenetic regulation of gene expression and possibly emerged as one of the most crucial histone marks.

All above-mentioned studies, however, provided only associative evidence of H4K8ac involvement in transcription showing modest albeit statistically significant overlaps between H4K8ac-bound genetic loci and transcriptionally deregulated genes [21, 39, 40]. Here, we wanted to establish direct links between H4K8ac and transcription, and evaluate these in context of the neighboring H4 PTMs (H4K5ac, H4K12ac and H4K16ac) detected in *P. falciparum*. For this, we suppressed the studied acetylations in *P. falciparum* cell lines that (over) expressed H4 genes with mutations at lysines 5, 8, 12, 16. The hypoacetylation of histone H4 resulted in broad transcriptional changes characterized by upregulation of genes involved in cell proliferation and downregulation of genes of host–parasite interactions and antigenic presentation/variation. Accordingly, there was an increase in multiplication rate of *P. falciparum* parasites when H4K8ac or all 4 H4 acetylations were suppressed. On the other hand, H4K8 exhibited maximum level of hyperacetylation induced by HDAC inhibitors compared to all other euchromatin marks; however, this hyperacetylation is highly transient. ChIP-Seq results showed that H4K8ac shifts its position back-and-forth between the 5′IGRs in the early and late IDC stages to the 5′ORF in the middle IDC stage. Besides its euchromatin-linked roles, H4K8ac also functions in the context of heterochromatin where it is involved in activation of the major antigenic variation gene family (*var*).

Results

Dominant-negative transgenic lines for H4 acetylations

In the first step, we created dominant-negative *P. falciparum* transgenic parasite lines transfected with pBcamR plasmid containing HA-tagged H4 genes in which the targeted lysine (K) residues (K5, K8, K12, K16) where replaced with arginine (R) (Additional file 1). In total, we generated five transgenic lines with the individual mutations (H4K5R, H4K8R, H4K12R, H4K16R) or with all 4 lysines changed to arginine (H4ac4R) and two additional control lines with either HA-tag only (HA) or wild-type HA-tagged H4 gene (H4-HA). As expected, there was an increase in the plasmid copy number and expression of the transgenic H4 proteins in *P. falciparum* grown in 10 μg/ml of blasticidin compared to 2.5 μg/ml in all parasite lines (Additional file 2: Figure S1a and Fig. 1a, respectively). Cell lysate fractionation showed that the transgenic H4 proteins could be solubilized only by high salt extraction of the nuclear fraction (Fig. 1b). More importantly, the intensities of HA-tagged H4 proteins were found to be, 3.12-fold higher for H4K8R, 2.6-fold higher for H4K12R and 3.6-fold higher for H4K16R compared to endogenous H4 protein in the transfectants at 10 μg/ml of blasticidin (Fig. 1a and Additional file 2: Figure S1h). This indicates partial but significantly higher displacement of endogenous proteins by the HA-tagged H4 protein. Transgenic proteins were also detected in the acid-extracted histone fractions and immuno-fluorescence microscopy of H4 K8R confirmed its nuclear localization (Additional file 2: Figure S1b). Taken together, these results demonstrate that the episomally expressed H4 proteins are targeted to the *P. falciparum* nucleosomes, presumably displacing the endogenous H4 protein and by that suppressing the respective acetylations. Hence, these transgenic lines provide an experimental tool to investigate the role of H4 lysine residues in epigenetic regulation of gene expression during *P. falciparum* IDC.

Functional implication of the dominant-negative inhibition of H4

Next, we carried out transcriptome analyses of the transfectants grown in the presence of 10 μg/ml of blasticidin. By comparing the transcriptional profiles of the H4 K/R mutants with those from both controls at three stages of the IDC, we observed broad transcriptional changes (Fig. 1c). Remarkably, there appears to be strong similarities between the transcriptional changes induced by the dominant-negative effect of all five H4K/R mutants. Overall, there were higher number of genes that were differentially expressed in the ring and schizont compared to the trophozoite stages. The transcription patterns of the dominant-negative transfectants was not or negatively correlated to HA and positively correlated to each other when normalized to H4-HA (Additional file 2: Figure S1c). On the other hand, there were essentially no transcriptional changes between the unmutated H4-HA and transfectant with the empty vector (HA) also grown at 10 μg/ml of blasticidin (Additional file 2: Figure S1d). Altogether, these results indicate that the acetylations at the 4 lysine residues of H4 act in a similar way possibly reflecting a synergistic and/or complimentary function in transcription during the *P. falciparum* IDC. Nonetheless, there were subtle differences within the individual profiles with H4ac4R clustering closer with H4K8R in the ring and trophozoite stages (Pearson's correlation coefficient 0.75 and 0.71, respectively) compared to the other 3 mutants (Fig. 1c; Additional file 2: Figure S1c). Given that

H4ac4R shows consistently the strongest effect on transcription, this may suggest a key role of H4K8ac in this process.

Functional analyses of the induced transcriptional changes showed a high consistency in the altered mRNA levels for all enriched pathways for all four H4 acetylations (Fig. 1d; Additional file 3: Table S1, Additional file 4: Figure S2). Moreover, similar pathways were deregulated in the H4K8R transfectant at the schizont stage when the blasticidin concentration was increased from 2.5 to 10 μg/ml whereas H4-HA grown at 2.5 and 10 μg/ml of blasticidin induced very few transcriptional changes (Additional file 2: Figure S1e), confirming that the transcriptional deregulation is a direct result of dominant-negative expression of the mutated H4 protein. Namely, the over-expression of the H4 mutants caused upregulation of a wide range of genes involved in cellular pathways associated with growth and replication (e.g., proliferation) of *P. falciparum* during the IDC. These include *genes encoding factors of the assembly of cytoplasmic and the organellar ribosomes, glycolysis, DNA replication* and *nuclear encoded genes for apicoplast and mitochondrial protein transport* (annotated by MPMP; http://mpmp.huji.ac.il/) (Fig. 1d). In addition, there was an upregulation of two classes of stress response pathways including the *DNA repair machinery* and *thioredoxin metabolism*. While some pathways (such as DNA repair) were upregulated throughout the IDC, others were typically de-repressed during their transcriptional downturn. For example, the genes coding for 40S, 50S and 60S cytoplasmic ribosomal proteins normally expressed during rings [43] were upregulated in the trophozoite and schizont stages, whereas genes encoding for mitochondrial and apicoplast ribosomal proteins normally expressed in the schizonts were upregulated in the rings (Additional file 2: Figure S1f). This indicates that H4 acetylations facilitate transcriptional suppression of proliferation related genes in a life cycle-specific manner, hence regulating the growth of parasite. This occurs in all three major developmental stages of the IDC in which the transgenic histones were expressed at the protein levels (Additional file 2: Figure S1g). Interestingly, the disruptions of H4K8ac or the 4 acetylations lead to significant increases in the parasite multiplication rate (MR) by up to 1.3-fold and twofold, respectively (Fig. 1e). This is most likely a direct result of the induced levels of the mutated H4 proteins as no alterations in MR were observed at 2.5 μg/ml of blasticidin in all transfectant parasites lines. Surprisingly, this was not associated with an increased number of newly formed merozoites in the segmented schizonts measured by light microscopy (Fig. 1f). Moreover, both the mutant and non-mutant transfectants have an identical length of the IDC, measured by transcriptomics analysis (Additional

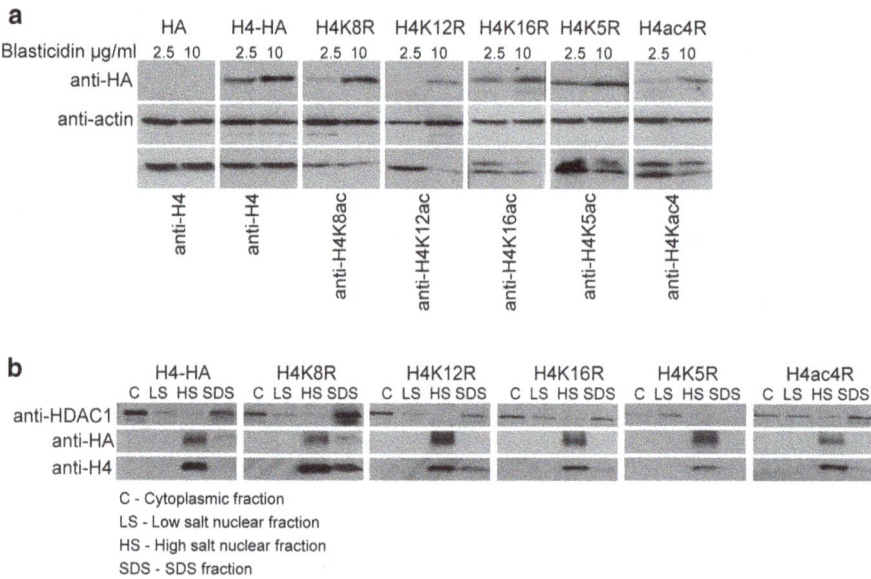

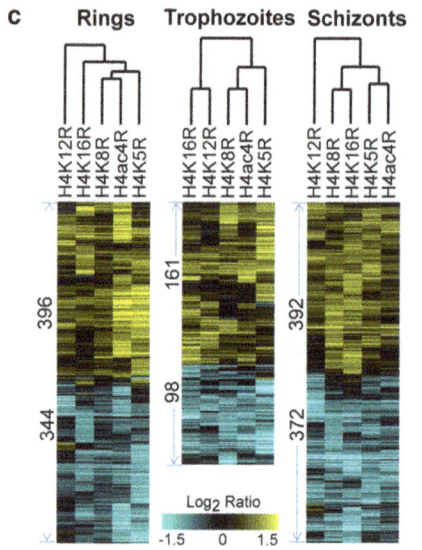

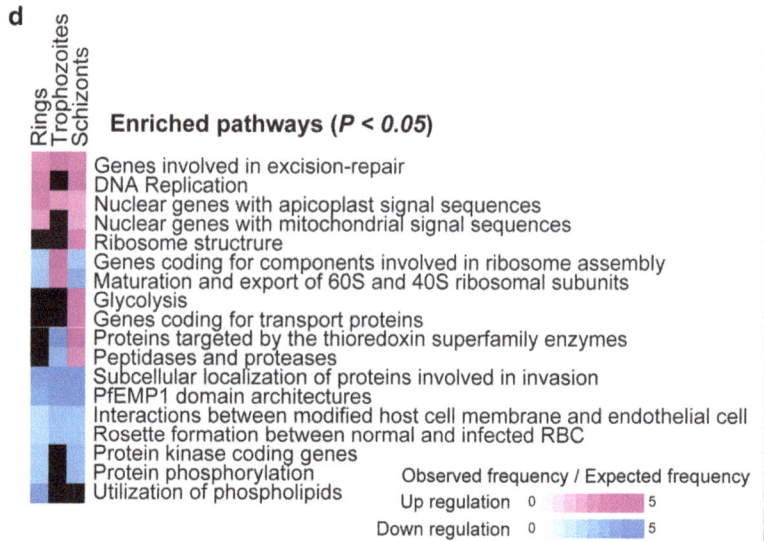

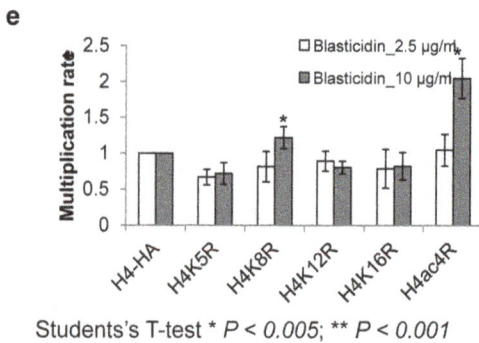

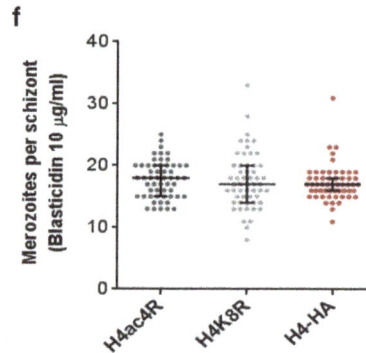

(See figure on previous page.)
Fig. 1 Histone H4 acetylation mutants. **a** Western blot analysis using total protein from the transfectants grown at 2.5 and 10 µg/ml blasticidin to check the presence of HA-tagged H4 proteins using antibodies against HA. Histone antibodies were used as positive control to confirm the expression of endogenous histones and actin was used as a loading control. **b** Western blot analysis carried out from cytoplasmic, low salt nuclear, high salt nuclear and total SDS fractions to confirm nuclear localization of HA-tagged H4 proteins in the transfectants. **c** Microarray expression analysis for transfectants grown at 10 µg/ml blasticidin. *Heat maps* represent the differentially expressed genes in the transfectants at each stage (Student's t test, $P < 0.05$). Data shown is $\log_2$ ratio of average of biological triplicates, normalized to H4-HA transfectant. **d** Functional pathways significantly up or downregulated in the differentially expressed gene sets (data from **c**) of the transfectants. The *scale* shows the fold enrichment of number of genes in the specific pathways [based on hypergeometric test ($P < 0.05$)]. **e** Effect of H4 mutations on multiplication. Cells were grown in 2.5 or 10 µg/ml blasticidin and FACS using Hoechst stain was carried out to count the number of new rings at each invasion cycle for 6 cycles. **f** The scatter plot of merozoite number in the segmented schizonts in H4ac4R, H4K8R and H4-HA transfectants determined by Giemsa stain light microscopy

file 5: Table S5). This suggests that the increased MR is likely facilitated by a higher fraction of viable merozoite with elevated levels of infectivity. It will be interesting to explore this possibility in future studies.

The most predominant functional groups enriched among the downregulated genes were factors of host–parasite interactions annotated as *PfEMP1 domain architecture, rosette formation between normal and infected RBC* and *interactions between modified host cell membrane and endothelial cells* (Fig. 1d). In particular, we observed downregulation of essentially all members of two gene families implicated in antigenic variation and immune evasion, *var* and *rifin*. The other important downregulated pathway consisted of genes involved in *merozoite invasion*, which includes merozoite surface proteins, rhoptry associated proteins, reticulocyte binding protein homologs and erythrocyte binding antigens (Additional file 3: Table S1). In contrast, downregulation of these genes was independent of the developmental stage, typically occurring throughout the IDC (Fig. 1d; Additional file 4: Figure S2). We validated that the 40S and 50S cytoplasmic ribosomal proteins were upregulated by, 1.7 to 2-fold, and, 1.7- to 4.3-fold, in trophozoite and schizont stages, respectively. On the other hand, invasion-related genes, RAP2 and RAP3, and, MSP and RAP3, were confirmed to be downregulated showing a fold change of <1 compared to wild-type parasites in trophozoite and schizont stages, respectively (Additional file 2: Figure S1i). Taken together, these results suggest that the 4 lysine residues of the H4 tail play a central role in regulating several key biological functionalities by which the parasite adapts to its environment. Specifically, the H4 acetylations facilitate an inverse regulatory relationship between growth and multiplication on one-side and host–parasite interactions on the other.

H4K8ac is the dynamic component of the H4 tetra-acetylation moiety

Next, we wished to compare the responsiveness of the four H4 lysine acetylations to HDAC inhibition in a broader context of *P. falciparum* chromatin remodeling

[21, 39, 40]. Highly synchronized trophozoites were treated with IC_{90} concentrations of Trichostatin A (TSA) or apicidin (50 and 70 nM, respectively) for 6 h followed by removal of the drug and culturing for another 2, 4 and 6 h, similar to our previous study [44]. Out of the 4 H4 acetylations, H4K8ac and (to a lesser degree) H4K16ac responded to both HDAC inhibitors, while H4K5 and H4K12 acetylation levels were unchanged (Fig. 2a). Both HDAC inhibitors induced acetylation at other euchromatin marks including H3K23ac, H3K56ac and to lesser degree H3K9ac whereas euchromatin-linked (tri)methylations of H3K4 and H4K20 were unresponsive. Finally, there was a dramatic increase in the signal intensity using an antibody against H4 tetra-acetylation (H4ac4). This likely corresponds to the changes mainly at H4K8ac and (less so) H4K16ac, given that the levels of H4K5ac and H4K12ac appear to be constitutive. As mentioned above, H4K8ac is the most dynamic euchromatin mark whose occupancy is tightly correlated with transcriptional activity, while H4K16ac showed only moderate-to-low levels of occupancy changes and is virtually uncoupled from transcription [21]. This suggests that out of the two dynamic H4 acetylations, H4K8ac may play a pivotal role in regulation of gene expression that is ultimately mediated by the H4ac4 epigenetic moiety. Hence, in the following parts of our study, we focus on H4K8ac as a major regulatory factor in gene expression during the *P. falciparum* IDC.

The effect of both TSA and apicidin on all the responsive histone acetylations appears to be transient as most of the histone modifications returned to their basal levels over 6 h after the compound removal (Fig. 2a). This is consistent with our previous study showing a transient effect of HDAC inhibitors on *P. falciparum* transcriptome [44]. To explore this phenomenon further, we carried out a "treatment/wash-off" time course experiment where the highly synchronized trophozoites were first exposed to 50 nM of TSA and subsequently samples were collected 0, 0.5, 2 and 4 h after the drug removal (Fig. 2b). These samples were used for transcriptomic and ChIP-on-chip analysis of H4K8ac occupancy,

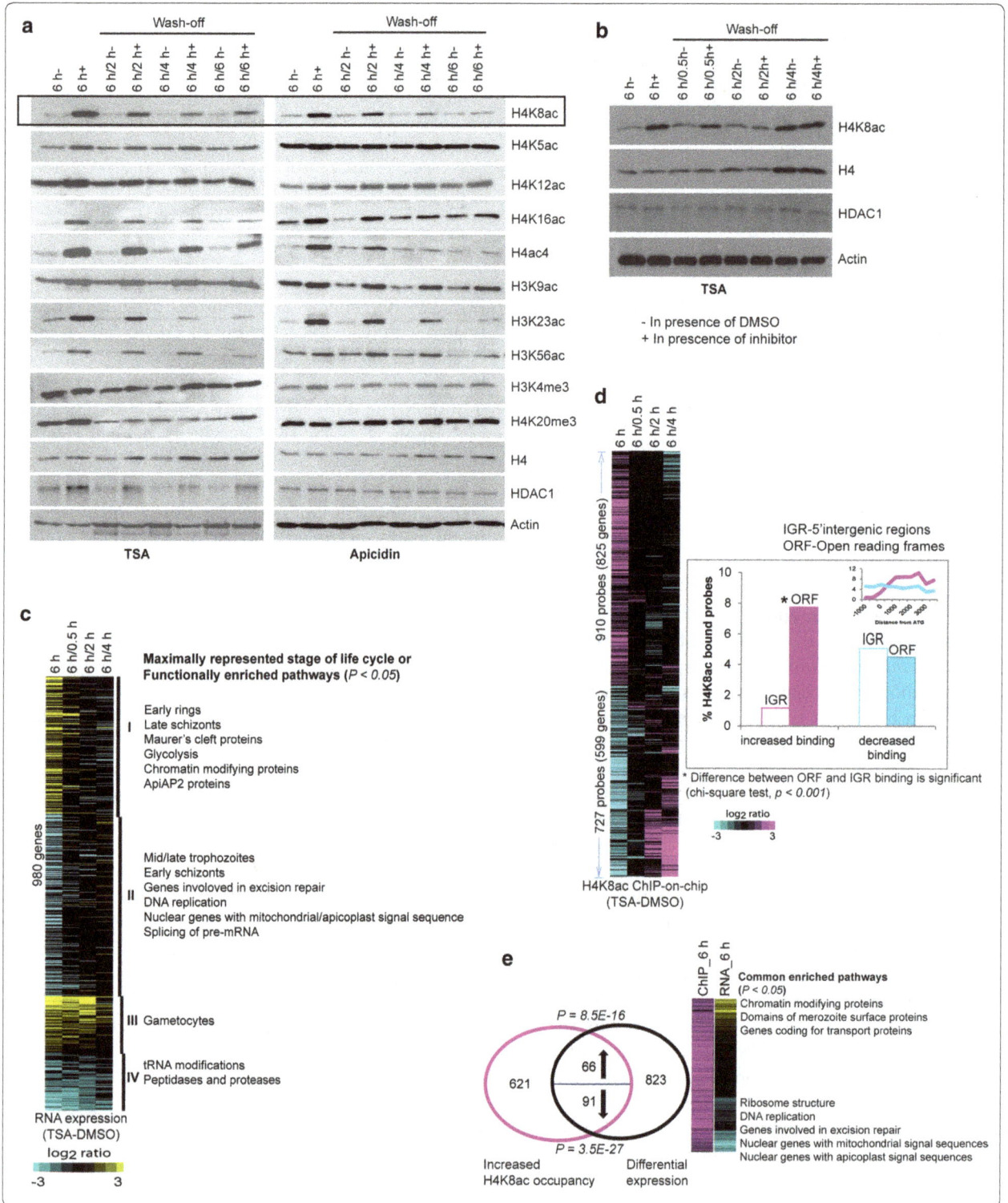

a

Wash-off   Wash-off

H4K8ac
H4K5ac
H4K12ac
H4K16ac
H4ac4
H3K9ac
H3K23ac
H3K56ac
H3K4me3
H4K20me3
H4
HDAC1
Actin

TSA   Apicidin

b

Wash-off

H4K8ac
H4
HDAC1
Actin

TSA

- In presence of DMSO
+ In presence of inhibitor

c

**Maximally represented stage of life cycle or
Functionally enriched pathways** (*P* < 0.05)

Early rings
Late schizonts
I   Maurer's cleft proteins
Glycolysis
Chromatin modifying proteins
ApiAP2 proteins

Mid/late trophozoites
Early schizonts
II   Genes involved in excision repair
DNA replication
Nuclear genes with mitochondrial/apicoplast signal sequence
Splicing of pre-mRNA

III Gametocytes

IV   tRNA modifications
Peptidases and proteases

RNA expression
(TSA-DMSO)
log2 ratio
-3   3

d

IGR-5'intergenic regions
ORF-Open reading frames

H4K8ac ChIP-on-chip
(TSA-DMSO)

* Difference between ORF and IGR binding is significant
(chi-square test, *p* < 0.001)

log2 ratio
-3   3

e

P = 8.5E-16

621   66   823
           91

P = 3.5E-27

Increased   Differential
H4K8ac occupancy   expression

Common enriched pathways
(*P* < 0.05)
Chromatin modifying proteins
Domains of merozoite surface proteins
Genes coding for transport proteins

Ribosome structure
DNA replication
Genes involved in excision repair
Nuclear genes with mitochondrial signal sequences
Nuclear genes with apicoplast signal sequences

(See figure on previous page.)
Fig. 2 Effect of HDAC inhibitors on *P. falciparum*. **a** Western blot analysis to check the effect of HDAC inhibitors on histone modifications. Tropho-zoite stage parasites were cultured for 6 h with DMSO (6 h−), 70 nM of Trichostatin A (TSA) or 75 nM of Apicidin (6 h+). The drug was subsequently washed off, and the cells were grown for another 2 h (6 h/2 h), 4 h (6 h/4 h) or 6 h (6 h/6 h). The antibodies used for each blot are shown on the right. **b** Similar experiment was repeated with TSA treatment for 6 h, subsequent wash off and growth for 0.5 h (6 h/0.5 h), 2 h (6 h/2 h) and 4 h (6 h/4 h). Total protein was extracted from the respective samples for western blot analysis using antibodies against histone H4 or H4K8ac or Pf HDAC1. **c** Microarray expression analysis of TSA treated cells. TSA or DMSO treatment for 6 h was done in biological triplicates and subsequent wash-off experiments were done in duplicates. The *heat map* (*left panel*) shows average $\log_2$ ratios of replicates (TSA minus DMSO) of the genes differentially expressed after 6 h TSA exposure (Student's *t* test, $P < 0.05$). *Right panel* represents the parasite stage whose genes are maximally expressed as well as the functionally enriched MPM pathways in each of the clusters. **d** ChIP-on-chip using H4K8ac antibody to assess differential binding upon TSA treatment. All experiments were done in at least duplicates. The *heat map* on the *left* shows average $\log_2$ ratios of replicates (TSA minus DMSO) of the probes differentially acetylated after 6-h TSA exposure (Wilcoxon rank sum test, $P < 0.05$). *Graph* on the *right* shows the percentage of probes that are differentially acetylated upon TSA treatment for 6 h. *Inset graph* shows the differential occupancy across the gene body. **e** *Venn diagram* represents the overlap between differential expression and increased H4K8 occupancy in the ORFs due to TSA. Binomial distribution was used to assign *P* values for up or downregulated genes overlapping with H4K8ac hyperacetylation. *Heat map* on the *right* represents the fold change in H4K8ac occupancy or RNA expression of the hyperacetylated genes. Functional pathways (MPM) significantly enriched in the up- or downregulated gene sets are shown

simultaneously. In this experimental setting, TSA caused a dramatic deregulation of the IDC transcriptional cascade with a broad downregulation of trophozoite-specific genes and upregulation of genes normally expressed in other stages (early ring and early schizonts) (Fig. 2c). This deregulation is largely reversible such that even 30 min after the drug removal, the majority of the transcripts returned to their original levels. However, there were two gene clusters whose TSA-induced transcriptional change was stable for at least 2 h. These include genes of gametocyte and merozoite stages among the upregulated and *tRNA modifications* and *peptidases and proteases* among the downregulated genes (Fig. 2c; Additional file 6: Table S2). In future studies, it will be interesting to understand the molecular mechanisms that underline both the reversible and irreversible transcriptional changes induced by the HDAC inhibitors. The chromosomal occupancy of H4K8ac appeared also highly sensitive to TSA at least 1637 genetic loci corresponding to 1424 genes (Fig. 2d). There is a remarkable asymmetry in the distribution of the TSA-induced H4K8ac occupancy that is significantly enriched within the ORFs ($P < 0.001$) compared to the 5′IGRs (Fig. 2d inset). There was a modest but significant overlap between the TSA-induced acetylation and altered transcription. The increased H4K8ac within the ORFs coincided with increased expression of 66 genes (hypergeometric test, $P = 8.5E-16$) and decreased expression of 91 genes (hypergeometric test, $3.5E-27$) (Fig. 2e; Additional file 7: Table S3). The upregulated gene set is functionally enriched for several pathways such as *chromatin modifying proteins* and *merozoite surface proteins* and genes with maximum expression in gametocytes and merozoite stages. The downregulated gene set upon H4K8ac hyperacetylation included factors of *DNA replication, excision repair* and *ribosome structure*. This effect

is opposite to the H4 hypoacetylation where genes of the proliferation pathways were upregulated, and merozoite surface proteins were downregulated (Fig. 1d). This further supports our model in which H4K8ac plays a central, regulatory role within the H4ac4 moiety.

H4K8ac regulates gene expression within both euchromatin and heterochromatin

Next, we wished to explore the distribution of the H4K8R mutant protein within the *P. falciparum* chromatin and correlate these to the induced transcriptional changes to find key regions of its epigenetic activity. For this we compared ChIP-on-chip results from the mutant H4K8R and unmutated H4-HA using an anti-HA antibody (Fig. 3a). Here, we observed variable occupancy patterns across the *P. falciparum* chromosomes, which suggests of the transgenic histones, displaced the endogenous histones partially with some regions being fully protected while others tolerating the mutant histone to a higher degree. Overall, there was a significant skew of the H4K8R binding to the 5′IGRs in rings and trophozoites while in schizonts, the H4K8R and H4-HA exhibit a more similar distribution. This is in sharp contrast to the TSA-induced hyperacetylation that occurs almost exclusively at the ORF of the genes (see Fig. 2d). Moreover, we observed that the increased (hypoacetylation driving) occupancy of H4K8R coincided significantly with differential expression including both up and downregulated genes. The upregulated gene set was enriched for nuclear *encoded genes responsible for apicoplast/mitochondrial import, DNA replication* and *repair machinery*. On the other hand, the downregulated genes with enhanced H4K8R occupancy included genes encoding the two main subtelomeric gene families involved in antigenic variations (*var* and *rifin*). This is particularly evident to the *var* genes where the enhanced H4K8R occupancy

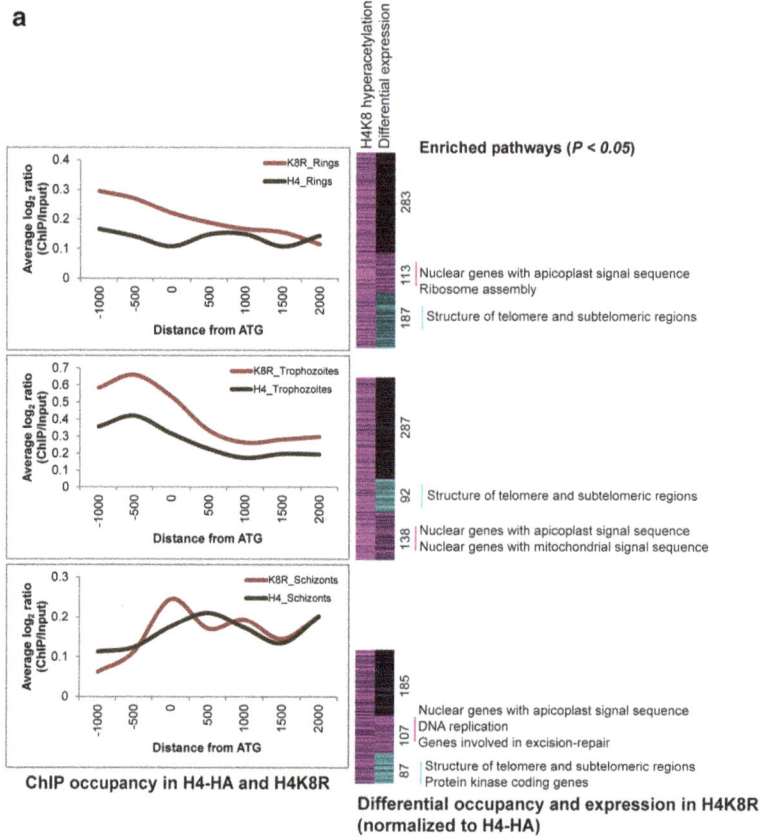

a

ChIP occupancy in H4-HA and H4K8R

Differential occupancy and expression in H4K8R
(normalized to H4-HA)

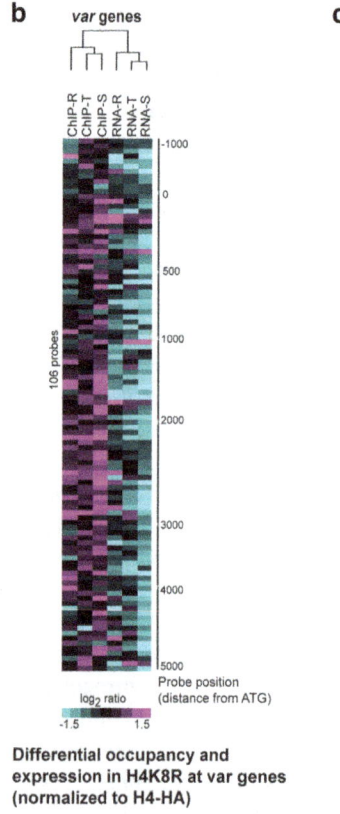

b

var genes

Differential occupancy and
expression in H4K8R at var genes
(normalized to H4-HA)

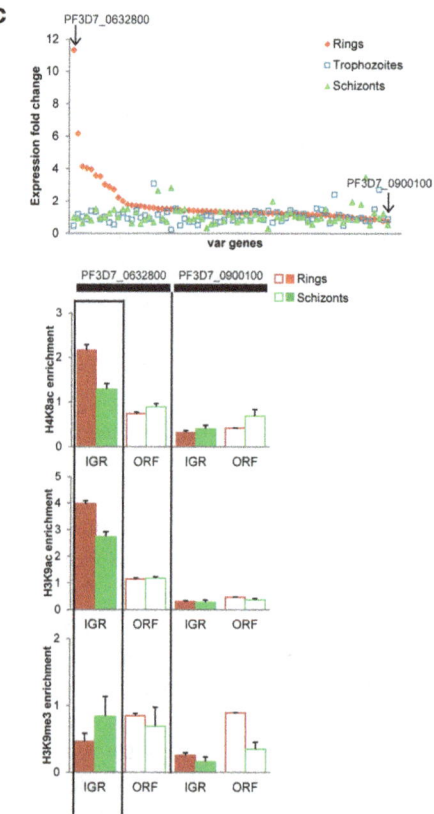

c

(See figure on previous page.)

Fig. 3 Euchromatin- and heterochromatin-linked H4K8ac. **a** ChIP-on-chip for H4-HA and H4K8R transfectants using anti-HA antibody. All microarrays were done in triplicates, and differential binding was assessed by Wilcoxon rank sum test ($P < 0.05$). *Graphs* on the *left* show average ChIP/Input log$_2$ ratio of all the probes binned according to their distance from the start codon. *Heat maps* on *right* depict the probes with increased binding and change in expression in the corresponding genes in H4K8R transfectants (compared to H4-HA). Functional pathways (MPMP) significantly enriched in the up or downregulated gene sets are shown. **b** Analysis of *var* genes linked heterochromatin in H4K8R. *Heat map* represents the differential binding at *var* gene loci and differential RNA expression of corresponding genes (H4K8R normalized to H4-HA transfectants). The distance from ATG marked on the *right* depicts the microarray probe position with respect to gene start. **c** Histone marks at active/inactive *var* gene in 3D7 clone. *Graph* on *top* represents microarray gene expression values of *var* genes in one of the 3d7 clones. The highest and least expressed *var* genes are marked with the *arrow*. The lower 3 *graphs* show the ChIP enrichment by qPCR of the most dominantly expressed *var* gene and the least expressed *var* gene (input subtracted Ct values normalized to ORF of PF3D7_1240300). Primers were designed within 1000 bp on each side of the gene start. The results of the other 3 clones are shown in Additional file 4: Figure S3

coincides with dramatically decreased transcript levels in all three stages of the IDC (Fig. 3b). This implies that the H4K8R-driven hypoacetylation at the *var* ORFs suppresses their transcription. This is surprising given that the subtelomeric regions of the *P. falciparum* chromosomes are in the heterochromatin state associated with typical heterochromatin factors such as H3K9me3 and heterochromatin binding protein HP1 [15, 35]. Our result suggests that H4K8ac, one of the main euchromatin markers in *P. falciparum*, also contributes to *var* gene regulation.

Given that the ChIp-on-chip measurements (Fig. 3b) were carried out with a parasite culture that were not selected for a single *var* gene expression, it was impossible to discern if the detected H4K8ac effect occurs specifically at the dominant transcript or all *var* genes. To distinguish between these two possibilities, we examined several isogenic clones derived by serial dilution of an in vitro culture of the *P. falciparum* 3D7 strain that exhibit a single var gene expression. These clones reflect the mutually exclusive expression with a single dominant transcript expressed at significantly higher levels in the ring stage, compared to the rest of the family (Fig. 3c; Additional file 8: Figure S3a). Using chromatin immunoprecipitation, we show that in 3 of the 4 clones, H4K8ac along with H3K9ac, is over-enriched at the intergenic region of the dominant *var* gene in the ring stage (Fig. 3c; Additional file 8: Figure S3b). On the other hand, H3K9me3 was enriched in the intergenic region of the dominant *var* transcript during the schizont stage, while in the ring stage this modification was mostly present in the ORF (Fig. 3c; Additional file 8: Figure S3b). Taken together, these results indicate that H4K8ac contributes to transcriptional regulation not only in euchromatin but also within a heterochromatin context. Consistent with its linking role between growth and host–parasite interaction (see above), this modification participates in induction (and possibly mutual expression) of the main antigenic factor in *P. falciparum*, the *var* gene family.

Genome-wide acetylation of H4K8-bound DNA follows transcriptional dynamics during the IDC

Next, we carried out chromatin immunoprecipitation (ChIP)-coupled high-throughput sequencing (ChIP-Seq) throughout the *P. falciparum* IDC to investigate the chromosomal distribution of H4K8ac with high resolution. H4K8ac-immunoprecipitated DNA from highly synchronized parasites at the ring (8–12 hpi), trophozoite (24–28 hpi) and schizont (34–38 hpi) stages was subjected to massively parallel sequencing to obtain an average genome coverage ranging from 18–29X and 296–476X using MiSeq and HiSeq, respectively (Additional file 1). The normalized ChIP over input tag counts calculated for 2 kb region around the start codon (from 1 kb upstream to 1 kb downstream) of each gene showed correlation > 0.8 between MiSeq and HiSeq runs (Additional file 9: Figure S4a). Figure 4a shows examples of read counts spread across 3 genes highly expressed in ring, trophozoite and schizont stage, respectively. A total of 4648, 6306 and 3515 peaks were obtained for the ring, trophozoite and schizont stage HiSeq samples, respectively (Fig. 4b; Additional file 10: Table S4). In all 3 IDC developmental stages, we could observe H4K8ac occupancy peaks that were fully within the ORF or 5′IGRs, while others overlapped both. The identified H4K8ac occupancy peaks reside either within or in intergenic regions of 52, 92 and 68% genes in the ring, trophozoite and schizont stages, respectively (Fig. 4b). These results suggest that H4K8ac is predominant in the trophozoite stage where its occupancy peaks associate with the largest number of genes. There is a higher overall sequence coverage generated by the ChIP-Seq reads in the 5′IGRs followed by the peaks overlapping both 5′IGRs and ORFs and that followed by peaks exclusive to the ORFs (Fig. 4c). This is in good agreement with the previous ChIP-on-chip results showing that the majority of H4K8ac is associated with 5′IGRs with lesser occupancy within ORF [21]. We confirm this by ChIP-on-chip measurements with the identical same samples used for the RNA-Seq and showed that H4K8ac exhibits the highest occupancy

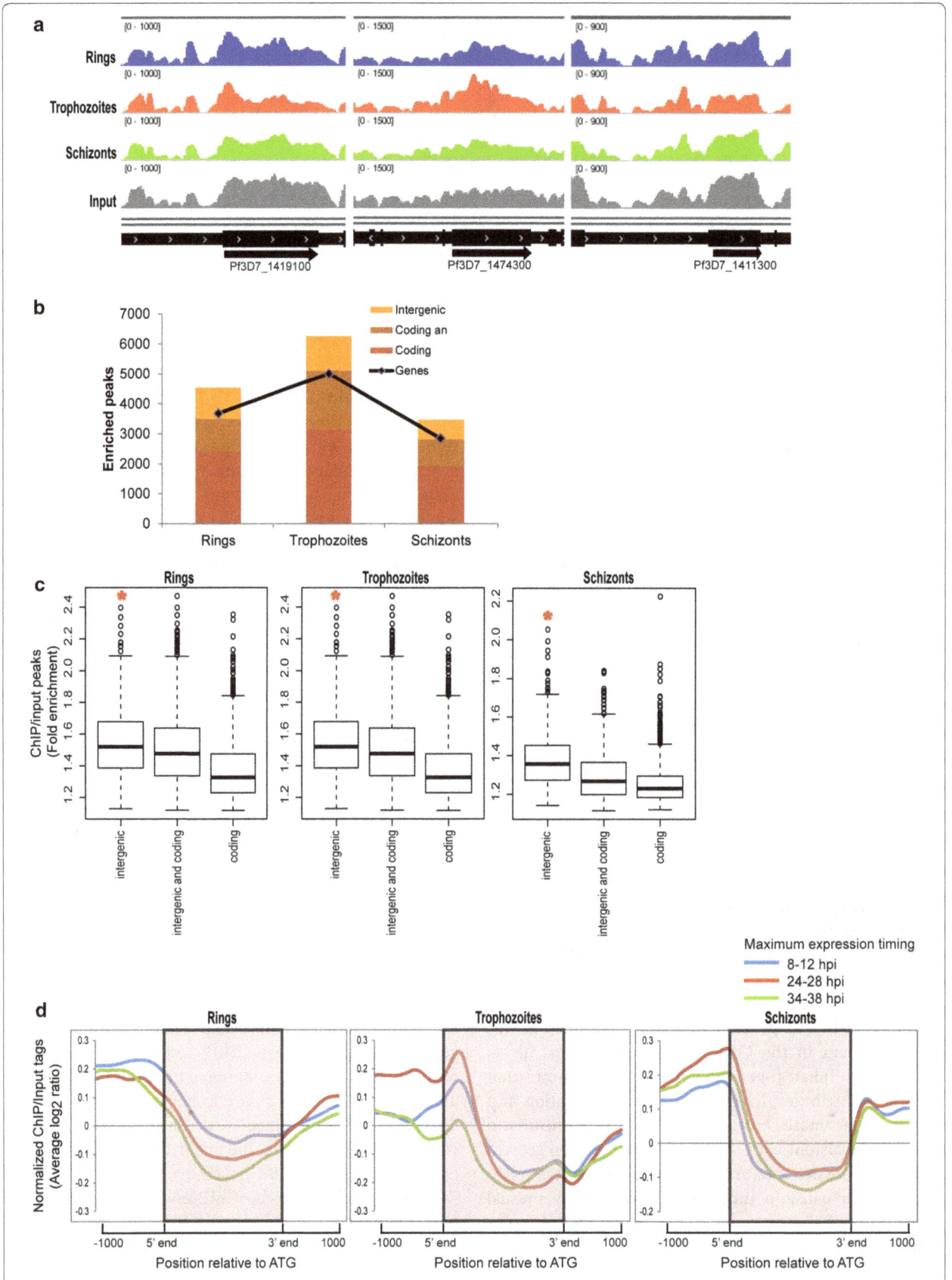

within 5'IGRs with the highest signal being detected in the trophozoite stage (Additional file 9: Figure S4b).

To evaluate this further, we analyzed the H4K8ac ChIP-Seq results in the context of transcriptional stage-specificity for each gene (Fig. 4d). Indeed, in the first part of the IDC, the ring and trophozoite stages, H4K8ac showed maximum occupancy, predominantly at the 5'IGRs (1 kb upstream of the translational start) of transcriptionally active genes in both IDC stages, respectively. However, in schizonts, H4K8ac remains predominantly at the trophozoite-specific genes and increases at schizont genes only marginally. Strikingly, there is a significant shift in the positioning of H4K8ac along the gene structures throughout the IDC. In the ring stage, H4K8ac is mainly associated with relatively distal 5'IGRs, while in the trophozoite stage, there is a narrow peak distributions at the translational start sites, skewed slightly toward the 5' termini of the ORFs (Fig. 4d). In the schizont stage, H4K8ac appears to "retreat" back to the more distal 5'IGRs, but a significant proportion of its occupancy still overlaps with the translational start sites. There is also a considerable rise in H4K8ac occupancy at the 3'IGRs in schizonts compared to the trophozoites (Fig. 4d, left two panels). Interestingly, in spite of the temporal shifts and the changing levels of the H4K8ac, the overall gene-based positional profiles of the H8K8ac were highly similar for all genes in each stage regardless of their transcriptional stage-specificity. Taken together, these results provide further evidence that H4K8ac is associated with transcriptional regulation of *P. falciparum* genes during the IDC, particularly regulating genes expressed during the mid-sections of the IDC, the trophozoite stage. At this stage, H4K8ac may contribute to transcriptional activation by repositioning its occupancy from 5'IGRs into the 5'termini of the ORFs.

Discussion

It is now clear that in *P. falciparum*, gene regulation, in some way, is responsive to the dynamic pattern of histone marks that undergoes multitude of changes through the asexual life cycle [4, 20]. The unique plasticity of the distribution of histone marks during *P. falciparum* development is unseen in other eukaryotes. Here, we have demonstrated an important role for the histone H4 acetylations, especially at lysine 8 position, to be crucial in gene regulation in *P. falciparum*. It was previously shown that the activating histone marks are abundant and widely distributed in *P. falciparum* genome, but only some coincide with the transcriptional dynamics. However, H4K8ac plays a unique role in *P. falciparum* biology that is distinct from H3K9ac and the other activating histone marks. Although all of them are marks of transcriptionally active euchromatin in *P. falciparum*, correlation of H4K8ac with RNA polymerase II occupancy [38] and response of H4K8ac to HDAC inhibitors and DNA damaging agents is unique [39, 40]. The results presented here highlight the highly dynamic character of H4K8ac in the genome and provide first direct evidence about its importance for transcriptional reprogramming during the *P. falciparum* IDC. A certain threshold level of H4K8ac is likely required in both euchromatin and heterochromatin to allow the correct transcriptional regulation that employs both activating and suppressing factors. Here, we wish to argue that H4K8ac is the key dynamic component of the tetra-acetylation state of histone 4 (H4ac4) that ultimately functions as a main epigenetic effector, presumably via binding to trans-factors ("chromatin readers"). Although little is known about chromatin readers in *P. falciparum*, these may include proteins with conserved chromatin binding domains such as the bromodomain carrying proteins PfBDP1 and PfBDP2 that were recently shown to regulate transcription of genes of merozoite invasion [37]. Alternatively, these may involve other unique *Plasmodium* genes that were proposed to also function as chromatin readers despite of an apparent absence of conserved chromatin binding domains such as Pf14-3-3 that was shown to bind phosphorylated H3 [45].

Clonally variant multigene (CVM) families impart phenotypic diversity to *P. falciparum*, which is essential for its survival and immune evasion. There is ample evidence that single active *var* gene is enriched with H3K9ac and H3K4me3 [23, 25]. The association of H4ac with the active *var* gene was seen in a ChIP-assay followed by dot blot [46] but has never been confirmed in a more systematic way. In addition, PfMYST, the *P. falciparum* histone acetyl transferase, which acetylates histone H4 at K5, 8, 12 and 16, occupies active *var* promoter [47]. We demonstrate that H4K8 acetylation, apart from euchromatin, is another putative component of the heterochromatin environment, showing over-enrichment at the upstream region of the dominant *var* gene along with H3K9ac. In one of the studies, clonally variant multigene family genes showed a similar occupancy profile across the IDC to ring stage genes for all histone marks (including H3K9ac) except H4ac and H3Kme3 serving as the activation and repressive marks, respectively, at CVM genes [22]. This shows that association of H3K9ac and H4K8ac at the *var* genes or other clonally variant multigene family genes might have distinct roles. While both H3K9ac and H4K8ac are required for the mutually exclusive expression of dominant members, H4K8ac is required to overall regulate the expression of CVM genes.

The most significant observation made by our study is the inverse transcriptional relationship between the proliferation genes and genes of host–parasite interaction governed by H4ac4 and H4K8ac, specifically. The increased MR in case of H4K8R and H4ac4R mutants supports the model of H4K8ac playing the role of main regulator in this process. The enhanced MR is likely a reflection of the increased number of invasive merozoites and/or their viability, presumably boosted by the upregulation of the proliferation-linked genes. This observation tempts us to speculate that the *Plasmodium* cells have the ability to regulate their growth and (inversely) antigenic presentation both of which is controlled epigenetically. This is consistent with a previously conceived model in which the asexually growing *Plasmodium* parasite is in a rapidly proliferating cellular state [48], which upon casual triggering undergoes a cascade of events leading to developmental stage transitions. These triggers include several factors of physiologic environment to which the parasite can respond in its natural conditions within human host [1]. All these factors were previously shown to trigger epigenetic factors of gene expression presumably affecting the parasite physiology as a response. In 2 of our previous studies, artemisinin-resistant parasites were marked by overexpression of genes belonging to ribosome assembly and maturation [49, 50]. Upregulation of invasion-related rhoptry and merozoite surface proteins is seen during reduced expression of multidrug resistant genes Pfcrt and Pfmdr1 [51]. As such, the reduced expression of these genes during H4 hypoacetylation might be mimicking a putative resistant mechanism leading to better survival of

the parasites. HDAC inhibitor-induced H4K8ac hyperacetylation is also mimicked when the cells are treated with artemisinin [40]. In light of these results, it will be interesting to investigate the potential role of H4K8ac in parasite's adaptation to its natural host during both individual- and population-level infections.

Targeting chromatin remodeling and mechanisms of epigenetic regulation of transcription is believed to have a high potential for development of new malaria intervention strategies. TSA and other HDAC inhibitors have been previously evaluated as anti-malaria drugs due to their promising activity and selectivity seen in vitro and in vivo against *Plasmodium* parasites [52]. In accordance with the role of H4K8ac to be the major hallmark of regulation among the other histone H4 modifications, it was also found to be the most responsive modification to HDAC inhibitors in *P. falciparum*. The transient hyperacetylation was copied in expression as well as H4K8 binding to DNA. The return of hyperacetylation to pre-treatment conditions after drug removal supports the notion that these drugs are fast metabolized in the cell followed by decrease in their effective concentration [53], indicating the reversible effect to be a more general property of HDAC inhibitors. However, the irreversible effect on some histone acetylations tempts us to speculate that these two types of events are mediated by distinct histone modifications potentially generated by two distinct HDAC enzymes: PfHDAC1 for the reversible changes of euchromatin [44, 54] and the PfHDA2 for the irreversible changes involved in the gametocyte conversion [55]. Our hypothesis is supported by the fact that the small number of genes that remained de-repressed even after TSA removal belonged to gametocyte and merozoite stages. Our results confirm the role of histone modifications in exo-erythrocytic stages, and HDAC inhibitors have shown to cause hyperacetylation of *P. falciparum* gametocyte histone proteins [56]. Overall, our study opens avenue to explore HDAC inhibitors and other compound which target histone acetylations especially H4K8ac for developing novel drugs against multiple stages of malaria parasites.

Conclusions

Plasmodium falciparum exhibits unique molecular mechanisms controlling chromatin remodeling as well as mechanisms that link epigenetic markers with transcription. Targeting of these molecular mechanisms has a high potential for development of new malaria intervention strategies. With our efforts to look at the combinatorial effects of histone marks on expression patterns, we can hope to decipher the histone code underlying complex regulatory make-up of these parasites and identify important elements of disease development. Our studies demonstrate the unique role of H4K8ac in maintaining the chromatin environment in *P. falciparum*. A threshold level of this acetylation most likely is required to keep the euchromatin or heterochromatin in a transcriptionally active state.

Methods
Parasite culture and drug treatments
P. falciparum strain 3D7 cells were cultured and synchronized under standard conditions [57]. Parasitemia was determined either by microscopic counting or cell sorting in the flow cytometer LSR Fortessa X-20 (BD Biosciences) using Hoechst 33342 (Sigma) [58].

Drug treatments
Plasmodium falciparum cells were treated with IC$_{90}$ values of apicidin (70 nM) or TSA (50 nM) at 5% parasitemia and 2% hematocrit for 6 h at 20-24 hpi. DMSO was used as a vehicle control. For wash-off experiments, drug-treated cells were pelleted down and washed with RPMI at least twice before culturing in fresh media without drug for specified time intervals and subsequently harvested for protein extraction, RNA extraction or chromatin immunoprecipitation. All experiments were done in either duplicates or triplicates.

Plasmid construction and transfection
Histone H4 gene was mutated such that the targeted lysine (K) residues (K5, K8, K12, K16) were replaced with arginine (R). These amino acid exchanges were previously shown to preserve the positive charge but prevent acetylation [59, 60]. For this, plasmid pBCamR-3HA [35] was modified to create HA-tagged version of wild-type or mutated histone H4 gene in all constructs. Mutations in H4 gene were introduced by amplifying complete H4 gene of *P. falciparum* using forward primer carrying the respective change in nucleotide sequence (see Additional file 1). Unmodified H4 or H4 carrying mutations were cloned upstream of 3HA at *Bam* H1/*Nhe* 1 sites of pBcamR-3HA to make a fusion with the 3HA tag epitope and transfected into *P. falciparum* using the blasticidin-driven regulatable transgene expression systems [61]. *Plasmodium falciparum* strain 3D7 was transfected as described [62] to carry episomal copies of the plasmids in presence of selection marker blasticidin. Once the transfectants were stably maintained, the concentration of blasticidin was increased from 2.5 to 10 µg/ml in a dominant-negative selection system, which allowed the dominant expression of HA-tagged H4 protein.

Protein extraction and immunodetection
For total protein, parasitized RBCs were lysed with 0.1% saponin and washed 2–3 times with PBS. Parasite pellets were resuspended in Laemmli SDS sample buffer,

incubated at 100 °C for 10 min and centrifuged at maximum speed for 10 min to recover the supernatant.

Nuclear fractionation was done as described [35]. Briefly, saponin-lysed parasite pellets were incubated in cell lysis buffer CLB (20 mM HEPES (pH7.9), 10 mM KCl, 1 mM EDTA, 1 mM EGTA, 0.65% NP-40, 1 mM DTT, protease inhibitors (Complete TM, Roche Diagnostics)) for 5 min on ice. Nuclei were pelleted at 5000 rpm, washed twice with CLB and digested with 300U MNase (Fermentas) in digestion buffer DB (20 mM Tris-HCl, pH7.5, 15 mM NaCl, 60 mM KCl, 1 mM CaCl2, 5 mM MgCl2, 5 mM MnCl2, 300 mM sucrose, 0.4% NP-40, 1 mM DTT, protease inhibitors) for 20 min at 37 °C. Soluble low salt nuclear fractions were recovered by centrifugation for 10 min at 13,000 rpm. Remaining nuclear debris was washed twice in DB and resuspended in high salt buffer HSB (20 mM HEPES (pH 7.9), 800 mM KCl, 1 mM EDTA, 1 mM EGTA, 1 mM DTT, protease inhibitors) by vortexing for 20 min at 4 °C. High salt nuclear fraction was recovered; after centrifugation for 5 min at 13,000 rpm, the high salt nuclear fraction was saved. The insoluble pellet was solubilized in SDS extraction buffer [2%SDS, 10 mM Tris-HCl (pH 7.5)] by vortexing for 20 min at room temperature.

Acid extraction of histones was modified from the original protocol [63]. Briefly, the insoluble nuclear pellet containing DNA and histones was treated overnight with 0.25 M HCl at 4 °C. The acid-extractable protein was precipitated with trichloroacetic acid (TCA), washed in ice-cold acetone, and resuspended in Laemmli sample buffer.

For immunoblotting, identical amount of total protein lysates separated by 12% SDS-PAGE were transferred onto nitrocellulose membrane. Western blot analyses were carried out using primary antibodies probed against the core histone modifications (or HA from Sigma) obtained from Millipore, Upstate and horseradish peroxidase-conjugated secondary antibody from GE Healthcare. Polyclonal PfHDAC1 anti-serum [39], and actin (Millipore) were used as loading controls. Enhanced chemiluminescence kit was used for detection according to manufacturer's instructions (Santa Cruz biotechnology, INC).

Immunofluorescence was carried out as described [64] using primary antibodies against HA (1:200) or H4K8ac (1:1000) and fluorophore-conjugated secondary antibodies (Invitrogen). Slides were visualized under Carl Zeiss LSM 510 Confocal Laser Scanning Microscope.

RNA preparation
Parasitized RBC pellets containing synchronized parasites from different stages were harvested and stored at −80 °C. RNA extraction, cDNA synthesis and

amplification of cDNA were carried out as illustrated [65].

Chromatin immunoprecipitation

Formaldehyde cross-linked chromatin was sonicated for 8 cycles or 20 cycles in case of ChIP-on-chip or ChIP-Seq, respectively, and immunoprecipitated as described [21].

Microarray hybridizations and data analysis

Hybridization

Random amplification of immunoprecipitated DNA as well as input DNA for microarray was carried out for 30 cycles as described [65]. Equal amounts of Cy5 and Cy3 labeled ChIP and input DNA, respectively, were hybridized on *P. falciparum* custom arrays containing 5402 50-mer intergenic oligonucleotide probes and 10,416 70-mer ORF probes representing 5343 coding genes [66]. For RNA analysis, cDNA labeled with Cy5 was mixed with equal amounts of Cy3-labeled cDNA made from RNA from 3D7 parasites of all stages. The hybridizations were done on arrays containing *P. falciparum* ORF probes only. All hybridizations were performed on the Agilent hybridization system, and microarray scanning was done using Power Scanner (Tecan, Austria). Data were acquired using GenePix Prov6.0 software (Axon Instruments, USA).

Data analysis

The microarray data were normalized using the Limma package of R [67]. Briefly, LOWESS normalization was applied to all spots on each array followed by quantile normalization between arrays. Spots with flags >0 and median foreground intensity >1.5-fold median background intensity for either channel were included. The normalized log2 ratio of Cy5 versus Cy3 channel was used to present the fold change. The gene expression ratio was calculated by averaging the ratios for all probes mapping to the ORF of a gene.

Functional analyses were carried out to calculate functionally enriched pathways in a given dataset (using the functional gene annotation by MPMP [68]). Over-representation of pathways as compared to their respective frequency in the genome was calculated based on hypergeometric test ($P < 0.05$).

Quantitative real-time PCR

For ChIP validation, qPCR was carried out on immunoprecipitated and input DNA samples. ChIP enrichment was calculated by using the ΔCt method (Ct of immunoprecipitated target gene — Ct of input target gene) where Ct is the threshold cycle. All PCRs were done in duplicates or triplicates. Validation of invasion and proliferation genes was done by reverse transcribing RNA from H4-HA and H4K8R parasites (rings, trophozoites and schizonts) into cDNA. All qPCRs were performed using either Applied Biosystems SYBR Select Mix or SYBR Green PCR Master Mix (Bio-Rad) according to the manufacturer's instructions.

High-throughput ChIP-Seq

Library preparation

Purified DNA from H4K8ac immunoprecipitated chromatin of ring, trophozoite and schizont stages as well as sonicated genomic input DNA (schizont stage input DNA) was used to prepare ChIP-Seq libraries. Libraries for ChIP-Seq were prepared using NEBNext Ultra DNA library prep kit from Illumina according to manufacturer's instructions with a few modifications in amplification. Libraries were amplified for 3 PCR cycles (15 s at 98 °C, 30 s at 55 °C, 30 s at 62 °C) followed by 7–9 PCR cycles (15 s at 98 °C, 30 s at 63 °C, 30 s at 65 °C) using KAPA HiFi HotStart Ready Mix (Kapa Biosystems, Woburn, MA). 150-bp paired end reads were obtained using HiSeq 2500 (Illumina).

Data processing

Sequence reads were aligned to the *P. falciparum* genome (downloaded from Gene DB, Jul2015) using BWA, version 0.7.12 [69]. After adapter trimming using skewer version, 0.1.125 [70], and removal of PCR duplicates, uniquely aligned reads (mapping quality >20, the probability of incorrect mapping <0.01) were extracted out for further study.

Data visualization

The processed data were visualized using integrative genomics viewer (IGV) interface [71]. ChIP-Seq profiles for gene models were generated using total reads counts. For this, total read coverage at 50-bp intervals was determined for regions spanning 1 kb upstream of start codon and 1 kb downstream of stop codon. Read coverage within ORF was calculated in 30 bins with the interval size approximate to 50 bp (the average length of genes in the whole genome is 1500 bp). Finally, for every library, the read coverage of each bin was normalized by the corresponding read coverage of that bin in the control input library.

Peak calling

The enrichment of histone occupancies was determined by peak calling algorithm built in MACS, version 2.1 [72] applied with parameters—nomodel-q 0.25—broad—slocal 500.

Additional files

Additional file 1. (a) Vector map (b) ChIP-Seq coverage (c) Primer sequences used in this study (d) Supplemental figure and table legends.

Additional file 2: Figure S1. *P. falciparum* transgenic lines for mutations in H4 acetylations (related to Fig. 1).

Additional file 3: Table S1. Genes differentially expressed in the transfectants (related to Fig. 1).

Additional file 4. Figure S2. Differentially expressed pathways in the transfectants (related to Fig. 1).

Additional file 5. Table S5. Parasite age estimation from RNA expression data of the transfectants (related to Fig. 1).

Additional file 6. Table S2. Genes that remain deregulated after removal of TSA (related to Fig. 2).

Additional file 7. Table S3. Overlap between H4K8ac binding and expression changes during TSA treatment (related to Fig. 2).

Additional file 8. Figure S3. Histone marks at dominant var genes (related to Fig. 3).

Additional file 9. Figure S4. ChIP coupled to high throughput sequencing (related to Fig. 4).

Additional file 10. Table S4. MACS identified peaks for H4K8ac ChIP-Seq (related to Fig. 4).

Abbreviations

ac: acetylation; me: methylation; IDC: intraerythrocytic developmental cycle; PTM: post transcriptional modification; ORF: open reading frame; IGR: intergenic region; HDAC: histone deacetylase inhibitor; TSA: trichostatin A; ChIP: chromatin immunoprecipitation; bp: base pair.

Authors' contributions

APG and ZB conceived and designed the experiments and wrote the manuscript. APG, JT, MK and AP performed the experiments. APG and LZ analyzed the data. All authors read and approved the final manuscript.

Acknowledgements

The authors are thankful to Dr. Sachel Mok for providing 3D7 clones, Prof. Till Voss for sharing pBCamR-3HA plasmid and Chin Wai Ho for technical assistance.

Competing interests

The authors declare that they have no competing interests.

Funding

This work is funded by the Tier 2 grant of the Ministry of Education Singapore; Grant Number MOE2013-T2-1-055.

References

1. Merrick CJ, Duraisingh MT. Epigenetics in *Plasmodium*: what do we really know? Eukaryot Cell. 2010;9:1150–8.
2. Ay F, Bunnik EM, Varoquaux N, Vert JP, Noble WS, Le Roch KG. Multiple dimensions of epigenetic gene regulation in the malaria parasite *Plasmodium falciparum*: gene regulation via histone modifications, nucleosome positioning and nuclear architecture in *P. falciparum*. BioEssays. 2015;37:182–94.
3. Cui L, Miao J. Chromatin-mediated epigenetic regulation in the malaria parasite *Plasmodium falciparum*. Eukaryot Cell. 2010;9(8):1137–49.
4. Duffy MF, Selvarajah SA, Josling GA, Petter M. Epigenetic regulation of the *Plasmodium falciparum* genome. Brief Funct Genomics. 2014;13:203–16.
5. Hoeijmakers WA, Stunnenberg HG, Bartfai R. Placing the *Plasmodium falciparum* epigenome on the map. Trends Parasitol. 2012;28:486–95.
6. Salcedo-Amaya AM, Hoeijmakers WA, Bartfai R, Stunnenberg HG. Malaria: could its unusual epigenome be the weak spot? Int J Biochem Cell Biol. 2010;42:781–4.
7. Voss TS, Bozdech Z, Bartfai R. Epigenetic memory takes center stage in the survival strategy of malaria parasites. Curr Opin Microbiol. 2014;20:88–95.
8. Batugedara G, Lu XM, Bunnik EM, Le Roch KG. The role of chromatin structure in gene regulation of the human malaria parasite. Trends Parasitol. 2017;33:364–77.
9. Kirchner S, Power BJ, Waters AP. Recent advances in malaria genomics and epigenomics. Genome Med. 2016;8:92.
10. Gupta AP, Bozdech Z. Epigenetic landscapes underlining global patterns of gene expression in the human malaria parasite, *Plasmodium falciparum*. Int J Parasitol. 2017;47(7):399–407.
11. Miao J, Fan Q, Cui L, Li J. The malaria parasite *Plasmodium falciparum* histones: organization, expression, and acetylation. Gene. 2006;369:53–65.
12. Coulson RM, Hall N, Ouzounis CA. Comparative genomics of transcriptional control in the human malaria parasite *Plasmodium falciparum*. Genome Res. 2004;14:1548–54.
13. Baum J, Papenfuss AT, Mair GR, Janse CJ, Vlachou D, Waters AP, Cowman AF, Crabb BS, de Koning-Ward TF. Molecular genetics and comparative genomics reveal RNAi is not functional in malaria parasites. Nucleic Acids Res. 2009;37:3788–98.
14. Lopez-Rubio JJ, Mancio-Silva L, Scherf A. Genome-wide analysis of heterochromatin associates clonally variant gene regulation with perinuclear repressive centers in malaria parasites. Cell Host Microbe. 2009;5:179–90.
15. Salcedo-Amaya AM, van Driel MA, Alako BT, Trelle MB, van den Elzen AM, Cohen AM, Janssen-Megens EM, van de Vegte-Bolmer M, Selzer RR, Iniguez AL, et al. Dynamic histone H3 epigenome marking during the intraerythrocytic cycle of *Plasmodium falciparum*. Proc Natl Acad Sci USA. 2009;106:9655–60.
16. Lemieux JE, Kyes SA, Otto TD, Feller AI, Eastman RT, Pinches RA, Berriman M, Su XZ, Newbold CI. Genome-wide profiling of chromosome interactions in *Plasmodium falciparum* characterizes nuclear architecture and reconfigurations associated with antigenic variation. Mol Microbiol. 2013;90:519–37.
17. Trelle MB, Salcedo-Amaya AM, Cohen AM, Stunnenberg HG, Jensen ON. Global histone analysis by mass spectrometry reveals a high content of acetylated lysine residues in the malaria parasite *Plasmodium falciparum*. J Proteome Res. 2009;8:3439–50.
18. Saraf A, Cervantes S, Bunnik EM, Ponts N, Sardiu ME, Chung DW, Prudhomme J, Varberg JM, Wen Z, Washburn MP, et al. Dynamic and combinatorial landscape of histone modifications during the intraerythrocytic developmental cycle of the malaria parasite. J Proteome Res. 2016;15:2787–801.
19. Coetzee N, Sidoli S, van Biljon R, Painter H, Llinas M, Garcia BA, Birkholtz LM. Quantitative chromatin proteomics reveals a dynamic histone post-translational modification landscape that defines asexual and sexual *Plasmodium falciparum* parasites. Sci Rep. 2017;7:607.
20. Bartfai R, Hoeijmakers WA, Salcedo-Amaya AM, Smits AH, Janssen-Megens E, Kaan A, Treeck M, Gilberger TW, Francoijs KJ, Stunnenberg HG. H2A.Z demarcates intergenic regions of the *Plasmodium falciparum* epigenome that are dynamically marked by H3K9ac and H3K4me3. PLoS Pathog. 2010;6:e1001223.
21. Gupta AP, Chin WH, Zhu L, Mok S, Luah YH, Lim EH, Bozdech Z. Dynamic epigenetic regulation of gene expression during the life cycle of malaria parasite *Plasmodium falciparum*. PLoS Pathog. 2013;9:e1003170.
22. Karmodiya K, Pradhan SJ, Joshi B, Jangid R, Reddy PC, Galande S. A comprehensive epigenome map of *Plasmodium falciparum* reveals unique mechanisms of transcriptional regulation and identifies H3K36me2 as a global mark of gene suppression. Epigenet Chromatin. 2015;8:32.
23. Petter M, Lee CC, Byrne TJ, Boysen KE, Volz J, Ralph SA, Cowman AF, Brown GV, Duffy MF. Expression of *P. falciparum* var genes involves exchange of the histone variant H2A.Z at the promoter. PLoS Pathog. 2011;7:e1001292.
24. Jiang L, Mu J, Zhang Q, Ni T, Srinivasan P, Rayavara K, Yang W, Turner

L, Lavstsen T, Theander TG, et al. PfSETvs methylation of histone H3K36 represses virulence genes in *Plasmodium falciparum*. Nature. 2013;499:223–7.

25. Lopez-Rubio JJ, Gontijo AM, Nunes MC, Issar N, Hernandez Rivas R, Scherf A. 5′ Flanking region of var genes nucleate histone modification patterns linked to phenotypic inheritance of virulence traits in malaria parasites. Mol Microbiol. 2007;66:1296–305.

26. Hoeijmakers WA, Flueck C, Francoijs KJ, Smits AH, Wetzel J, Volz JC, Cowman AF, Voss T, Stunnenberg HG, Bartfai R. *Plasmodium falciparum* centromeres display a unique epigenetic makeup and cluster prior to and during schizogony. Cell Microbiol. 2012;14:1391–401.

27. Petter M, Selvarajah SA, Lee CC, Chin WH, Gupta AP, Bozdech Z, Brown GV, Duffy MF. H2A.Z and H2B.Z double-variant nucleosomes define intergenic regions and dynamically occupy var gene promoters in the malaria parasite *Plasmodium falciparum*. Mol Microbiol. 2013;87(6):1167–82.

28. Fraschka SA, Henderson RW, Bartfai R. H3.3 demarcates GC-rich coding and subtelomeric regions and serves as potential memory mark for virulence gene expression in *Plasmodium falciparum*. Sci Rep. 2016;6:31965.

29. Bunnik EM, Polishko A, Prudhomme J, Ponts N, Gill SS, Lonardi S, Le Roch KG. DNA-encoded nucleosome occupancy is associated with transcription levels in the human malaria parasite *Plasmodium falciparum*. BMC Genomics. 2014;15:347.

30. Kensche PR, Hoeijmakers WA, Toenhake CG, Bras M, Chappell L, Berriman M, Bartfai R. The nucleosome landscape of *Plasmodium falciparum* reveals chromatin architecture and dynamics of regulatory sequences. Nucleic Acids Res. 2016;44:2110–24.

31. Westenberger SJ, Cui L, Dharia N, Winzeler E. Genome-wide nucleosome mapping of *Plasmodium falciparum* reveals histone-rich coding and histone-poor intergenic regions and chromatin remodeling of core and subtelomeric genes. BMC Genomics. 2009;10:610.

32. Ponts N, Harris EY, Prudhomme J, Wick I, Eckhardt-Ludka C, Hicks GR, Hardiman G, Lonardi S, Le Roch KG. Nucleosome landscape and control of transcription in the human malaria parasite. Genome Res. 2010;20:228–38.

33. Ponts N, Harris EY, Lonardi S, Le Roch KG. Nucleosome occupancy at transcription start sites in the human malaria parasite: a hard-wired evolution of virulence? Infect Genet Evol. 2011;11:716–24.

34. Brancucci NM, Bertschi NL, Zhu L, Niederwieser I, Chin WH, Wampfler R, Freymond C, Rottmann M, Felger I, Bozdech Z, Voss TS. Heterochromatin protein 1 secures survival and transmission of malaria parasites. Cell Host Microbe. 2014;16:165–76.

35. Flueck C, Bartfai R, Volz J, Niederwieser I, Salcedo-Amaya AM, Alako BT, Ehlgen F, Ralph SA, Cowman AF, Bozdech Z, et al. *Plasmodium falciparum* heterochromatin protein 1 marks genomic loci linked to phenotypic variation of exported virulence factors. PLoS Pathog. 2009;5:e1000569.

36. Perez-Toledo K, Rojas-Meza AP, Mancio-Silva L, Hernandez-Cuevas NA, Delgadillo DM, Vargas M, Martinez-Calvillo S, Scherf A, Hernandez-Rivas R. *Plasmodium falciparum* heterochromatin protein 1 binds to tri-methylated histone 3 lysine 9 and is linked to mutually exclusive expression of var genes. Nucleic Acids Res. 2009;37:2596–606.

37. Josling GA, Petter M, Oehring SC, Gupta AP, Dietz O, Wilson DW, Schubert T, Langst G, Gilson PR, Crabb BS, et al. A *Plasmodium falciparum* bromodomain protein regulates invasion gene expression. Cell Host Microbe. 2015;17:741–51.

38. Rai R, Zhu L, Chen H, Gupta AP, Sze SK, Zheng J, Ruedl C, Bozdech Z, Featherstone M. Genome-wide analysis in *Plasmodium falciparum* reveals early and late phases of RNA polymerase II occupancy during the infectious cycle. BMC Genomics. 2014;15:959.

39. Chaal BK, Gupta AP, Wastuwidyaningtyas BD, Luah YH, Bozdech Z. Histone deacetylases play a major role in the transcriptional regulation of the *Plasmodium falciparum* life cycle. PLoS Pathog. 2010;6:e1000737.

40. Gupta DK, Patra AT, Zhu L, Gupta AP, Bozdech Z. DNA damage regulation and its role in drug-related phenotypes in the malaria parasites. Sci Rep. 2016;6:23603.

41. Wang Z, Zang C, Rosenfeld JA, Schones DE, Barski A, Cuddapah S, Cui K, Roh TY, Peng W, Zhang MQ, Zhao K. Combinatorial patterns of histone acetylations and methylations in the human genome. Nat Genet. 2008;40:897–903.

42. Magraner-Pardo L, Pelechano V, Coloma MD, Tordera V. Dynamic remodeling of histone modifications in response to osmotic stress in

Saccharomyces cerevisiae. BMC Genomics. 2014;15:247.

43. Bozdech Z, Llinas M, Pulliam BL, Wong ED, Zhu J, DeRisi JL. The transcriptome of the intraerythrocytic developmental cycle of *Plasmodium falciparum*. PLoS Biol. 2003;1:E5.

44. Andrews KT, Gupta AP, Tran TN, Fairlie DP, Gobert GN, Bozdech Z. Comparative gene expression profiling of *P. falciparum* malaria parasites exposed to three different histone deacetylase inhibitors. PLoS ONE. 2012;7:e31847.

45. Dastidar EG, Dzeyk K, Krijgsveld J, Malmquist NA, Doerig C, Scherf A, Lopez-Rubio JJ. Comprehensive histone phosphorylation analysis and identification of Pf14-3-3 protein as a histone H3 phosphorylation reader in malaria parasites. PLoS ONE. 2013;8:e53179.

46. Freitas-Junior LH, Hernandez-Rivas R, Ralph SA, Montiel-Condado D, Ruvalcaba-Salazar OK, Rojas-Meza AP, Mancio-Silva L, Leal-Silvestre RJ, Gontijo AM, Shorte S, Scherf A. Telomeric heterochromatin propagation and histone acetylation control mutually exclusive expression of antigenic variation genes in malaria parasites. Cell. 2005;121:25–36.

47. Miao J, Fan Q, Cui L, Li X, Wang H, Ning G, Reese JC, Cui L. The MYST family histone acetyltransferase regulates gene expression and cell cycle in malaria parasite *Plasmodium falciparum*. Mol Microbiol. 2010;78:883–902.

48. Salcedo-Sora JE, Caamano-Gutierrez E, Ward SA, Biagini GA. The proliferating cell hypothesis: a metabolic framework for *Plasmodium* growth and development. Trends Parasitol. 2014;30:170–5.

49. Mok S, Imwong M, Mackinnon MJ, Sim J, Ramadoss R, Yi P, Mayxay M, Chotivanich K, Liong KY, Russell B, et al. Artemisinin resistance in *Plasmodium falciparum* is associated with an altered temporal pattern of transcription. BMC Genomics. 2011;12:391.

50. Mok S, Ashley EA, Ferreira PE, Zhu L, Lin Z, Yeo T, Chotivanich K, Imwong M, Pukrittayakamee S, Dhorda M, et al. Drug resistance. Population transcriptomics of human malaria parasites reveals the mechanism of artemisinin resistance. Science. 2015;347:431–5.

51. Adjalley SH, Scanfeld D, Kozlowski E, Llinas M, Fidock DA. Genome-wide transcriptome profiling reveals functional networks involving the *Plasmodium falciparum* drug resistance transporters PfCRT and PfMDR1. BMC Genomics. 1090;2015:16.

52. Andrews KT, Tran TN, Fairlie DP. Towards histone deacetylase inhibitors as new antimalarial drugs. Curr Pharm Des. 2012;18(24):3467–79.

53. Waterborg JH. Plant histone acetylation: in the beginning. Biochim Biophys Acta. 2011;1809:353–9.

54. Patel V, Mazitschek R, Coleman B, Nguyen C, Urgaonkar S, Cortese J, Barker RH, Greenberg E, Tang W, Bradner JE, et al. Identification and characterization of small molecule inhibitors of a class I histone deacetylase from *Plasmodium falciparum*. J Med Chem. 2009;52:2185–7.

55. Coleman BI, Skillman KM, Jiang RH, Childs LM, Altenhofen LM, Ganter M, Leung Y, Goldowitz I, Kafsack BF, Marti M, et al. A *Plasmodium falciparum* histone deacetylase regulates antigenic variation and gametocyte conversion. Cell Host Microbe. 2014;16:177–86.

56. Trenholme K, Marek L, Duffy S, Pradel G, Fisher G, Hansen FK, Skinner-Adams TS, Butterworth A, Ngwa CJ, Moecking J, et al. Lysine acetylation in sexual stage malaria parasites is a target for antimalarial small molecules. Antimicrob Agents Chemother. 2014;58:3666–78.

57. Trager W, Jensen JB. Human malaria parasites in continuous culture. 1976. J Parasitol. 2005;91:484–6.

58. Malleret B, Claser C, Ong AS, Suwanarusk R, Sriprawat K, Howland SW, Russell B, Nosten F, Renia L. A rapid and robust tri-color flow cytometry assay for monitoring malaria parasite development. Sci Rep. 2011;1:118.

59. Zhang W, Bone JR, Edmondson DG, Turner BM, Roth SY. Essential and redundant functions of histone acetylation revealed by mutation of target lysines and loss of the Gcn5p acetyltransferase. EMBO J. 1998;17:3155–67.

60. Dion MF, Altschuler SJ, Wu LF, Rando OJ. Genomic characterization reveals a simple histone H4 acetylation code. Proc Natl Acad Sci USA. 2005;102:5501–6.

61. Epp C, Raskolnikov D, Deitsch KW. A regulatable transgene expression system for cultured *Plasmodium falciparum* parasites. Malar J. 2008;7:86.

62. Nkrumah LJ, Muhle RA, Moura PA, Ghosh P, Hatfull GF, Jacobs WR Jr, Fidock DA. Efficient site-specific integration in *Plasmodium falciparum* chromosomes mediated by mycobacteriophage Bxb1 integrase. Nat Methods. 2006;3:615–21.

63. Longhurst HJ, Holder AA. The histones of *Plasmodium falciparum*: identification, purification and a possible role in the pathology of malaria. Parasitology. 1997;114(Pt 5):413–9.

64. Luah YH, Chaal BK, Ong EZ, Bozdech Z. A moonlighting function of *Plasmodium falciparum* histone 3, mono-methylated at lysine 9? PLoS ONE. 2010;5:e10252.

65. Bozdech Z, Mok S, Gupta AP. DNA microarray-based genome-wide analyses of Plasmodium parasites. Methods Mol Biol. 2013;923:189–211.

66. Hu G, Llinas M, Li J, Preiser PR, Bozdech Z. Selection of long oligonucleotides for gene expression microarrays using weighted rank-sum strategy. BMC Bioinform. 2007;8:350.

67. Edwards D. Non-linear normalization and background correction in one-channel cDNA microarray studies. Bioinformatics. 2003;19:825–33.

68. Ginsburg H. Progress in in silico functional genomics: the malaria metabolic pathways database. Trends Parasitol. 2006;22:238–40.

69. Li H. Exploring single-sample SNP and INDEL calling with whole-genome de novo assembly. Bioinformatics. 2012;28:1838–44.

70. Jiang H, Lei R, Ding SW, Zhu S. Skewer: a fast and accurate adapter trimmer for next-generation sequencing paired-end reads. BMC Bioinform. 2014;15:182.

71. Robinson JT, Thorvaldsdottir H, Winckler W, Guttman M, Lander ES, Getz G, Mesirov JP. Integrative genomics viewer. Nat Biotechnol. 2011;29:24–6.

72. Zhang Y, Liu T, Meyer CA, Eeckhoute J, Johnson DS, Bernstein BE, Nusbaum C, Myers RM, Brown M, Li W, Liu XS. Model-based analysis of ChIP-Seq (MACS). Genome Biol. 2008;9:R137.

9

NuA4 histone acetyltransferase activity is required for H4 acetylation on a dosage-compensated monosomic chromosome that confers resistance to fungal toxins

Hironao Wakabayashi[1][*][iD], Christopher Tucker[1], Gabor Bethlendy[2,3], Anatoliy Kravets[1], Stephen L. Welle[4,5], Michael Bulger[5], Jeffrey J. Hayes[1] and Elena Rustchenko[1]

Abstract

Background: The major human fungal pathogen *Candida albicans* possesses a diploid genome, but responds to growth in challenging environments by employing chromosome aneuploidy as an adaptation mechanism. For example, we have shown that *C. albicans* adapts to growth on the toxic sugar L-sorbose by transitioning to a state in which one chromosome (chromosome 5, Ch5) becomes monosomic. Moreover, analysis showed that while expression of many genes on the monosomic Ch5 is altered in accordance with the chromosome ploidy, expression of a large fraction of genes is increased to the normal diploid level, presumably compensating for gene dose.

Results: In order to understand the mechanism of the apparent dosage compensation, we now report genome-wide ChIP-microarray assays for a sorbose-resistant strain harboring a monosomic Ch5. These data show a significant chromosome-wide elevation in histone H4 acetylation on the mCh5, but not on any other chromosome. Importantly, strains lacking subunits of the NuA4 H4 histone acetyltransferase complex, orthologous to a complex previously shown in *Drosophila* to be associated with a similar gene dosage compensation mechanism, did not show an increase in H4 acetylation. Moreover, loss of NuA4 subunits severely compromised the adaptation to growth on sorbose.

Conclusions: Our results are consistent with a model wherein chromosome-wide elevation of H4 acetylation mediated by the NuA4 complex plays a role in increasing gene expression in compensation for gene dose and adaption to growth in a toxic environment.

Keywords: *Candida albicans*, NuA4, H4 and H3 acetylation, Chromosome 5 monosomy

Background

Candida albicans is an opportunistic fungal pathogen that is part of the normal microbial community of the digestive tract and genitalia of humans. *C. albicans* does no harm to the healthy host; however, it causes life-threatening infections in severely immunocompromised patients. *C. albicans* normally possesses a diploid genome organized in eight pairs of chromosomes, but uses reversible loss or gain of an entire chromosome or a large part of chromosomes to survive in toxic environments that would otherwise kill cells or prevent their propagation [reviewed in 1, 2]. For example, the loss of one chromosome 5 (Ch5) or duplication of the right arm of Ch5 confers laboratory resistance to the anti-fungal caspofungin [3, 4]. The loss of one Ch5 also confers laboratory resistance to the anti-fungal flucytosine, as well as resistance to toxic sugar L-sorbose, which kills *C. albicans* in a manner similar to caspofungin or other front-line drugs from the echinocandin class [reviewed in 3]. In addition, changes in Ch5 ploidy are frequently found in fluconazole-resistant clinical isolates of *C. albicans*

*Correspondence: Hironao_Wakabayashi@urmc.rochester.edu
[1] Department of Biochemistry and Biophysics, University of Rochester Medical Center, Rochester, NY, USA
Full list of author information is available at the end of the article

[reviewed in 5, 6]. In light of data pointing to the importance of Ch5 in *C. albicans* drug resistance, there is a growing need to better understand the control of Ch5 ploidy and regulation of genes on this chromosome.

Previous studies of changes in transcription of Ch5 genes associated with the loss of one Ch5 in a mutant [Sor125(55)] adapted to growth on the toxic sugar sorbose (Sou+) revealed expression of many genes on the monosomic Ch5 decreased twofold, conforming to the loss of DNA copy number. However, surprisingly, expression of ~ 40% of gene transcripts from monosomic Ch5 corresponded to the normal disomic or near disomic levels [7]. In addition to monosomy of Ch5, this mutant acquired duplication of a chimeric Ch4/7b, resulting in trisomy of this chromosome (Fig. 1a), and facilitating the Sou+ phenotype [8]. Similar to Ch5, our data show that while expression of many genes on the trisomic Ch4/7b

increased 1.5-fold, as expected from gene copy number, many other genes were expressed at the normal disomic or near disomic levels (C. Tucker and E. Rustchenko, unpublished observation). These observations led us to propose that in *C. albicans*, transcriptional compensation for gene dose serves to facilitate the formation and maintenance of aneuploid chromosome states that are required for survival in adverse environments [7]. We have previously reported multiple genes encoding negative regulators of laboratory resistance to echinocandins or sorbose that are scattered across Ch5 [9]. While the final number of regulatory genes is yet to be determined, some are involved in cell wall biosynthesis, while others important to the Sou+ and drug susceptibility phenotypes are subjects of antisense regulation [9–11]. However, despite these advancements, little is known about the formation and maintenance of the monosomic Ch5 state.

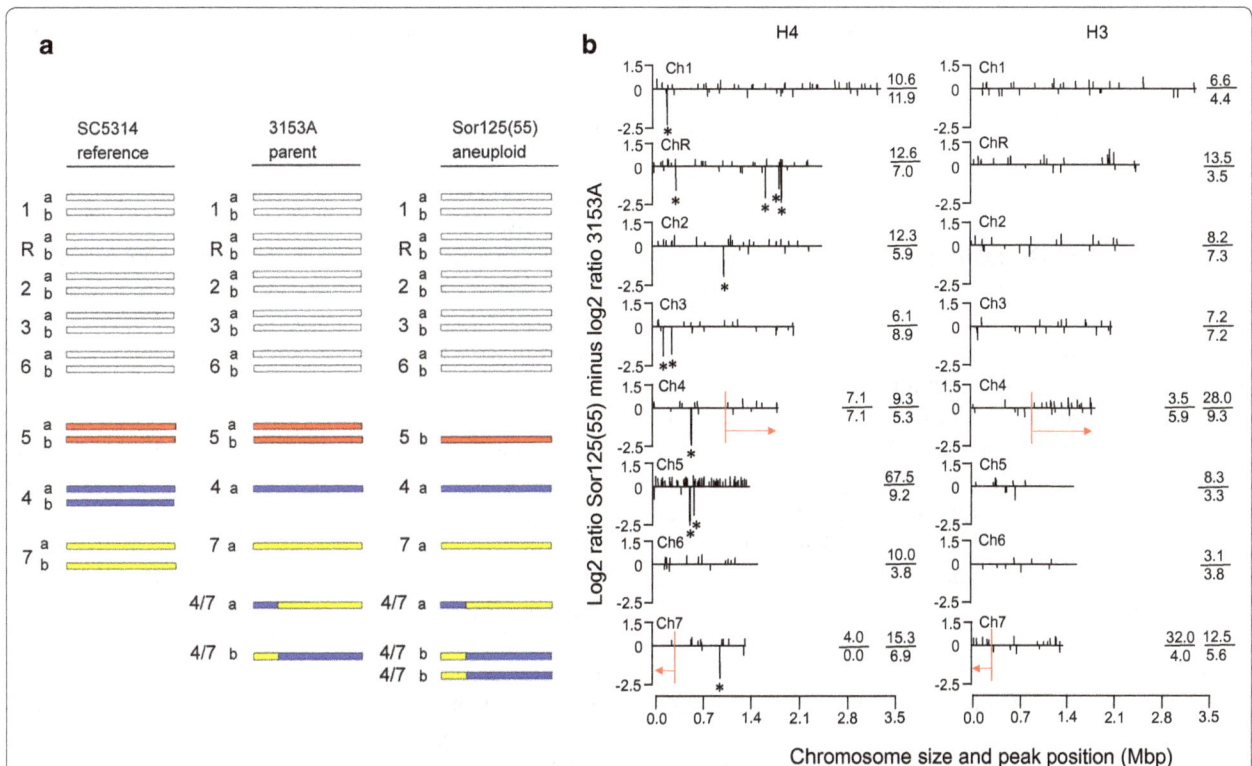

Fig. 1 Schematic chromosome patterns and ChIP-Chip results. **a** Horizontal bars represent the individual chromosomes of the parental strain 3153A, its aneuploid derivative Sor125(55), and the reference strain SC5314, as indicated. Chromosomes are designated from 1 to 7 and R, on the left, as their sizes decrease from top to bottom. For the chromosome sizes, see [8]. ChR refers to the chromosome containing a cluster of tandemly repeated rDNA units. Homologous chromosomes are indicated with "a" and "b." [Adopted from 8]. **b** Graphic presentation of ChIP-Chip results showing acetylation of histone H4 and histone H3, as indicated, on the chromosomes of the mutant Sor125(55), compared to the parental strain 3153A. Each chromosome is presented with a single graph. The X-axis indicates the position of the probes on each chromosome in the reference strain SC5314, as annotated in genomic assembly 21 in CGD. The Y-axis shows the averaged log2 ratio Sor125(55) minus log2 ratio 3153A (see "Methods"). The horizontal red arrows on Ch 4 and Ch7 indicate the trisomic regions due to duplication of Ch4/7b (see Fig. 1a). The numbers on the right indicate the density of positive (top) and negative (bottom) peaks of acetylation on each chromosome, while the second values on the right for Ch4 and Ch7 indicate the density of positive and negative peaks within the trisomic regions. Asterisks indicate large negative peaks denoting apparent loss of H4 acetylation in Sor125(55)

The regulation of gene expression on the monosomic Ch5 in *C. albicans* is possibly analogous to the well-described and essential dosage compensation found on the single male X chromosome in *Drosophila melanogaster*. In this organism, compensation requires histone H4 acetylation mediated by the male-specific lethal (MSL) complex, which includes the histone acetyltransferase (HAT) subunit MOF [12]. MSL acetylation of H4 is associated with twofold upregulation of almost all genes on the single X chromosome. The orthologous complex in *C. albicans* is termed the nucleosome acetyltransferase of H4 (NuA4) complex, in which the Esa1p subunit harbors the catalytic activity [13]. The NuA4 complex is evolutionarily conserved from yeast to humans and is known to be a key regulator of transcription [14, 15]. In *Saccharomyces cerevisiae*, NuA4 is responsible for the acetylation of nucleosomal histone H4 at lysine K5, K8, and K12 and, to a lesser extent, at K16. In *C. albicans*, NuA4 contributes mainly to acetylation of H4 K5 and H4 K12, whereas the SAS HAT complex contributes to acetylation of H4 K16 [16]. In support, deletion of a gene encoding the NuA4 catalytic subunit Esa1p in *C. albicans* significantly diminished acetylation of H4 K5 and H4 K12 [16]. Also, deleting Nbn1p, which comprises part of the NuA4 catalytic module, diminished overall H4 acetylation [17].

Considering the well-characterized model of dosage compensation mechanisms operative on the single X chromosome of *D. melanogaster*, we wished to determine whether elevated H4 acetylation is similarly associated with the aforementioned widespread transcriptional upregulation found on the monosomic Ch5 in the *C. albicans* sorbose-resistant mutant [7]. Indeed, we found increased H4 acetylation exists across the monosomic Ch5, while, in contrast, increased H3 acetylation is found within the trisomic Ch4/7b region. Importantly, we show that NuA4 subunits are required for elevated H4 acetylation in strains bearing a monosomic Ch5 and for efficient adaption to growth of *C. albicans* in the presence of the toxic sorbose.

Results

Acetylation of H4 and H3 on aneuploid chromosomes in *C. albicans*

We first asked whether histone H4 acetylation patterns across the genome are altered in the well-characterized sorbose-resistant strain Sor125(55) containing a monosomic Ch5 and a trisomic Ch4/7b, compared to the normal diploid parental strain (Fig. 1a). We performed chromatin immunoprecipitation on chip (ChIP-Chip) using a pan-H4-acetylation antibody with hybridization to our custom tiling arrays (see "Methods"). Analysis of the data shows sporadic changes in acetylation genome wide, with both increases and decreases apparent at

many loci (Fig. 1b, note that the majority of loci have changes that do not exceed a threshold and are not plotted). However, we observed a generalized and much larger fivefold to tenfold increase in H4 acetylation on the monosomic Ch5 than that found on all other chromosomes. All chromosomes showed similar amounts of decreased H4 acetylation.

We then asked whether the greatly elevated level of histone acetylation we observe on the monosomic Ch5 is specific to histone H4 or a more generalized feature of nucleosomes in the mutant strain by examining genome-wide acetylation of histone H3, which is also commonly increased in association with transcription. We performed ChIP-Chip using an antibody specific for H3 acetylated at the transcription-related sites K9 and K14 and found that H3 acetylation on Ch5 appeared similar to that found on all other disomic chromosomes, exhibiting a sporadic pattern of increases and decreases at various loci in the mutant compared to the parental strain. In contrast, we detected approximately a twofold to tenfold overall increase in H3 acetylation localized to the trisomic Ch4/7b region of the genome (Fig. 1b).

We also noticed a number of large negative peaks in the H4 acetylation plot, indicating loci exhibiting significant decreases in this modification in the mutant compared to parental cells (marked with asterisks in Fig. 1b, note that many of these peaks are too closely positioned to be distinguished in this figure, but are listed in Additional file 1: Table S1). There were 43 such peaks scattered throughout the genome, associated with 27 open reading frames (ORFs) based on the criteria of being positioned either within 1 kb of an ORF or within of an ORF. The function of the majority of the peak-associated genes is not known, thus understanding the phenotypic outcome associated with these changes in acetylation awaits further analysis.

The NuA4 complex plays a role in elevated H4 acetylation on the monosomic Ch5

As described above, elevated histone acetylation associated with dosage compensation of the single X chromosome in male flies is due to the HAT MOF that acetylates histone H4 [12]. To determine whether the orthologous NuA4 complex in *C. albicans* is involved in the large increase in acetylation observed across Ch5 in sorbose-resistant [Sor125(55)] cells, we initially employed three *C. albicans* strains, each with the NuA4 complex disrupted. Specifically, we employed strain Nbn1 (*nbn1* −/−) in which both alleles of the *NBN1* (orf19.878) gene encoding the Nbn1p subunit of the catalytic NuA4 HAT module are deleted. We also used the strain Eaf3 (*eaf3* −/−), in which both alleles of orf19.2660, encoding Eaf3p, an ortholog of yeast *EAF3* within the targeting module of the *S. cerevisiae* NuA4 complex, are deleted. Finally, we employed strain Esa1 (*esa1 −/+*), in which one allele of the essential *ESA1* (orf19.5416) gene encoding the catalytic subunit of the NuA4 complex is disrupted (Table 1). PCR and Southern blot analyses confirmed disruption of one *ESA1* allele and deletion of both alleles of *NBN1* and *EAF3* in these strains (data not shown).

We first tested how loss of the NuA4 subunits affected the growth of Esa1, Nbn1 or Eaf3 cells, as compared to their parent strains. We spread approximately 3000 colony-forming units (cfu) of each strain on plates with YPD universal rich medium and, when micro-colonies appeared, measured number of cells in colonies in approximately 4-h intervals. We found that the loss of Nbn1p from the catalytic module of the NuA4 complex

greatly affected the growth, whereas the loss of Eaf3p from the targeting module or disruption of one copy of *ESA1* encoding the NuA4 catalytic subunit had little effect on growth (Fig. 2). The Esa1 strain was not included in further experiments because the effect of disruption of a single copy of *ESA1* might not result in NuA4 insufficiency [for example, see 9].

We next assessed whether disruptions of the NuA4 complex altered histone acetylation levels upon transition of Nbn1 or Eaf3 cells from disomic to monosomic state of Ch5. We generated monosomic Ch5 mutants from the NuA4+ parental strain DAY286 and the NuA4 deletion mutants Nbn1 and Eaf3 (Table 1) using our well-established method of plating cells on L-sorbose medium [18] (see "Methods"). We confirmed loss of one Ch5 in

Table 1 *Candida albicans* strains used in this study

Strain	Genotype	Source
BWP17	*ura3-Δ::imm434/ura3-Δ::imm434, arg4-Δ::hisG/arg-Δ::hisG, his1-Δ::hisG/his1-Δ::hisG*	Tag-Module collection
Esa1	Same as BWP17, but *esa1-Δ/ESA1*	Same as above
DAY286	Same as BWP17, but *ura3-Δ::imm434/ura3-Δ::imm434, ARG4::URA3::arg4-Δ::hisG/arg-Δ::hisG, his1-Δ::hisG/his1-Δ::hisG*	FGSC[a]
DAY286-1	Same as DAY286, but Ch5 monosomy	This study
Nbn1	Same as DAY286, but *nbn1-Δ/nbn1-Δ*	FGSC
Nbn1-1	Same as Nbn1, but Ch5 monosomy	This study
Nbn1-2	Same as Nbn1, but Ch5 monosomy	This study
Nbn1-3	Same as Nbn1, but Ch5 monosomy	This study
Eaf3	Same as DAY286, but *eaf3-Δ/eaf3-Δ*	FGSC
Eaf3-1	Same as Eaf3, but Ch5 monosomy	This study
3153A	Laboratory strain, normal diploid	[8, 27]
Sor125(55) a.k.a. Sor55	Same as 3153A, but Ch5 monosomy, *MTLa*, and Ch4/7b trisomy[b]	[8, 27]

[a] Fungal Genomic Strains Center

[b] Both Sor125(55) and its parental strain, 3153A, harbor two hybrid chromosomes resulting from a reciprocal exchange between a Ch4 and a Ch7 of the reference strain SC5314 [8], which we term Ch4/7a and Ch4/7b. In Sor125(55), Ch4/7b is duplicated; therefore, genes on this hybrid chromosome, together with the corresponding portions of the intact Ch4 and Ch7, have a copy number of three [8]

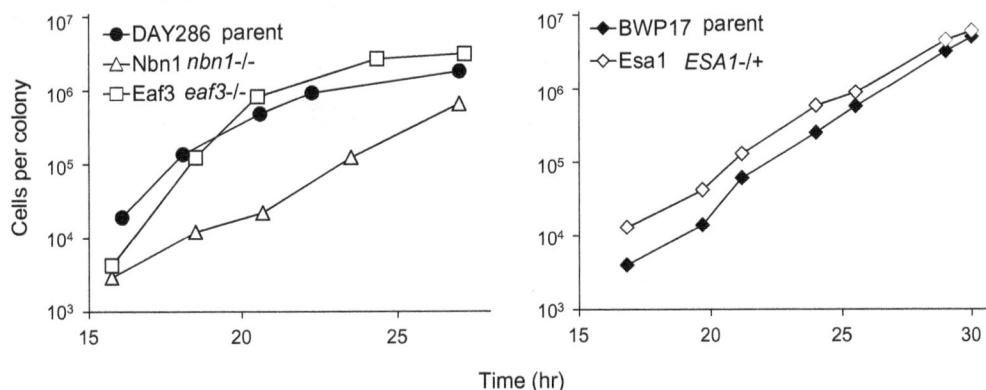

Fig. 2 Growth curves of strains with a disrupted NuA4 complex. Shown are Nbn1 (*nbn1 −/−*) and Eaf3 (*eaf3 −/−*) versus parental DAY286 (left) and Esa1 (*ESA1 −/+*) versus parental BWP17 (right), as indicated. See Table 1 for the strains' relationship and genotype

derived mutants by PCR with primers for *MTL* loci, residing on Ch5 [7]. We also examined the chromosome banding pattern in one mutant from each parental strain by pulse-field gel electrophoresis (PFGE) (Additional file 2: Figure S1). PFGE analysis showed that ChR carrying tandem repeats of ribosomal DNA (rDNA) units exhibited some instability in the Nbn1 monosomic Ch5 mutant as expected, since changes of ChR lengths frequently occur due to changes in the number of rDNA units in the cluster of tandemly assembled rDNA units [19]. Otherwise, chromosome banding patterns did not show any other alterations in mutant strains other than monosomy of Ch5, common to all mutants. It is important to note that adaptation of *C. albicans* to growth on sorbose as the sole source of carbon almost exclusively involves a transition to a monosomic Ch5 state without other chromosome alterations, in multiple genetic backgrounds (> 95% incidence) [3, reviewed in 1, 2, 20], and this method is now used by many laboratories.

Having established the monosomic Ch5 strains in the desired parental and deletion backgrounds, we next employed Western blot analysis with either pan-acetyl H4 antibody or antibodies, specific for acetylation at H4 K5, H4 K12, or H4 K16 to compare acetylation in these strains. Importantly, we found a significant increase in total acetylation of H4 in the Sor125(55) strain that harbors a monosomic Ch5, compared to the parent 3153A (Fig. 3a, also see Additional file 3: Figure S2), consistent with our ChIP analysis (Fig. 1b). Likewise, we observed

a significant total increase in H4 acetylation in the newly generated monosomic Ch5 mutant DAY286-1 versus its parent DAY286 (Fig. 3a, also see Additional file 3: Figure S2). Increased acetylation was also detectable with antibodies specific for H4 K5ac or K12ac [Sor125(55)], or H4 K5ac (DAY286-1). These data are in good agreement with our genome-wide ChIP analyses and indicate that the large elevation of H4 acetylation specific to Ch5 in strains monosomic for Ch5 can be observed in blots of histones isolated from whole cells.

We next compared H4 acetylation in parental cells lacking subunits from the NuA4 HAT and strains harboring a monosomic Ch5 derived from these cells. In a strain lacking the Nbn1 subunit from the catalytic module of NuA4, we found no difference in overall H4 acetylation compared to the parental DAY286 (Fig. 3b, black bar) and no significant difference was detected with antibodies specific for single acetylation events in the H4 tail (Fig. 3b, gray, hatched and white bars). Importantly, in contrast to Ch5 monosomic strains with a fully functional NuA4 complex, we found no increase in H4 acetylation associated with transition of the Nbn1 deletion strain to a monosomic Ch5 state (Fig. 3a, Nbn1-1, black bars). See results for two more independent Ch5 monosomic derivatives of Nbn1 in Additional file 4: Figure S3. The result was similar whether comparing pan-acetylation of H4 or specific acetylation at individual H4 lysines (Fig. 3a, Nbn1-1, gray, hatched, and white bars). Likewise, in cells lacking the Eaf3p subunit from the targeting

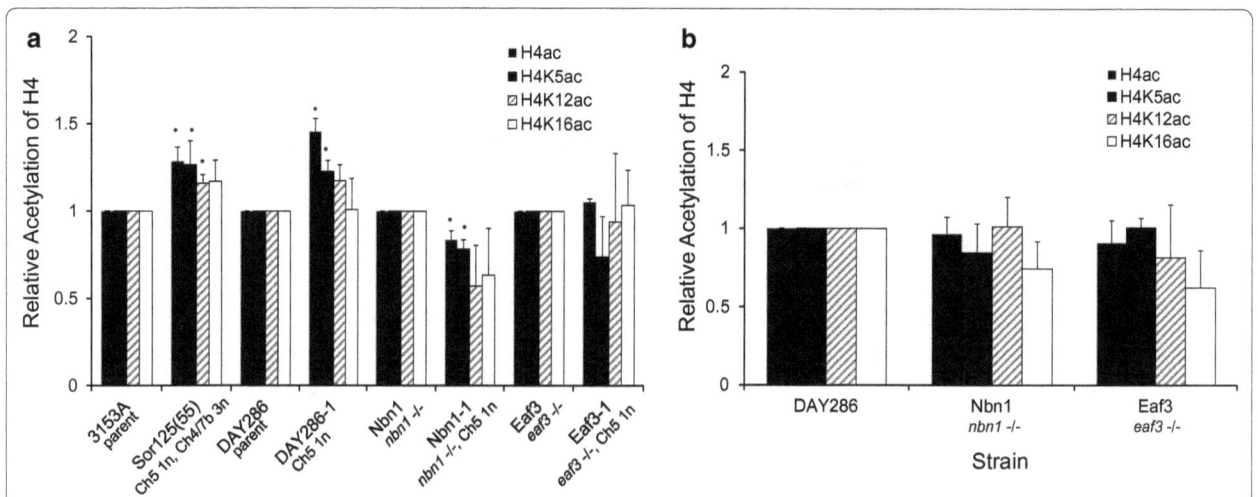

Fig. 3 Increase in H4 acetylation associated with adaptation of monosomic Ch5 state requires an intact NuA4 complex. **a** Comparison of total H4 acetylation in parental *C. albicans* strains and monosomic Ch5 strains derived as described in the tex. Shown is total H4 acetylation in: (i) the parental strain (3153A) and the monosomic Ch5 mutant Sor125(55); (ii) the parental strain DAY286 and the monosomic Ch5 mutant DAY286-1; (iii) parental strain Nbn1 (*nbn −/−*) and the monosomic Ch5 mutant Nbn1-1; and (iv) the parental stran Eaf3 (*eaf −/−*) and the monosomic Ch5 mutant Eaf3-1. **b** Total H4 acetylation in deletion mutants Nbn1 and Eaf3 compared to their parental strain DAY286. H4 acetylation was determined by Western blot with the indicated antibodies and results normalized to the parental strain response for each set (see text). Data are averaged from three independent experiments. Asterisks indicate statistical significance of the difference in acetylation between mutants and parentals with *p* value < 0.05 calculated according to Student's *t* test

module of NuA4, no increase in overall acetylation was observed upon transition from a disomic to monosomic Ch5 state (Fig. 3a, compare Eaf3 and Eaf3-1). Our results indicate that the elevated H4 acetylation observed in mutants with monosomic Ch5 is due to the activity of the NuA4 complex.

Role of the NuA4 complex in formation of the monosomic Ch5

To determine whether the NuA4 complex has a role in formation of the monosomic Ch5 state, we determined whether the rates of production of Sou⁺ cells monosomic for Ch5 are affected by deletion of Nbn1 or Eaf3 subunits compared to the parental strain DAY286. The experiments were conducted, as previously reported [18]. Production of Sou⁺ colonies representing Sou⁺ mutants was observed for each strain by spreading from 2×10^5 to 6×10^5 of the Sou⁻ cells on L-sorbose plates in three independent experiments. The appearance of Sou⁺ colonies was recorded daily and is presented as an averaged accumulation curve (Fig. 4a). Survival of Sou⁻ cells was assessed from sorbose plates by daily transfer of entire agar disks to YPD plates, incubation for 3 days, and counting the number of grown colonies [see 18 for more details]. The daily survival of Sou⁻ cells of each strain is presented in Additional file 5: Figure S4. The daily rate of production of Sou⁺ mutants was calculated by dividing the number of Sou⁺ colonies appearing daily by number of cells that were viable 4 days earlier, the predicted day of the mutational events leading to transformation (Fig. 4b).

[For example, we considered that mutational events that occurred in parental DAY286 cells on day 0 would result in appearance of visible Sou⁺ colonies on day 4 (Fig. 4a)]. We found greatly diminished rates of Sou⁺ mutants' production with strains Nbn1 and Eaf3 lacking NuA4 subunits versus DAY286 with a normal NuA4 complex. This result shows that efficient adaptation to sorbose by formation of the monosomic Ch5 depends on the NuA4 complex.

Discussion

In this work, we show that a chromosome-wide increase in acetylation of histone H4 occurs on the single Ch5 of the *C. albicans* sorbose-resistant mutant Sor125(55), on which ~ 40% of genes are upregulated to disomic or near disomic levels, thus compensating for decreased gene dosage. Such an increase in H4 acetylation seems to be a general attribute of the monosomic Ch5 state, as it also occurred on the single Ch5 in the mutant DAY286-1, which has a different genetic background. Increased H4 acetylation, however, does not occur on the trisomic Ch4/7b in the Sor125(55) mutant, on which multiple genes are downregulated to disomic or near disomic levels, again compensating for increased gene dosage (E. Rustchenko, unpublished observation). Instead, a chromosome-wide increase in acetylation of histone H3 occurs on the trisomic Ch4/7b, but does not occur on the monosomic Ch5 or any disomic chromosome. Thus, unique epigenetic signals occur on chromosomes with different aneuploidies, consistent with

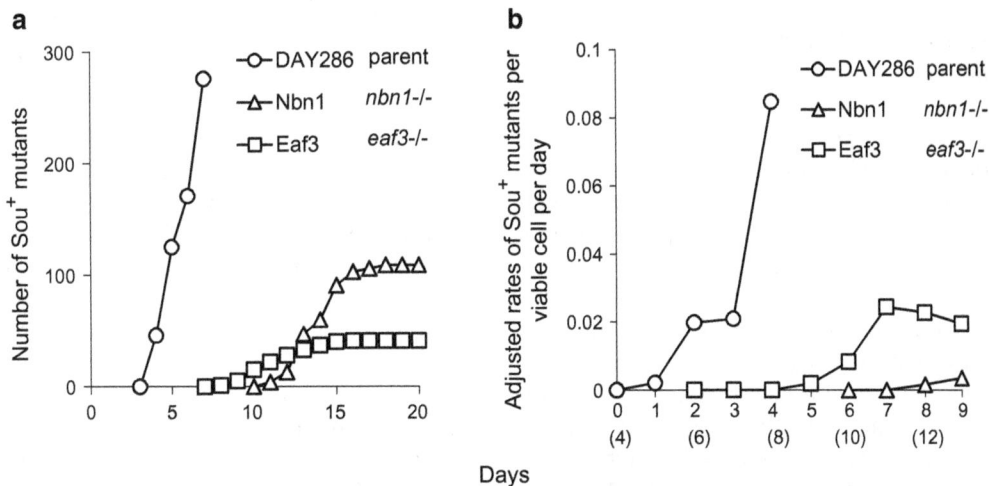

Fig. 4 Daily accumulation of Sou⁺ colonies (**a**) and adjusted rates of generation of Sou⁺ mutants per viable cell per day (**b**). Shown are deletion strains Nbn1 (*nbn1* −/−) and Eaf3 (*eaf3* −/−), as well as their parental strain DAY286, as indicated. Note that in (**b**), both the deduced time of formation of the mutations (top number) and the time of the appearance of the corresponding Sou⁺ colonies (bottom number, in parentheses) are presented. See Results section for more explanations. Note the delayed and diminished production of Sou⁺ colonies and diminished mutant rates by Nbn1 and Eaf3

distinct transcriptional responses. We provide evidence that the bulk of the increased acetylation of histone H4 is due to activity of the NuA4 HAT complex, orthologous to the enzyme complex previously implicated in dosage compensation on the single male X chromosome in *Drosophila*. While the enzymes responsible for the observed increase in H3 acetylation have yet to be identified, previous work in *C. albicans* suggests that the Rtt109 or SAGA/ADA complexes might be responsible [21, 22].

Interestingly, disruption of a single allele of the essential *ESA1* gene encoding the catalytic subunit of the NuA4 complex or lack of the Eaf3p subunit from targeting module did not have a significant effect on strain viability. These results suggest that sufficient Esa1p is produced in the hemizygous Esa1 strain for cell survival and Eaf3p is not critical for survival. Also, whole-cell Western blotting showed that strains lacking NuA4 subunits Nbn1p or Eaf3p did not exhibit reduced H4 acetylation attributable to NuA4. However, Wang et al. [16] reported that deletion of the NuA4 catalytic subunit Esa1p resulted in a significant decrease in the acetylation level of *C. albicans* nucleosomal H4. Importantly, deletion of either the Nbn1p or Eaf3p subunits of the NuA4 complex eliminated the characteristic increase in H4 acetylation observed on the monosomic Ch5 in DAY286-1 and compromised the ability of the deletion strains to undergo transformation to a monosomic Ch5 state. Our results support a model wherein a major pathway to generation of sorbose-resistant *C. albicans* mutants involves adaptation of a monosomic Ch5 state, with a concurrent chromosome-wide increase in H4 acetylation to facilitate a twofold upregulation of the large proportion of Ch5 genes required for normal cellular homeostasis.

We find consistency in the fact that there are epigenetic changes on both aneuploid chromosomes in Sor125(55), monosomic Ch5 and accompanying trisomic Ch4/7b. Although the details of gene regulation on Ch4/7b are less understood, and will be the subject of future study, the increase in acetylation of H3 is likely to be the outcome of a mechanism distinct than that responsible for upregulating expression of single-copy genes on Ch5. This mechanism might be involved in upregulation of a subset of genes on the trisomic Ch4/7b in conjunction with the increase in gene dosage associated with sorbose adaptation, concomitant with compensatory downregulation of many other genes to the disomic level. Given that mutants were preserved immediately after generation, we believe that the increase in H3 and H4 acetylation we detected is directly coupled with generation of aneuploid states and not with an unspecified instability due to, for example, how cells were maintained and handled (see "Methods").

Our results suggest NuA4 as a novel drug target to reduce the viability of strains resistant to anti-fungals. We have shown that a significant fraction of sorbose-resistant mutants with a monosomic Ch5 also exhibit decreased susceptibility to the anti-candidal agent caspofungin [3]. In addition, approximately a quarter of mutants that adapted to caspofungin upon direct exposure to this drug transitioned to a monosomic Ch5 state, which defines the decreased drug susceptibility [4]. Interestingly of over 20 HATs in *S. cerevisiae*, only Esa1p, the catalytic subunit of NuA4, is essential for viability [23]. In *C. albicans* NuA4, four subunits (Esa1p, Nbn1p, Eaf6p, and Epl1p) comprise the catalytic core, while eight, including Eaf3p, make up a structural/targeting module [24]. Given its role in resistance, drugs that act upon the NuA4 complex in *Candida*, but not in humans could provide a pathway to combating *C. albicans* infections.

Conclusions

Our data imply that acetylation associated with the monosomic Ch5 requires critical subunits of the *C. albicans* NuA4 HAT complex, suggesting a mechanism similar to that operative on the single X chromosome in *Drosophila* males, and setting a precedent for a fundamental cellular process appearing earlier in evolution. Recently, transcriptional dosage compensation was reported for sets of genes on different trisomic or higher ploidy aneuploid chromosomes in wild or laboratory strains of *S. cerevisiae* [25]. However, to our knowledge, no report is available regarding the state of histone acetylation on aneuploid chromosomes in *S. cerevisiae*. It will be interesting in future experiments to determine how NuA4 HAT activity contributes to the generation of monosomic Ch5 strains and the role of increase in H3 acetylation on Ch4/7b.

Methods
Strains, media, and growth conditions

The parental strain BWP17 [26] and a derivative strain Esa1 in which one copy of the gene *ESA1* encoding a catalytic subunit of NuA4 HAT complex is disrupted were supplied from the Tag-Module collection, University of Toronto. The parental strain DAY286 [26], a derivative strain Nbn1, lacking the *NBN1* gene encoding a subunit from the catalytic module of the NuA4 complex, and a derivative strain Eaf3, lacking the *EAF3* (orf19.2660) gene encoding a subunit identified as a component of the targeting module of yeast *S. cerevisiae* NuA4, were supplied by Fungal Genetics Stock Center (FGSC), University of Missouri, Kansas City (http://www.fgsc.net/candida/FGSCcandidaresources.htm). All the above strains cannot utilize sorbose, Sou⁻, and die on medium in which toxic sorbose is available as the only carbon source [27].

Also, we used the Sou$^-$ strain 3153A and its Sou$^+$ derivative Sor125(55) [7] (see Fig. 1a for schematics of chromosome ploidy of these strains). See Table 1 for strains used in this work.

Yeast extract/pepton/dextrose (YPD), synthetic dextrose (SD), L-sorbose, or sorbitol media were previously described [11, 28]. Media were supplemented with uridine (50 μg/ml), histidine (20 μg/ml), or arginine (10 μg/ml) when needed, as previously indicated [28]. Strains were preserved as − 70 °C stocks immediately after generation of mutant colonies and handled such that to avoid accumulation of unspecified mutations and chromosome instability due to, for example, long-term propagation, the use of aged cultures or exposure to low temperatures [see 29, 30 for more].

Generation and analysis of Ch5 monosomic mutants

Mutants that lost one Ch5 were generated by plating cells on solid synthetic medium in which L-sorbose is substituted for glucose as described [3, 18]. Intact chromosomes were prepared and precisely separated using contour-clamped homogenous electrophoretic field (CHEF) version of PFGE, as described [30].

ChIP-Chip analysis

We performed chromatin immunoprecipitation (ChIP) as described previously [31]. Briefly, cells were transferred from − 70 °C stocks to sorbitol master plates as a streak and incubated overnight, and approximately 3500 cfu were plated on each sorbitol plate to generate independent colonies. Upon reaching approximately 1×10^5 cells per colony, cells were collected and frozen at − 70 °C. Next, cells were removed from − 70 °C and proteins and DNA were formaldehyde cross-linked, cells were lysed, and whole-cell extract was prepared, and chromatin was sonicated with a Biorupter™ sonicator (Diagenode, Denville, NJ) to fragments from approximately 200 to 500 bp and then incubated with antibody specifically reactive to H4 acetylated at lysines K5, K8, K12, and K16 (pan-acetylation) (catalog #04-557, Upstate Biotechnology/Millipore, Billerica, MA), or antibody specifically reactive to H3 acetylated at K9 and K14 (catalog #39140, Active Motif, Carlsbad, CA) and immunoprecipitation performed. The cross-linking was reversed and ChIP and input control samples purified and amplified with WGA2 GenomePlex Complete Whole Genome Amplification (WGA) kit from Sigma-Aldrich Corporation (St. Louis, MO). Custom-designed tiling microarrays were generously provided by NimbleGen Inc. (Madison, WI) (http://www.nimblegen.com). Each microarray contained 710,907 probes that permitted up to four matches within the genome. Of all probes, approximately 20,000 matched the Ch5 sequences. The probes were in situ synthesized 50-mers with an average probe spacing of 35 bp on each strand tiled in an unbiased fashion across the entire genome. Empty features on the array were filled with randomly generated probes that had G + C content and length comparable to the C. albicans probes as controls for non-specific binding.

Array hybridizations were performed by Roche NimbleGen. Each array was hybridized with a mixture of ChIP-selected DNA fragments labeled with Cy5 and control input DNA from the same strain labeled with Cy3. We obtained the data from two independent experiments each for H4 and H3 acetylation. Signal intensities were scanned and normalized, and the ratio of ChIP DNA/Input DNA for either Sor125(55) or 3153A and difference between log2 ratios Sor125(55)/Input DNA and log2 ratios 3153A/Input DNA for either H4 or H3 acetylation were determined and are presented independently for each repeat in Additional file 6: Table S2. Raw data are available at GEO with the accession number GSE81684. An analysis was performed using an algorithm written in Python and plotted as the position of the probes on each chromosome in the reference strain SC5314, as annotated in genomic assambly 21 in CGD, versus the averaged log2 ratio Sor125(55) minus log2 ratio 3153A. Thus, the graphs reflect the difference in acetylation with data plotted above or below the X-axis reflecting increased or decreased acetylation in aneuploid cells compared to that in disomic parental cells, respectively. Thus, a score of +1 reflects a twofold increase in acetylation at a particular locus in the mutant compared to the parental strain. To minimize random noise on the plots, which obscured real effects due to the high probe density, log2 ratio differences were counted as significant and plotted only if the following criteria were met: mean log2 ratio difference of 0.2 units or greater (in the same direction) over three consecutive probes and $p < 0.1$ for all three consecutive probes. The use of three consecutive probes for this analysis is based on the fact that the average size of the immunoprecipitated DNA fragments was similar to the size of a genomic region covered by three probes. Thus, each dot corresponds to the location of three consecutive probes on the chromosome that satisfied a high stringency statistical filter.

Western blot analysis

Candida albicans cells were seeded on YPD plates (3000 cfu/plate) and incubated at 37 °C. Cells were collected in wash buffer containing 20 mM Tris, pH 8, 0.1 M NaCl, 0.5 mM dithiothreitol, 5 mM EDTA and washed three times in this buffer. Cells were suspended in lysis buffer containing 20 mM Tris, pH 8, 0.1 M NaCl, 0.1% NP-40, 0.5 mM dithiothreitol, 5 mM EDTA, 10 mM Sodium butyrate, and Protease Inhibitor Mini Tablets (Thermo Scientific, Rockford, IL), then microglass beads (Biospec Products, Bartlesville, OK) added, and cell lysed in a BeadBeater (Mini-Beadbeater-1, Biospec Products) at 20 s duration with 10 ~ 1-min intervals on ice. The lysed cells were mixed with 0.4 N HCl to a final concentration of 0.2 N, incubated on ice for 30 min with occasional vortexing, and centrifuged at 14,000 rpm for 10 min at 4 °C. Supernatants were collected, and 1/5 volume of 100% trichloroacetic acid was added, and after 30 min of incubation on ice, the samples were centrifuged at 14,000 rpm for 10 min at 4 °C and supernatants were discarded. The pellets were resuspended with ice-cold acetone containing 0.05 N HCl and centrifuged at 14,000 rpm for 10 min at 4 °C. The pellets were rinsed with ice-cold acetone lacking HCl twice more than resuspended with 20 mM Tris, pH 8.

Histone preparations were divided into 1 μg samples and were separated on a series of 15% SDS-PAGE gels and then transferred to PVDF membrane (0.2-μm pore size, Thermo Scientific) for Western blots; thus, H4 pan-acetylation and total H4 blots were determined for same histone preparation with the same amount of histone H4. Each Western blot analysis was conducted from three independent cell cultures. The membranes were probed with anti-histone H4, anti-pan-acetylated H4, anti-histone H4acK5, anti-histone H4acK8, anti-histone H4acK12, or anti-histone H4acK16 antibody (catalog ##39270, 39244, 39584, 39173, 39166, or 39168, respectively, Active Motif, Carlsbad, CA) followed by the incubation with ALP-labeled secondary antibody (anti-rabbit IgG linked to alkaline phosphatase, ImmunoReagents, Inc, Raleigh, NC) and detected by chemifluorescence (ECF substrate; GE Healthcare, Piscataway, NJ, USA). Fluorescence signals were detected using Molecular Imager Gel Doc XR + system (Bio-Rad, Hercules, CA), and the band densities and background corrections were quantified by Image Lab software (Bio-Rad). Acetylation levels were obtained as a density value of each histone H4 acetylation divided by density of H4. Then, acetylation levels of mutant versus parent were calculated. Data were subjected to statistical analysis (Student's *t* test).

Additional files

Additional file 1: Table S1. Positive and negative peaks of H3 or H4 acetylation presented for individual chromosomes of *Candida albicans*.

Additional file 2: Figure S1. Chromosome separation with PFGE of *C. albicans* mutants that adapted to utilize toxic L-sorbose. Names of the mutants and their parental strains are indicated on a top. Top gel shows precise separation of three smallest chromosomes 7, 6, and 5, as indicated on the right, while longer chromosomes are compressed in a top portion of the gel. Note that each of these chromosomes is presented by two bands, because homologous chromosomes in each pair are not of the equal size. Bottom gel shows precise separation of chromosomes 4, 3, 2, 1, and R, as indicated on the right. Note the lack of one chromosome 5 in the mutants. Also shown are chromosomes of the *Saccharomyces cerevisiae* strain 867 that serve as markers of *C. albicans* chromosomes.

Additional file 3: Figure S2. Example of Western blot with histone H4 antibodies of *C. albicans* strains as indicated on the top. Antibodies are indicated on the left. Purified histone extract from each strain was prepared and subjected to electrophoresis on 15% polyacrylamide gels followed by Western blot analysis (Methods).

Additional file 4: Figure S3. Histone H3 acetylation. (A) Example of Western blot with histone H3 antibodies of *C. albicans* strains as indicated on the top. For details, see the legend of Fig. S2. (B) Relative amount of H3 acetylation calculated from three independent Western blot analyses.

Additional file 5: Figure S4. The survival of DAY286, Nbn1 (*nbn1* −/−), and Eaf3 (*eaf3* −/−) Sou⁻ cells on L-sorbose medium. Daily survival rate was measured according to (18).

Additional file 6: Table S2. ChIP data (http://www.ncbi.nlm.nih.gov/geo/query/acc.cgi?acc=GSE81684).

Abbreviations
ChIP: chromatin immunoprecipitation; HAT: histone acetyltransferase; ORF: open reading frame; CHEF: contour-clamped homogenous electrophoretic field; PFGE: pulse-field gel electrophoresis; PVDF: polyvinylidene difluoride.

Authors' contributions
ER, MB, and JH conceived and designed the experiments and wrote the manuscript. HW, GB, and AK performed the experiments. SLW, ChT, and HW analyzed the data. All authors read and approved the final manuscript.

Author details
[1] Department of Biochemistry and Biophysics, University of Rochester Medical Center, Rochester, NY, USA. [2] Roche Diagnostics Corporation, Indianapolis, IN, USA. [3] Present Address: Parabase Genomics, Dorchester, MA, USA. [4] Department of Medicine, University of Rochester Medical Center, Rochester, NY, USA. [5] Department of Pediatrics, Center for Pediatric Biochemical Research, University of Rochester Medical Center, Rochester, NY, USA.

Acknowledgements
We thank the Tag-Module collection in the University of Toronto and the FGSC for deletion strains. We thank Cheeptip Benyajati for the critical reading of the manuscript.

Competing interests
The authors declare that they have no competing interests.

Funding

This work was supported in part by the University of Rochester Funds to ER, as well as partially supported by National Institutes of Health Grants Number 1R01AI110764 to ER, R01GM52426 to JH, and R01DK070687 to MB.

References

1. Rustchenko E. Chromosome instability in *Candida albicans*. FEMS Yeast Res. 2007;7:1–11.
2. Rustchenko E. Specific chromosome alterations of *Candida albicans*: mechanisms for adaptation to pathogenicity. In: Nombela C, Cassel GH, Baquero F, Gutierrez-Fuentes JA, editors. Evolutionary biology of bacterial and fungal pathogens. Washington: ASM Press; 2008. p. 197–212.
3. Yang F, Kravets A, Bethlendy G, Welle S, Rustchenko E. Chromosome 5 monosomy of *Candida albicans* controls susceptibility to various toxic agents, including major antifungals. Antimicrob Agents Chemother. 2013;57:5026–36.
4. Yang F, Zhang L, Wakabayashi H, Myers J, Jiang Y, Cao Y, Jimenez-Ortigosa C, Perlin DS, Rustchenko E. Tolerance to caspofungin in *Candida albicans* is associated with at least three distinctive mechanisms that govern expression of *FKS* Genes and cell wall remodeling. Antimicrob Agents Chemother. 2017;61:e00071-17. doi:10.1128/AAC.00071-17.
5. Kwon-Chung KJ, Chang YC. Aneuploidy and drug resistance in pathogenic fungi. PLoS Pathog. 2012;8:e1003022.
6. Ford ChB, Funt JM, Abbey D, Issi L, Guiducci C, Martinez DA, Delorey T, Li BY, White ThC, Cuomo Ch, Rao RP, Berman J, Thompson DA, Regev A. The evolution of drug resistance in clinical isolates of *Candida albicans*. eLIFE. 2014. doi:10.7554/eLife.00662.
7. Kravets A, Qin H, Ahmad A, Bethlendy G, Gao Q, Rustchenko E. Widespread occurrence of dosage compensation in *Candida albicans*. PLoS ONE. 2010;5:e10856.
8. Kravets A, Yang F, Bethlendy G, Sherman F, Rustchenko E. Adaptation of *Candida albicans* to growth on sorbose via monosomy of chromosome 5 accompanied by duplication of another chromosome carrying a gene responsible for sorbose utilization. FEMS Yeast Res. 2014;14:708–13.
9. Suwunnakorn S, Wakabayashi H, Rustchenko E. Chromosome 5 of human pathogen *Candida albicans* carries multiple genes for negative control of caspofungin and anidulafungin susceptibility. Antimicrob Agents Chemother. 2016;60:7457–67.
10. Kabir M, Ahmad A, Greenberg J, Wang Y-K, Rustchenko E. Loss and gain of chromosome 5 controls growth of *Candida albicans* on sorbose due to dispersed redundant negative regulators. Proc Natl Acad Sci USA. 2005;102:12147–52.
11. Ahmad A, Kravets A, Rustchenko E. Transcriptional regulatory circuitries in the human pathogen *Candida albicans* involving sense-antisense interactions. Genetics. 2012;190:537–47.
12. Hallacli E, Akhtar A. X chromosomal regulation in flies: when less is more. Chromosome Res. 2009;17:603–19.
13. Eisen A, Utley RT, Nourani A, Allard S, Schmidt P, Lane WS, Lucchesi JC, Côté J. The yeast NuA4 and Drosophila MSL complexes contain homologous subunits important for transcription regulation. J Biol Chem. 2001;276:3484–91.
14. Doyon Y, Côté J. The highly conserved and multifunctional NuA4 HAT complex. Curr Opin Genet Dev. 2004;14:147–54.
15. Doyon Y, Selleck W, Lane WS, Tan S, Côté J. Structural and functional conservation of the NuA4 histone acetyltransferase complex from yeast to humans. MCB. 2004;24:1884–96.
16. Wang X, Chang P, Ding J, Chen J. Distinct and redundant roles of the two MYST histone acetyltransferases Esa1 and Sas2 in cell growth and morphogenesis of *Candida albicans*. Eucaryot Cell. 2013;12:438–49.
17. Lu Y, Su Ch, Mao X, PalaRaniga P, Liu H, Chen J. Efg1-mediated recruitment of NuA4 to promoters is required for hypha-specific Swi/Snf binding and activation in *Candida albicans*. Mol Biol Cell. 2008;19:4260–72.
18. Janbon G, Sherman F, Rustchenko E. Appearance and properties of L-sorbose-utilizing mutants of *Candida albicans* obtained on a selective plate. Genetics. 1999;153:653–64.
19. Rustchenko EP, Curran TM, Sherman F. Variations in the number of ribosomal DNA units in morphological mutants and normal strains of *Candida albicans* and in normal strains of *Saccharomyces cerevisiae*. J Bacteriol. 1993;175:7189–99.
20. Rustchenko E. *Candida albicans* adaptability to environmental challenges by means of specific chromosomal alterations. In: Pandalai SG, editor. Recent research developments in bacteriology, vol. 1. Trivandrum: Transworld Research Network; 2003. p. 91–102.
21. Lopez da Rosa J, Kaufman PD. Chromatin-mediated *Candida albicans* virulence. Biochim Biophys Acta. 2012;1819:349–55.
22. Sellam A, Askew C, Epp E, Lavoi H, Whiteway M, Nantel A. Genome-wide mapping of the coactivator Ada2p yields insight into the functional roles of SAGA/ADA complex in *Candida albicans*. Mol Biol Cell. 2009;20:2389–400.
23. Mitchell L, Lambert JP, Gerdes M, Al-Madhoun AS, Skerjanc IS, Figeys D, Baetz K. Functional dissection of the NuA4 histone acetyltransferase reveals its role as a genetic hub and that Esf1 is essential for complex integrity. Mol Cell Biol. 2008;28:2244–56.
24. Chittuluru JR, Chaban Y, Monnet-Saksouk J, Carrozza M, Sapountzi V, Selleck W, Huang J, Utley RT, Cramet M, Allard S, Cai G, Workman JL, Fried MG, Tan S, Côté J, Asturias FJ. Structure and nucleosome interaction of the yeast NuA4 and piccolo-NuA4 histone acetyltransferase complexes. Nat Struct Mol Biol. 2011;18:1196–203.
25. Hose J, Mun Yong C, Sardi M, Wang Z, Newton MA, Gasch AP. Dosage compensation can buffer copy-number variation in wild yeasts. eLIFE. 2015. doi:10.7554/eLife.05462.001.
26. Davis DA, Bruno VM, Loza L, Filler SG, Mitchell AP. *Candida albicans* Mds3p, is a conserved regulator of pH responses and virulence identified through insertional mutagenesis. Genetics. 2002;162:1573–81.
27. Janbon G, Sherman F, Rustchenko E. Monosomy of a specific chromosome determines L-sorbose utilization: a novel regulatory mechanism in *Candida albicans*. Proc Natl Acad Sci USA. 1998;95:5150–5.
28. Sherman F. Getting started with yeast. Methods Enzymol. 2003;350:3–41.
29. Rustchenko-Bulgac E. Variations of *Candida albicans* electrophoretic karyotypes. J Bacteriol. 1991;173:6586–96.
30. Ahmad A, Kabir MA, Kravets A, Andaluz E, Larriba G, Rustchenko E. Chromosome instability and unusual features of some widely used strains of *Candida albicans*. Yeast. 2008;25:433–48.
31. Drouin S, Robert F. Genome-wide location analysis of chromatin-associated proteins by ChIP on CHIP: controls matter. 2015; in press. https://www.researchgate.net/publication/268401136_Genome-wide_Location_Analysis_of_Chromatin-associated_Proteins_by_ChIP_on_CHIP_Controls_Matter

10

Dynamics of 5-methylcytosine and 5-hydroxymethylcytosine during pronuclear development in equine zygotes produced by ICSI

Sonia Heras, Katrien Smits, Catharina De Schauwer and Ann Van Soom[*]

Abstract

Background: Global epigenetic reprogramming is considered to be essential during embryo development to establish totipotency. In the classic model first described in the mouse, the genome-wide DNA demethylation is asymmetric between the paternal and the maternal genome. The paternal genome undergoes ten-eleven translocation (TET)-mediated active DNA demethylation, which is completed before the end of the first cell cycle. Since TET enzymes oxidize 5-methylcytosine to 5-hydroxymethylcytosine, the latter is postulated to be an intermediate stage toward DNA demethylation. The maternal genome, on the other hand, is protected from active demethylation and undergoes replication-dependent DNA demethylation. However, several species do not show the asymmetric DNA demethylation process described in this classic model, since 5-methylcytosine and 5-hydroxymethylcytosine are present during the first cell cycle in both parental genomes. In this study, global changes in the levels of 5-methylcytosine and 5-hydroxymethylcytosine throughout pronuclear development in equine zygotes produced in vitro were assessed using immunofluorescent staining.

Results: We were able to show that 5-methylcytosine and 5-hydroxymethylcytosine both were explicitly present throughout pronuclear development, with similar intensity levels in both parental genomes, in equine zygotes produced by ICSI. The localization patterns of 5-methylcytosine and 5-hydroxymethylcytosine, however, were different, with 5-hydroxymethylcytosine homogeneously distributed in the DNA, while 5-methylcytosine tended to be clustered in certain regions. Fluorescence quantification showed increased 5-methylcytosine levels in the maternal genome from PN1 to PN2, while no differences were found in PN3 and PN4. No differences were observed in the paternal genome. Normalized levels of 5-hydroxymethylcytosine were preserved throughout all pronuclear stages in both parental genomes.

Conclusions: In conclusion, the horse does not seem to follow the classic model of asymmetric demethylation as no evidence of global DNA demethylation of the paternal pronucleus during the first cell cycle was demonstrated. Instead, both parental genomes displayed sustained and similar levels of methylation and hydroxymethylation throughout pronuclear development.

Keywords: Horse, Pronucleus, Epigenetic reprogramming, 5-Methylcytosine, 5-Hydroxymethylcytosine, DNA methylation, DNA hydroxymethylation, Active demethylation

*Correspondence: Ann.VanSoom@UGent.be
Department of Reproduction, Obstetrics and Herd Health, Faculty
of Veterinary Medicine, Ghent University, 9820 Merelbeke, Belgium

Background

During mammalian development, two major waves of epigenetic reprogramming take place, one during germ line differentiation and the second during preimplantation embryo development. The epigenetic reprogramming during embryo development is of major importance because it is considered as being essential to establish a totipotent state [1]. The methylation of the fifth carbon of cytosine, 5-methylcytosine (5mC), was the first epigenetic modification discovered in DNA and, hence, is nowadays the best studied DNA epigenetic modification. It plays a key role in gene expression regulation, X chromosome inactivation, gene imprinting and the control of endogenous retrotransposons [2]. Based on studies in the mouse, the genome-wide DNA demethylation during embryo development, which excludes imprinted genes and retrotransposons, is proposed to be asymmetric between the maternal and the paternal genome [3, 4]. As such, the complete demethylation of the paternal DNA is achieved before the end of the first cell cycle by active demethylation. In contrast, the demethylation of the maternal genome is a passive process associated with cell divisions [5]. This epigenetic reprogramming during preimplantation embryo development, however, is not strictly conserved between species. While this asymmetric global demethylation pattern is conserved in rat [2, 6] and human [7], some species such as rabbit [8], pig [9], goat [10] and sheep [11] failed to follow this model. In these species, no DNA demethylation is observed during the first cell cycle, regardless of the parental origin of the genome. Other species, such as cattle, show an intermediate pattern with partial demethylation of the paternal pronucleus (pPN) in the zygote [11].

In 2009, a new modified form of cytosine, 5-hydroxymethylcytosine (5hmC), was identified. This new modification is generated by the oxidation of 5mC by the ten-eleven translocation (TET) enzymes [12, 13], which are able to further oxidize 5hmC into 5-formylcytosine (5fC) and 5-carboxylcytosine (5caC) [12]. As such, TET3 is considered to be the initiator of active DNA demethylation in the paternal genome. The protein STELLA (PGC7), on the other hand, protects the maternal genome from active demethylation by binding to the dimethylated histone H3 lysine 9 (H3K9me2) and excluding TET3 which results in preventing the oxidation of 5mC to 5hmC [14]. The fact that 5hmC was considered then a DNA demethylation transient, was supported by the observed complementary patterns of 5mC and 5hmC in the pPN during pronuclear development in mouse, rabbit and cattle, in which a reduction in 5mC levels is accompanied by an increase in 5hmC levels [15]. Though, the presence of high 5hmC levels in several tissues, including the nervous system, indicates that this epigenetic DNA modification plays its own epigenetic role [13] being involved in chromatin and transcription regulation [16, 17].

The biological implication of this lack of interspecies conservation of global asymmetric DNA demethylation is currently unknown. Moreover, the contradictory results observed within the same species using different immunofluorescent staining protocols [16–19] have questioned the model as well. In this study, we aimed to gain further insight into the interspecies conservation of 5mC and 5hmC patterns during the first cell cycle. To this end, we characterized for the first time the dynamics of 5mC and 5hmC throughout pronuclear development in equine zygotes produced by ICSI, independently in the maternal (mPN) and the paternal (pPN) pronucleus, using an immunofluorescent staining protocol optimized previously [20, 21].

Results

Zygote classification according to pronuclear morphology

In this study, the dynamic patterns of 5mC and 5hmC were analyzed independently for the pPN and the mPN during pronuclear development in in vitro-produced equine zygotes. To this end, the zygotes ($n = 141$) were classified into different stages according to size, position and conformation of the pronuclei. Due to the lipid-rich cytoplasm of equine oocytes and zygotes, immunostaining was used to identify the following pronuclear stages: (1) PN0: decondensing sperm head and meiosis II finished with the second polar body extruded and the chromosomes which start to decondense, (2) PN1: the decondensed DNA of the mPN and pPN is forming two small pronuclei, (3) PN2: the pronuclei are increasing in size and start to migrate toward the center, (4) PN3: the pronuclei reached their maximum size and are in apposition and (5) PN4: the pronuclei are in apposition and display an area with a fibrillary aspect.

To obtain all pronuclear stages, equine zygotes were collected at five time points ranging from 8 to 23 h after fertilization using intracytoplasmic sperm injection (ICSI). The distribution of the pronuclear and cleavage stages among the different collection time points is illustrated in Fig. 1. As the oocytes were fertilized using ICSI, the moment of penetration of the spermatozoon was timed precisely. Nevertheless, a lot of variability in the developmental stage of the embryos within each collection time point was observed, indicating that the activation of the oocyte and further pronuclear formation in the individual embryos were not evolving synchronously.

The oocyte activation rate increased with time in culture (h). The shortest activation rate was observed at 8 h after ICSI, with less than 40% of the injected oocytes activated, while 75.5% was activated at 23 h after ICSI, i.e.,

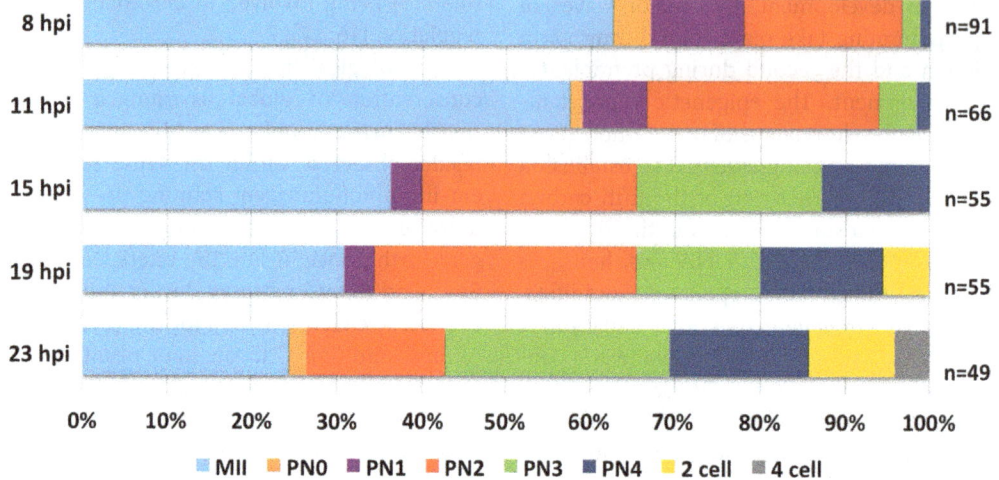

Fig. 1 Distribution of ICSI-produced equine embryos among developmental stages in each time point of collection. The time points of collection are measured in hours after ICSI (hpi). The oocytes in metaphase II (MII) did not activate after the injection of the spermatozoon, which was found intact in the cytoplasm of the oocyte. The number of injected oocytes collected at each time point is indicated in the *right*. Degenerated oocytes and zygotes were excluded from the analysis

the last time point of collection. At 8 h after ICSI, most zygotes were identified as PN1 (11%) and PN2 (18%) stages. At 11 h after ICSI, the majority of zygotes were at PN2 stage (27%), with only 8% and 3% at PN1 and PN3, respectively. At 15 h after ICSI, PN2 (26%) was again the most observed stage although 22% of the collected zygotes were in PN3. At 19 h after ICSI, zygotes were still mostly in PN2 (31%), while 15% of the zygotes were in either PN3 or PN4. At 23 h after ICSI, only 16% of the zygotes were in PN2 and PN3 became the most common stage (27%). Within these last two time points of collection, i.e., 19 and 23 h after ICSI, 5% and 10% of the embryos already reached the two-cell stage, respectively. Furthermore, 23 h after ICSI, 4% of the embryos were at the four-cell stage. PN1 might be considered as a very transitory stage as it was only observed in 19 of the 141 zygotes included in the study, while PN2, the stage in which the pronuclei grow and migrate, was present at all collection time points and can be considered as the longest stage with 59 of the 141 zygotes observed in that stage.

Determination of the parental pronuclear origin and DNA content

In the mouse, the relative pronuclear size is used to determine the parental origin of the pronuclei [22, 23]. However, the relative size of the pronuclei in the horse is variable and does not differ between the mPN and pPN, as confirmed in this study (Fig. 2). Therefore, the asymmetric pattern of the tri-methylation of histone H3 lysine 9 (H3K9me3) was used to accurately differentiate between the mPN and pPN. This histone modification is

in the horse only present in the mPN throughout pronuclear development, as described previously [20].

After determination of the stage and parental origin of the pronuclei, the total DNA fluorescence was calculated for each pronucleus of each zygote by multiplying the fluorescence intensity by its corresponding pronuclear area. When the total DNA fluorescence between the different pronuclear stages was compared independently for the mPN and the pPN (Fig. 3), a significant increase was observed between PN1 and PN2 in the mPN (p

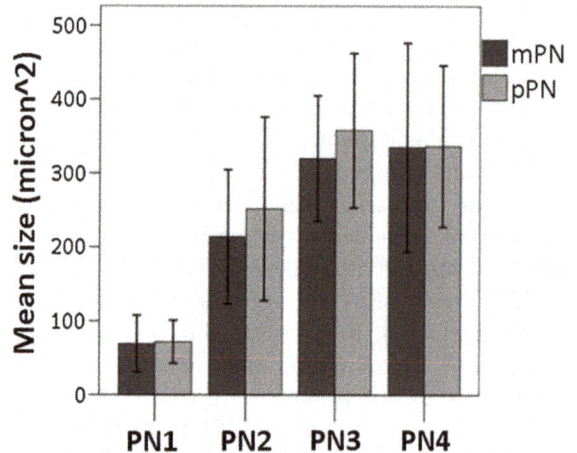

Fig. 2 Evolution of the pronuclear size during development. No differences were found in mean pronuclear size (at the largest diameter) between the mPN and pPN in the four pronuclear stages. PN1 $n = 19$, PN2 $n = 59$, PN3 $n = 37$, PN4 $n = 26$. *Error bars* represent ± 1 SD

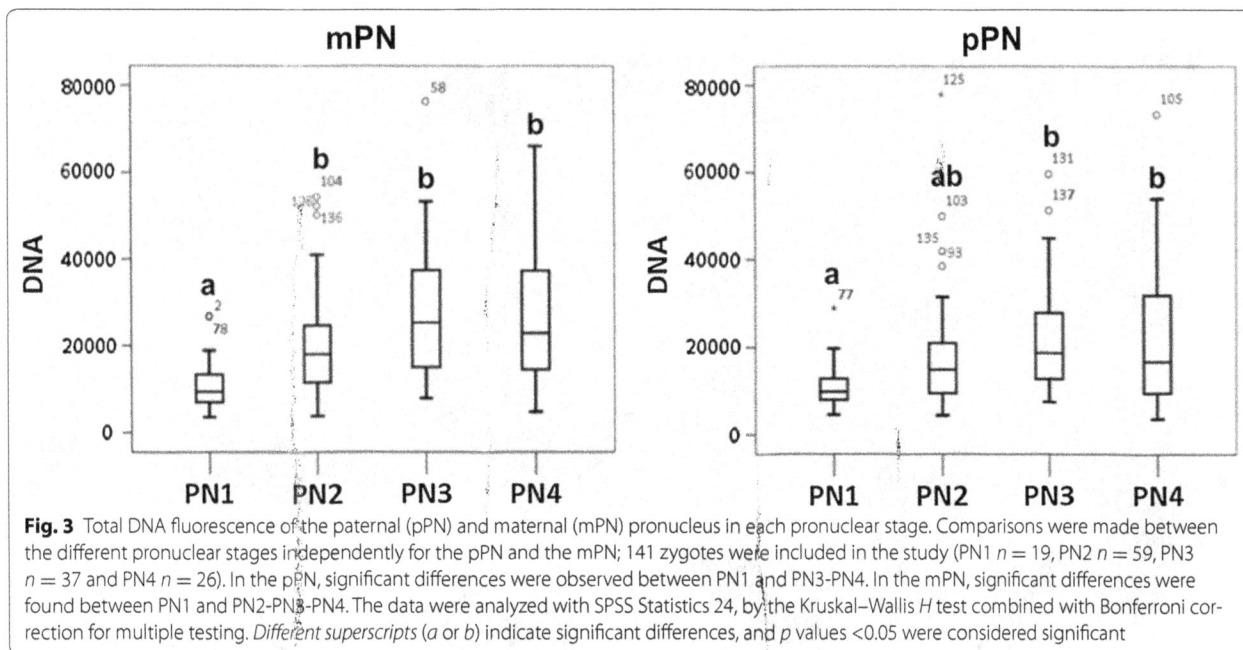

Fig. 3 Total DNA fluorescence of the paternal (pPN) and maternal (mPN) pronucleus in each pronuclear stage. Comparisons were made between the different pronuclear stages independently for the pPN and the mPN; 141 zygotes were included in the study (PN1 $n = 19$, PN2 $n = 59$, PN3 $n = 37$ and PN4 $n = 26$). In the pPN, significant differences were observed between PN1 and PN3-PN4. In the mPN, significant differences were found between PN1 and PN2-PN3-PN4. The data were analyzed with SPSS Statistics 24, by the Kruskal–Wallis H test combined with Bonferroni correction for multiple testing. *Different superscripts (a or b) indicate significant differences, and p* values <0.05 were considered significant

value = 0.01) and between PN1 and PN3 in the pPN (p value = 0.01), which is indicative for DNA replication. In the rabbit, it has been demonstrated that DNA replication occurs during the migration and enlargement of the pronuclei (PN2 and PN3 of their classification) by injecting DIG-11dUTP in the zygotes [24]. Consequently, the total fluorescence of 5mC or 5hmC was divided by its corresponding total DNA fluorescence, as performed by Reis Silva et al. [24], to correct for the DNA replication that occurs during pronuclear development [25].

Dynamics of DNA methylation during pronuclear development

The global DNA methylation pattern of 60 ICSI-produced equine zygotes was analyzed throughout pronuclear development (PN1: $n = 6$, PN2: $n = 25$, PN3: $n = 18$, PN4: $n = 11$).

5mC was highly present in both the mPN and the pPN and no differences in intensity could be determined at a glance between the mPN and the pPN (Fig. 4). Interestingly, the distribution of 5mC in the pronuclei was rather heterogeneous with higher concentrations in certain regions of the DNA (Fig. 5a).

The dynamics of the normalized 5mC fluorescence (5mC/DNA) between the different pronuclear stages showed a significant increase between PN1 and PN2 in the mPN (p value = 0.022) (Fig. 6). In the pPN, on the other hand, no significant differences in normalized 5mC fluorescence were observed between pronuclear stages.

Regarding total 5mC fluorescence, a significant increase was found between PN1 and PN2, PN3, and PN4 in the mPN (p value = 0.015, 0.003 and 0.011, respectively) and between PN1 and PN2 in the pPN (p value = 0.032). The total DNA fluorescence increased significantly between PN1 and PN3 (p value = 0.023) in the mPN, but the increase in DNA fluorescence observed in the pPN when all zygotes included in the study were taken into account (5mC + 5hmC: $n = 141$) (Fig. 3) did not reach statistical significance when only the 5mC-stained pPN ($n = 60$) was considered (Fig. 6). Finally, higher normalized 5mC fluorescence was observed in the mPN than in the pPN at PN3, with a paternal/maternal ratio of 0.83 (p value = 0.046).

Dynamics of DNA hydroxymethylation during pronuclear development

To evaluate the global hydroxymethylation pattern in equine zygotes, 81 zygotes were produced using ICSI and distributed in four pronuclear stages (PN1: $n = 13$, PN2: $n = 34$, PN3: $n = 19$, PN4: $n = 15$).

In all the zygotes analyzed, 5hmC was highly present throughout pronuclear development in both the pPN and mPN, with no observable differences depending on the parental pronuclear origin (Fig. 7). In contrast to the distribution of 5mC, 5hmC was very homogeneously distributed in the DNA (Fig. 5b). After quantifying the fluorescence intensity, no significant differences were observed in the normalized

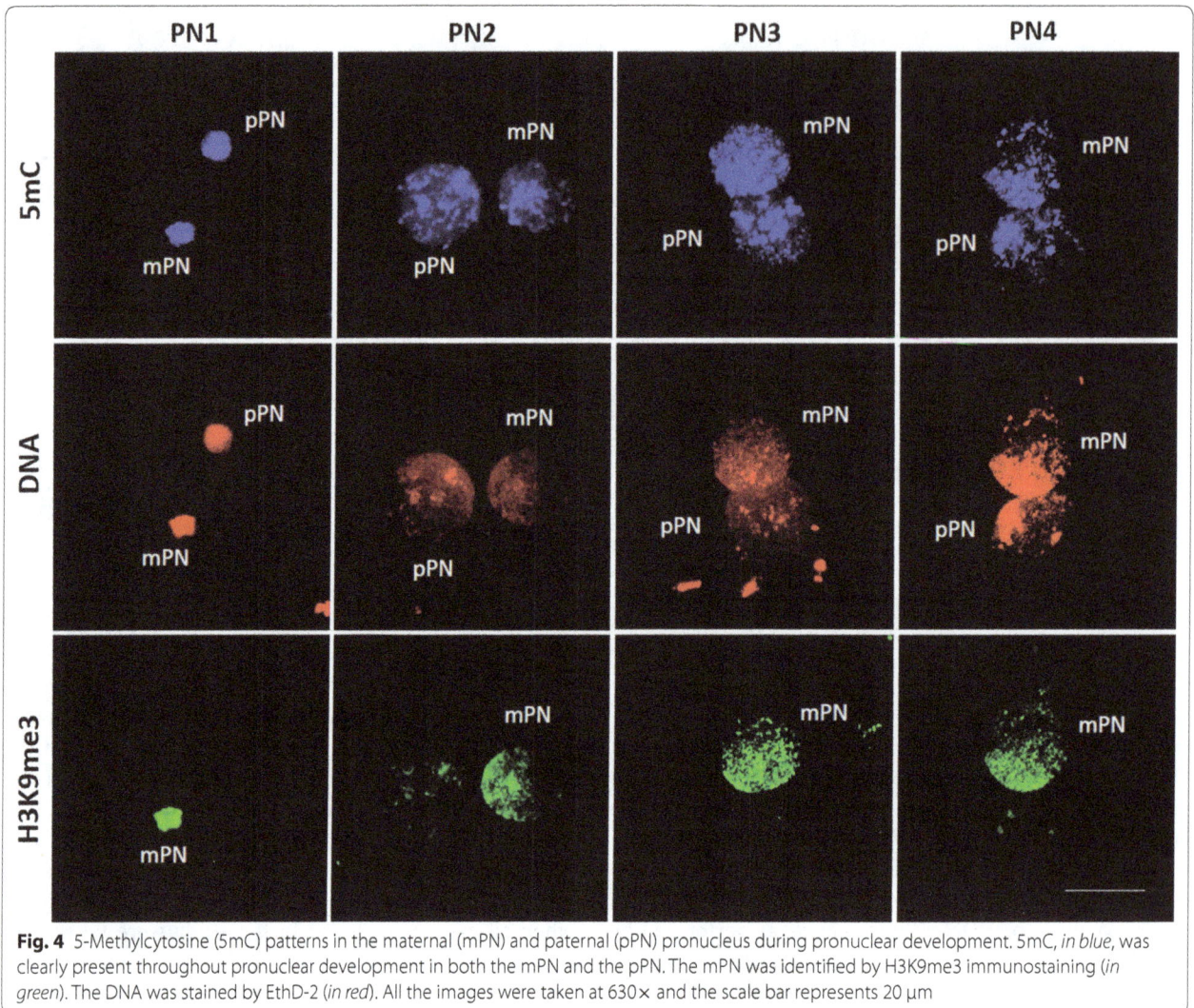

Fig. 4 5-Methylcytosine (5mC) patterns in the maternal (mPN) and paternal (pPN) pronucleus during pronuclear development. 5mC, *in blue*, was clearly present throughout pronuclear development in both the mPN and the pPN. The mPN was identified by H3K9me3 immunostaining (*in green*). The DNA was stained by EthD-2 (*in red*). All the images were taken at 630× and the scale bar represents 20 µm

5hmC levels (5hmC/DNA) between pronuclear stages regardless the parental origin of the pronuclei (Fig. 8), indicating that the ratio of 5hmC to DNA is relatively constant throughout pronuclear development. The total 5hmC fluorescence showed a significant increase between PN1 and PN2, PN3 and PN4 (p value = 0.025, 0.009 and 0.004, respectively) in the mPN, and between PN1 and PN3 (p value = 0.008) in the pPN (Fig. 8). Analogously, the total DNA fluorescence between pronuclear stages also increased significantly between PN1 and PN3-PN4 in both the mPN (p value = 0.001) and the pPN (p value = 0.01 and 0.002, respectively). These results indicate that the increase in 5hmC is associated with the increase in DNA. Significantly lower levels of normalized 5hmC were found in the pPN compared to the mPN in PN2, with a paternal/maternal ratio of 0.87 (p value = 0.027). The

opposite situation was observed in PN3, with significantly higher levels of normalized 5hmC in the pPN compared to the mPN and a paternal/maternal ratio of 1.25 (p value = 0.027).

Discussion

In this study, we analyzed for the first time the global dynamic patterns of 5mC and 5hmC independently for the pPN and the mPN during pronuclear development in equine zygotes produced by ICSI.

To evaluate the dynamic changes in 5mC and 5hmC during pronuclear development, equine zygotes were collected at five different time points within the first 23 h after ICSI. Within this time frame, we were able to assess the whole pronuclear development as the embryo stages collected ranged from injected oocytes, which were not activated yet to four-cell embryos.

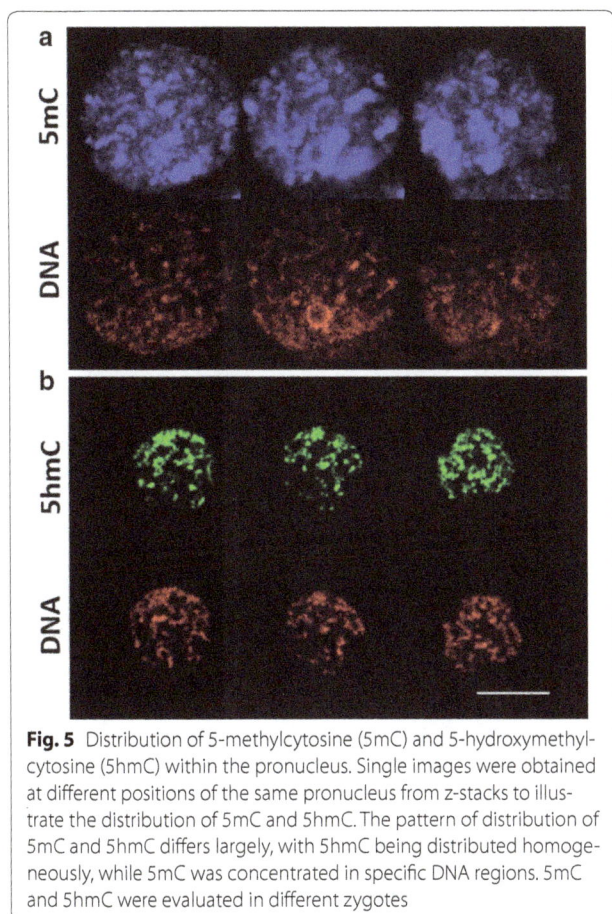

Fig. 5 Distribution of 5-methylcytosine (5mC) and 5-hydroxymethyl-cytosine (5hmC) within the pronucleus. Single images were obtained at different positions of the same pronucleus from z-stacks to illustrate the distribution of 5mC and 5hmC. The pattern of distribution of 5mC and 5hmC differs largely, with 5hmC being distributed homogeneously, while 5mC was concentrated in specific DNA regions. 5mC and 5hmC were evaluated in different zygotes

The presence of 5mC and 5hmC was evaluated by immunofluorescent staining in individual embryos. Although this is a very powerful tool to study the global changes in 5mC and 5hmC levels, the results are highly influenced by the protocol used, with the proper exposure of the epitopes and the optimal incubation time and concentration of the primary antibodies being of major importance [18]. For example, it has been demonstrated in mouse zygotes that a reduction in the concentration of 5hmC primary antibody resulted in a reduced signal in the mPN [18], and the incubation with 5hmC primary antibody for longer than 1 h at room temperature or shorter than 6 h at 4 °C resulted in higher levels in the pPN than in the mPN, while these differences were no longer observed when saturation binding conditions were used [17]. Additionally, when epitopes are retrieved using the traditional acid-based method, an active demethylation in mouse zygotes is observed resulting in complete DNA demethylation of the pPN as early as 8 h after fertilization [4, 11, 15, 23, 26]. However, if a new method to retrieve the 5mC epitope in mice is used, which combines the traditional acid treatment with a short tryptic digestion, high levels of 5mC in both pronuclei throughout

pronuclear development are observed instead of the active demethylation of the pPN [22].

To overcome this bias, we optimized the immunofluorescent staining in equine zygotes in a previous study, in which the traditional acid treatment (4 N HCl) was compared to the new epitope retrieval method (4 N HCl + tryptic digestion) to evaluate the dynamics of 5mC and 5hmC and no significant differences between both methods were observed [27]. The new epitope retrieval method was chosen in this case because it reduces the acid treatment time by half. Furthermore, the optimal concentration and incubation time for the primary antibodies were determined.

The different accessibility of 5mC and 5hmC antibodies to the DNA depending on its parental origin in the mouse, evidenced by the asymmetric pattern of this marks observed between the mPN and pPN after short incubation with the primary antibodies or different epitope retrieval methods, indicates a different chromatin conformation depending on its parental origin. Indeed, the paternal DNA undergoes massive changes during pronuclear development, which might explain the different conformation. In the horse, however, this possible asymmetric chromatin conformation does not affect the binding of 5mC and 5hmC antibodies.

During the first cell cycle, DNA replication takes place [25] and the size and complexity of the pronuclei vary enormously, which renders fluorescence intensity quantification even more challenging. In order to minimize these influences, we carefully classified the collected zygotes into four different pronuclear stages, based on the size and the conformation of their pronuclei. Additionally, the normalized fluorescence of 5mC or 5hmC was calculated for each pronucleus to correct for DNA replication.

In this study, we found no evidence of active DNA demethylation of the pPN in the horse. 5mC was highly present in both the mPN and pPN throughout pronuclear development. In the mPN, we found only increased normalized 5mC levels between PN1 and PN2, but no differences were further found with PN3 and PN4, reflecting stable normalized 5mC levels from PN1 to PN4. These observations are in contrast to the classic model of global methylation erasure of the pPN during the first cell cycle as established in mouse [4] and conserved in human [7, 11] and rat [6, 28]. However, they are in line with the results obtained in other species including rabbits [8, 24], sheep [11], pigs [9] and goats [10], where no loss of methylation was reported neither in the pPN or the mPN during the first cell cycle.

At first, 5hmC was considered as being only an intermediate form for DNA demethylation through TET oxidation [12]. This hypothesis was supported by the

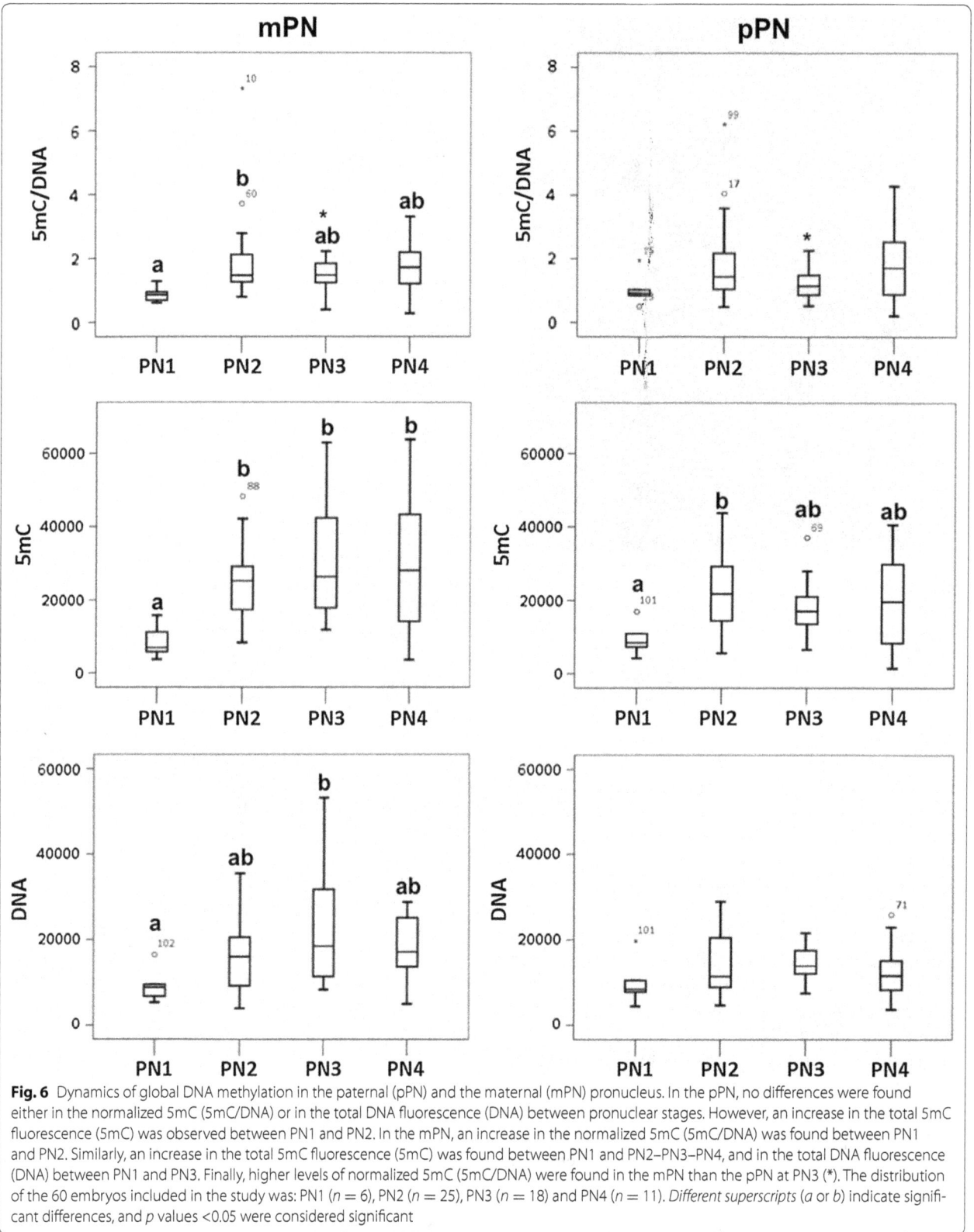

Fig. 6 Dynamics of global DNA methylation in the paternal (pPN) and the maternal (mPN) pronucleus. In the pPN, no differences were found either in the normalized 5mC (5mC/DNA) or in the total DNA fluorescence (DNA) between pronuclear stages. However, an increase in the total 5mC fluorescence (5mC) was observed between PN1 and PN2. In the mPN, an increase in the normalized 5mC (5mC/DNA) was found between PN1 and PN2. Similarly, an increase in the total 5mC fluorescence (5mC) was found between PN1 and PN2–PN3–PN4, and in the total DNA fluorescence (DNA) between PN1 and PN3. Finally, higher levels of normalized 5mC (5mC/DNA) were found in the mPN than the pPN at PN3 (*). The distribution of the 60 embryos included in the study was: PN1 ($n = 6$), PN2 ($n = 25$), PN3 ($n = 18$) and PN4 ($n = 11$). *Different superscripts* (*a* or *b*) indicate significant differences, and *p* values <0.05 were considered significant

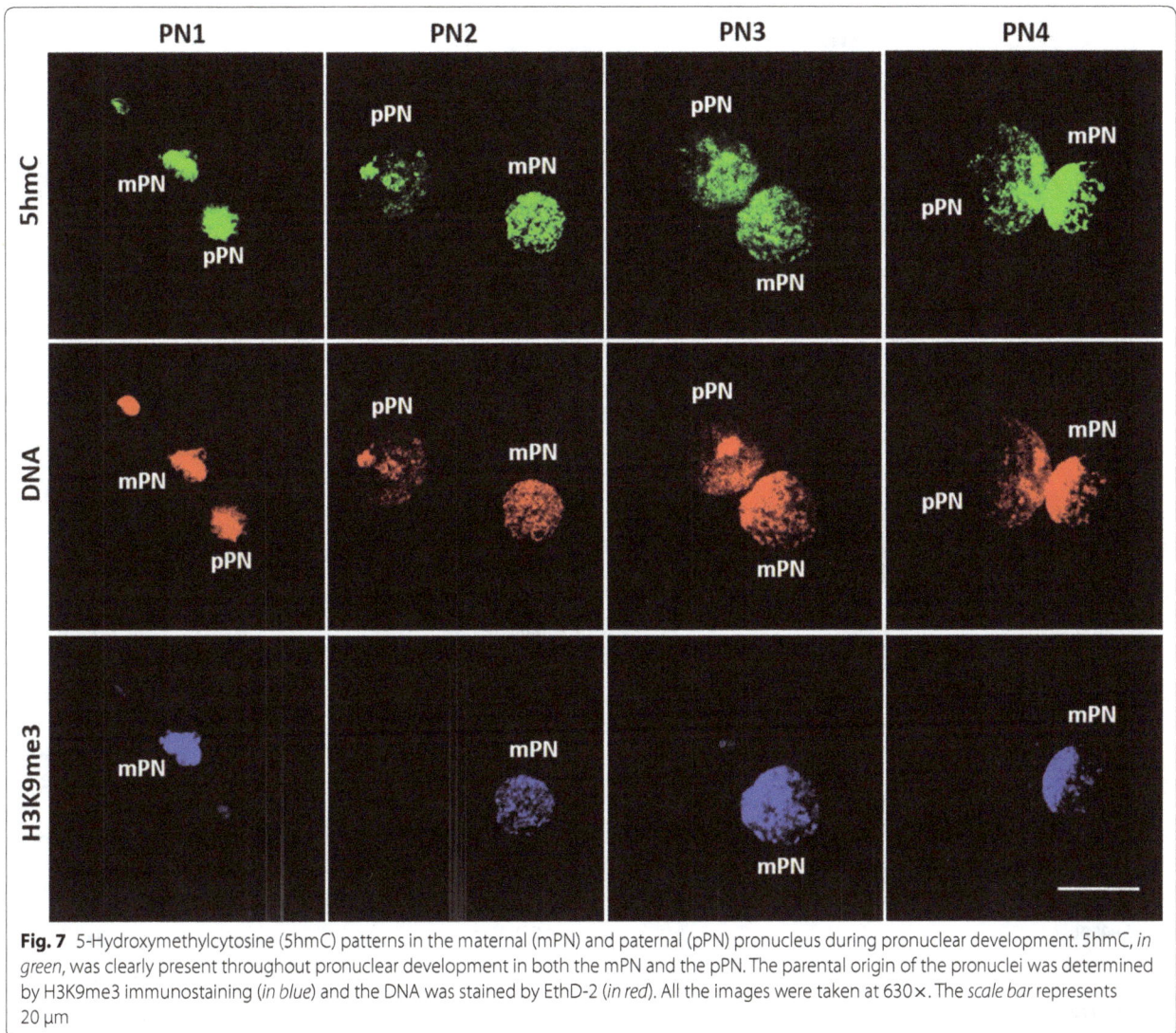

Fig. 7 5-Hydroxymethylcytosine (5hmC) patterns in the maternal (mPN) and paternal (pPN) pronucleus during pronuclear development. 5hmC, *in green*, was clearly present throughout pronuclear development in both the mPN and the pPN. The parental origin of the pronuclei was determined by H3K9me3 immunostaining (*in blue*) and the DNA was stained by EthD-2 (*in red*). All the images were taken at 630×. The *scale bar* represents 20 μm

observed inverse patterns between 5mC and 5hmC reported in mouse, rabbit and bovine zygotes, where increasing 5hmC levels in the pPN was accompanied by decreasing 5mC levels [5, 15, 29]. Nevertheless, it must be mentioned that some studies in mouse, using different immunostaining protocols, were not able to confirm this inverse pattern [16, 17]. Also in this study, this inverse relationship between 5mC and 5hmC was not observed, and 5mC as well as 5hmC coexisted in both the mPN and pPN throughout pronuclear development in the horse. Even though 5mC and 5hmC levels were not studied within the same embryo, and consequently, we were not able to correlate the dynamics of both marks, we did observe a completely different distribution pattern. While 5mC seemed more concentrated in some DNA regions, 5hmC was homogeneously distributed in the pronuclear DNA. This different distribution pattern of 5mC and 5hmC has also been observed in cleavage stage embryos in the mouse, especially in the emerging pluripotent lineage where the presence of 5mC was restricted to some intensely stained foci [30]. This asymmetric pattern might be a consequence of the strong association of 5mC to heterochromatin, while 5hmC is preferentially associated with euchromatin [30]. Moreover, this finding also supports the growing evidence that 5hmC plays its own epigenetic role [16, 17, 31, 32].

It has been previously reported in mice, rabbit and cattle [5, 15, 16] that 5hmC is only present (or present in a consistent higher level) in the pPN compared to its maternal counterpart. This was not observed in the horse, where high levels of 5hmC were found in both mPN and pPN. In this study, higher levels of 5hmC were demonstrated in

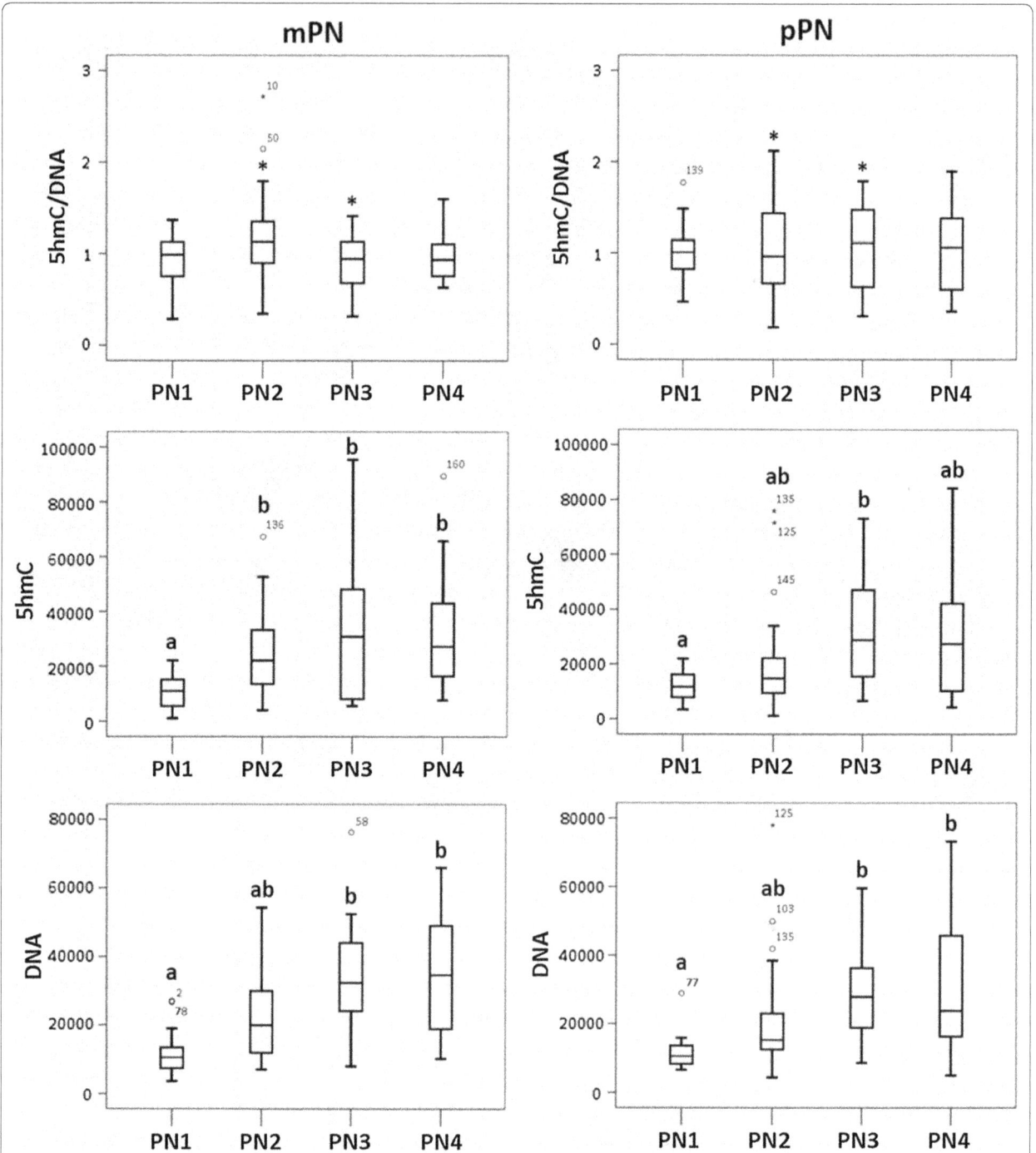

Fig. 8 Dynamics of global DNA hydroxymethylation in the paternal (pPN) and the maternal (mPN) pronucleus. No differences in the normalized levels of 5hmC (5hmC/DNA) were observed between the pronuclear stages in pPN or the mPN. However, the levels of normalized 5hmC were significantly higher in the maternal PN2 than its paternal counterpart and in the paternal PN3 than its maternal counterpart (*). A significant increase in the levels of total 5hmC (5hmC) was found in the pPN between PN1 versus PN3 and in the mPN between PN1 vs. PN2–PN3–PN4. Additionally, a significant increase in the total DNA fluorescence (DNA) was observed between PN1 and PN3–PN4, in both pPN and mPN. The distribution of the 81 embryos included in the study was: PN1 ($n = 13$), PN2 ($n = 34$), PN3 ($n = 19$) and PN4 ($n = 15$). *Different superscripts (a or b)* indicate significant differences, and *p* values <0.05 were considered significant

the maternal PN2 compared to the paternal counterpart and vice versa in PN3: This is probably due to an intrinsic artifact of the fluorescence intensity quantification rather than to a biological explanation, especially when considering the high complexity of the pronuclei. Additionally, no differences were observed in normalized 5hmC levels between pronuclear stages regardless of the parental origin of the pronuclei in the horse, evidencing a stable presence of 5hmC during pronuclear development. This persistent level of 5hmC throughout pronuclear development in both pPN and mPN has also been reported in mouse [17] and human [19].

In cattle, a strong association between the pattern of H3K9me3 and of 5mC in the pPN was reported, where during pronuclear development, the presence of both marks simultaneously increased in the pPN [33]. It was hypothesized that de novo H3K9 methylation directed the de novo DNA methylation, and as such, in cattle, global demethylation of the pPN is followed by a gradual H3K9 methylation and finally DNA methylation [33]. In the horse, this association was not observed. Instead, 5mC levels remained high and constant during pronuclear development in the pPN, with no evidence of demethylation, while H3K9me3 was never present in the pPN, as previously demonstrated in the horse [20] and rabbit zygotes [24].

In conclusion, there is no evidence of global demethylation, nor active nor passive, in the horse during the first cell cycle. The normalized 5mC levels were high and stable during pronuclear development, showing no evidence of passive demethylation associated with DNA replication in the PN regardless their parental origin. Moreover, no evidence of active demethylation (replication independent and TET mediated) was observed as stable and non-complementary patterns of normalized 5mC and 5hmC were present in both PN during pronuclear development. However, it is important to keep in mind that the embryos used in this study were produced in vitro by ICSI. In some species, such as rabbit [24] and rat [28], the loss of methylation observed by immunostaining in the pPN of in vivo-derived zygotes, was reduced after in vitro culture [24], IVF [28] and more dramatically after ICSI [28]. In sheep [11] and cattle [34], on the other hand, no differences in DNA methylation patterns were observed between in vivo and IVF [11, 34] or ICSI [34] zygotes using the same immunostaining technique. In mouse, reduced loss of methylation of the pPN was observed after IVF when the acid-based epitope retrieval was used for immunostaining. When the acid-tryptic-based epitope retrieval was used instead, no differences in DNA methylation pattern were observed between in vivo-derived and in vitro-produced mouse zygotes, with both mPN and pPN showing high levels of DNA methylation

[22]. Whether the results of this study apply for equine in vivo zygotes as well remains to be investigated. The collection of equine zygotes in vivo is challenging, since the horse is a monovulatory species that not responds to superovulation and as such, only one zygote at a time could be on average surgically collected. On the other hand, the ICSI protocol used in this study yields a 20% blastocyst rate, which is in line with the results obtained by other groups [35]. Additionally, healthy foals have been born from embryos produced using this protocol. Consequently, an active demethylation of the pPN during first cell cycle might not be essential to obtain normal, healthy offspring in the horse. In line with this observation, the use of round spermatid injection (ROSI) in mice leads to similar high levels of 5mC in the pPN and the mPN, with the zygotes being still able to develop to term [36]. This indicates that active demethylation of the pPN is not essential for normal embryo development in the mouse as well.

Conclusions

In the present study, we described, for the first time in the horse, the global dynamics of DNA methylation and hydroxymethylation throughout pronuclear development, independently in the paternal and maternal genome.

No evidence of active demethylation of the paternal genome was found in the horse. Instead, 5mC and 5hmC coexisted at high levels in both parental genomes throughout pronuclear development. Moreover, 5mC and 5hmC displayed different distribution patterns in the pronuclei with 5hmC homogeneously distributed, while 5mC was more clustered within certain DNA regions. This suggests that both marks provide their own epigenetic information. Considering these results, global DNA demethylation during the first cell division might not be essential for embryo development in the horse.

Methods

Equine in vitro embryo production

Equine zygotes were produced in vitro by piezo drill-assisted ICSI as described previously [37]. Briefly, ovaries were collected from slaughtered mares and processed within 4 h. Cumulus oocyte complexes (COCs) were aspirated from follicles larger than 5 mm using a 16-gauge needle attached to a vacuum pump (-100 mm Hg) and matured in groups of maximum 30 in 500 µL of Dulbecco's modified Eagle medium nutrient mixture F-12 (DMEM/F12; Gibco)-based maturation medium [38] for minimum 25 h at 38.5 °C in a humidified atmosphere of 5% CO_2 in air. After maturation, COCs were denuded by gentle pipetting in 0.05% bovine hyaluronidase (Sigma-Aldrich) for two minutes, followed by further pipetting in HEPES-buffered TCM199 medium (Gibco) supplemented with 10% fetal bovine serum (FBS). Only oocytes with an extruded polar body were used for piezo drill-assisted ICSI. Frozen and fresh sperm of two different stallions was used for ICSI; after Percoll gradient centrifugation, the sperm was washed and kept in calcium-free TALP and manipulated in 9% polyvinylpyrrolidone (Sigma-Aldrich) in phosphate-buffered saline (PBS, Gibco). All manipulations were performed on the heated stage (37 °C) of an inverted microscope; a progressively motile sperm was immobilized by piezo pulses and subsequently injected into the cytoplasm of a mature oocyte using a piezo drill. The injected oocytes were cultured in groups of 10–15 in 20 μl drops of DMEM-F12 supplemented with 10% FBS (Greiner Bio-One) and 50 μg/ml Gentamycin (Sigma) at 38.2 °C in a humidified atmosphere of 5% CO_2, 5% O_2 and 90% N_2. Presumptive zygotes were collected after 8, 11, 15, 19 and 23 h of culture in order to obtain all the pronuclear stages.

Immunofluorescent staining

After collection, presumptive zygotes were vortexed for 1 min to remove any remaining cumulus cells, fixed in 4% paraformaldehyde (PFA; Sigma-Aldrich) for 20 min at room temperature and kept in 2% PFA at 4 °C for a maximum of four days until immunostaining was performed. The zygotes were subsequently washed in PBS containing 0.5% bovine serum albumin (BSA, Sigma-Aldrich) for 1 h at room temperature. After washing, they were permeabilized with 0.5% Triton X-100 (Sigma-Aldrich) and 0.05% Tween 20 (Sigma-Aldrich) in PBS for 1 h at room temperature, and washed three times for 5 min in 0.5% BSA in PBS. Zygotes were blocked in PBS containing 2% BSA for 1 h at room temperature and washed three times for 2 min in 0.5% BSA in PBS. After washing, they were incubated with the primary antibody rabbit anti-H3K9me3 (1:100, Active Motif) in 2% BSA in PBS overnight at 4 °C. Simultaneously, four zygotes were incubated with the non-immune control antibody rabbit IgG (0.01 mg/ml; Rockland 011-0102) in 2% BSA in PBS overnight at 4 °C. Zygotes which served as negative control remained in 2% PBS in BSA without adding any primary antibody. Next, all zygotes, including the non-immune and the negative control, were washed three times for 10 min in 0.5% BSA in PBS, post-fixed in 4% PFA for 25 min and washed again three times in 0.5% BSA in PBS for 10 min. Epitope retrieval was performed as described previously [20] by treating the zygotes with 4 N HCL (Sigma-Aldrich) for 30 min at room temperature and with 100 mM Tris–HCl (pH 8.5; Sigma-Aldrich) for 10 min at room temperature, followed by three times washing with 5% BSA in PBS for 5-min and 20-s treatment with 0.25% (w/v) trypsin (Sigma-Aldrich T4799) at 37 °C. Tryptic digestion was stopped by incubating the zygotes with 30% goat serum (Gibco) in PBS for 2 min at room temperature and subsequently washing them three times with 0.5% BSA in PBS for 5 min. After epitope retrieval, zygotes were treated with 1 mg/mL RNase A (Affymetrix) for 30 min at 37 °C to avoid the binding of the nuclear stain, ethidium homodimer 2 (EthD-2; Molecular Probes), to RNA [39]. Subsequently, they were washed three times in 0.5% BSA in PBS for 5 min and incubated with the nuclear stain 0.5 nM EthD-2 in 0.5% BSA in PBS for 30 min at room temperature. After four washing steps of 2 min in 0.5% BSA in PBS, zygotes were incubated in blocking solution overnight at 4 °C. Subsequently, zygotes previously incubated with rabbit anti-H3K9me3 were incubated with either mouse anti-5mC (0.01 mg/ml; EpiGentek A-1014-050) or mouse anti-5hmC (0.01 mg/ml; Active Motif 39999) primary antibodies in 30% goat serum in PBS overnight at 4 °C. At the same time, zygotes previously incubated with rabbit IgG control antibody were incubated with mouse IgG control antibody (0.01 mg/ml; Sigma-Aldrich) in 30% goat serum in PBS overnight at 4 °C. Zygotes used as negative control were placed in 30% goat serum in PBS without the addition of primary antibodies. After incubation with the primary antibodies, zygotes were washed and serially incubated with the two secondary antibodies in 30% goat serum in PBS, for 1 h at room temperature each. For the H3K9me3-5mC immunostaining, goat anti-rabbit FITC (0.02 mg/ml; Life Technologies) and goat anti-mouse Alexa Fluor 405 (0.02 mg/ml; Abcam) secondary antibodies were used, respectively. For the H3K9me3-5hmC immunostaining, goat anti-rabbit Alexa Fluor 405 (0.02 mg/ml; Abcam) and goat anti-mouse Alexa Fluor 488 (0.02 mg/ml; Abcam) secondary antibodies were used, respectively. After four washes of 5 min in 0.5% BSA in PBS, two zygotes were mounted per slice in 24.7 mg/ml 1,4-diazabicyclo[2.2.2]octane (DABCO; Sigma-Aldrich) in 90% glycerol and 10% PBS, to prolong the lifetime of the dyes.

Fluorescence microscopy and image analysis

To avoid fading, evaluation of the embryos was performed 2 days after immunostaining was completed, using a Leica TSC SPE-II confocal microscope (Leica, Belgium) with an ACS APO 63× oil immersion objective (Leica) and laser lines at 405, 488 and 561 nm wavelengths.

For each wavelength, single images were taken independently of the equatorial cross section of each pronucleus. Additionally, digital optical sections of the area containing both pronuclei were taken for each wavelength using Z-series acquisition feature every 0.5 μm.

Quantitative analysis of fluorescence intensities was performed using the ImageJ software.

The area of the equatorial cross section of each pronucleus was manually outlined and measured in the single images. After subtracting the background, the mean fluorescence intensity was measured for 5mC, 5hmC and EthD-2. The mean fluorescence intensities were then multiplied by the pronuclear areas to obtain the total fluorescence of 5mC, 5hmC and EthD-2. Finally, for each pronucleus, the total fluorescence of 5mC or 5hmC was divided by the total fluorescence of EthD-2 to obtain the normalized fluorescence.

When a small part of the mPN and pPN was overlapping, the overlapping area was not measured. Additionally, when most of the area of the pronuclei was overlapping and the analysis of both pronuclei could not be made independently, the zygotes were excluded from the study.

Statistical analysis

Four replicates were performed for 5mC and 5hmC. The Kruskal–Wallis H test combined with Bonferroni correction for multiple testing was used to compare the total and normalized fluorescence between the different pronuclear stages, independently for the mPN and pPN. The paired samples t test was used to compare the size and the normalized fluorescence between the mPN and the pPN in each pronuclear stage. All the analyses were performed with SPSS Statistics 24 and p values <0.05 were considered significant.

Abbreviations

TET: ten-eleven translocation enzymes; 5mC: 5-methylcytosine; 5hmC: 5-hydroxymethylcytosine; 5fC: 5-formylcytosine; 5caC: 5-carboxylcytosine; H3K9me2: histone H3 lysine 9 di-methylation; ICSI: intracytoplasmic sperm injection; mPN: maternal pronucleus; pPN: paternal pronucleus; H3K9me3: histone H3 lysine 9 tri-methylation; ROSI: round spermatid injection; COCs: cumulus oocytes complexes; DMEM/F12: Dulbecco's modified Eagle medium nutrient mixture F-12; FBS: fetal bovine serum; PBS: phosphate-buffered saline; PFA: paraformaldehyde; BSA: bovine serum albumin; EthD-2: ethidium homodimer 2; DABCO: 1,4-diazabicyclo[2.2.2]octane.

Authors' contributions

SH was involved in conception and design, sample collection, immunofluorescent staining, confocal imaging and manuscript writing; KS performed the ICSI; KS, CDS and AVS participated in critical review of the manuscript. All authors read and approved the final manuscript.

Acknowledgements

The authors thank Petra Van Damme for her excellent technical assistance.

Competing interests

The authors declare that they have no competing interests.

Funding
This work was supported by a grant awarded to SH from the "Agentschap voor Innovatie door Wetenschap en Technologie" (IWT) Grant Number 111438. Experimental work was supported by a GOA Grant—Project No. 01G01112—Pathways to pluripotency and differentiation in embryos and embryonic stem cells.

References

1. Reik W, Dean W, Walter J. Epigenetic reprogramming in mammalian development. Science. 2001;293(1089–109):3. doi:10.1126/science.1063443.
2. Dean W, Santos F, Stojkovic M, Zakhartchenko V, Walter J, Wolf E, et al. Conservation of methylation reprogramming in mammalian development: aberrant reprogramming in cloned embryos. Proc Natl Acad Sci USA. 2001;98:13734–8. doi:10.1073/pnas.241522698.
3. Oswald J, Engemann S, Lane N, Mayer W, Olek A, Fundele R, et al. Active demethylation of the paternal genome in the mouse zygote. Curr Biol. 2000;10:475–8.
4. Mayer W, Niveleau A, Walter J, Fundele R, Haaf T. Demethylation of the zygotic paternal genome. Nature. 2000;403:501–2. doi:10.1038/35000654.
5. Iqbal K, Jin SG, Pfeifer GP, Szabo PE. Reprogramming of the paternal genome upon fertilization involves genome-wide oxidation of 5-methylcytosine. Proc Natl Acad Sci USA. 2011;108:3642–7. doi:10.1073/pnas.1014033108.
6. Zaitseva I, Zaitsev S, Alenina N, Bader M, Krivokharchenko A. Dynamics of DNA-demethylation in early mouse and rat embryos developed in vivo and in vitro. Mol Reprod Dev. 2007;74:1255–61. doi:10.1002/mrd.20704.
7. Xu Y, Zhang JJ, Grifo JA, Krey LC. DNA methylation patterns in human tripronucleate zygotes. Mol Hum Reprod. 2005;11:167–71. doi:10.1093/molehr/gah145.
8. Shi W, Dirim F, Wolf E, Zakhartchenko V, Haaf T. Methylation reprogramming and chromosomal aneuploidy in in vivo fertilized and cloned rabbit preimplantation embryos. Biol Reprod. 2004;71:340–7. doi:10.1095/biolreprod.103.024554.
9. Jeong YS, Yeo S, Park JS, Koo DB, Chang WK, Lee KK, et al. DNA methylation state is preserved in the sperm-derived pronucleus of the pig zygote. Int J Dev Biol. 2007;51:707–14. doi:10.1387/ijdb.072450yj.
10. Hou J, Lei TH, Liu L, Cui XH, An XR, Chen YF. DNA methylation patterns in in vitro-fertilised goat zygotes. Reprod Fertil Dev. 2005;17:809–13.
11. Beaujean N, Hartshorne G, Cavilla J, Taylor J, Gardner J, Wilmut I, et al. Non-conservation of mammalian preimplantation methylation dynamics. Curr Biol. 2004;14:R266–7. doi:10.1016/j.cub.2004.03.019.
12. Tahiliani M, Koh KP, Shen Y, Pastor WA, Bandukwala H, Brudno Y, et al. Conversion of 5-methylcytosine to 5-hydroxymethylcytosine in mammalian DNA by MLL partner TET1. Science. 2009;324:930–5. doi:10.1126/science.1170116.
13. Kriaucionis S, Heintz N. The nuclear DNA base 5-hydroxymethylcytosine is present in Purkinje neurons and the brain. Science. 2009;324:929–30. doi:10.1126/science.1169786.
14. Szabo PE, Pfeifer GP. H3K9me2 attracts PGC7 in the zygote to prevent Tet3-mediated oxidation of 5-methylcytosine. J Mol Cell Biol. 2012;4:427–9. doi:10.1093/jmcb/mjs038.
15. Wossidlo M, Nakamura T, Lepikhov K, Marques CJ, Zakhartchenko V, Boiani M, et al. 5-Hydroxymethylcytosine in the mammalian zygote is linked with epigenetic reprogramming. Nat Commun. 2011;2:241. doi:10.1038/ncomms1240.
16. Salvaing J, Aguirre-Lavin T, Boulesteix C, Lehmann G, Debey P, Beaujean N. 5-Methylcytosine and 5-hydroxymethylcytosine spatiotemporal profiles in the mouse zygote. PLoS ONE. 2012;7:e38156. doi:10.1371/journal.pone.0038156.
17. Li Y, O'Neill C. 5′-Methylcytosine and 5′-hydroxymethylcytosine each provide epigenetic information to the mouse zygote. PLoS ONE. 2013;8:e63689. doi:10.1371/journal.pone.0063689.
18. Salvaing J, Li Y, Beaujean N, O'Neill C. Determinants of valid measurements of global changes in 5′-methylcytosine and 5′-hydroxymethylcytosine by immunolocalisation in the early embryo. Reprod Fertil Dev. 2014. doi:10.1071/rd14136.

19. Petrussa L, Van de Velde H, De Rycke M. Similar kinetics for 5-methylcytosine and 5-hydroxymethylcytosine during human preimplantation development in vitro. Mol Reprod Dev. 2016;83:594–605. doi:10.1002/mrd.22656.

20. Heras S, Smits K, Leemans B, Van Soom A. Asymmetric histone 3 methylation pattern between paternal and maternal pronuclei in equine zygotes. Anal Biochem. 2015;471:67–9. doi:10.1016/j.ab.2014.11.005.

21. Heras S, Forier K, Rombouts K, Braeckmans K, Van Soom A. DNA counterstaining for methylation and hydroxymethylation immunostaining in bovine zygotes. Anal Biochem. 2014;454:14–6. doi:10.1016/j.ab.2014.03.002.

22. Li Y, O'Neill C. Persistence of cytosine methylation of DNA following fertilisation in the mouse. PLoS ONE. 2012;7:e30687. doi:10.1371/journal.pone.0030687.

23. Santos F, Hendrich B, Reik W, Dean W. Dynamic reprogramming of DNA methylation in the early mouse embryo. Dev Biol. 2002;241:172–82. doi:10.1006/dbio.2001.0501.

24. Reis Silva AR, Adenot P, Daniel N, Archilla C, Peynot N, Lucci CM, et al. Dynamics of DNA methylation levels in maternal and paternal rabbit genomes after fertilization. Epigenetics. 2011;6:987–93. doi:10.4161/epi.6.8.16073.

25. Hyttel P, Greve T, Callesen H. Ultrastructural aspects of oocyte maturation and fertilization in cattle. J Reprod Fertil Suppl. 1989;38:35–47.

26. Barton SC, Arney KL, Shi W, Niveleau A, Fundele R, Surani MA, et al. Genome-wide methylation patterns in normal and uniparental early mouse embryos. Hum Mol Genet. 2001;10:2983–7.

27. Heras S, Smits K, Van Soom A. Optimization of DNA methylation immunostaining in equine zygotes produced after ICSI. *EpiConcept COST Workshop* 2014:1.

28. Yoshizawa Y, Kato M, Hirabayashi M, Hochi S. Impaired active demethylation of the paternal genome in pronuclear-stage rat zygotes produced by in vitro fertilization or intracytoplasmic sperm injection. Mol Reprod Dev. 2010;77:69–75. doi:10.1002/mrd.21109.

29. Zhang P, Su L, Wang Z, Zhang S, Guan J, Chen Y, et al. The involvement of 5-hydroxymethylcytosine in active DNA demethylation in mice. Biol Reprod. 2012;86:104. doi:10.1095/biolreprod.111.096073.

30. Li Y, Seah MK, O'Neill C. Mapping global changes in nuclear cytosine base modifications in the early mouse embryo. Reproduction. 2016;151:83–95. doi:10.1530/rep-15-0207.

31. Hahn MA, Qiu R, Wu X, Li AX, Zhang H, Wang J, et al. Dynamics of 5-hydroxymethylcytosine and chromatin marks in mammalian neurogenesis. Cell Rep. 2013;3:291–300. doi:10.1016/j.celrep.2013.01.011.

32. Iurlaro M, Ficz G, Oxley D, Raiber EA, Bachman M, Booth MJ, et al. A screen for hydroxymethylcytosine and formylcytosine binding proteins suggests functions in transcription and chromatin regulation. Genome Biol. 2013;14:R119. doi:10.1186/gb-2013-14-10-r119.

33. Park JS, Jeong YS, Shin ST, Lee KK, Kang YK. Dynamic DNA methylation reprogramming: active demethylation and immediate remethylation in the male pronucleus of bovine zygotes. Dev Dyn. 2007;236:2523–33. doi:10.1002/dvdy.21278.

34. Abdalla H, Hirabayashi M, Hochi S. Demethylation dynamics of the paternal genome in pronuclear-stage bovine zygotes produced by in vitro fertilization and ooplasmic injection of freeze-thawed or freeze-dried spermatozoa. J Reprod Dev. 2009;55:433–9.

35. Galli C, Colleoni S, Duchi R, Lagutina I, Lazzari G. Developmental competence of equine oocytes and embryos obtained by in vitro procedures ranging from in vitro maturation and ICSI to embryo culture, cryopreservation and somatic cell nuclear transfer. Anim Reprod Sci. 2007;98:39–55. doi:10.1016/j.anireprosci.2006.10.011.

36. Polanski Z, Motosugi N, Tsurumi C, Hiiragi T, Hoffmann S. Hypomethylation of paternal DNA in the late mouse zygote is not essential for development. Int J Dev Biol. 2008;52:295–8. doi:10.1387/ijdb.072347zp.

37. Smits K, Govaere J, Hoogewijs M, Piepers S, Van Soom A. A pilot comparison of laser-assisted vs piezo drill ICSI for the in vitro production of horse embryos. Reprod Domest Anim. 2012;47:e1–3. doi:10.1111/j.1439-0531.2011.01814.x.

38. Galli C, Colleoni S, Duchi R, Lagutina I, Lazzari G. Developmental competence of equine oocytes and embryos obtained by in vitro procedures ranging from in vitro maturation and ICSI to embryo culture, cryopreservation and somatic cell nuclear transfer. Anim Reprod Sci. 2007;98:39–55. doi:10.1016/j.anireprosci.2006.10.011.

39. Suzuki T, Fujikura K, Higashiyama T, Takata K. DNA staining for fluorescence and laser confocal microscopy. J Histochem Cytochem. 1997;45:49–53.

11

Histone peptide microarray screen of chromo and Tudor domains defines new histone lysine methylation interactions

Erin K. Shanle[1†], Stephen A. Shinsky[2,3,6†], Joseph B. Bridgers[2], Narkhyun Bae[4], Cari Sagum[4], Krzysztof Krajewski[2], Scott B. Rothbart[5], Mark T. Bedford[4*] and Brian D. Strahl[2,3*]

Abstract

Background: Histone posttranslational modifications (PTMs) function to regulate chromatin structure and function in part through the recruitment of effector proteins that harbor specialized "reader" domains. Despite efforts to elucidate reader domain–PTM interactions, the influence of neighboring PTMs and the target specificity of many reader domains is still unclear. The aim of this study was to use a high-throughput histone peptide microarray platform to interrogate 83 known and putative histone reader domains from the chromo and Tudor domain families to identify their interactions and characterize the influence of neighboring PTMs on these interactions.

Results: Nearly a quarter of the chromo and Tudor domains screened showed interactions with histone PTMs by peptide microarray, revealing known and several novel methyllysine interactions. Specifically, we found that the CBX/HP1 chromodomains that recognize H3K9me also recognize H3K23me2/3—a poorly understood histone PTM. We also observed that, in addition to their interaction with H3K4me3, Tudor domains of the Spindlin family also recognized H4K20me3—a previously uncharacterized interaction. Several Tudor domains also showed novel interactions with H3K4me as well.

Conclusions: These results provide an important resource for the epigenetics and chromatin community on the interactions of many human chromo and Tudor domains. They also provide the basis for additional studies into the functional significance of the novel interactions that were discovered.

Keywords: Chromatin, Histone methylation, Chromodomain, Tudor domain, Peptide microarray

Background

Dynamic regulation of chromatin structure is linked to the regulation of all DNA-templated processes including gene expression [1, 2]. Histone posttranslational modifications (PTMs) represent one of the major mechanisms for regulating the chromatin landscape [3, 4]. Histone PTMs include acetylation, phosphorylation, and methylation among many others, and these different chemical signatures exert their effects on chromatin structure both in a type- and site-specific manner [5–8]. One of the primary mechanisms by which histone PTMs alter chromatin structure is via the recruitment of effector proteins that contain specialized "reader" domains that specifically recognize different histone PTMs [5, 9, 10]. These effectors may be transcription factors or additional chromatin-modifying machinery, and often their functions define the downstream consequences associated with distinct histone PTMs [10–12].

Histone lysine (K) methylation represents one of the major histone PTMs and over 170 methyllysine readers are thought to exist in humans [13]. Lysine methylation can take the form of mono-, di-, or trimethylation, and

*Correspondence: mtbedford@mdanderson.org;
brian_strahl@med.unc.edu
†Erin K. Shanle and Stephen A. Shinsky contributed equally to this work
[2] Department of Biochemistry and Biophysics, The University of North Carolina, Chapel Hill, NC 27599, USA[4] Department of Epigenetics and Molecular Carcinogenesis, The University of Texas MD Anderson Cancer Center, Smithville, TX 78957, USA
Full list of author information is available at the end of the article

often each state is associated with distinct genomic locations and unique functional outcomes [14–16]. The major sites of K-methylation on histone H3 are K4, K9, K27, K36, and K79, while K20 is the predominant K-methylation site on histone H4 [14, 15, 17, 18]. Interestingly, histone K-methylation is associated with both transcriptionally permissive and transcriptionally repressive states of chromatin, dependent on the site and degree of methylation. For instance, monomethylation of histone H4 at Lys 20 (H4K20me1) is associated with actively transcribed regions, while trimethylation of H3K9 (H3K9me3) is associated with transcriptionally silent chromatin [19–23]. This suggests that methyllysine readers are specific both for the degree of methylation and for the sequence context surrounding the methylated residue. However, some sites of K-methylation, such as H3K9 and H3K27, share a common ARKS sequence context, suggesting that some reader domains may recognize multiple sites and may require additional factors for specificity. Indeed, many chromatin modifiers and transcriptional machinery contain multiple reader domains that simultaneously engage multiple histone PTMs in a form of "cross-talk" [24]. Furthermore, the residues flanking methyllysine hotspots are often subject to PTMs that may permit or impede binding of methyllysine readers. For example, phosphorylation of serine 10 of H3 (H3S10p) inhibits binding of H3K9me3-specific reader domains, and cis-histone tail H3K9me3/S10p has been observed in cells [25, 26]. Similarly, phosphorylation of H3T3, and to a lesser degree H3T6, impedes binding of H3K4me3-specific reader domains, while symmetrical or asymmetrical dimethylation of H3R2 (H3R2me2s/a) has little effect on certain domain interactions [27], but can block or promote others [28–30].

Known methyllysine reader domains include the plant homeodomain (PHD) fingers, the bromo-adjacent homology (BAH) domains, and the "Royal family" domains [17, 31, 32]. The Royal family of methyllysine readers is conserved throughout eukaryotic evolution and includes the chromo-, Tudor-, PWWP-, and malignant brain tumor (MBT)-structural domains [33]. Members of this family are known to interact with multiple different sites of K-methylation on histones and other proteins. Structural and mechanistic studies of the heterochromatin protein 1 (HP1) chromodomain provided some of the first insights into the molecular mechanism of methyllysine recognition. A co-crystal structure of the HP1 chromodomain bound to an H3K9me3 peptide revealed that a hydrophobic region composed of three conserved aromatic residues stabilizes the interaction with the methyllysine side chain via cation–π and hydrophobic interactions [34]. Additional structure–function studies showed that this "aromatic cage" is a general feature of many methyllysine readers even outside the Royal family [35]. However, the

site- and methyl-state specificity for many members of the Royal family is unclear. Moreover, several of the known members of this family are uncharacterized for their interactions with histone PTMs.

While multiple histone K-methylation sites are known, advances in mass spectrometry have increased the number of known sites [36]. However, it is unclear whether the current repertoire of known methyllysine readers, including members of the Royal family, also interacts with these newly identified sites or whether there are distinct readers for each site. In this investigation, we set out to use a high-throughput proteomics approach utilizing customized histone peptide microarrays to survey a large number of human chromo and Tudor domains for their binding to various histone PTMs. We identified several new histone PTM–reader domain interactions for previously characterized readers and identified a subset of H3K9-methyl readers that are also capable of reading H3K23 methylation—a newly identified yet poorly understood histone PTM. The results of our survey will facilitate future studies characterizing the structure–function relationships and biological consequences of these interactions.

Results

We selected 31 chromodomains and 39 Tudor or Tudor-like domains to screen using our histone peptide microarray platform. Several additional protein reader domains were also screened, for a total of 83 proteins screened by peptide microarray (Additional file 1: Table S1). Our peptide microarray platform consists of nearly 300 biotinylated histone peptides harboring unique PTMs at one or more residues immobilized to a streptavidin-coated glass surface (Additional file 2: Table S2) [37, 38]. The arrays were probed in duplicate with purified GST-tagged protein domains, and binding to specific PTMs was detected using fluorescently labeled antibodies (see Methods). Twenty-two of the 83 selected domains showed positive hits with one or more histone peptides on the arrays. As shown in Table 1, many of the known binding targets for these protein domains were detected, as well as several novel interactions. Based on these observations, we focused on further characterizing (1) the novel interaction between Spindlin1 (SPIN1) triple Tudor domain and H4K20me3 and (2) the interaction between several chromodomains and H3K23me2/3.

SPIN1 is a transcriptional coactivator that contains a triple Tudor domain known to interact with H3K4me3—a mark associated with actively transcribed genes [39–41]. As expected, the SPIN1 triple Tudor and several other known H3K4me-binding domains, including the PHD domain of Taf3 and the tandem Tudor domains of JMJD2A and SGF29, strongly interacted with H3K4me2/3 (Fig. 1a; Additional file 3: Figure S1, and Additional file 4: Figure S2). Interestingly, phosphorylation of H3T3 inhibited binding of

Table 1 Summary of reader domain interactions identified via histone peptide microarrays

Protein	Domain	Top array hits	Known interactions	References
HP1β/CBX1	Chromo	H3K9me1/2/3, H3K23me1/2/3	H3K9me1/2/3, H3K23me1/2/3	[21, 22, 46, 52]
HP1γ/CBX3	Chromo	*H3K9me1/2/3, H3K23me3*[a]	H3K9me2/3	[21, 22, 52, 74]
HP1α/CBX5	Chromo	*H3K9me1/2/3, H3K23me3*[a]	H3K9me1/2/3	[21, 48, 52]
CDYL1b	Chromo	*H3K9me1/2/3, H3K27me2/3*	H3K9me1/2/3, H3K27me1/2/3	[75]
CDYL2	Chromo	*H3K9me2/3, H3K27me3*	H3K9me1/2/3, H3K27me1/2/3	[76]
CHD1	Chromo (2)	H3K4me2/3	H3K4me3	[54, 55]
CHD7	Chromo (2)	H4me0	H3K4me1/2/3	[56]
CHD9	Chromo (2)	H4me0		
MPP8	Chromo	*H3K9me1/2/3, H3K23me2/3*	H3K9me1/2/3, H3K23me1/2/3	[46, 77, 78]
53BP1	Tudor (2)	*H3K4me2*, H3K18me2[a], H3K36me2[a], *H4K20me1/2*	H3K4me2, H4K20me1/2	[42, 52]
JMJD2A	Tudor (2)	*H3K4me2/3*, H3K18me3[a], H3K9me3, *H4K20me2/3*	H3K4me3, H4K20me2/3	[51, 52]
PHF20	Tudor (2)	H3K9me2/3, H4K8me1	H3K9me2, H3K27me2, H3K36me2	[35]
UHRF1	Tudor-like	H3K9me2/3	H3K9me2/3	[79, 80]
PHF1	Tudor	H3K36me3	H3K36me2/3	[81–84]
SGF29	Double Tudor	*H3K4me3*	H3K4me1/2/3	[85]
SPIN1	Triple Tudor	*H3K4me2/3*, H3K18me3, *H4K20me3*[a]	H3K4me2/3	[40]
TDRD2	Tudor (extended)	H3K4me3		
TDRD3	Tudor	H3R2me2a, H3R8me2a	pan-Rme2a	[53]
GLP	ANK	H3K9me1/2	H3K9me1/2	[86]
ING2	PHD	H3K4me2/3	H3K4me2/3	[87, 88]
TAF3	PHD	*H3K4me2/3*	H3K4me2/3	[89]
L3MBTL1	MBT (3)	H3K4me2, H3K9me2, H4K8me1	H3K4me1/2, H3K9me1/2, H3K27me1/2, H4K20me1/2	[90–92]

Italics, validated by peptide pull-down assays

[a] Novel interactions

SGF29 tandem Tudor and Taf3 PHD, but did not interfere with SPIN1 triple Tudor or JMJD2A tandem Tudor interaction with H3K4me3. In contrast, all of the H3K4me3 reader domains, including CHD1 chromodomain, accommodated H3K4me3 with neighboring H3R2me2a. Furthermore, the chromodomains of CHD7 and CHD9 showed robust interaction with both the unmodified H3 and H4 N-terminal tails, while the chromodomain of CHD1 showed interaction with the H4 tail but not with the unmodified H3 tail (Fig. 1a; Additional file 5: Figure S3, Additional file 6: Figure S4, and Additional file 7: Figure S5). In addition, the tandem Tudor of 53BP1 preferentially bound H4K20me1/2 as previously described (Fig. 1a) [42].

Surprisingly, the SPIN1 triple Tudor domain also bound H4K20me3-containing peptides (Fig. 1b, c). This novel interaction was highly reproducible on the peptide microarrays and was confirmed by in-solution peptide pull-down assays (Additional file 3: Figure S1A and S1B). We further validated the specificity of the interaction using peptide pull-down experiments with full-length SPIN1 protein and tested the possibility that other Spindlin family members (SPIN2B, SPIN3, and SPIN4) interact with H4K20me3. Indeed, full-length SPIN1 protein

preferentially bound H4K20me3 and showed less interaction with mono- or dimethylated H4K20 (Fig. 1d). Like SPIN1, full-length SPIN2B, SPIN3, and SPIN4 also showed preference for trimethylated H3K4 and H4K20 (Fig. 1d). In order to test whether SPIN1 family members interact with H4K20me3 peptides in the context of the cellular milieu, we transfected 293T cells with constructs containing individual SPIN1 family proteins fused with green fluorescent protein (GFP) and performed peptide pull-down experiments with whole cell lysates. The results from these assays mimic the results obtained using purified proteins, further suggesting that SPIN1 family proteins preferentially interact with H3K4me3 and H4K20me3 in cells (Fig. 1e; Additional file 3: Figure S1C). Notably, while we were unable to purify SPIN2A to high enough quality for use in in vitro pull-down assays using purified proteins, we were able to assess SPIN2A binding to H3K4me3/H4K20me3 peptides using lysates from transfected cells. These results show that, like other SPIN1 family members, SPIN2A is capable of interacting with both H3K4me3 and H4K20me3 (Fig. 1e).

Previous crystallographic analysis of the interaction between SPIN1 triple Tudor and H3K4me3 revealed

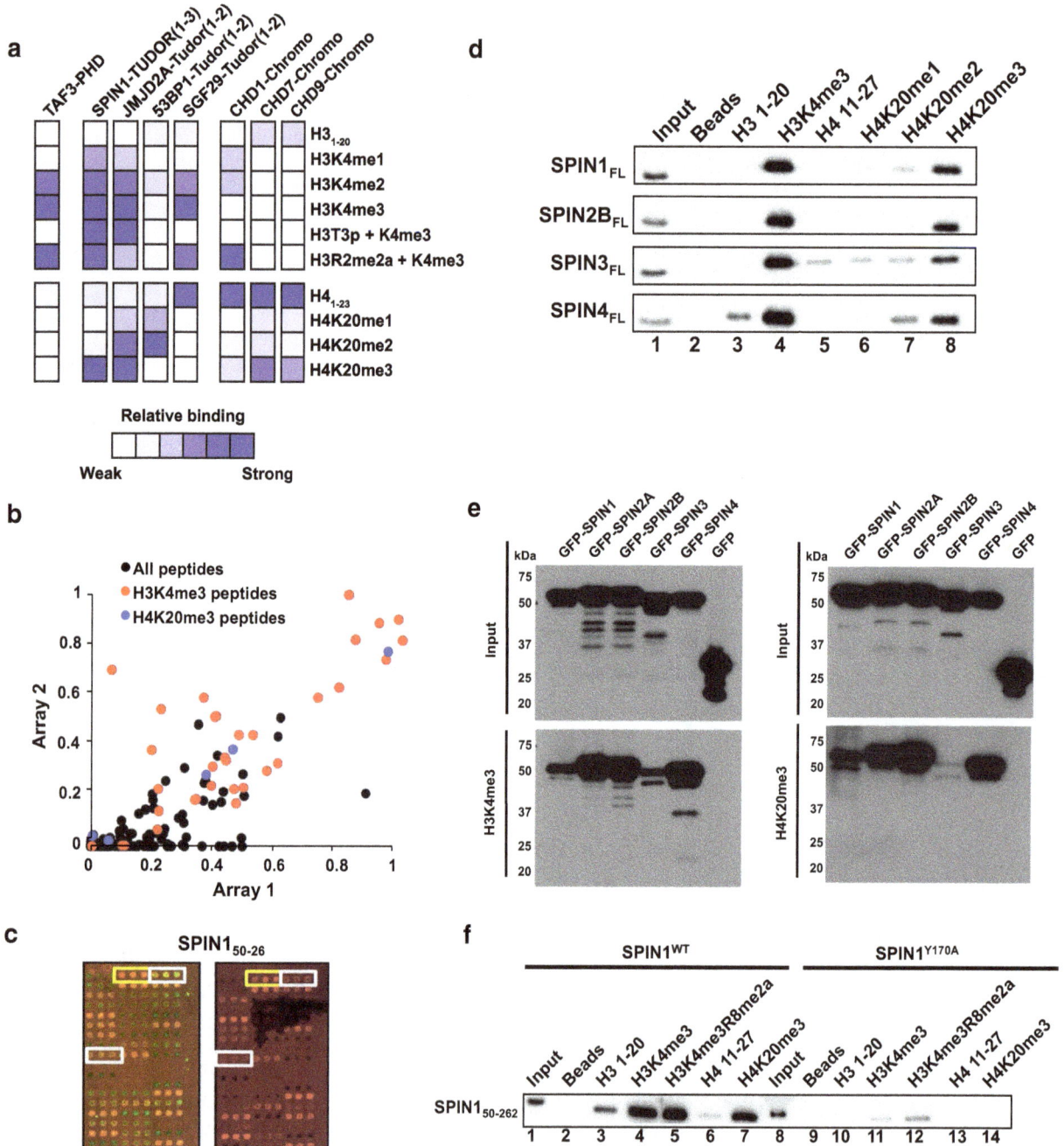

Fig. 1 SPIN1 triple Tudor domain interacts with H4K20me3 as well as H3K4me3. **a** Heat map showing the relative binding detected for each of the indicated domains on the peptide microarray platform. Data represent the average of two independent arrays relative to the most intense binding signal within the indicated set of peptides. **b** Scatter plot of the relative binding of SPIN1 triple Tudor domain from two independent peptide arrays. H3K4me3-containing peptides are shown in red and H4K20me3-containing peptides are shown in *blue*. All other peptides are shown in *black*. **c** Representative picture of a section of the peptide microarray for SPIN1 triple Tudor domain. The *left panel* shows both the *green* (peptide) and *red* (protein binding) fluorescent channels, while the *right panel* depicts only the red fluorescence channel for clarity. Positive antibody controls are outlined in *white* and the positive interaction with the H4K20me3 peptide is outlined in *yellow*. Full array images are shown in Additional file 3: Figure S1. **d** Western blot results of peptide pull-down experiments with purified full-length Spindlin family members. The input is shown in *Lane 1* and the corresponding bound fraction is shown in *Lanes 2–8*. **e** Western blot results of peptide pull-down experiments with whole cell lysates derived from transiently transfected HEK 293T cells (GFP-SPIN1, 2A, 2B, 3 and 4). **f** Western blot results of peptide pull-down experiments with purified SPIN1 wild type (SPIN1^WT) or aromatic cage mutant in the second Tudor domain (SPIN1^Y170A) demonstrating a loss of H3K4me3 and H4K20me3 in the mutant

an aromatic cage in the second Tudor domain of the SPIN1 triple Tudor that coordinates the trimethylated lysine residue [41]. In order to determine whether the interaction with H4K20me3 occurs through the same aromatic cage, we generated a tyrosine 170 to alanine (Y170A) mutation in the aromatic cage of SPIN1 and tested the interaction with H4K20me3 using peptide pull-down analyses. As shown in Fig. 1f, the Y170A variant showed minimal interaction with both H3K4me3 and H4K20me3. The SPIN1 Y170A mutation was previously shown to disrupt H3K4me3 binding [43]; thus, our results suggest that H4K20me3 is coordinated in the same aromatic pocket. Additional structural and functional studies show that H3R8me2a enhances the binding of SPIN1 to H3K4me3-containing peptides [43]. The methylated Arg8 side chain is coordinated in a hydrophobic pocket in the first Tudor domain of SPIN1, while the H3K4me3 residue is coordinated in the second Tudor domain [43]. In order to confirm that the Y170A mutant specifically disrupts the second Tudor domain, we tested the interaction of SPIN1 with a peptide containing the double H3K4me3/H3R8me2a modification. Both wild-type SPIN1 and the Y170A variant showed enhanced

interaction with the H3K4me3/H3R8me2a peptide compared to the H3K4me3 peptide, suggesting specific disruption of the second Tudor domain in the Y170A variant (Fig. 1f). Together, these results indicate that the second Tudor domain of SPIN1 can bind both H3K4me3 and H4K20me3 in a mutually exclusive manner.

The second novel interaction we detected occurred between several chromodomains and H3K23me peptides. As shown in Fig. 2a, the chromodomains of CBX1, CBX3, CBX5, CDYL2, and MPP8 showed strong interaction with H3K9me3, even in the context of neighboring arginine 8 asymmetrical or symmetrical dimethylation. In all cases, H3S10phos inhibited the interaction with H3K9me3 and there was a clear preference for H3K9me3 over H3K27me3. In addition to the well-characterized interactions with H3K9me3, we also observed binding to H3K23me-containing peptides. To validate these interactions, we performed peptide pull-down experiments with H3K9, H3K23, and K3K27 peptides with varied degrees of methylation. As shown in Fig. 2b, we observed that CBX1, CBX3, CBX5, and MPP8 chromodomains interacted with H3K9me1/2/3 peptides and H3K23me2/3 peptides, but showed minimal interactions

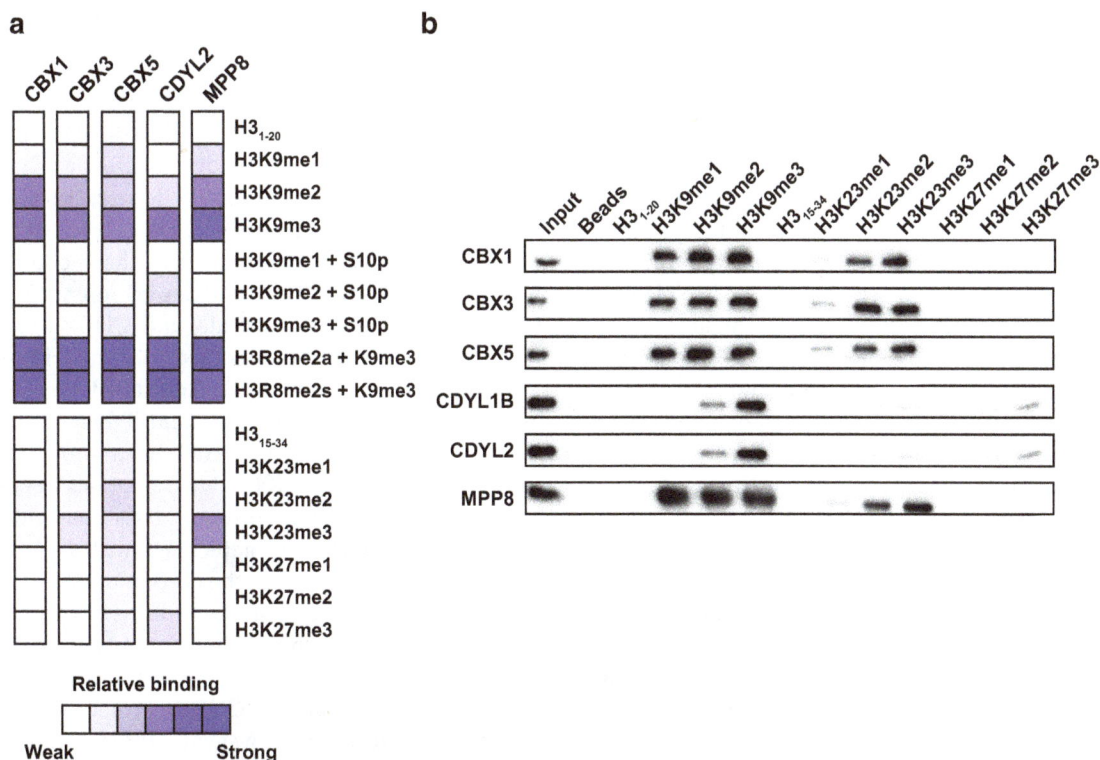

Fig. 2 Chromodomains interact with H3K23me2/3 in addition to H3K9me1/2/3. **a** Heat map showing the relative binding detected for each of the indicated domains on the peptide microarray platform. Data represent the average of two independent arrays relative to the most intense binding signal within the indicated set of peptides. **b** Western blot results of peptide pull-downs performed with the indicated GST-tagged domain and histone peptide. The input is shown in *Lane 1* and the bound fraction is shown in *Lanes 2–13*

with H3K27me1/2/3 peptides. CDYL1B and CDYL2 showed preference for H3K9me2/3 peptides and very weak interactions with H3K23me peptides and H3K27me peptides. These results suggest that members of the CBX family of H3K9me3 reader domains, which have high selectivity for H3K9 over a very similar sequence motif at H3K27 (i.e., ARKS), are robust readers of H3K23 methylation, thus implicating H3K23 methylation, along with H3K9me, in the silencing functions of these domains.

Discussion

The aim of this study was to create a valuable resource of chromo and Tudor reader domains for their interactions and cross-talk between histone PTMs. This work was facilitated by the use of a high-throughput approach employing peptide microarrays containing nearly 300 biotinylated histone peptides harboring up to five PTMs on each peptide (Additional file 2: Table S2). While several other histone peptide microarray platforms have been described [44–47], there are several notable features of our peptide array platform that aided the current study. These include highly purified peptides of lengths greater than 20 amino acids, along with each peptide being spotted multiple times by multiple pins to provide a robust number of data points that gave us high confidence in the interactions (and changes in these interactions by neighboring PTMs) that we observed.

Our survey of histone reader domains is one of the largest screens for histone PTM–reader domain interactions to date. We expressed and purified 83 protein domains, including 31 chromodomains and 39 Tudor or Tudor-like domains. We screened each domain in duplicate, and 22 domains exhibited consistent, reproducible binding to histone peptides on our arrays. The majority of the protein domains we tested, however, did not exhibit binding to histone peptides (see full list of domains screened in Additional file 1: Table S1). There are several possible explanations for this. First, our previous observations suggest that binding affinities weaker than approximately 30 μM are typically beyond the limit of detection for this platform [37]. It is notable that many reader domains exhibit weak interactions with histone peptides, which may account for a substantial number of negatives in our screen. For example, the chromodomain of CBX2 has been shown to bind H3K9me3 and H3K27me2 peptides with a binding affinity of ~40 μM via fluorescence polarization [48], which would explain why this chromodomain failed to show PTM interactions as compared to the other CBX domains. Second, we screened several protein domains with unknown histone PTM binding targets. For example, the Tudor domains of TDRD1 and TDRD2 are known to interact with methylated Piwi proteins [49, 50], but there are no known methyl-histone binding targets

known to date. Similarly, TDRD4, TDRD9, and several other TDRD family members have no known methyl-histone binding targets, and it is possible that these Tudor domains do not interact with histones. Third, the recombinant protein domains we expressed and purified may require additional sequences from their respective proteins that are needed for histone PTM binding and are not present in the domains we designed. Indeed, the single Tudor domain of PHF20 was negative on our arrays, but the tandem Tudor domain interacted with H3K9me2, as previously shown [35]. In addition, the domains we purified may require interaction with other proteins in order to bind histones. Finally, it is possible that the conditions we used in this high-throughput approach were not amenable to binding for some proteins.

Of the 31 Tudor or Tudor-like domains we screened, several known interactions were detected on our arrays (Table 1). Both 53BP1 and JMJD2A tandem Tudor domains showed binding to H3K4me and H4K20me peptides as previously shown [42, 51, 52]. Some novel interactions were also detected on the peptide arrays, such as binding to H3K18me, but further experiments need to be performed to validate these findings. TDRD3 Tudor domain specifically recognized asymmetrically dimethylated peptides, as previously shown [53], but our results suggest that this Tudor domain has broad affinity for Rme2a-containing peptide (Additional file 8: Figure S6 and Table 1). Of the 39 chromodomains we surveyed, nearly 20% interacted with modified histone peptides. Many of these interactions are well characterized, such as binding to H3K9me peptides by the CBX family of proteins [21, 22, 52]. We also observed interaction between CHD1 chromodomain and H3K4me3 as previously described [54, 55]. Intriguingly, the chromodomains of CHD1, CHD7, and CHD9 all showed interactions with unmodified histone H4 peptide, and CHD7 and CHD9 also interacted with unmodified histone H3 peptide (Additional file 5: Figure S3, Additional file 6: Figure S4, Additional file 5: Figure S5, and Table 1). Although CHD9 chromodomains have not been shown to bind methylated histones, H3K4me1/2/3 peptides were shown to competitively disrupt histone interactions with purified histones, which is in agreement with the idea that the H3N terminus can bind the chromodomain of CHD9 [56]. It should be noted that unmodified histone H3 peptides were not tested in these experiments, but based on our findings, we speculate the K4 unmodified peptide would have also competed CHD9 chromodomain interaction given our results show general H3 binding without preference to the H3K4 methyl state.

Due to the large scope of this microarray screen, we focused on validating only a subset of interactions by peptide pull-down experiments. The two most significant and novel interactions we uncovered were with H3K23me and H4K20me peptides. First, we observed that several chromodomains interacted with H3K23me, in addition to the known H3K9me targets. MPP8 and CBX1 chromodomains have been shown to interact with H3K23me peptides at low micromolar binding affinities [46]. We confirmed these results in our screen and observed that additional chromodomains, CBX3 and CBX5, also interact with H3K23me2/3. In our study, we observed preferential binding to di- and trimethylated states of H3K23. All three methylation states of H3K23 have been detected in cells [46], and a recent report suggests that H3K23me3 colocalizes with H3K27me3 and plays a role in protecting heterochromatin from double-strand DNA breaks during meiosis [57]. It is possible that the dual modification may provide an even better binding substrate for these chromodomains, which will be interesting to explore in future biological studies. In contrast, the CDY family members that exhibited binding on our arrays (CDYL1B and CDYL2) showed preference for H3K9me3 and did not interact with H3K23me peptides. Notably, a recent report suggested that H3K23 methylation regulates levels of H3K36 methylation by recruiting the H3K36 demethylase, KDM4B, via its double Tudor domain [58]. However, none of the Tudor domains surveyed here showed interaction with H3K23 methylation.

We also focused on the novel interaction between the SPIN1 Tandem Tudor domain and H4K20me2/3, which we validated by peptide pull-down experiments and demonstrated with other Spindlin family members (Fig. 1d). In the context of full-length protein, SPIN1 shows remarkable preference for trimethylation at H3K4 and H4K20. SPIN2B and SPIN3 show similar selectivity, while SPIN4 seems to accommodate both di- and trimethylated H4K20me3. SPIN1 is composed of three homologous Tudor domains denoted I, II, and III [39]. The second Tudor domain (II) is composed of the aromatic residues Phe 141, Trp 151, Tyr 170, and Tyr 177, which together form the aromatic cage that coordinates the methyl lysine [41]. Our observation that the Tyr170Ala variant loses interaction with both H3K4me3 and H4K20me3 suggests that domain II is responsible for recognizing both modifications. This further suggests that SPIN1 interacts with H3K4me3 and H4K20me3 at different times and/or different genomic locations, possibly under different cellular conditions.

Lysine 20 is the predominant methylation site on H4, and this modification is important for development in higher eukaryotes (reviewed in [59]). H4K20 methylation is associated with regulating transcription, the DNA damage response, and cell cycle progression [59]. Multiple H4K20 methyl readers have been described, and many of them contain a Tudor or Tandem Tudor domain

like SPIN1 [59]. These readers of H4K20me are thought to mediate the cellular roles ascribed to H4K20 methylation. For instance, the Tandem Tudor domain of 53BP1 is required for localization of 53BP1 to sites of DNA damage where it acts as a mediator of DNA damage signaling and repair [42, 60, 61]. Interestingly, different levels of H4K20 methylation are associated with different effects on transcription. For instance, H4K20me1 is associated with active transcriptional states of chromatin, while H4K20me3 is associated with transcriptionally silent chromatin regions [19, 62, 63]. Our data suggest that SPIN1 and other Spindlin family members are capable of recognizing both a transcriptional activation modification (i.e., H3K4me3) and a transcriptional silencing modification (i.e., H4K20me3). This result is, to the best of our knowledge, the first example of a reader domain that can read both activating and deactivating histone PTMs.

As H4K20 methylation is important in several cellular processes, aberrant H4K20 methylation is observed in several cancers and mutations in H4K20 methyl reader domains have been described in human developmental disorders [64–66]. Furthermore, SPIN1 is overexpressed in several varieties of malignant tumors and upregulation of SPIN1 is known to increase cellular proliferation and cause chromosomal instability and abnormal mitosis [67–71]. The interaction of SPIN1 with H4K20me2/3 may play a part in mediating some of the roles associated with H4K20 methylation, but additional in vivo studies are needed to determine the biological importance of this interaction. Chemical probes that inhibit the interaction of SPIN1 with H3K4me3 peptides in vitro have been described [72, 73]. Our work suggests that these probes could also be useful tools for characterizing the SPIN1-H4K20me interaction in vivo.

Conclusions

This high-throughput screen aimed to determine the histone PTM binding targets of known and putative chromo and Tudor reader domains to create a valuable resource for future studies of these domains. Our survey encompassed the majority of human chromo and Tudor domains, and uncovered known and unknown histone PTM interactions. Of the many hits we observed, we focused on two novel interactions: (1) chromodomain recognition of H3K23me and (2) recognition of H4K20me3 by the Spindlin family of proteins. Future work will be needed to uncover the importance of these interactions in vivo.

Methods
Protein expression and purification
Codon optimized constructs were synthesized and cloned into pGEX-4T-1 expression vectors (GE

Healthcare) by Biomatik. Proteins were expressed in sol-uBL21 (DE3) (Amsbio) grown in Terrific Broth II media (MP Biomedicals). After culturing at 37 °C until an OD_{600} of ~0.6, cells were chilled for 30 min at 4 °C before induction with 1 mM IPTG for 20 h at 16 °C. Cells were harvested by centrifugation and pellets were flash frozen in liquid nitrogen. For purification, thawed cell pellets were resuspended in binding buffer (50 mM Tris pH 7.5, 250 mM NaCl, 4 mM DTT, 10% glycerol) supplemented with a protease inhibitor cocktail (Roche), 0.1 mM phenylmethane sulfonyl fluoride (PMSF), 0.5 mg/ml chicken egg lysozyme (Sigma), and 0.2% (v/v) Triton X-100. After incubation on ice for 45 min, cells were lysed by sonication and clarified by centrifugation. Lysates were incubated with glutathione agarose (Pierce) and then washed with 10 bed volumes of binding buffer. Bound protein was eluted with elution buffer (50 mM Tris pH 8.0, 250 mM NaCl, 4 mM DTT, 10% glycerol, 10 mM reduced glutathione) and then dialyzed against 3 l of binding buffer at 4 °C. Samples were concentrated by centrifugation and protein concentration and purity were determined by Bradford Assay (BioRad) and SDS-PAGE, respectively.

Peptide microarrays

The peptide microarrays were generated and assayed as described previously [37, 38], except that the arrays contained four triplicate spots of each peptide. Briefly, GST-tagged proteins were diluted to 0.5–2 µM in phosphate-buffered saline (PBS) supplemented with 0.1% (v/v) Tween-20 (PBST) and 5% (w/v) bovine serum albumin (BSA, EMD Millipore Omnipure Fraction V) and incubated with peptide microarrays overnight at 4 °C. Arrays were washed three times with PBS and then probed with an anti-GST antibody (EpiCypher Inc.; Cat. No. 13-0022) diluted to 1:1000 in PBST + 5% BSA. Arrays were washed again 3× with PBS and then probed with an Alexa Fluor 647-conjugated anti-rabbit antibody at 1:10,000 (ThermoFisher). Arrays were imaged using a Typhoon Scanner and protein binding was determined as previously described [37, 38]. The average signal intensity for each peptide was normalized to the most intense binding within an array, and normalized binding was averaged for at least two independent replicates for each protein. Heat maps of relative binding were generated using JavaTree View (version 1.16r4) after normalizing the relative binding within the subset of peptides selected for the heat map.

In-solution peptide pull-down assays

For pull-down experiments using purified proteins, a total of 50 pmol of GST-tagged protein was incubated with 500 pmols of biotinylated histone peptide for 1 h at 4 °C in peptide binding buffer (50 mM Tris pH 8.0,

300 mM NaCl, 0.1% NP-40). Following incubation, the protein–peptide mixture was incubated with streptavidin-coated magnetic beads (Pierce) and pre-equilibrated with peptide binding buffer, for 1 additional hour at 4 °C. The beads were washed three times with peptide binding buffer, and bound complexes were eluted with 1x SDS loading buffer, then resolved via SDS-PAGE and transferred to a PVDF membrane. The membrane was probed with an anti-GST antibody (EpiCypher Inc.; Cat. No. 13-0022) diluted 1:4000 in PBST supplemented with 5% (w/v) BSA (Sigma).

For pull-down experiments using cell lysates, HEK 293T cells were transiently transfected with GFP-SPIN1, 2A, 2B, 3, and 4 using polyethylenimine according to manufacturer's instructions. Cells were lysed in ice-cold mild lysis buffer (50 mM Tris HCl pH 7.5, 150 mM NaCl, 0.1% NP-40, 5 mM EDTA, 5 mM EGTA, 15 mM MgCl2) containing protease inhibitor cocktail (Roche). Thirty microliter of streptavidin agarose beads (Millipore) was pre-washed with binding buffer and incubated with 10 µg of biotinylated histone peptides for 2 h with rocking at 4 °C. The beads were then washed three times with 500 µl binding buffer to remove unbound peptide. The peptide–streptavidin agarose mix was then incubated overnight with the whole cell lysates and rocked at 4 °C. After three washes with 500 µl binding buffer, 30 µl of 2× SDS loading buffer was added to the beads and boiled. The samples were subjected to SDS-PAGE and Western blotting analysis using polyclonal GFP antibody (Santa Cruz Biotech, 1:3000).

Additional files

Additional file 1: Table S1. Protein domains screened by histone peptide microarray.

Additional file 2: Table S2. Histone peptide library used for peptide microarrays.

Additional file 3: Figure S1. SPIN1 Tudor domain interaction with H3K4me3 and H4K20me3. A) Representative array images of SPIN1 Tudor domain showing peptide binding indicated in red (right panel). The peptide tracer is shown in green (left panel). Positive antibody controls are outlined in white. B) Western blot results of peptide pull-down experiments with SPIN1 tandem Tudor domain. The input is shown in Lane 1 and the corresponding bound fraction is shown in Lanes 2–8. C) Western blot results of peptide pull-down experiments with whole cell lysates derived from transiently transfected HEK 293T cells (GFP-SPIN1, 2A, 2B, 3 and 4).

Additional file 4: Figure S2. Peptide microarray data for SGF29 and Taf3. Representative array images of A) SGF29 double Tudor domain and Taf3 PHD domain showing peptide binding indicated in red (right panel). The peptide tracer is shown in green (left panel). Positive antibody controls are outlined in white.

Additional file 5: Figure S3. CHD1 chromodomain histone peptide microarray. A) Representative array images of CHD1 chromodomain showing peptide binding indicated in red (right panel). The peptide tracer is shown in green (left panel). Positive antibody controls are outlined in white. B) Scatter plot of the relative binding of CHD1 chromodomain from two independent peptide arrays. All modified and unmodified H4 (1–23) peptides are shown in blue, and H3K4me2/3-containing peptides are shown in red. All other peptides are shown in black. C) Relative binding to the indicated histone peptides from two representative arrays. Data were normalized to the most intense binding and the average and standard deviation of triplicate spots is shown.

Additional file 6: Figure S4. CHD7 chromodomain histone peptide microarray. A) Representative array images of CHD7 chromodomain showing peptide binding indicated in red (right panel). The peptide tracer is shown in green (left panel). Positive antibody controls are outlined in white. B) Scatter plot of the relative binding of CHD7 chromodomain from two independent peptide arrays. All modified and unmodified H4 (1–23) peptides are shown in red. All other peptides are shown in black. C) Relative binding to the indicated histone peptides from one representative array. Data were normalized to the most intense binding and the average and standard deviation of triplicate spots is shown.

Additional file 7: Figure S5. CHD9 chromodomain histone peptide microarray. A) Representative array images of CHD9 chromodomain showing peptide binding indicated in red (right panel). The peptide tracer is shown in green (left panel). Positive antibody controls are outlined in white. B) Scatter plot of the relative binding of CHD9 chromodomain from two independent peptide arrays. All modified and unmodified H4 (1–23) peptides are shown in red. All other peptides are shown in black. C) Relative binding to the indicated histone peptides from one representative array. Data were normalized to the most intense binding and the average and standard deviation of triplicate spots is shown.

Additional file 8: Figure S6. TDRD3 Tudor domain histone peptide microarray. A) Representative array images of TDRD3 Tudor domain showing peptide binding indicated in red (right panel). The peptide tracer is shown in green (left panel).

Abbreviations
BAH: bromo-adjacent homology; GST: glutathione S-transferase; MBT: malignant brain tumor; PBS: phosphate-buffered saline; PHD: plant homeodomain; PTM: posttranslational modification.

Authors' contributions
EKS, SAS, MTB, and BDS designed the study. EKS, SAS, JBB, and SBR performed the experiments. NB and CS generated the recombinant effector domains for the screening. KK created peptides and contributed to the fabrication and design of the peptide arrays. EKS and SAS analyzed the data and wrote the manuscript. All authors read and approved the final manuscript.

Author details
[1] Department of Biological and Environmental Sciences, Longwood University, Farmville, VA 23909, USA. [2] Department of Biochemistry and Biophysics, The University of North Carolina, Chapel Hill, NC 27599, USA. [3] Lineberger Comprehensive Cancer Center, The University of North Carolina School of Medicine, Chapel Hill, NC 27599, USA. [4] Department of Epigenetics and Molecular Carcinogenesis, The University of Texas MD Anderson Cancer Center, Smithville, TX 78957, USA. [5] Center for Epigenetics, Van Andel Research Institute, Grand Rapids, MI 49503, USA. [6] Present Address: Roy & Diana Vagelos Laboratories, Department of Chemistry, University of Pennsylvania, Philadelphia, PA, USA.

Acknowledgements
We would like to acknowledge Ian Tsun and Lucas Aponte for help with protein purification and peptide pull-down experiments.

Competing interests
BDS and MTB acknowledge being co-founders of EpiCypher, Inc.

Funding
EKS was supported in part through Longwood University and by the NIGMS IRACDA Grant K12-GM000678. SAS was funded through a UNC Lineberger Cancer Center Postdoctoral Fellowship. BDS is supported by NIH Grant GM110058. MTB is supported by NIH Grant DK062248. SBR is supported by NIH Grant CA181343.

References
1. Luger K, Hansen JC. Nucleosome and chromatin fiber dynamics. Curr Opin Struct Biol. 2005;15:188–96.
2. Li G, Reinberg D. Chromatin higher-order structures and gene regulation. Curr Opin Genet Dev. 2011;21:175–86.
3. Berger SL. The complex language of chromatin regulation during transcription. Nature. 2007;447:407–12.
4. Bannister AJ, Kouzarides T. Regulation of chromatin by histone modifications. Cell Res. 2011;21:381–95.
5. Gardner KE, Allis CD, Strahl BD. Operating on chromatin, a colorful language where context matters. J Mol Biol. 2011;409:36–46.
6. Strahl BD, Allis CD. The language of covalent histone modifications. Nature. 2000;403:41–5.
7. Rothbart SB, Strahl BD. Interpreting the language of histone and DNA modifications. Biophys Acta (BBA) Gene Regul Mech. 2014;1839:627–43.
8. Kouzarides T. Chromatin modifications and their function. Cell. 2007;128:693–705.
9. Musselman CA, Lalonde ME, Cote J, Kutateladze TG. Perceiving the epigenetic landscape through histone readers. Nat Struct Mol Biol. 2012;19:1218–27.
10. Taverna SD, Li H, Ruthenburg AJ, Allis CD, Patel DJ. How chromatin-binding modules interpret histone modifications: lessons from professional pocket pickers. Nat Struct Mol Biol. 2007;14:1025–40.
11. Daniel JA, Pray-Grant MG, Grant PA. Effector proteins for methylated histones: an expanding family. Cell Cycle. 2005;4:919–26.
12. Bottomley MJ. Structures of protein domains that create or recognize histone modifications. EMBO Rep. 2004;5:464–9.
13. Herold JM, Ingerman LA, Gao C, Frye SV. Drug discovery toward antagonists of methyl-lysine binding proteins. Curr Chem Genomics. 2011;5:51–61.
14. Martin C, Zhang Y. The diverse functions of histone lysine methylation. Nat Rev Mol Cell Biol. 2005;6:838–49.
15. Lachner M, O'Sullivan RJ, Jenuwein T. An epigenetic road map for histone lysine methylation. J Cell Sci. 2003;116:2117–24.
16. Black JC, Van Rechem C, Whetstine JR. Histone lysine methylation dynamics: establishment, regulation, and biological impact. Mol Cell. 2012;48:491–507.
17. Patel DJ. A structural perspective on readout of epigenetic histone and DNA methylation marks. Cold Spring Harb Perspect Biol. 2016;8:a018754.
18. Moore KE, Gozani O. An unexpected journey: lysine methylation across the proteome. Biochim Biophys Acta (BBA) Gene Regul Mech. 1839;2014:1395–403.
19. Barski A, Cuddapah S, Cui K, Roh T-Y, Schones DE, Wang Z, et al. High-resolution profiling of histone methylations in the human genome. Cell. 2007;129:823–37.
20. Li Z, Nie F, Wang S, Li L. Histone H4 Lys 20 monomethylation by histone methylase SET8 mediates Wnt target gene activation. Proc Natl Acad Sci. 2011;108:3116–23.
21. Bannister AJ, Zegerman P, Partridge JF, Miska EA, Thomas JO, Allshire RC, et al. Selective recognition of methylated lysine 9 on histone H3 by the HP1 chromo domain. Nature. 2001;410:120–4.
22. Lachner M, O'Carroll D, Rea S, Mechtler K, Jenuwein T. Methylation of histone H3 lysine 9 creates a binding site for HP1 proteins. Nature. 2001;410:116–20.
23. Schotta G, Ebert A, Krauss V, Fischer A, Hoffmann J, Rea S, et al. Central role of Drosophila SU (VAR) 3–9 in histone H3-K9 methylation and heterochromatic gene silencing. EMBO J. 2002;21:1121–31.
24. Suganuma T, Workman JL. Crosstalk among histone modifications. Cell. 2008;135:604–7.
25. Hirota T, Lipp JJ, Toh B-H, Peters J-M. Histone H3 serine 10 phosphorylation by Aurora B causes HP1 dissociation from heterochromatin. Nature. 2005;438:1176–80.

26. Fischle W, Tseng BS, Dormann HL, Ueberheide BM, Garcia BA, Shabanowitz J, et al. Regulation of HP1-chromatin binding by histone H3 methylation and phosphorylation. Nature. 2005;438:1116–22.

27. Gatchalian J, Gallardo CM, Shinsky SA, Ospina RR, Liendo AM, Krajewski K, et al. Chromatin condensation and recruitment of PHD finger proteins to histone H3K4me3 are mutually exclusive. Nucleic Acids Res. 2016;44:6102–12.

28. Iberg AN, Espejo A, Cheng D, Kim D, Michaud-Levesque J, Richard S, et al. Arginine methylation of the histone H3 tail impedes effector binding. J Biol Chem. 2008;283:3006–10.

29. Migliori V, Müller J, Phalke S, Low D, Bezzi M, Mok WC, et al. Symmetric dimethylation of H3R2 is a newly identified histone mark that supports euchromatin maintenance. Nat Struct Mol Biol. 2012;19:136–44.

30. Hyllus D, Stein C, Schnabel K, Schiltz E, Imhof A, Dou Y, et al. PRMT6-mediated methylation of R2 in histone H3 antagonizes H3 K4 trimethylation. Genes Dev. 2007;21:3369–80.

31. Yap KL, Zhou M-M. Keeping it in the family: diverse histone recognition by conserved structural folds. Crit Rev Biochem Mol Biol. 2010;45:488–505.

32. Shi X, Kachirskaia I, Walter KL, Kuo J-HA, Lake A, Davrazou F, et al. Proteome-wide analysis in Saccharomyces cerevisiae identifies several PHD fingers as novel direct and selective binding modules of histone H3 methylated at either lysine 4 or lysine 36. J Biol Chem. 2007;282:2450–5.

33. Maurer-Stroh S, Dickens NJ, Hughes-Davies L, Kouzarides T, Eisenhaber F, Ponting CP. The Tudor domain "Royal Family": tudor, plant agenet, chromo, PWWP and MBT domains. Trends Biochem Sci. 2003;28:69–74.

34. Jacobs SA, Khorasanizadeh S. Structure of HP1 chromodomain bound to a lysine 9-methylated histone H3 tail. Science. 2002;295:2080–3.

35. Adams-Cioaba MA, Li Z, Tempel W, Guo Y, Bian C, Li Y, et al. Crystal structures of the Tudor domains of human PHF20 reveal novel structural variations on the Royal Family of proteins. FEBS Lett. 2012;586:859–65.

36. Washburn MP, Zhao Y, Garcia BA. Reshaping the chromatin and epigenetic landscapes with quantitative mass spectrometry. Mol Cell Proteom. 2016;15:753–4.

37. Rothbart SB, Krajewski K, Strahl BD, Fuchs SM. Peptide microarrays to interrogate the "histone code". Methods Enzymol. 2012;512:107.

38. Fuchs SM, Krajewski K, Baker RW, Miller VL, Strahl BD. Influence of combinatorial histone modifications on antibody and effector protein recognition. Curr Biol. 2011;21:53–8.

39. Zhao Q, Qin L, Jiang F, Wu B, Yue W, Xu F, et al. Structure of human Spindlin1 Tandem tudor-like domains for cell cycle regulation. J Biol Chem. 2007;282:647–56.

40. Wang W, Chen Z, Mao Z, Zhang H, Ding X, Chen S, et al. Nucleolar protein Spindlin1 recognizes H3K4 methylation and stimulates the expression of rRNA genes. EMBO Rep. 2011;12:1160–6.

41. Yang N, Wang W, Wang Y, Wang M, Zhao Q, Rao Z, et al. Distinct mode of methylated lysine-4 of histone H3 recognition by tandem tudor-like domains of Spindlin1. Proc Natl Acad Sci. 2012;109:17954–9.

42. Botuyan MV, Lee J, Ward IM, Kim J-E, Thompson JR, Chen J, et al. Structural basis for the methylation state-specific recognition of histone H4-K20 by 53BP1 and Crb2 in DNA repair. Cell. 2006;127:1361–73.

43. Su X, Zhu G, Ding X, Lee SY, Dou Y, Zhu B, et al. Molecular basis underlying histone H3 lysine–arginine methylation pattern readout by Spin/Ssty repeats of Spindlin1. Genes Dev. 2014;28:622–36.

44. Bock I, Kudithipudi S, Tamas R, Kungulovski G, Dhayalan A, Jeltsch A. Application of Celluspots peptide arrays for the analysis of the binding specificity of epigenetic reading domains to modified histone tails. BMC Biochem. 2011;12:48.

45. Su Z, Boersma MD, Lee J-H, Oliver SS, Liu S, Garcia BA, et al. ChIP-less analysis of chromatin states. Epigenetics Chromatin. 2014;7:7.

46. Liu H, Galka M, Iberg A, Wang Z, Li L, Voss C, et al. Systematic identification of methyllysine-driven interactions for histone and nonhistone targets. J Proteome Res. 2010;9:5827–36. doi:10.1021/pr100597b.

47. Bua DJ, Kuo AJ, Cheung P, Liu CL, Migliori V, Espejo A, et al. Epigenome microarray platform for proteome-wide dissection of chromatin-signaling networks. PLoS ONE. 2009;4:e6789.

48. Bernstein E, Duncan EM, Masui O, Gil J, Heard E, Allis CD. Mouse polycomb proteins bind differentially to methylated histone H3 and RNA and are enriched in facultative heterochromatin. Mol Cell Biol. 2006;26:2560–9.

49. Chen C, Jin J, James DA, Adams-Cioaba MA, Park JG, Guo Y, et al. Mouse Piwi interactome identifies binding mechanism of Tdrkh Tudor domain to arginine methylated Miwi. Proc Natl Acad Sci. 2009;106:20336–41.

50. Vagin VV, Wohlschlegel J, Qu J, Jonsson Z, Huang X, Chuma S, et al. Proteomic analysis of murine Piwi proteins reveals a role for arginine methylation in specifying interaction with Tudor family members. Genes Dev. 2009;23:1749–62.

51. Huang Y, Fang J, Bedford MT, Zhang Y, Xu R-M. Recognition of histone H3 lysine-4 methylation by the double tudor domain of JMJD2A. Science. 2006;312:748–51.

52. Kim J, Daniel J, Espejo A, Lake A, Krishna M, Xia L, et al. Tudor, MBT and chromo domains gauge the degree of lysine methylation. EMBO Rep. 2006;7:397–403.

53. Yang Y, Lu Y, Espejo A, Wu J, Xu W, Liang S, et al. TDRD3 is an effector molecule for arginine-methylated histone marks. Mol Cell. 2010;40:1016–23.

54. Sims RJ, Chen C-F, Santos-Rosa H, Kouzarides T, Patel SS, Reinberg D. Human but not yeast CHD1 binds directly and selectively to histone H3 methylated at lysine 4 via its tandem chromodomains. J Biol Chem. 2005;280:41789–92.

55. Flanagan JF, Mi L-Z, Chruszcz M, Cymborowski M, Clines KL, Kim Y, et al. Double chromodomains cooperate to recognize the methylated histone H3 tail. Nature. 2005;438:1181–5. doi:10.1038/nature04290.

56. Schnetz MP, Bartels CF, Shastri K, Balasubramanian D, Zentner GE, Balaji R, et al. Genomic distribution of CHD7 on chromatin tracks H3K4 methylation patterns. Genome Res. 2009;19:590–601.

57. Papazyan R, Voronina E, Chapman JR, Luperchio TR, Gilbert TM, Meier E, et al. Methylation of histone H3K23 blocks DNA damage in pericentric heterochromatin during meiosis. Elife. 2014;3:e02996.

58. Su Z, Wang F, Lee J-H, Stephens KE, Papazyan R, Voronina E, et al. Reader domain specificity and lysine demethylase-4 family function. Nat Commun. 2016;7:13387.

59. van Nuland R, Gozani O. Histone H4 lysine 20 (H4K20) methylation, expanding the signaling potential of the proteome one methyl moiety at a time. Mol Cell Proteom. 2016;15:755–64.

60. Tuzon CT, Spektor T, Kong X, Congdon LM, Wu S, Schotta G, et al. Concerted activities of distinct H4K20 methyltransferases at DNA double-strand breaks regulate 53BP1 nucleation and NHEJ-directed repair. Cell Rep. 2014;8:430–8.

61. Sanders SL, Portoso M, Mata J, Bähler J, Allshire RC, Kouzarides T. Methylation of histone H4 lysine 20 controls recruitment of Crb2 to sites of DNA damage. Cell. 2004;119:603–14.

62. Mikkelsen TS, Ku M, Jaffe DB, Issac B, Lieberman E, Giannoukos G, et al. Genome-wide maps of chromatin state in pluripotent and lineage-committed cells. Nature. 2007;448:553–60.

63. Vakoc CR, Sachdeva MM, Wang H, Blobel GA. Profile of histone lysine methylation across transcribed mammalian chromatin. Mol Cell Biol. 2006;26:9185–95.

64. Behbahani TE, Kahl P, von der Gathen J, Heukamp LC, Baumann C, Gütgemann I, et al. Alterations of global histone H4K20 methylation during prostate carcinogenesis. BMC Urol. 2012;12:1.

65. Van Den Broeck A, Brambilla E, Moro-Sibilot D, Lantuejoul S, Brambilla C, Eymin B, et al. Loss of histone h4k20 trimethylation occurs in preneoplasia and influences prognosis of non-small cell lung cancer. Clin Cancer Res. 2008;14:7237–45.

66. Kuo AJ, Song J, Cheung P, Ishibe-Murakami S, Yamazoe S, Chen JK, et al. The BAH domain of ORC1 links H4K20me2 to DNA replication licensing and Meier-Gorlin syndrome. Nature. 2012;484:115–1.

67. Franz H, Greschik H, Willmann D, Ozretić L, Jilg CA, Wardelmann E, et al. The histone code reader SPIN1 controls RET signaling in liposarcoma. Oncotarget. 2015;6:4773.

68. Jiang F, Zhao Q, Qin L, Pang H, Pei X, Rao Z. Expression, purification, crystallization and preliminary X-ray analysis of human spindlin1, an ovarian cancer-related protein. Protein Pept Lett. 2006;13:203–5.

69. Wang J-X, Zeng Q, Chen L, Du J-C, Yan X-L, Yuan H-F, et al. SPINDLIN1 promotes cancer cell proliferation through activation of WNT/TCF-4 signaling. Mol Cancer Res. 2012;10:326–35.

70. Zhang P, Cong B, Yuan H, Chen L, Lv Y, Bai C, et al. Overexpression of spindlin1 induces metaphase arrest and chromosomal instability. J Cell Physiol. 2008;217:400–8.

71. Yuan H, Zhang P, Qin L, Chen L, Shi S, Lu Y, et al. Overexpression of SPIN-DLIN1 induces cellular senescence, multinucleation and apoptosis. Gene. 2008;410:67–74.

72. Robaa D, Wagner T, Luise C, Carlino L, McMillan J, Flaig R, et al. Identification and structure-activity relationship studies of small-molecule inhibitors of the methyllysine reader protein Spindlin1. ChemMedChem. 2016;11:2327–38.

73. Wagner T, Greschik H, Burgahn T, Schmidtkunz K, Schott A-K, McMillan J, et al. Identification of a small-molecule ligand of the epigenetic reader protein Spindlin1 via a versatile screening platform. Nucleic Acids Res. 2016;44:e88.

74. Kaustov L, Ouyang H, Amaya M, Lemak A, Nady N, Duan S, et al. Recognition and specificity determinants of the human Cbx chromodomains. J Biol Chem. 2011;286:521–9.

75. Franz H, Mosch K, Soeroes S, Urlaub H, Fischle W. Multimerization and H3K9me3 binding are required for CDYL1b heterochromatin association. J Biol Chem. 2009;284:35049–59.

76. Fischle W, Franz H, Jacobs SA, Allis CD, Khorasanizadeh S. Specificity of the chromodomain Y chromosome family of chromodomains for lysine-methylated ARK (S/T) motifs. J Biol Chem. 2008;283:19626–35.

77. Kokura K, Sun L, Bedford MT, Fang J. Methyl-H3K9-binding protein MPP8 mediates E-cadherin gene silencing and promotes tumour cell motility and invasion. EMBO J. 2010;29:3673–87.

78. Chang Y, Horton JR, Bedford MT, Zhang X, Cheng X. Structural insights for MPP8 chromodomain interaction with histone H3 lysine 9: potential effect of phosphorylation on methyl-lysine binding. J Mol Biol. 2011;408:807–14.

79. Nady N, Lemak A, Walker JR, Avvakumov GV, Kareta MS, Achour M, et al. Recognition of multivalent histone states associated with heterochromatin by UHRF1 protein. J Biol Chem. 2011;286:24300–11.

80. Rothbart SB, Krajewski K, Nady N, Tempel W, Xue S, Badeaux AI, et al. Association of UHRF1 with methylated H3K9 directs the maintenance of DNA methylation. Nat Struct Mol Biol. 2012;19:1155–60.

81. Musselman CA, Avvakumov N, Watanabe R, Abraham CG, Lalonde M-E, Hong Z, et al. Molecular basis for H3K36me3 recognition by the Tudor domain of PHF1. Nat Struct Mol Biol. 2012;19:1266–72.

82. Ballaré C, Lange M, Lapinaite A, Martin GM, Morey L, Pascual G, et al. Phf19 links methylated Lys36 of histone H3 to regulation of Polycomb activity. Nat Struct Mol Biol. 2012;19:1257–65.

83. Cai L, Rothbart SB, Lu R, Xu B, Chen W-Y, Tripathy A, et al. An H3K36 methylation-engaging Tudor motif of polycomb-like proteins mediates PRC2 complex targeting. Mol Cell. 2013;49:571–82.

84. Qin S, Guo Y, Xu C, Bian C, Fu M, Gong S, et al. Tudor domains of the PRC2 components PHF1 and PHF19 selectively bind to histone H3K36me3. Biochem Biophys Res Commun. 2013;430:547–53.

85. Bian C, Xu C, Ruan J, Lee KK, Burke TL, Tempel W, et al. Sgf29 binds histone H3K4me2/3 and is required for SAGA complex recruitment and histone H3 acetylation. EMBO J. 2011;30:2829–42.

86. Collins RE, Northrop JP, Horton JR, Lee DY, Zhang X, Stallcup MR, et al. The ankyrin repeats of G9a and GLP histone methyltransferases are mono- and dimethyllysine binding modules. Nat Struct Mol Biol. 2008;15:245–50.

87. Pena PV, Davrazou F, Shi X, Walter KL, Verkhusha VV, Gozani O, et al. Molecular mechanism of histone H3K4me3 recognition by plant homeodomain of ING2. Nature. 2006;442:100–3.

88. Shi X, Hong T, Walter KL, Ewalt M, Michishita E, Hung T, et al. ING2 PHD domain links histone H3 lysine 4 methylation to active gene repression. Nature. 2006;442:96–9.

89. Vermeulen M, Mulder KW, Denissov S, Pijnappel WW, van Schaik FM, Varier RA, et al. Selective anchoring of TFIID to nucleosomes by trimethylation of histone H3 lysine 4. Cell. 2007;131:58–69.

90. Ruthenburg AJ, Li H, Patel DJ, Allis CD. Multivalent engagement of chromatin modifications by linked binding modules. Nat Rev Mol Cell Biol. 2007;8:983–94.

91. Min J, Allali-Hassani A, Nady N, Qi C, Ouyang H, Liu Y, et al. L3MBTL1 recognition of mono-and dimethylated histones. Nat Struct Mol Biol. 2007;14:1229–30.

92. Trojer P, Li G, Sims RJ, Vaquero A, Kalakonda N, Boccuni P, et al. L3MBTL1, a histone-methylation-dependent chromatin lock. Cell. 2007;129:915–28.

12

SMYD5 regulates H4K20me3-marked heterochromatin to safeguard ES cell self-renewal and prevent spurious differentiation

Benjamin L. Kidder[1,2,3]*, Gangqing Hu[3], Kairong Cui[3] and Keji Zhao[3]*

Abstract

Background: Epigenetic regulation of chromatin states is thought to control the self-renewal and differentiation of embryonic stem (ES) cells. However, the roles of repressive histone modifications such as trimethylated histone 4 lysine 20 (H4K20me3) in pluripotency and development are largely unknown.

Results: Here, we show that the histone lysine methyltransferase SMYD5 mediates H4K20me3 at heterochromatin regions. Depletion of SMYD5 leads to compromised self-renewal, including dysregulated expression of OCT4 targets, and perturbed differentiation. SMYD5-bound regions are enriched with repetitive DNA elements. Knockdown of SMYD5 results in a global decrease of H4K20me3 levels, a redistribution of heterochromatin constituents including H3K9me3/2, G9a, and HP1α, and de-repression of endogenous retroelements. A loss of SMYD5-dependent silencing of heterochromatin nearby genic regions leads to upregulated expression of lineage-specific genes, thus contributing to the decreased self-renewal and perturbed differentiation of SMYD5-depleted ES cells.

Conclusions: Altogether, these findings implicate a role for SMYD5 in regulating ES cell self-renewal and H4K20me3-marked heterochromatin.

Keywords: Embryonic stem cells, SMYD5, H4K20me3, Repetitive DNA, LTR, LINE, Pluripotent, Epigenetics, Chromatin, Heterochromatin, Genomics, RNA-Seq, ChIP-Seq, Self-renewal, Gene expression, Embryoid body, Differentiation, Histone methyltransferase

Background

Compared to the extensive studies on active histone modifications, limited investigations have been performed on heterochromatic markers. Heterochromatic domains are generally inaccessible to DNA binding factors and transcriptionally silent [1]. Large regions of heterochromatin can be found around chromosomal structures such as centromeres and telomeres, while smaller domains are interspersed throughout the genome [2]. Heterochromatin plays a critical role in gene expression during development and differentiation [3] and is also involved in maintaining genome integrity by stabilizing repetitive DNA sequences throughout the genome by inhibiting recombination between homologous DNA repeats [4]. Heterochromatin is associated with H3K9 and H4K20 methylation. ESET/Setdb1 and LSD1, which control the methylation status of H3 lysine 9, are important for silencing of endogenous retroviruses (ERVs) in ES cells and during early embryogenesis [5, 6], suggesting critical roles of H3K9 methylation in ES cells. However, it remains unclear how H4K20me3 is regulated and how it contributes to the repression of endogenous retroelements in ES cells.

H4K20 methylation marks, which are evolutionarily conserved from yeast (*S. pombe*) to humans [7], have been implicated in having diverse cellular functions

*Correspondence: Benjamin.kidder@wayne.edu; keji.zhao@nih.gov
[3] Systems Biology Center, National Heart, Lung and Blood Institute, National Institutes of Health, Bethesda, MD, USA
Full list of author information is available at the end of the article

including the formation of heterochromatin, gene regulation and repression of transcription [8], DNA damage repair [9, 10], DNA replication [11], chromosome condensation [12], and genome stability [10, 13]. Although H4K20me1 is found in active genes [14, 15], H4K20me3 is associated with the formation of pericentric heterochromatin by sequential methylation of H4K20me1 and H4K20me2 by Suv420h1 or Suv420h2, respectively [10, 16]. H4K20me3 marks repress transcription of repetitive elements [10, 17, 18]. SMYD5 has recently been shown to be a histone methyltransferase that mediates H4K20me3 modification in Drosophila and mouse primary macrophage cells [19]. However, the role of SMYD5 in mouse development, ES cell self-renewal and differentiation, and regulation of heterochromatin has not been fully elucidated.

In this study, we show that the H4K20me3 methyltransferase, SMYD5, targets H4K20me3 in heterochromatin regions containing retroelements and facilitates HP1α binding. Our results suggest that SMYD5 represses lineage-specific genes and thus contributes to the maintenance of ES cell lineage.

Results

SMYD5 regulates ES cell self-renewal

Using RNA-Seq assays, we found that SMYD5 is highly expressed in ES cells and downregulated upon differentiation (see Additional file 1: Figure S1A) [20]. To study the role of SMYD5 in ES cell function, we knocked down *Smyd5* with lentiviral particles encoding three different short hairpin RNAs (shRNAs) (see Additional file 1: Figure S1B). Depletion of SMYD5 resulted in a loss of normal ES cell colony morphology (see Additional file 1: Figure S1C), where shSmyd5 ES cell colonies became flattened and lost their tight cell–cell contact and became scattered at the colony periphery. The severity of phenotypes correlated with the knockdown efficiency (see Additional file 1: Figure S1C). To confirm the specificity

of the shRNA sequences, we performed a rescue experiment by overexpressing an shRNA-resistant version of wild-type (WT) SMYD5 or an shRNA-resistant enzymatically mutant version of SMYD5 (H315L and C317A) [19]. Our results show that control ES cells (short hairpin luciferase—shLuc and shLuc + WT) maintained their colony morphology and overexpression of wild-type SMYD5 in short hairpin Smyd5 (shSmyd5) ES cells (shSmyd5 + WT) restored the 3D colony morphology of the majority of colonies to an ESC-like phenotype (Fig. 1a, b). While 99% of shLuc ES cell colonies exhibited an ESC-like morphology, only 11% of shSmyd5 ES cells remained intact (Fig. 1a, b). However, 70% of shSmyd5 ES cells overexpressing wild-type SMYD5 displayed an ESC-like morphology (Fig. 1b). In addition, while overexpression of mutant SMYD5 decreased the number of intact ESC-like colonies in shLuc ES cells (shLuc + mut) (Fig. 1a, b), the number of intact shSmyd5 ES cells (shSmyd5 + mut) did not significantly change, demonstrating that SMYD5 is important for ES cell self-renewal. Moreover, alkaline phosphatase (AP) staining, a marker of undifferentiated ES cells, was mostly absent in shSmyd5 ES cells or shSmyd5 ES cell colonies overexpressing mutant SMYD5 relative to control (shLuc) ES cells (Fig. 1c, d). However, AP staining was restored in 80% of shSmyd5 ES cells overexpressing wild-type SMYD5, further demonstrating that SMYD5 is important for ES cell self-renewal. In addition, we observed wild-type levels of SMYD5 expression in shSmyd5 ES cells overexpressing wild-type SMYD5 (Fig. 1e).

By comparing the global gene expression profiles of shSmyd5 and shLuc ES cells using RNA-Seq, we found 1616 genes differentially expressed (DE) at least twofold, and 4235 genes differentially expressed (DE) at least 1.5 fold, including the pluripotency regulators Oct4/*Pou5f1*, *Nanog*, and *Tbx3* (Fig. 1f), as exemplified by the UCSC genome browser tracks (Fig. 1g). Genes differentially expressed at least 1.5 fold were used for downstream

(See figure on next page.)
Fig. 1 SMYD5 regulates ES cell self-renewal. **a** Bright-field microscopy of ES cells infected with shLuc or shSmyd5 lentiviral particles and wild-type (WT) SMYD5 or an enzymatically mutant (mut) version of SMYD5 (H315L and C317A) lentiviral particles and stably selected with puromycin and G418. **b** ES cell colonies were scored by morphology. The percentage of colonies with an ES-like morphology (compact and round vs. flattened) are represented as mean ± SEM. *P* values were calculated using a *t* test. **c** Alkaline phosphatase (AP) staining of ES cells. **d** ES cells were scored by AP staining. The percentage of AP positive colonies is represented as mean ± SEM. *p* values were calculated using a *t* test. **e** Quantitative RT-PCR (Q-RT-PCR) expression of SMYD5 using primers for three different regions of the SMYD5 coding region. **f** Scatter plot of RNA sequencing (RNA-Seq) gene expression analysis between shLuc and shSmyd5 ES cells. Log2 adjusted differentially expressed genes are plotted. Genes whose expression is greater than twofold (shLuc vs shSmyd5) and with an RPKM > 1 (reads per kilo bases of exon model per million reads) and FDR < 0.001 are shown in *black*. **g** UCSC genome browser view of differential expression of self-renewal genes in shSmyd5 and shLuc control ES cells. **h** Q-RT-PCR analysis of expression of Smyd5 and self-renewal genes in shLuc and shSmyd5 ES cells. **i** Gene set enrichment analysis (GSEA) [22] of differentially expressed genes in Smyd5 knockdown ES cells relative to undifferentiated and differentiated embryoid bodies (EBs). **j** Gene ontology (GO) functional annotation of differentially expressed genes analyzed using DAVID [23]. **k** Mouse gene atlas expression analysis evaluated using Network2Canvas [66] demonstrates that lineage and ES cell genes are misexpressed in shSmyd5 ES cells. Each node (*square*) represents a gene list (shLuc vs shSmyd5 DE genes) associated with a gene-set library (mouse gene atlas). The brightness (*white*) of each node is determined by its *p* value

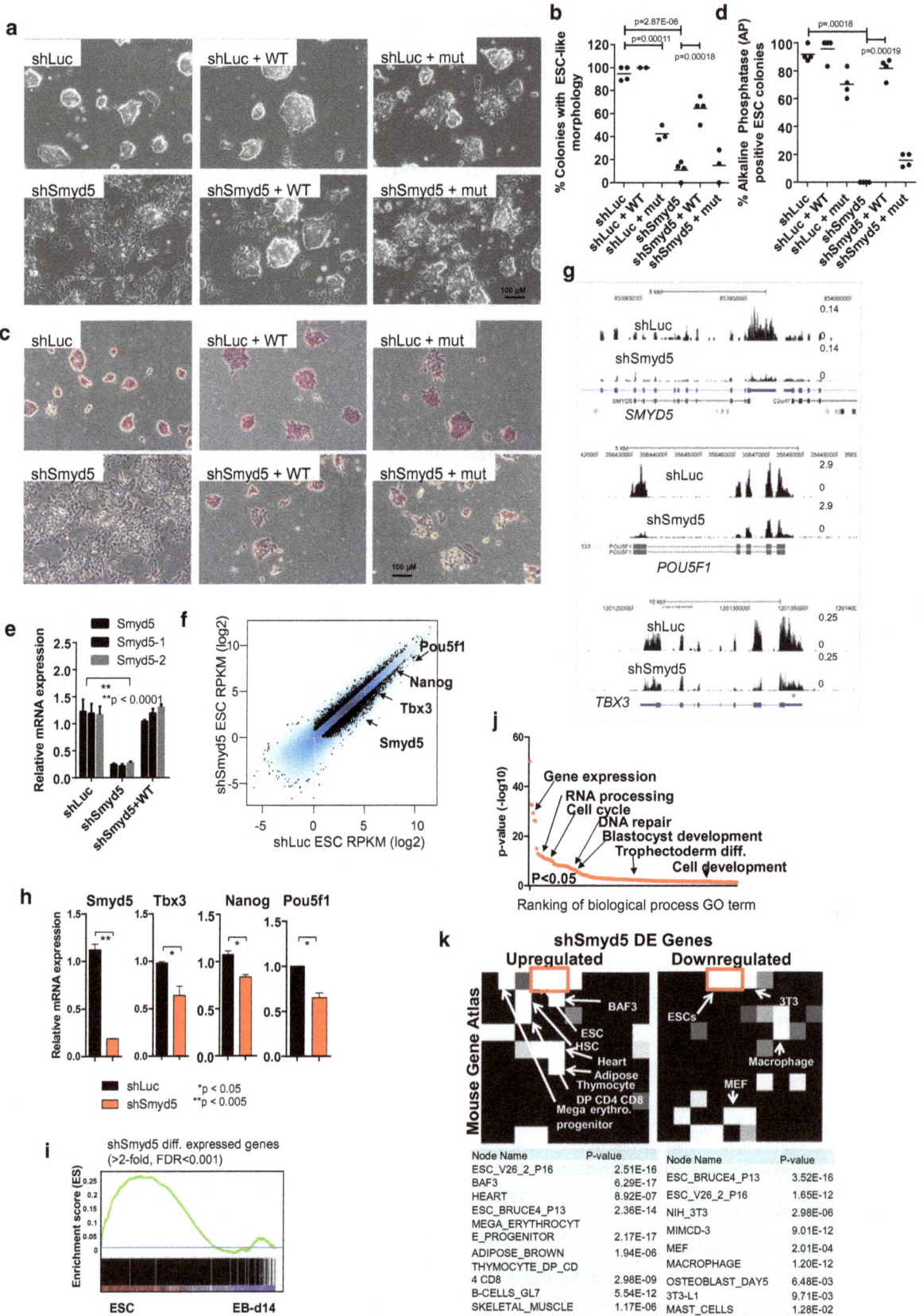

analyses. Moreover, several stem cell genes, including *Esrrb* and *Tbx3*, were downregulated in *Smyd5* knockdown ES cells. Because ESRRB and TBX3 occupy promoter regions of other stem cell genes, including *Oct4* and *Nanog*, a reduction in their expression may further influence the self-renewal state of *Smyd5* knockdown ES cells. We confirmed the expression of several self-renewal genes using Q-RT-PCR (Fig. 1h). We then compared these DE genes with global expression data from ES cells and embryoid body (EB) differentiated cells [20], which emulates early embryo development [21] (see "Methods"), using gene set enrichment analysis (GSEA) [22]. This analysis shows that differentially expressed genes are enriched in ES cells (Fig. 1i), suggesting that a loss of SMYD5 impacts ES cell function. Moreover, DAVID [23] gene ontology (GO) terms enriched in DE genes include gene expression, cell cycle, RNA processing, DNA repair, blastocyst development, trophectoderm differentiation, and cell development (Fig. 1j). We also found that expression of many of the DE genes between shLuc and shSmyd5 ES cells is not only enriched in ES cells (Fig. 1k), but expression of many upregulated genes in shSmyd5 ES cells is also enriched in committed lineages, suggesting that expression of both lineage genes and ES cell regulators is impacted by depleting SMYD5. These results implicate a role for SMYD5 in repressing expression of lineage-specific genes in ES cells.

Because *Pou5f1* and *Nanog* are downregulated in shSmyd5 ES cells, we compared genes bound by OCT4, SOX2, and NANOG in ES cells [24] and genes misexpressed in shSmyd5 ES cells. We found that 34% of the DE genes are bound by OCT4 in ES cells (Fig. 2a), and OCT4, SOX2, and NANOG also co-occupied a number of SMYD5-regulated genes (Fig. 2a), suggesting that depletion of SMYD5 leads to perturbation of the core ES cell transcriptional circuitry.

Altered differentiation of SMYD5-depleted ES cells

To investigate the function of SMYD5 during ES cell differentiation, we induced EB formation of shLuc and shSmyd5 ES cells in the absence of leukemia inhibitory factor (LIF). EB formation, which involves a change in culture condition from 2D to 3D, is a suitable assay for evaluating ES cell differentiation because it recapitulates embryogenesis [21, 25]. shLuc and shSmyd5 ES cells were cultured in the absence of LIF on low-attachment dishes to induce EB differentiation over two weeks. While shLuc ES cells formed mainly circular or globular EB structures containing a primitive endoderm (PE) layer during early differentiation (day 6) (Fig. 2b, top), shSmyd5 ES cells formed complex structures containing bulges lined with a PE layer (Fig. 2b, top). The PE layer forms in vivo by differentiation of cells located on the surface of the ICM

facing the blastocoel [25]. This pattern was further pronounced at day 8 of EB differentiation (Fig. 2b, bottom). We also utilized teratoma formation assays to evaluate the in vivo differentiation ability of SMYD5-depleted ES cells into cells of the three germ layers (ectoderm, mesoderm, and endoderm). Hematoxylin and eosin (H&E) staining confirmed the presence of complex structures and more advanced differentiation in shSmyd5 EBs (Fig. 2c). A further evaluation of differentiation using teratoma formation revealed that while shSmyd5 and shLuc teratomas are both able to give rise to cells of the three germ layers including ectoderm (keratinized epithelium, epidermis), mesoderm (mesenchymal cells, muscle, adipocytes), and endoderm (glandular epithelium) (Fig. 2d), shSmyd5 teratomas had a greater presence of glandular endodermal cells.

Knockdown of SMYD5 leads to accelerated gene expression changes during ES cell differentiation

To identify gene expression defects caused by depletion of SMYD5, we evaluated DE genes during differentiation of shLuc and shSmyd5 ES cells (Fig. 3a) using RNA-Seq. K-means clustering ($k = 20$) was used to identify patterns of gene expression variability (Fig. 3a). These results highlight clusters of genes such as self-renewal genes, including *Nanog* (Fig. 3a, b; see Additional file 1: Figure S1D), that were rapidly downregulated in shSmyd5 EBs, and lineage-specific genes, such as *Sox17*, *Afp*, *Gata6*, and *Tbx5* (Fig. 3a, b; see Additional file 1: Figure S1D), that were differentially expressed between shSmyd5 and shLuc EBs. Because Sox17 is a transcription factor that is expressed in endodermal lineages, a driver of extraembryonic endoderm transcriptional programs, and is important in antagonizing expression of *Pou5f1* and *Nanog* during differentiation [26], its upregulation during differentiation of SMYD5-depleted ES cells may explain the presence of complex structures involving primitive endoderm. To compare the transcriptomes of shSmyd5 EBs relative to shLuc EBs, we used principal component analysis (PCA), which showed that shSmyd5 EBs progress through the first two components at an altered trajectory (Fig. 3c), suggesting that knockdown of *Smyd5* leads to perturbed differentiation.

To further address the phenomenon of altered differentiation, we built a predictive model to determine the probability of expression changes due to chance or altered differentiation following differentiation of shSmyd5 EBs. Expression changes that indicate altered differentiation include genes that show altered downregulation or upregulation during EB differentiation. Our findings demonstrate that the percentage of upregulated or downregulated genes outpaced the expected (Fig. 3d, see "Methods"), demonstrating that shSmyd5 EBs exhibit

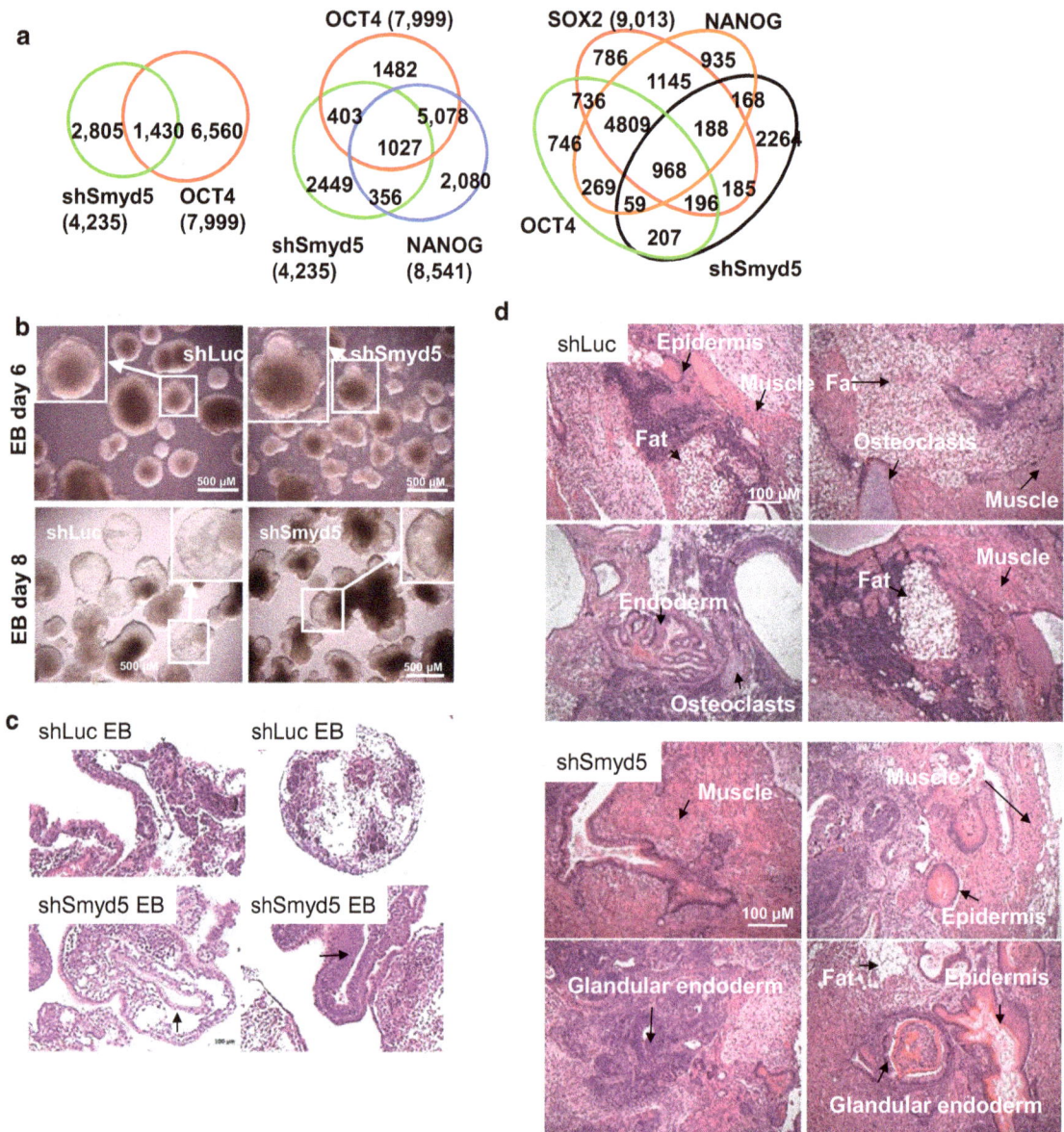

Fig. 2 Altered differentiation of SMYD5-depleted ES cells. **a** Venn diagrams showing overlap between differentially expressed genes in shSmyd5 and shLuc ES cells and genes bound by OCT4, OCT4, and NANOG, or OCT4, SOX2, and NANOG. **b** Embryoid body (EB) formation shows abnormal differentiation of shSmyd5 ES cells. **c** Hematoxylin and eosin (H&E) histological sections of shLuc and shSmyd5 day 9 EBs. The *arrowheads* depict altered and advanced differentiation of shSmyd5 EBs relative to control (shLuc) EBs. The *bottom left panel* shows an EB with an atypical internal epithelial-like structure; the *bottom right panel* shows a thick epithelial-like layer. **d** H&E histological sections of teratomas generated from shLuc and shSmyd5 ES cells injected into SCID–beige mice. Tumors were harvested 4–6 weeks post-injection and evaluated using standard H&E histological methods. Transmitted white-light microscopy of sectioned teratomas. Heterogeneous differentiation of shLuc and shSmyd5 ES cells into ectoderm (keratinized epidermal cells), mesoderm (muscle and mesenchymal cells, adipocytes), and endoderm (glandular structures)

altered differentiation. The x-axis represents genes that are upregulated or downregulated by alpha-fold from ES cells (d0 of differentiation) to day 6 EBs, or from ES cells to day 10 EBs.

GSEA was then used to investigate the expression state of DE genes during differentiation in the absence of SMYD5. Our results demonstrate that ES cell-enriched genes are differentially expressed during early differentiation (Fig. 3e, top graphs) while EB-enriched genes are differentially expressed later in differentiation (Fig. 3e, bottom graphs). DAVID GO analysis revealed that many developmental GO terms were overrepresented during

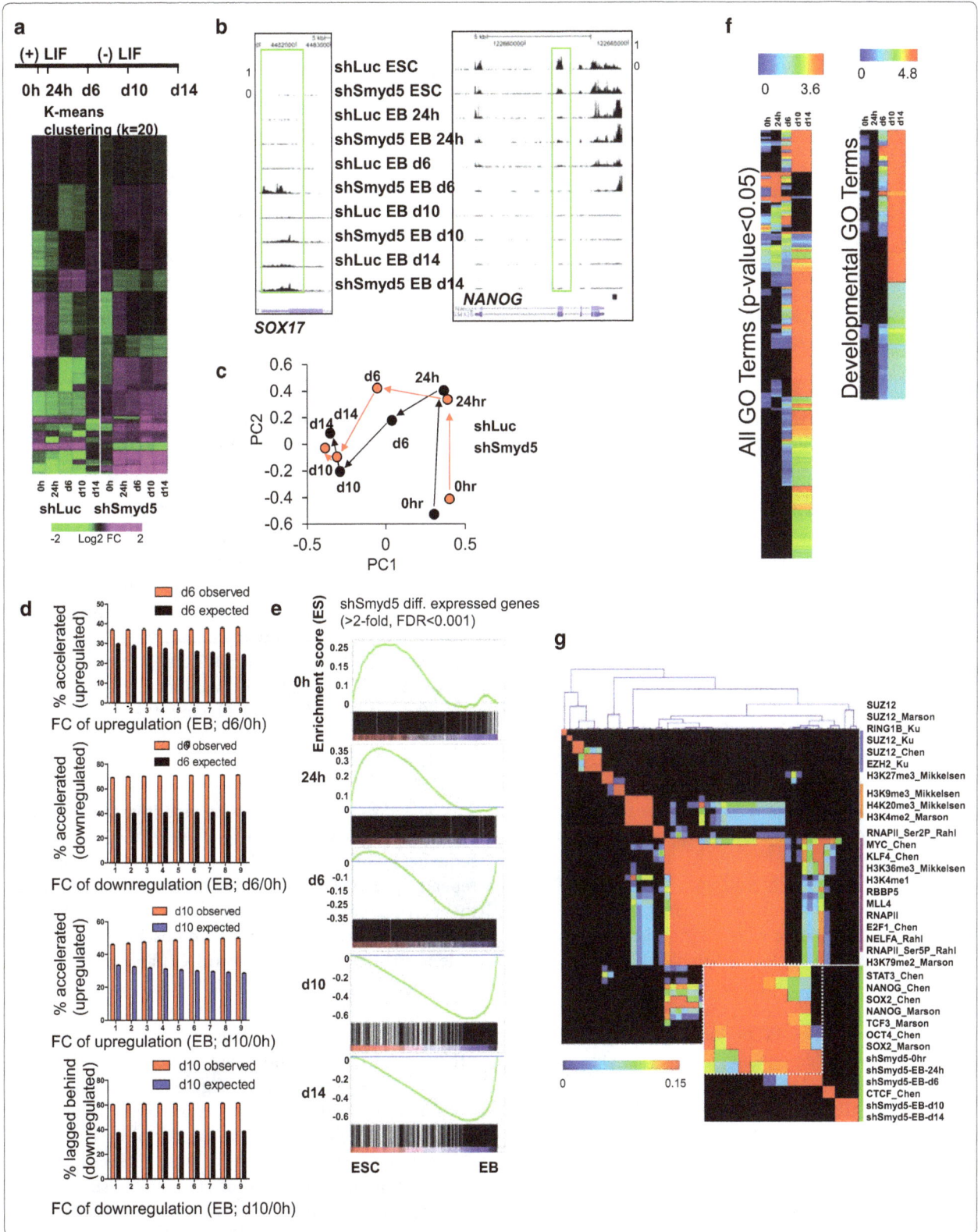

(See figure on previous page.)

Fig. 3 Transcriptome analysis reveals altered differentiation of SMYD5-depleted ES cells. **a** K-means clustering analysis of RNA-Seq data from shLuc and shSmyd5 ES cells differentiated without LIF for 14 days. The experimental design is shown on *top*. Differentially expressed genes (>twofold; RPKM > 1) clustered according to k-means. **b** Custom tracks of RNA-Seq data in the UCSC genome browser for undifferentiated and differentiated shLuc control and shSmyd5 ES cells. **c** Principal component analysis (PCA) of differentially expressed genes during EB differentiation of shSmyd5 and shLuc ES cells. **d** Prediction of differentially expressed genes due to chance or altered differentiation. The percentage of genes that lag behind during differentiation of shSmyd5 ES cells is less than expected. *Top each bar* represents a group of genes upregulated by at least alpha-fold (*X* axis) from ESC (0 h) to EB day 6 in the control cells. The percentage of genes with expression values that follow the order: EB day 6 (shLuc) > EB day 6 (shSmyd5) > ES cell is calculated (observed; *red bars*); *error bars* are generated by bootstrapping. The expression values of all genes are randomly shuffled independently for EB day 6 (shLuc), EB day 6 (shSmyd5), and ES cells and are repeated many times to give the means and standard deviations for the expectations (expected; *blue bars*). The *red bars* represent observed data. *Bottom each bar* represents a group of genes upregulated by at least alpha-fold (*X* axis) from ESC (0 h) to EB day 10 in the control cells. The percentage of genes with expression values that follow the order: EB day 10 (shLuc) > EB day 10 (shSmyd5) > ES cell is calculated (observed; *red bars*); *error bars* are generated by bootstrapping. The expression values of all genes are randomly shuffled independently for EB day 10 (shLuc), EB day 10 (shSmyd5), and ES cells and are repeated many times to give the means and standard deviations for the expectations (expected; *blue bars*). The *red bars* represent observed data. **e** Gene set enrichment analysis (GSEA) of differentially expressed genes during differentiation of shSmyd5 ES cells relative to ES cells and day 14 EBs. **f** DAVID gene ontology analysis of differentially expressed genes between shLuc and shSmyd5 ES cells and during EB differentiation. The hierarchical clustering heat map on the right shows enrichment of developmental GO terms. **g** Correlation matrix of differentially expressed (DE) genes during shSmyd5 ES cell differentiation with promoter binding of transcription factors and epigenetic modifiers. Heat map generated by evaluating pair-wise affinities between differentially expressed (DE) genes during shLuc and shSmyd5 EB differentiation using RNA-Seq datasets generated from this study (0, 24 h, 6, 10, 14 days) and published ChIP-Seq data [14, 24, 27–29, 67]. AutoSOME [68] was used to generate pair-wise affinity values

later EB differentiation (Fig. 3f), suggesting that SMYD5 regulates developmental genes during differentiation.

Because depletion of SMYD5 resulted in the differential expression of many OCT4, SOX2, and NANOG targets (Fig. 2a), we evaluated whether DE genes during shSmyd5 EB differentiation were also occupied by ES cell-enriched transcription regulators or marked by histone modifications using public datasets [14, 24, 27–29] (Fig. 3g). In shSmyd5 ES cells and during early EB formation (24 h), we observed a strong correlation between DE genes and binding of ES cell-enriched factors OCT4, SOX2, NANOG, and STAT3 (Fig. 3g), suggesting that depletion of SMYD5 leads to the dysregulated expression of pluripotency-regulator targets during early EB differentiation.

SMYD5 mediates H4K20me3 modification in ES cells

To evaluate the genome-wide distribution of SMYD5 in ES cells, we utilized biotin-mediated ChIP-Seq (bioChIP-Seq) [30, 31] and FLAG ChIP-Seq as described in the materials and methods. Using this approach, we observed a high overlap of SMYD5 binding using these two approaches (see Additional file 2: Figure S2A) by evaluating the density of SMYD5 at "Spatial Clustering for Identification of ChIP-Enriched Regions" (SICER) islands (see "Methods") (Fig. 4a) as well as by heat maps (Fig. 4b; see Additional file 2: Figure S2B) and average profiles (Fig. 4c; see Additional file 2: Figure S2C).

Because SMYD5 has been shown to deposit H4K20me3 marks [19], we evaluated global levels of H4K20 methylation in *Smyd5* knockdown ES cells using western blotting. Our results show that depletion of SMYD5 led to decreased H4K20me3, but not H4K20me2 or H4K20me1

(Fig. 4d; see Additional file 2: Figure S2D), demonstrating that SMYD5 confers H4K20me3 methyltransferase activity. We also observed a restoration of H4K20me3 levels in shSmyd5 ES cells overexpressing an shRNA-resistant version of SMYD5 (Fig. 4d, bottom right). Moreover, our ChIP-Seq data indicated that a majority of SMYD5 occupied regions (79%) contain H4K20me3 marks (Fig. 4e), where SMYD5 binding is significantly enriched in H4K20me3 islands (Fig. 4f). Overall, these results demonstrate that SMYD5 occupies chromatin regions containing H4K20me3.

SMYD5 and H4K20me3 co-occupy repetitive DNA elements

Since H4K20me3 is known to be enriched in repetitive sequences [14], we analyzed the enrichment of DNA repeats in SMYD5 islands. To this end, we evaluated the percent coverage of H4K20me3 or SMYD5 peaks that overlap a repeat element. We observed enrichment of long interspersed elements (LINE) and long-terminal repeat (LTR) elements in H4K20me3 (see Additional file 3: Figure S3A, top) and SMYD5 regions (see Additional file 3: Figure S3A, bottom), respectively. Similarly, LINE and LTR elements were enriched in H3K9me3 regions (see Additional file 3: Figure S3A, middle panel), consistent with the co-localization of H3K9me3 and H4K20me3 on chromatin. Enrichment of LINE and LTR elements was markedly higher at these regions relative to random genomic sequences of comparable size and frequency, and H3K4me3 regions, which were used as controls, demonstrating that SMYD5 and H4K20me3 islands are enriched at LINE and LTR repetitive DNA elements.

We also evaluated the percent coverage of LINE and LTR sequences for all SMYD5, H4K20me3, and

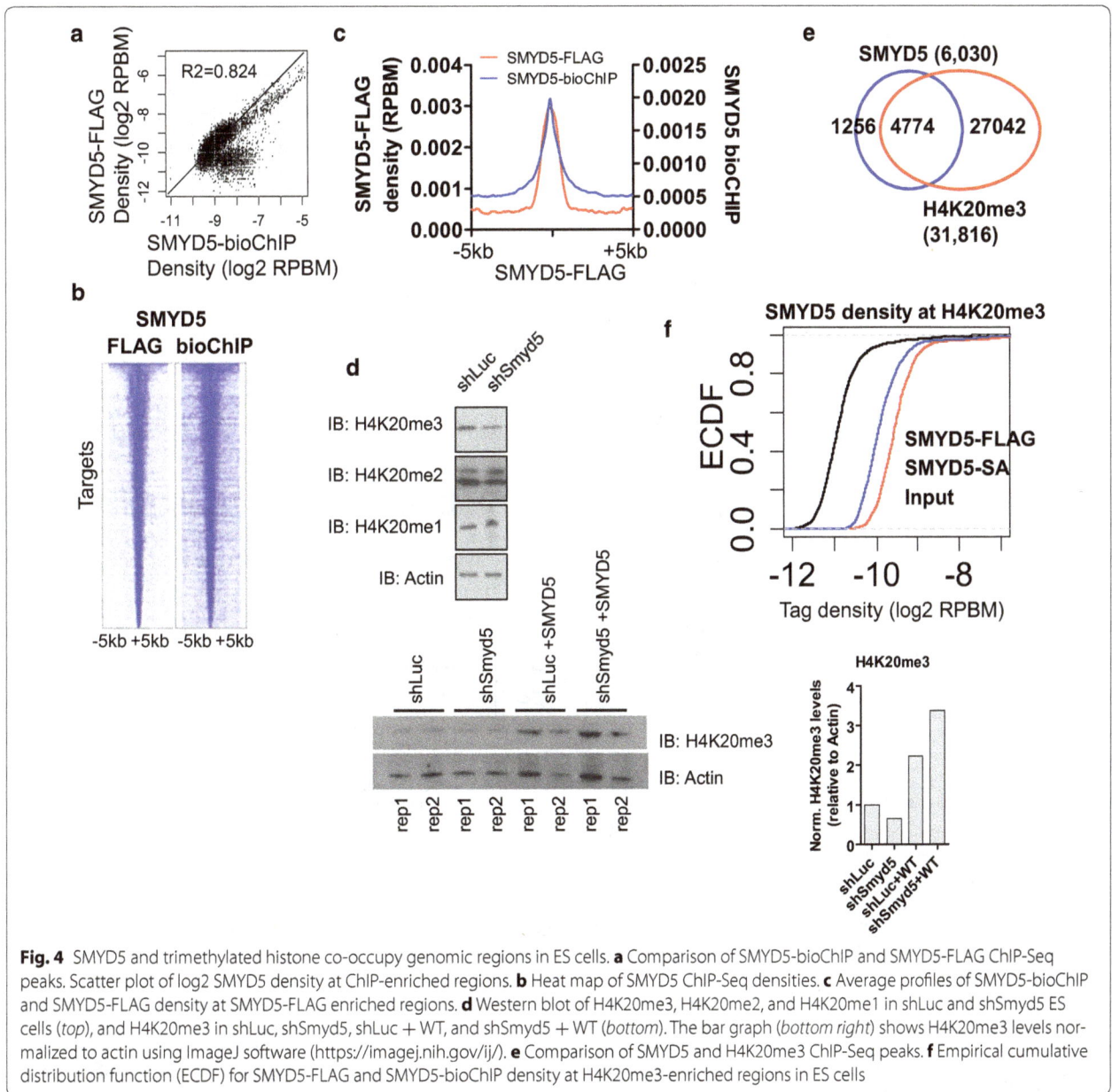

Fig. 4 SMYD5 and trimethylated histone co-occupy genomic regions in ES cells. **a** Comparison of SMYD5-bioChIP and SMYD5-FLAG ChIP-Seq peaks. Scatter plot of log2 SMYD5 density at ChIP-enriched regions. **b** Heat map of SMYD5 ChIP-Seq densities. **c** Average profiles of SMYD5-bioChIP and SMYD5-FLAG density at SMYD5-FLAG enriched regions. **d** Western blot of H4K20me3, H4K20me2, and H4K20me1 in shLuc and shSmyd5 ES cells (*top*), and H4K20me3 in shLuc, shSmyd5, shLuc + WT, and shSmyd5 + WT (*bottom*). The bar graph (*bottom right*) shows H4K20me3 levels normalized to actin using ImageJ software (https://imagej.nih.gov/ij/). **e** Comparison of SMYD5 and H4K20me3 ChIP-Seq peaks. **f** Empirical cumulative distribution function (ECDF) for SMYD5-FLAG and SMYD5-bioChIP density at H4K20me3-enriched regions in ES cells

H3K9me3 islands. Using this approach, we observed enrichment of LINE and LTR sequences in H4K20me3, SMYD5, and H3K9me3 regions relative to random genomic regions (Fig. 5a).

To investigate which repeat family members are enriched in H4K20me3, H3K9me3, or SMYD5 regions, we evaluated the percentage of peaks that overlap a repeat family element. We observed enrichment of ERVK (LTR class) and L1 (LINE class) family repetitive elements in H4K20me3 (see Additional file 3: Figure S3B, top), H3K9me3 (see Additional file 3: Figure S3B, middle), and SMYD5 regions (see Additional file 3: Figure S3B, bottom), respectively. We also evaluated the percent

coverage of ERVK and L1 sequences for all SMYD5, H4K20me3, and H3K9me3 islands. Using this approach, we observed enrichment of ERVK and L1 sequences in H4K20me3, H3K9me3, and SMYD5 regions relative to random genomic regions (Fig. 5b), To gain further insight into which specific repeats are enriched in SMYD5, H4K20me3, and H3K9me3 regions, we evaluated the percentage of peaks that overlap repeat subfamilies within the LINE or LTR repeat class. Of the hundreds of annotated repeat subfamilies that we surveyed, only a few repeat subfamilies were found to be enriched within these regions (see Additional file 3: Figure S3C). While several L1Md repeats (L1Md_T, L1Md_A) are enriched

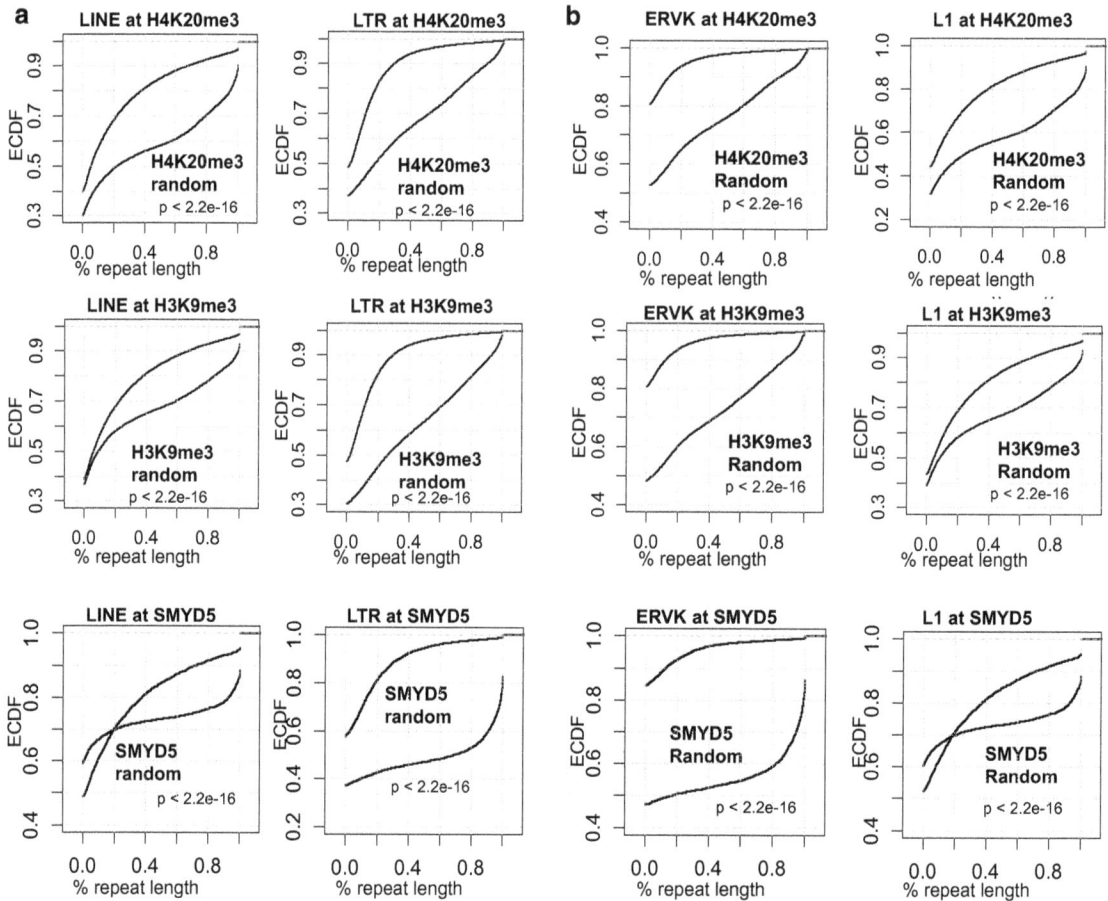

Fig. 5 SMYD5 and H4K20me3 occupy repetitive DNA elements in ES cells. Comparison of H4K20me3, H3K9me3, and SMYD5 enriched sequences and annotated repetitive sequences (http://www.repeatmasker.org). Empirical cumulative distribution (ECDF) for the percent coverage of a **a** LINE or LTR repeat class or **b** family member (L1 or ERVK) across all H4K20me3 (*top*) or H3K9me3 (*middle*) islands, or SMYD5 regions (*bottom*) relative to random genomic regions (*black*). Y-axis shows the percentage of genes that exhibit a percent repeat length less than the value specified by the x-axis. A *line* shifted to the right means a systematic increase in the percent coverage of a repeat element in ChIP-Seq peaks relative to random genomic sequences. p value for all <2.2e−16 (Kolmogorov–Smirnov test)

in H4K20me3, H3K9me3, and SMYD5 occupied regions, two repeats (L1Md_F and L1Md_F2) were only enriched in H4K20me3 and H3K9me3 marked regions (see Additional file 3: Figure S3C).

By clustering repeat subfamilies by their enrichment in SMYD5, H4K20me3, or H3K9me3 regions, our results further demonstrate that a few LINE/LTR repeat subfamilies are enriched within H4K20me3 (see Additional file 4: Figure S4A), H3K9me3 (see Additional file 4: Figure S4B), and SMYD5 regions (see Additional file 4: Figure S4C).

Knockdown of SMYD5 leads to decreased levels of H4K20me3, H3K9me3/2, and HP1α binding

To further directly test whether SMYD5 regulates H4K20me3 in ES cells, we investigated the global distribution of H4K20me3 in shSmyd5 ES cells using ChIP-Seq. Our results revealed that 12,358 islands showed a decrease in H4K20me3 levels in shSmyd5 ES cells (FDR < 0.001, fold-change >1.5) (Fig. 6a). A comparison of average H4K20me3 profiles around H4K20me3 peaks (Fig. 6b, left panel), and a boxplot (Fig. 6b, right panel), also showed global decreases in H4K20me3 levels in shSmyd5 ES cells compared with shLuc ES cells ($p < 2.2e−16$). Similarly, the levels of H4K20me3 at SMYD5-enriched regions also decreased in shSmyd5 ES cells ($p < 2.2e−16$) (Fig. 6c; see Additional file 5: Figure S5), consistent with a role for SMYD5 in regulating H4K20me3 on chromatin.

To determine whether both H4K20me3 and SMYD5 are simultaneously present at the same genomic regions, we performed Re-ChIP, also termed sequential ChIP [32], by immunoprecipitating ES cell chromatin first with an H4K20me3 or a FLAG antibody (for SMYD5), and second with a FLAG or H4K20me3 antibody, respectively. We then performed reChIP-PCR and found that each was significantly enriched relative to the control, Nanog promoter, which is not enriched with SMYD5 or H4K20me3 (see Additional file 6: Figure S6A). For example, H4K20me3 + SMYD5 reChIP levels were enriched >10–25 fold (relative to control; Nanog promoter) at H4K20me3/SMYD5 co-occupied regions, and SMYD5 + H4K20me3 reChIP levels were also elevated ~6–11 fold (relative to the control) at H4K20me3/SMYD5 marked regions (see Additional file 6: Figure S6B-C), demonstrating that a subset of H4K20me3 marked regions contain SMYD5 marks.

To test whether SMYD5-mediated H4K20me3 affects H3K9 methylation at heterochromatin regions, we evaluated the global distributions of H3K9me3 and H3K9me2 in *Smyd5* knockdown cells. These results show that levels decreased at a subset of H3K9me3 (Fig. 6d) and H3K9me2 islands (Fig. 6e) in shSmyd5 ES cells compared with shLuc ES cells. A comparison of average profiles and boxplots of H3K9me3 (Fig. 6f) or H3K9me2 (Fig. 6g) revealed decreased H3K9me3/2 enrichment, suggesting that the H4K20me3 signal may be important for the H3K9me3 modification in heterochromatin in ES cells. Our results show that 72% of H3K9me3 islands overlap with H4K20me3 islands, and 70% of SMYD5 islands overlap with H3K9me3 islands. In addition, 48% of SMYD5 islands exhibit decreased H3K9me3 levels (>1.5 fold-change, FDR < 0.001), while 36% of all H3K9me3 islands showed decreased levels in SMYD5-depleted ES cells. These results suggest that SMYD5-bound regions are more likely to exhibit decreased H3K9me3 levels relative to regions without SMYD5 binding in SMYD5-depleted cells.

Because H3K9 methylation is important for the recruitment of the heterochromatin protein, HP1 [33], we examined HP1α binding profiles in shSmyd5 ES cells using average profiles (Fig. 6h, left panel) and a boxplot (Fig. 6h, right panel). Our data indicate that HP1α levels decrease in shSmyd5 ES cells, suggesting that depletion

of SMYD5 leads to decreased heterochromatin. To test whether decreases in H3K9me3/2 and HP1α correlate with decreases in H4K20me3 levels, we compared their changes at H4K20me3 islands relative to input chromatin. These results show that H3K9me3/2 (Fig. 6i) and HP1α levels (Fig. 6j) decrease at H4K20me3 islands in shSmyd5 ES cells. We also surveyed H4K20me3 levels at H3K9me3 marked regions in SMYD5-depleted ES cells and observed greater decreases in H4K20me3 at regions where H3K9me3 decreased relative to regions where H3K9me3 levels did not change significantly ($p < 2.2e-16$) (Fig. 6k), suggesting that decreases in H4K20me3 are correlated with decreased H3K9me3 levels.

To investigate how SMYD5-mediated decreases in H4K20me3 affect H3K9me3 levels, we hypothesized that H3K9 methyltransferases may bind H4K20me3 marks. However, while we observed binding of G9a, HP1α, and ESET to H3K9me3/2 using an in vitro pull-down assay with biotinylated histone H4 and H3 peptides and nuclear extracts from ES cells, we did not observe binding to H4K20me3/2/1 (see Additional file 7: Figure S7A). To investigate whether SMYD5 directly interacts with heterochromatin constituents HP1α and G9a, we immunoprecipitated SMYD5 and performed immunoblotting using anti-HP1α, anti-G9a, and anti-SMYD5 antibodies. Our results show that SMYD5 binds to HP1α and G9a (Fig. 6l). Moreover, using ChIP-Seq analysis to survey global levels of G9a, we found that G9a levels decrease in shSmyd5 ES cells relative to shLuc ES cells (Fig. 6m). Our results also show that G9a occupies 40% of SMYD5 islands. Overall, these results suggest that SMYD5 interacts with heterochromatin proteins HP1α and G9a, and depletion of SMYD5 leads to decreased binding of HP1α and G9a. In addition, inspection of custom tracks on the UCSC genome browser revealed decreased levels of H4K20me3 and H3K9me3 at several SMYD5 occupied regions (Fig. 6n).

To investigate whether enrichment of repressive histone marks and associated chromatin factors decreases

(See figure on next page.)
Fig. 6 Depletion of SMYD5 leads to decreased H4K20me3, H3K9me3/2, and HP1α binding. **a** Change in the global distribution of H4K20me3 in shSmyd5 ES cells. **b** Average profile and *boxplot* of H4K20me3 ChIP-Seq tag density in shSmyd5 ES cells. **c** Average profiles of H4K20me3 at SMYD5-enriched regions in shLuc and shSmyd5 ES cells. **d**, **e** Changes in global distributions of **d** H3K9me3 (**e**) and H3K9me2 in shSmyd5 ES cells relative to shLuc ES cells. **f–h** Average profiles and *boxplot* of **f** H3K9me3, **g** H3K9me2, and **h** HP1α densities in shLuc and shSmyd5 ES cells. **i** Boxplot depicting density of H4K20me3, H3K9me3, and H3K9me2 at H4K20me3 islands. **j** Empirical cumulative distribution (ECDF) for the fold-change in density of HP1α in shSmyd5 ES cells. The *red line* shifted to the *left* of the input (*gray*) shows a systematic decrease in enrichment in shSmyd5 ES cells. *Boxplot* below shows HP1α density (log2 fold-change vs. input) in shLuc and shSmyd5 ES cells. **k** H4K20me3 density at regions at regions with decreased or unaltered H3K9me3 levels. **l** SMYD5 associates with HP1α and G9a. FLBIO-SMYD5 (biotinylated SMYD5 + BirA) or BirA (control) ES cells were used to immunoprecipitate SMYD5 protein with avidin-agarose beads. Immunoprecipitates were analyzed by immunoblotting with anti-HP1α, anti-G9a, and anti-SMYD5 antibodies. **m** Changes in the global distribution of G9a in shSmyd5 ES cells relative to shLuc ES cells. **n** Altered profiles of H4K20me3 at SMYD5-enriched regions (H2-Q1, H2-Q7) in shLuc and shSmyd5 ES cells

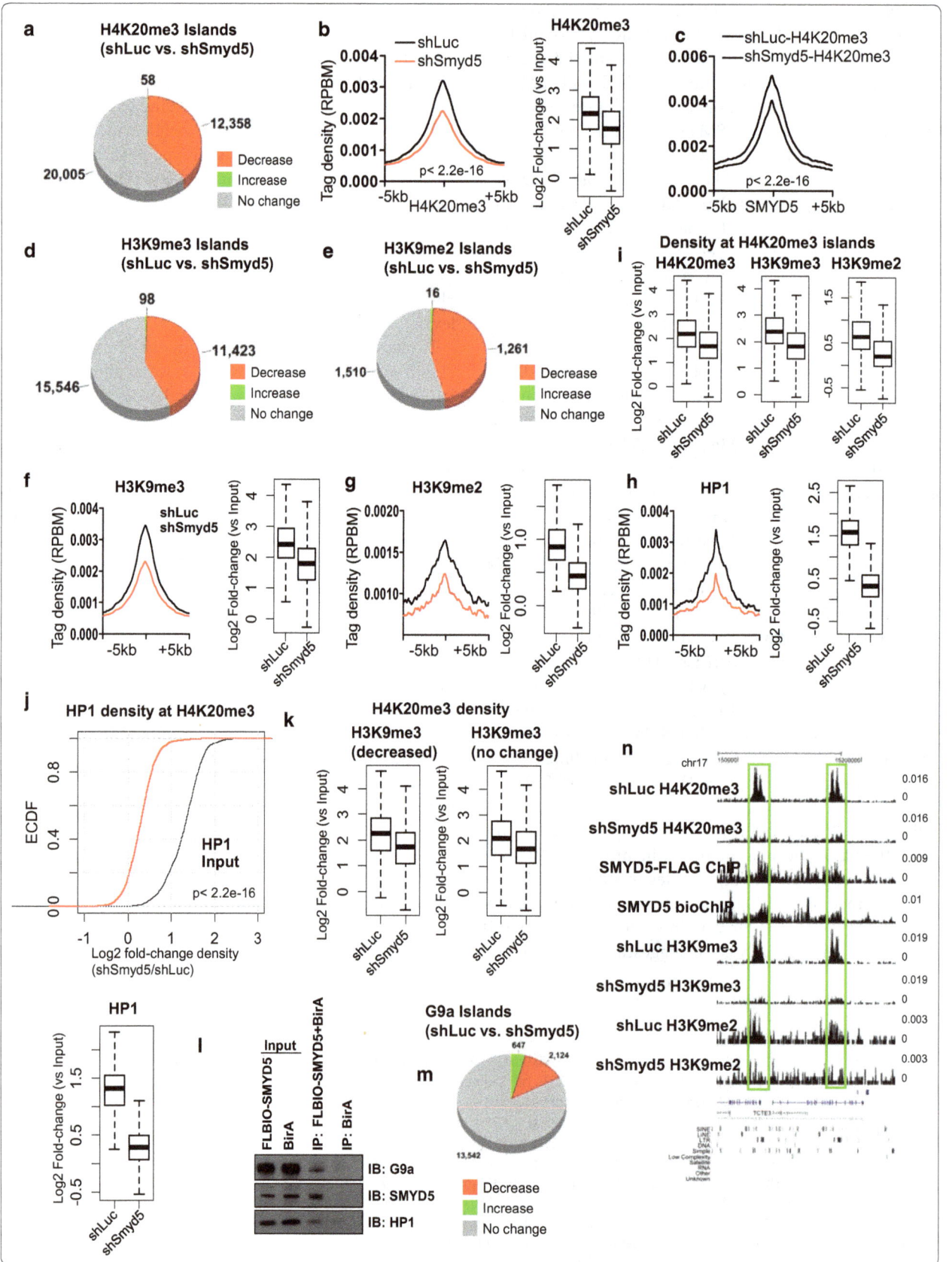

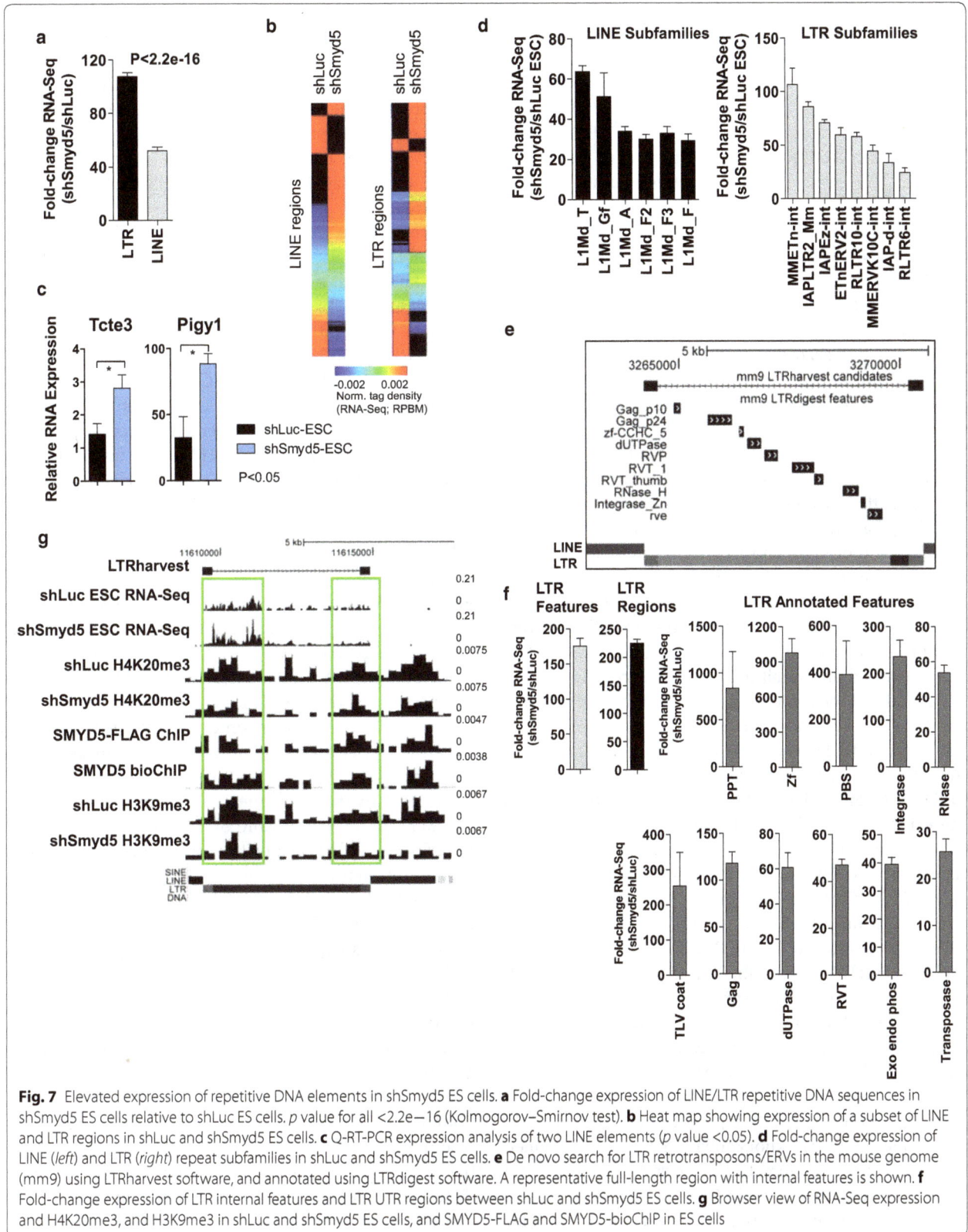

Fig. 7 Elevated expression of repetitive DNA elements in shSmyd5 ES cells. **a** Fold-change expression of LINE/LTR repetitive DNA sequences in shSmyd5 ES cells relative to shLuc ES cells. p value for all $<2.2e-16$ (Kolmogorov–Smirnov test). **b** Heat map showing expression of a subset of LINE and LTR regions in shLuc and shSmyd5 ES cells. **c** Q-RT-PCR expression analysis of two LINE elements (p value <0.05). **d** Fold-change expression of LINE (*left*) and LTR (*right*) repeat subfamilies in shLuc and shSmyd5 ES cells. **e** De novo search for LTR retrotransposons/ERVs in the mouse genome (mm9) using LTRharvest software, and annotated using LTRdigest software. A representative full-length region with internal features is shown. **f** Fold-change expression of LTR internal features and LTR UTR regions between shLuc and shSmyd5 ES cells. **g** Browser view of RNA-Seq expression and H4K20me3, and H3K9me3 in shLuc and shSmyd5 ES cells, and SMYD5-FLAG and SMYD5-bioChIP in ES cells

at LINE and LTR regions in the absence of SMYD5, we compared the densities of H4K20me3, H3K9me3, and HP1α at LINE (see Additional file 7: Figure S7B) and LTR (see Additional file 7: Figure S7C) regions. These results demonstrate that H4K20me3, H3K9me3, and HP1α levels decrease at LINE (see Additional file 7: Figure S7B) and LTR (see Additional file 7: Figure S7C) regions in shSmyd5 ES cells. We also observed enrichment of SMYD5 at LINE (see Additional file 7: Figure S7B) and LTR (see Additional file 7: Figure S7C) regions relative to Input DNA in ES cells. Overall, these results suggest that SMYD5 occupies LINE and LTR repetitive DNA elements and catalyzes H4K20me3 modifications, which are important for H3K9me3 modifications and HP1α binding in the heterochromatic regions.

Increased expression of repetitive DNA in SMYD5 depleted ES cells

To investigate whether SMYD5-mediated H4K20me3 plays a role in silencing of LINE, and LTR elements, we evaluated their expression in shLuc and shSmyd5 ES cells (Fig. 7a). These results revealed a marked increase in expression of LINE and LTR elements in shSmyd5 ES cells (Fig. 7a; $p < 2.2e-16$). We then applied a stringent filter to identify LINE and LTR regions (>0.003 RPBM tag density per site) and evaluated the RNA-Seq expression profile of this subset of LINE and LTR regions. Heat maps revealed an overall increase in expression of both LINE and LTR elements (Fig. 7b), suggesting that a loss of SMYD5 leads to global increases in expression of the LINE and LTR repetitive sequences. We confirmed the increased expression of several LINE regions, using Q-RT-PCR (Fig. 7c) and RNA-Seq. We also investigated the expression of LINE and LTR repeats in shLuc and shSmyd5 ES cells and observed de-repression of LTR repeat subfamilies, which are enriched in H4K20me3- and SMYD5-bound regions, including MMETn, IAPLTR2_Mm, IAPEz-int, ETnERV2-int, RLTR10-int, MMERK10C-int, IAP-d-int, RLTR6-int (MMETn, 107-fold increase; IAPLTR2_Mm, 86-fold increase; IAPEz-int, 71-fold increase), and de-repression of LINE repeat subfamilies, including L1Md_T, L1Md_Gf, L1Md_A, L1Md_F2, L1Md_F3, and L1Md_F in SMYD5-depleted ES cells (L1Md_T, 79-fold increase; L1Md_Gf, 51-fold increase; L1Md_T, 79-fold increase) (Fig. 7d).

Because we observed de-repression of LTR and LINE subfamilies in *Smyd5* knockdown ES cells, we reasoned that full-length, or intact, LTR retrotransposons and ERVs may be de-repressed in SMYD5-depleted ES cells. To test this possibility, we performed a de novo search for full-length LTR retrotransposons and ERVs in the mouse genome using LTRharvest software [34], which provides annotations of known LTR features. Using this approach,

we identified 11,394 full-length LTR retrotransposons/ERVs in the mouse genome. We then annotated these LTR regions using LTRdigest software [35] and identified 20,852 internal features including sequences encoding viral proteins such as gag and pol (Fig. 7e). An evaluation of the expression state of these LTR features revealed an overall increase in the expression of full-length (intact) LTR/ERV annotated features in SMYD5-depleted ES cells (Fig. 7f).

To investigate a relationship between the de-repression of LTRs/ERV regions and occupancy of SMYD5/H4K20me3, we evaluated the overlap between LTR regions and SMYD5/H4K20me3 occupancy. Using this method, we found 872 regions which were occupied by SMYD5/H4K20me3 and contained LTR/ERV sequences. An examination of custom UCSC genome browser tracks revealed SMYD5 binding at a representative region containing LTR and LINE elements accompanied by decreased levels of H4K20me3, H3K9me3, and H3K9me2 in shSmyd5 ES cells relative to shLuc ES cells (Fig. 7g; see Additional file 8: Figure S8).

To investigate whether loss of SMYD5-dependent silencing of LTR/ERV elements leads to upregulated expression of nearby genes, which was observed in ESET/Setdb1 knockout ES cells [36], we first evaluated the number of upregulated genes in SMYD5 knockdown ES cells that contain LTR/ERV sequences within 10 kb of their TSS (see Additional file 9: Figure S9A-B). Annotation of these LTR/ERV elements revealed that they mainly reside in intronic and intergenic regions (see Additional file 9: Figure S9C). We then evaluated the expression state of LTR/ERV elements nearby differentially expressed genes. These results revealed an increase in expression of LTR/ERV elements in SMYD5 knockdown ES cells relative to control ES cells (see Additional file 9: Figure S9D). Moreover, we also observed decreased H4K20me3 levels at nearby islands in SMYD5 knockdown ES cells (see Additional file 9: Figure S9E), suggesting that SMYD5-dependent control of H4K20me3 supports the repression of LTR/ERV elements of nearby genes. We then investigated whether upregulated genes in shSmyd5 ES cells contain SMYD5 binding and LTR/LINE elements within 10 kb of their TSS are lineage-specific. Indeed, our results show that expression of lineage-specific genes bound by SMYD5 and containing LTR/LINE elements is upregulated in shSmyd5 ES cells (see Additional file 9: Figure S9F). These results suggest that SMYD5-dependent silencing of LTR/LINE elements represses expression of lineage-specific genes in ES cells. Overall, these results demonstrate that SMYD5 influences gene expression of nearby genes by silencing LTR/ERV elements.

Discussion

SMYD5 is important for ES cell self-renewal and differentiation

ES cell self-renewal is governed by networks of transcription factors, including OCT4, SOX2, NANOG, and TBX3 [27, 37, 38], and epigenetic regulators such as BRG1 [39, 40] and KDM5B [20, 41] that participate in regulating transcription of genes that promote self-renewal while repressing developmental genes. Disruption of these factors abrogates self-renewal leading to specific or mixed-lineage differentiation. While many studies have focused on the roles of chromatin modifying enzymes that regulate active marks such as H3K4 methylation [20, 41–43], or repressive histone marks such as H3K27 or H3K9 methylation [44–48], few regulators of the repressive histone mark H4K20me3 have been shown to be important for mouse development [10, 13], and none have been shown to be important for ES cell self-renewal. In this study, we have provided evidence that SMYD5, which mediates H4K20me3 marks, is a critical regulator of ES cell function. We found that knockdown of *Smyd5* resulted in decreased ES cell colony integrity and decreased expression of pluripotency regulators such as *Oct4*, *Nanog*, and *Tbx3*, demonstrating that depletion of SMYD5 leads to compromised self-renewal. However, modulation of OCT4 levels did not diminish the impact of depleting SMYD5 in ES cells and during differentiation.

We also observed perturbed differentiation of SMYD5-depleted ES cells, where a loss of SMYD5 resulted in abnormal EB differentiation including the formation of complex structures containing circular bulges lined with a PE, and expression of endodermal genes such as *Sox17*. The differential formation of endoderm between control and shSmyd5 cells was also visible in teratomas. Moreover, our results describing an important role for the H4K20 histone methyltransferase, SMYD5, in ES cell differentiation is in alignment with a previous study which demonstrated that depletion of Suv420h1/h2 histone methyltransferases leads to compromised differentiation of ES cells [49]. Combined, these findings suggest that H4K20 HMTases are important for ES cells differentiation.

SMYD5 regulates H4K20me3 at repetitive DNA elements in ES cells

Our results support a role for SMYD5 in regulating H4K20me3 in ES cells. These results are in alignment with a previous study which implicated a role for SMYD5 as a methyltransferase that deposits H4K20me3 marks in *Drosophila* and in macrophages [19]. We found that SMYD5 binds H4K20me3-enriched regions

and depletion of SMYD5 results in global decreases in H4K20me3 as evaluated by western blotting and ChIP-Seq. These results argue for a critical role of SMYD5 in regulating H4K20me3. Interestingly, our data indicated that depletion of SMYD5 also decreased levels of H3K9me3/2, G9a, and HP1α. Because H4K20me3 is known to co-localize with H3K9 methylation at heterochromatic regions [7, 50] and H3K9me3 is important for recruitment of HP1 and heterochromatin formation [51–54], it is plausible that a loss of SMYD5 and H4K20me3 may lead to decreased heterochromatin through delocalization of H3K9me3 and HP1. Along this line, HP1 isoforms have been shown to recruit Suv420h1/2, which also induce H4K20 methylation [17], suggesting that interplay between H4K20 methyltransferases, histone modifications, and HP1 proteins regulates heterochromatin. Our results showing that SMYD5 interacts with HP1α and G9a, and depletion of SMYD5 leads to decreased HP1α and G9a binding, and H3K9me3/2 levels, is in alignment with this model. Decreased H3K9me3/2 levels may be due to a disrupted interaction between SMYD5 with G9a in SMYD5-depleted ES cells, as G9a deposits H3K9 methylation marks and is involved in regulating H3K9me3 levels in vivo [55]. It is also possible that H4K20me3 may interact with Suv39h1 or Suv39h2 histone methyltransferases, which deposit H3K9 methylation. In this case, a disrupted interaction between H4K20me3 and Suv39 h enzymes may lead to decreased H3K9 methylation levels. While we observed decreased H3K9me3 at a subset of regions (40%) in SMYD5-depleted ES cells, the majority of H3K9me3 marked-regions (60%) were unaltered (Fig. 6d), suggesting that H3K9 methylation levels change at a subset of regions in SMYD5-depleted ES cells. We also observed decreased G9a levels at a subset of islands (14%) in SMYD5-depleted ES cells (Fig. 6m), and occupancy of G9a at a subset (40%) of SMYD5 islands.

Disruption of repressive chromatin constituents of heterochromatin may trigger localized decondensation of chromatin, thus leading to de-repressed transcription of the underlying DNA. Consistent with this possibility, we observed that a decrease in H4K20me3 at LINE and LTR repetitive DNA regions by depletion of SMYD5 was accompanied by decreased levels of the heterochromatin mark H3K9me3 and increased expression of LINE and LTR repetitive DNA elements. Moreover, in addition to observing a redistribution of H3K9me3 levels in shSmyd5 ES cells using ChIP-Seq, we performed H3K9me3 immunofluorescence analysis using shLuc and shSmyd5 ES cells (see Additional file 10: Figure S10A, B) and observed decreased H3K9me3 heterochromatin foci in shSmyd5 ES cells (see Additional file 10: Figure S10C), further suggesting that depletion of SMYD5 leads to a relaxed chromatin state. While we observed a co-occurrence of the repressive histone modifications H4K20me3 and H3K9me3 at LINE and LTR repetitive elements, the role for multiple heterochromatin-associated histone modifications at repetitive genomic regions is not fully known. A possible explanation for the co-occurrence is that H4K20me3 and H3K9me3 may serve as redundant markers to facilitate chromatin compaction and maintenance of heterochromatin. Another explanation is that H4K20me3 and H3K9me3 may interact with a broader set of repressors compared with H4K20me3 or H3K9me3 alone. As such, combinatorial marking by H4K20me3 and H3K9me3 may provide greater repressive abilities relative to H4K20me3 or H3K9me3. Moreover, H4K20me3 and H3K9me3 may facilitate interactions between histone modifying enzymes and heterochromatin constituents. Along this line, the H3K9 methyltransferase ESET/Setdb1 has been shown to interact with multiple repressors, including KAP1 and HP1, KAP1 has been shown to interact with ESET/Setdb1 and HP1 [56], G9a has been shown to interact with HP1, and our results demonstrate that the H4K20me3 methyltransferase SMYD5 interacts with G9a.

Conclusions

Results presented here describe a role for SMYD5 in regulating ES cell maintenance by silencing differentiation genes. Our model suggests that repetitive DNA elements recruit SMYD5 to the vicinity of differentiation genes, thus keeping them silenced. Depletion of SMYD5 relieves the silencing of these genes and thus induces differentiation.

Methods

ES cell culture

R1 ES cells were cultured as previously described with minor modifications [20, 41]. Briefly, R1 ES cells were cultured on irradiated MEFs in DMEM, 15% FBS media containing LIF (ESGRO) at 37 °C with 5% CO_2. For ChIP experiments, ES cells were cultured on gelatin-coated dishes in ES cell media containing 1.5 μM CHIR9901 (GSK3 inhibitor) for several passages to remove feeder cells. ES cells were passed by washing with PBS using serological pipets (sc-200278, sc-200280) and dissociating with trypsin. For self-renewal experiments in the absence of LIF, ES cells were cultured on gelatin-coated dishes in ES cell media without LIF and without feeders. For embryoid body (EB) formation, ES cells were cultured in low-attachment binding dishes to promote 3D formation in ES cell media without LIF. Alkaline phosphatase staining was performed using a kit from Millipore according to the manufacturer's instructions.

Establishment of SMYD5 expressing ES cells

R1 ES cells were nucleofected with the pEF1α-BirAV5-neo plasmid and stably selected in the presence of 300 μg/mL G418 for at least 5–7 days. Individual ES cell colonies were picked and screened for BirA expression using western blotting. An ES cell clone expressing high levels of BirA was used for the subsequent experiments. Next, Smyd5 cDNA was amplified from ES cell cDNA and cloned into the pEF1α-FLBIO-puro vector using the BamHI and XbaI sites. BirA ES cells were nucleofected with the pEF1α-FLBIO-Smyd5-puro plasmid and stably selected in the presence of 1 μg/mL puromycin and 200 μg/mL G418. Individual ES cells clones were picked and screened for SMYD5 expression using an anti-FLAG antibody and western blotting. BirA ES cells were used as a negative control for immunoprecipitation and western blotting experiments. For generation of ES cells overexpressing wild-type or mutant (H315L and C317A) SMYD5, SMYD5 was PCR-amplified from ES cell cDNA and cloned into the pCDH-neo lentiviral vector (System Biosciences).

Lentiviral infection

ES cells were transduced with lentiviral particles encoding shRNAs as described previously [20, 41]. Briefly, shRNA template DNA oligos were annealed and double-stranded shRNA templates were cloned into the BamH1/EcoRI digested pGreenPuro Vector (System Biosciences) according to the manufacture's protocol. To generate lentiviral particles, 293T cells were co-transfected with an envelope plasmid (pLP/VSVG), packaging vector (psPAX2), and an shRNA (shLuc or shSmyd5) or cDNA expression vector (SMYD5) using lipofectamine 2000. Twenty-four to 48 h posttransfection, the medium containing lentiviral particles was harvested, filtered, and used to infect ES cells. Twenty-four hours post-transduction, ES cells were stably selected in the presence of 1 μg/mL puromycin.

Teratoma and tumor formation

Teratoma formation was performed as previously described [20]. Briefly, ES cells were cultured on gelatin-coated dishes to remove feeder cells, dissociated into single cells, and 10^6 ES cells were injected subcutaneously into immunocompromised SCID–beige mice. After three to four weeks, mice were euthanized and teratomas were washed and fixed in 10% buffered formalin. Teratomas were then embedded in paraffin. Thin sections were cut and stained with hematoxylin and eosin (H&E) using standard techniques. All animals were treated in accordance with Institution Animal Care and Use Committee guidelines under current approved protocols at NHLBI.

Q-RT-PCR expression analysis

RNA isolation and Q-RT-PCR were performed as previously described with minor modifications [20]. Briefly, RNA isolation and Q-RT-PCR were performed as previously described with minor modifications. Total RNA was harvested from ES cells using an RNeasy Mini Kit or miRNeasy Mini Kit (Qiagen, Valencia, CA) and DNase treated using Turbo DNA-free (Ambion). Reverse transcription was performed using a Superscript III kit (Invitrogen, Carlsbad, CA). Q-RT-PCR was performed using TaqMan probes, or custom FAM-labeled probes, and primers and TaqMan Universal PCR Master Mix reagents (Applied Biosystems). Primers used for Q-RT-PCR with Roche Universal probes were designed using the Universal Probe Library Assay design Center (Roche).

Immunoflourescence analysis

ES cells were fixed with 4% paraformaldehyde for 15 min at room temperature, washed with 0.1% Triton X-100 (Sigma), and blocked in 1% BSA/0.01% Tween-20 for 30 min. Fixed cells were incubated with an anti-H3K9me3 antibody (ab8898) overnight at 4 °C in blocking buffer. The next day, the cells were washed with blocking buffer, and incubated with DAPI in 0.1% Triton X-100, washed with blocking buffer, and mounted in ProLong Gold antifade reagent (Invitrogen).

ChIP-Seq

ChIP-Seq experiments were performed as previously described with minor modifications [20, 41, 57]. The H4K20me3 antibody (07-463) and the HP1α antibody were obtained from Millipore. The polyclonal H4K20me2 (ab9052), H4K20me1 (ab9051), H3K9me3 (ab8898), and H3K9me2 (ab1220) antibodies were obtained from Abcam. For SMYD5-FLAG ChIP-Seq, the monoclonal anti-FLAG antibody was obtained from Sigma. For SMYD5-bioChIP-Seq, streptavidin (SA) beads were obtained from Invitrogen.

Briefly, ES cells were harvested and chemically cross-linked with 1% formaldehyde (Sigma) for 5–10 min at 37 °C and subsequently sonicated. Sonicated cell extracts were used for ChIP assays. ChIP-enriched DNA was end-repaired using the End-It DNA End-Repair kit (Epicentre), followed by addition of a single A nucleotide, and ligation of PE adapters (Illumina) or custom-indexed adapters. PCR was performed using Phusion High-Fidelity PCR master mix. ChIP libraries were sequenced on an Illumina HiSeq platform according to the manufacture's protocol.

Sequence reads were mapped to the mouse genome (mm9) using Bowtie2 [58]. To allow mapping to repetitive elements, we used the default mode of Bowtie2, which searches for multiple alignments, and reports

the best one based on the alignment score (MAPQ) (http://bowtie-bio.sourceforge.net/bowtie2/manual. shtml).

ChIP-Seq read-enriched regions were identified relative to Input DNA (sonicated chromatin) as previously described with minor modifications [59, 60]. Briefly, ChIP-Seq read-enriched regions (peaks) were identified relative to Input DNA using "Spatial Clustering for Identification of ChIP-Enriched Regions" (SICER) software [60] with a window size setting of 200 bps, a gap setting of 400 bps, and a FDR setting of 0.001. For a comparison of ChIP-enrichment between samples, a fold-change threshold of 1.5 and an FDR setting of 0.001 were used. For transcription factors (see Fig. 3g), the ChIP-Seq read-enriched peaks were called by MACS [61] with a p value setting of 0.00001. The RPBM measure (read per base per million reads) was used to quantify the density of histone modification, SMYD5 binding, and Input DNA at genomic regions from ChIP-Seq datasets. We have also applied the Kolmogorov–Smirnov test to obtain p value statistics and compare densities at genomic regions.

reChIP

reChIP, also termed sequential ChIP, was performed as previously described with minor modifications [32]. Cross-linked chromatin from ES cells was immunoprecipitated with antibodies against either H4K20me3 or FLAG (for SMYD5) as described above (see "ChIP-Seq"), except that chromatin was eluted in a TE solution containing 20 mM DTT, 500 mM NaCL, and 1% SDS at 37° for 20 min. The eluted DNA was diluted 50-fold, and a second round of immunoprecipitations was performed against the FLAG or H4K20me3 antibody as described above. PCR primers for evaluating reChIP were designed from the indicated genomic regions. Real-time PCR was performed using an Applied Biosystems OneStepPlus machine. For reChIP, 1 μL of reChIP DNA or 1 μL of Input DNA was used as a template, and relative enrichment was determined from a standard curve for each primer using a standard curve of Input DNA, and using the Nanog promoter (which does not contain enrichment of H4K20me3 or SMYD5) as a normalizer.

RNA-Seq analysis

RNA was harvested from ES cells and EBs as described above. RNA-Seq was performed as previously described [20, 41]. RNA was harvested from ES cells and EBs as described above. mRNA was purified using a Dynabeads mRNA purification kit (Invitrogen). Double-stranded cDNA was generated using a SuperScript Double-Stranded cDNA synthesis kit (Invitrogen). cDNA was end-repaired using the End-It DNA End-Repair kit (Epicentre), followed by addition of a single A nucleotide, and ligation of PE adapters (Illumina) or custom-indexed adapters. PCR was performed using Phusion High-Fidelity PCR master mix. RNA-Seq libraries were sequenced on Illumina GAIIX or HiSeq platforms according to the manufacture's protocol.

The RPKM measure (reads per kilo bases of exon model per million reads) proposed previously [62] was used to quantify the mRNA expression level of a gene from RNA-Seq datasets. Differentially expressed genes were identified using EdgeR (FDR < 0.001 and FC > 2) [63]. Genes with RPKM < 3 in both conditions in comparison were excluded from this analysis.

Prediction of differentially expressed genes due to chance or accelerated differentiation

Expression changes that indicate accelerated differentiation include genes that show accelerated downregulation or upregulation during EB differentiation. The percentage of genes that lag behind during differentiation of shSmyd5 ES cells is less than expected. Each bar represents a group of genes upregulated by at least alpha-fold (X axis) from ESC (0 h) to EB day 6 in the control cells. The percentage of genes with expression values that follow the order: EB day 6 (shLuc) > EB day 6 (shSmyd5) > ES cell is calculated (Observed; red bars); error bars are generated by bootstrapping. The expression values of all genes are randomly shuffled independently for EB day 6 (shLuc), EB day 6 (shSmyd5), and ES cells and are repeated many times to give the means and standard deviations for the expectations (Expected; blue bars). The red bars represent observed data.

Annotation of repetitive DNA sequences

Repetitive DNA sequence classes (e.g., LINE, LTR), families (L1, ERVK), and names (e.g., L1Md_T, IAPLTR1) for the mm9 reference genome were defined according to the annotations provided by the UCSC genome browser and RepeatMasker (http://www.repeatmasker.org), which uses curated libraries of repeats such as Repbase (http://www.girinst.org/repbase/).

In vitro pull-down assay

Nuclear extracts were prepared from ES cells using a standard high salt extraction protocol [64]. Briefly, cells were lysed by Dounce homogenizing in buffer A, washed, and nuclear proteins were extracted with buffer C. The salt concentration was diluted as described [65], and incubated with histone peptides (unmodified or modified) prebound to avidin beads overnight at 4 °C. Beads were washed, eluted, and analyzed by SDS-PAGE.

Additional files

Additional file 1: Figure S1. Depletion of SMYD5 leads to decreased self-renewal and altered differentiation. (**A**) RNA-Seq data of Smyd5 expression in ES cells and day 14 differentiated embryoid bodies (EBs; log2 RPKM). (**B**) Q-RT-PCR analysis of *Smyd5* expression in control (shLuc) ES cells and SMYD5 shRNA knockdown ES cells (shSmyd5-1, shSmyd5-2, and shSmyd5-3). (**C**) Bright-field microscopy of ES cells infected with shLuc (control) or shSmyd5 lentiviral particles (shSmyd5-1, shSmyd5-2, and shSmyd5-3) and stably selected with puromycin. (**D**) Q-RT-PCR expression analysis during shLuc and shSmyd5 EB differentiation.

Additional file 2: Figure S2. SMYD5 and H4K20me3 co-occupancy in ES cells. (**A**) Venn diagram showing comparison of SMYD5-FLAG and SMYD5-bioChIP ChIP-enriched peaks. (**B**) Heat map of SMYD5-FLAG and FLAG-bioChIP ChIP-Seq densities at FLAG-bioChIP intersecting regions. (**C**) Average profiles of SMYD5-bioChIP and SMYD5-FLAG density at FLAG-bioChIP intersecting enriched regions. (**D**) Western blot of H4K20me3 in shLuc and shSmyd5 ES cells.

Additional file 3: Figure S3. SMYD5 and H3K9me3 are enriched at LINE and LTR repeats in ES cells. (**A**) Repetitive DNA sequences (LINE and LTR) are enriched in H4K20me3, H3K9me3 and SMYD5 genomic sites. Comparison of H4K20me3, H3K9me3, and SMYD5 enriched sequences and annotated repetitive sequences (http://www.repeatmasker.org). The percentage of ChIP-enriched regions with at least 60% repeat length is shown. Note the predominance of LTR and LINE repetitive DNA sequences in ChIP-enriched islands. (**B**) Repeat subfamilies belonging to the LINE and LTR repetitive DNA sequence classes are enriched in H4K20me3, H3K9me3 and SMYD5 genomic sites. The percentage of ChIP-enriched regions with at least 60% repeat length is shown. (**C**) Repetitive DNA sequence family members L1 and ERVK are enriched in H4K20me3, H3K9me3 and SMYD5 genomic regions.

Additional file 4: Figure S4. Enrichment of LINE and LTR repeat subfamilies at SMYD5, H4K20me3, and H3K9me3 occupied regions. (**A, C**) Hierarchical clustering heat map showing the percent coverage of repeat subfamilies belonging to the LINE and LTR class within (**A**) H4K20me3 (**B**) H3K9me3, and (**C**) SMYD5 ChIP-enriched regions. Red indicates an elevated percent coverage of a repeat element. The X axis shows the H4K20me3, H3K9me3, and SMYD5 ChIP-peaks while the Y axis shows the LINE or LTR element name.

Additional file 5: Figure S5. H4K20me3 density at SMYD5 bound regions. (**A**) Empirical cumulative distribution for the density of H4K20me3 at SMYD5-enriched regions in shLuc and shSmyd5 ES cells. The boxplot shows the density of H4K20me3 at SMYD5-enriched regions (log2 fold-change vs. Input) in shLuc and shSmyd5 ES cells.

Additional file 6: Figure S6. Re-ChIP Validation of H4K20me3 and SMYD5 co-occupancy in ES cells. (**A**) Real-time PCR depicting the relative enrichment of H4K20me3/SMYD5 co-occupied or control (Nanog promoter) genomic sites after sequential immunoprecipitations with an anti-FLAG antibody (for SMYD5) and then an anti-H4K20me3 antibody, or an anti-H4K20me3 antibody and then an anti-FLAG antibody. (**B**) Fold-change enrichment of H4K20me3/SMYD5 co-occupied genomic sites relative to the control (Nanog promoter) site. Region #10 is also depicted in Fig. 7g. (**C**) Annotation of H4K20me3/SMYD5 regions using HOMER software [69].

Additional file 7: Figure S7. Depletion of SMYD5 leads to decreased H4K20me3, H3K9me3, and HP1α at LINE/LTR repeats. (**A**) Peptide pull-down assays using ES cell nuclear extracts and unmodified or modified H4/H3 peptides were performed and analyzed by immunoblotting with anti-HP1α, anti-G9a, and anti-ESET antibodies. (**B**) Empirical cumulative distribution for the density of H4K20me3 (top left panel), H3K9me3 (bottom left panel), HP1α (bottom right panel) at LINE regions in shLuc and shSmyd5 ES cells, and SMYD5-FLAG in ES cells (top right panel). (**C**) Empirical cumulative distribution for the density of H4K20me3 (top left panel), H3K9me3 (bottom left panel), HP1α (bottom right panel) at LTR regions in shLuc and shSmyd5 ES cells, and SMYD5-FLAG in ES cells (top right panel).

Additional file 8: Figure S8. Elevated expression of repetitive DNA elements in SMYD5 knockdown ES cells. Browser view of RNA-Seq expression and H4K20me3, and H3K9me3 in shLuc and shSmyd5 ES cells, and SMYD5-FLAG and SMYD5-bioChIP in ES cells. The green box highlights a genomic region enriched with H4K20me3 and other histone modifications in control (shLuc) ES cells.

Additional file 9: Figure S9. Upregulated genes in SMYD5 knockdown cells associated with LTR/ERV elements and decreased H4K20me3. Loss of SMYD5-dependent silencing of LTR/ERV elements influences the expression of nearby genes. **(A)** Number of differentially expressed (DE) genes between shLuc and shSmyd5 ES cells (fold-change >1.5, p value <0.05). **(B)** Expression of upregulated genes between shLuc and shSmyd5 ES cells ($p = 5.528e-13$) (log2 RPKM). **(C)** Annotation of LTR/ERV elements nearby DE genes in shLuc and shSmyd5 ES using HOMER software [69]. **(D)** Fold-change expression of LTR/ERV elements at DE genes (A-C) relative to total mRNA in shSmyd5 ES cells relative to shLuc ES cells. **(E)** Density of H4K20me3 marks nearby LTR/ERV element and within 10 kb of TSS of DE genes. **(F)** Mouse gene atlas expression analysis evaluated using Network2Canvas [66] demonstrates that lineage and ES cell genes are misexpressed in shSmyd5 ES cells. Each node (square) represents a gene list (shLuc vs shSmyd5 DE genes bound by SMYD5 and containing LTR/LINE element) associated with a gene-set library (mouse gene atlas). The brightness (white) of each node is determined by its p value.

Additional file 10: Figure S10. Decreased H3K9me3 heterochromatin foci in SMYD5-depleted ES cells. **(A, B)** Immunofluorescence staining of H3K9me3 in **(A)** shLuc and **(B)** shSmyd5 ES cells. Nuclei were stained with DAPI. **(C)** Quantitation of nuclear H3K9me3 heterochromatin foci using ImageJ software.

Abbreviations
ES cells: embryonic stem cells; H4K20me3: trimethylated histone 4 lysine 20; H3K9me3: trimethylated histone 3 lysine 9; SMYD5: set and mynd domain 5; LTR: long terminal repeat; LINE: long interspersed nuclear element; shRNA: short hairpin RNA; ERV: endogenous retrovirus; TF: transcription factor; HP1: heterochromatin protein 1; RNA-Seq: RNA sequencing; WT: wild-type; 3D: three dimensional; AP: alkaline phosphatase; DE: differentially expressed; Q-RT-PCR: quantitative real-time PCR; GSEA: gene set enrichment analysis; GO: gene ontology; LIF: leukemia inhibitory factor; EB: embryoid body; PE: primitive endoderm; ICM: inner cell mass; H&E: hematoxylin and eosin; PCA: principle component analysis; bioChIP: biotin-mediated ChIP; SICER: spatial clustering for identification of ChIP-enriched regions; ChIP-Seq: chromatin immunoprecipitation sequencing; reChIP: sequential ChIP; kb: kilobase; TSS: transcription start site; SCID: sever combined immunodeficiency; FDR: false discovery rate; MACS: model-based analysis for ChIP-Seq; RPKM: reads per kilo bases of exon model per million reads; RPBM: read per base per million reads.

Authors' contributions
BLK and KZ conceived the project. BLK performed the experiments and analyzed the data. GH contributed to the computational data analysis. KC contributed to the experiments. BLK and KZ wrote the paper. All authors read and approved the final manuscript.

Author details
[1] Department of Oncology, Wayne State University School of Medicine, Detroit, MI, USA. [2] Karmanos Cancer Institute, Wayne State University School of Medicine, Detroit, MI, USA. [3] Systems Biology Center, National Heart, Lung and Blood Institute, National Institutes of Health, Bethesda, MD, USA.

Acknowledgements
We thank Drs. Zhiyong Ding, Wenfei Jin, Ana Robles and Curt Harris for helpful discussions, Dr. Jianlong Wang for providing the pEF1a-BirAV5-neo and pEF1a-Flagbio(FLBIO)-puro vectors. This work utilized the Wayne State University High Performance Computing Grid for computational resources (https://www.grid.wayne.edu/) and the computational resources of the NIH HPC Biowulf cluster. (http://hpc.nih.gov). The DNA Sequencing Core, Light Microcopy Core, and the Transgenic Core facilities of NHLBI assisted with this work.

Competing interests
The authors declare that they have no competing interests.

Funding
This work was supported by the Division of Intramural Research of the National Heart, Lung and Blood Institute, Karmanos Cancer Institute, Wayne State University, and a grant from the National Heart, Lung and Blood Institute (1K22HL126842-01A1) awarded to BLK.

References
1. Grewal SI, Moazed D. Heterochromatin and epigenetic control of gene expression. Science. 2003;301(5634):798–802. doi:10.1126/science.1086887.
2. Grewal SI, Elgin SC. Heterochromatin: new possibilities for the inheritance of structure. Curr Opin Genet Dev. 2002;12(2):178–87.
3. Avner P, Heard E. X-chromosome inactivation: counting, choice and initiation. Nat Rev Genet. 2001;2(1):59–67. doi:10.1038/35047580.
4. Guarente L. Sir2 links chromatin silencing, metabolism, and aging. Genes Dev. 2000;14(9):1021–6.
5. Matsui T, Leung D, Miyashita H, Maksakova IA, Miyachi H, Kimura H, et al. Proviral silencing in embryonic stem cells requires the histone methyltransferase ESET. Nature. 2010;464(7290):927–31. doi:10.1038/nature08858.
6. Macfarlan TS, Gifford WD, Driscoll S, Lettieri K, Rowe HM, Bonanomi D, et al. Embryonic stem cell potency fluctuates with endogenous retrovirus activity. Nature. 2012;487(7405):57–63. doi:10.1038/nature11244.
7. Lachner M, Sengupta R, Schotta G, Jenuwein T. Trilogies of histone lysine methylation as epigenetic landmarks of the eukaryotic genome. Cold Spring Harb Symp Quant Biol. 2004;69:209–18. doi:10.1101/sqb.2004.69.209.
8. Karachentsev D, Sarma K, Reinberg D, Steward R. PR-Set7-dependent methylation of histone H4 Lys 20 functions in repression of gene expression and is essential for mitosis. Genes Dev. 2005;19(4):431–5. doi:10.1101/gad.1263005.
9. Botuyan MV, Lee J, Ward IM, Kim JE, Thompson JR, Chen J, et al. Structural basis for the methylation state-specific recognition of histone H4-K20 by 53BP1 and Crb2 in DNA repair. Cell. 2006;127(7):1361–73. doi:10.1016/j.cell.2006.10.043.
10. Schotta G, Sengupta R, Kubicek S, Malin S, Kauer M, Callen E, et al. A chromatin-wide transition to H4K20 monomethylation impairs genome integrity and programmed DNA rearrangements in the mouse. Genes Dev. 2008;22(15):2048–61. doi:10.1101/gad.476008.
11. Vermeulen M, Eberl HC, Matarese F, Marks H, Denissov S, Butter F, et al. Quantitative interaction proteomics and genome-wide profiling of epigenetic histone marks and their readers. Cell. 2010;142(6):967–80. doi:10.1016/j.cell.2010.08.020.
12. Beck DB, Oda H, Shen SS, Reinberg D. PR-Set7 and H4K20me1: at the crossroads of genome integrity, cell cycle, chromosome condensation, and transcription. Genes Dev. 2012;26(4):325–37. doi:10.1101/gad.177444.111.
13. Oda H, Okamoto I, Murphy N, Chu J, Price SM, Shen MM, et al. Monomethylation of histone H4-lysine 20 is involved in chromosome structure and stability and is essential for mouse development. Mol Cell Biol. 2009;29(8):2278–95. doi:10.1128/MCB.01768-08.

14. Mikkelsen TS, Ku M, Jaffe DB, Issac B, Lieberman E, Giannoukos G, et al. Genome-wide maps of chromatin state in pluripotent and lineage-committed cells. Nature. 2007;448(7153):553–60.

15. Barski A, Cuddapah S, Cui K, Roh TY, Schones DE, Wang Z, et al. High-resolution profiling of histone methylations in the human genome. Cell. 2007;129(4):823–37. doi:10.1016/j.cell.2007.05.009.

16. Sanders SL, Portoso M, Mata J, Bahler J, Allshire RC, Kouzarides T. Methylation of histone H4 lysine 20 controls recruitment of Crb2 to sites of DNA damage. Cell. 2004;119(5):603–14. doi:10.1016/j.cell.2004.11.009.

17. Schotta G, Lachner M, Sarma K, Ebert A, Sengupta R, Reuter G, et al. A silencing pathway to induce H3-K9 and H4-K20 trimethylation at constitutive heterochromatin. Genes Dev. 2004;18(11):1251–62. doi:10.1101/gad.300704.

18. Fodor BD, Shukeir N, Reuter G, Jenuwein T. Mammalian Su(var) genes in chromatin control. Annu Rev Cell Dev Biol. 2010;26:471–501. doi:10.1146/annurev.cellbio.042308.113225.

19. Stender JD, Pascual G, Liu W, Kaikkonen MU, Do K, Spann NJ, et al. Control of proinflammatory gene programs by regulated trimethylation and demethylation of histone H4K20. Mol Cell. 2012;48(1):28–38. doi:10.1016/j.molcel.2012.07.020.

20. Kidder BL, Hu G, Yu ZX, Liu C, Zhao K. Extended self-renewal and accelerated reprogramming in the absence of Kdm5b. Mol Cell Biol. 2013;33(24):4793–810. doi:10.1128/MCB.00692-13.

21. Kurosawa H. Methods for inducing embryoid body formation: in vitro differentiation system of embryonic stem cells. J Biosci Bioeng. 2007;103(5):389–98. doi:10.1263/jbb.103.389.

22. Subramanian A, Tamayo P, Mootha VK, Mukherjee S, Ebert BL, Gillette MA, et al. Gene set enrichment analysis: a knowledge-based approach for interpreting genome-wide expression profiles. Proc Natl Acad Sci USA. 2005;102(43):15545–50.

23. Dennis G Jr, Sherman BT, Hosack DA, Yang J, Gao W, Lane HC, et al. DAVID: database for annotation, visualization, and integrated discovery. Genome Biol. 2003;4(5):P3.

24. Marson A, Levine SS, Cole MF, Frampton GM, Brambrink T, Johnstone S, et al. Connecting microRNA genes to the core transcriptional regulatory circuitry of embryonic stem cells. Cell. 2008;134(3):521–33. doi:10.1016/j.cell.2008.07.020.

25. Coucouvanis E, Martin GR. Signals for death and survival: a two-step mechanism for cavitation in the vertebrate embryo. Cell. 1995;83(2):279–87.

26. Niakan KK, Ji H, Maehr R, Vokes SA, Rodolfa KT, Sherwood RI, et al. Sox17 promotes differentiation in mouse embryonic stem cells by directly regulating extraembryonic gene expression and indirectly antagonizing self-renewal. Genes Dev. 2010;24(3):312–26. doi:10.1101/gad.1833510.

27. Chen X, Xu H, Yuan P, Fang F, Huss M, Vega VB, et al. Integration of external signaling pathways with the core transcriptional network in embryonic stem cells. Cell. 2008;133(6):1106–17.

28. Rahl PB, Lin CY, Seila AC, Flynn RA, McCuine S, Burge CB, et al. c-Myc regulates transcriptional pause release. Cell. 2010;141(3):432–45. doi:10.1016/j.cell.2010.03.030.

29. Ku M, Koche RP, Rheinbay E, Mendenhall EM, Endoh M, Mikkelsen TS, et al. Genomewide analysis of PRC1 and PRC2 occupancy identifies two classes of bivalent domains. PLoS Genet. 2008;4(10):e1000242. doi:10.1371/journal.pgen.1000242.

30. Kim J, Chu J, Shen X, Wang J, Orkin SH. An extended transcriptional network for pluripotency of embryonic stem cells. Cell. 2008;132(6):1049–61.

31. Kim J, Cantor AB, Orkin SH, Wang J. Use of in vivo biotinylation to study protein–protein and protein–DNA interactions in mouse embryonic stem cells. Nat Protoc. 2009;4(4):506–17. doi:10.1038/nprot.2009.23.

32. Bernstein BE, Mikkelsen TS, Xie X, Kamal M, Huebert DJ, Cuff J, et al. A bivalent chromatin structure marks key developmental genes in embryonic stem cells. Cell. 2006;125(2):315–26.

33. Fischle W, Tseng BS, Dormann HL, Ueberheide BM, Garcia BA, Shabanowitz J, et al. Regulation of HP1-chromatin binding by histone H3 methylation and phosphorylation. Nature. 2005;438(7071):1116–22. doi:10.1038/nature04219.

34. Ellinghaus D, Kurtz S, Willhoeft U. LTRharvest, an efficient and flexible software for de novo detection of LTR retrotransposons. BMC Bioinformatics. 2008;9:18. doi:10.1186/1471-2105-9-18.

35. Steinbiss S, Willhoeft U, Gremme G, Kurtz S. Fine-grained annotation and classification of de novo predicted LTR retrotransposons. Nucleic Acids Res. 2009;37(21):7002–13. doi:10.1093/nar/gkp759.

36. Karimi MM, Goyal P, Maksakova IA, Bilenky M, Leung D, Tang JX, et al. DNA methylation and SETDB1/H3K9me3 regulate predominantly distinct sets of genes, retroelements, and chimeric transcripts in mESCs. Cell Stem Cell. 2011;8(6):676–87. doi:10.1016/j.stem.2011.04.004.

37. Ivanova N, Dobrin R, Lu R, Kotenko I, Levorse J, DeCoste C, et al. Dissecting self-renewal in stem cells with RNA interference. Nature. 2006;442(7102):533–8.

38. Tam WL, Lim CY, Han J, Zhang J, Ang YS, Ng HH, et al. T-cell factor 3 regulates embryonic stem cell pluripotency and self-renewal by the transcriptional control of multiple lineage pathways. Stem Cells. 2008;26(8):2019–31. doi:10.1634/stemcells.2007-1115.

39. Kidder BL, Palmer S, Knott JG. SWI/SNF-Brg1 regulates self-renewal and occupies core pluripotency-related genes in embryonic stem cells. Stem Cells. 2009;27(2):317–28. doi:10.1634/stemcells.2008-0710.

40. Ho L, Jothi R, Ronan JL, Cui K, Zhao K, Crabtree GR. An embryonic stem cell chromatin remodeling complex, esBAF, is an essential component of the core pluripotency transcriptional network. Proc Natl Acad Sci USA. 2009;106(13):5187–91. doi:10.1073/pnas.0812888106.

41. Kidder BL, Hu G, Zhao K. KDM5B focuses H3K4 methylation near promoters and enhancers during embryonic stem cell self-renewal and differentiation. Genome Biol. 2014;15(2):R32. doi:10.1186/gb-2014-15-2-r32.

42. Ang YS, Tsai SY, Lee DF, Monk J, Su J, Ratnakumar K, et al. Wdr5 mediates self-renewal and reprogramming via the embryonic stem cell core transcriptional network. Cell. 2011;145(2):183–97. doi:10.1016/j.cell.2011.03.003.

43. Glaser S, Schaft J, Lubitz S, Vintersten K, van der Hoeven F, Tufteland KR, et al. Multiple epigenetic maintenance factors implicated by the loss of Mll2 in mouse development. Development. 2006;133(8):1423–32. doi:10.1242/dev.02302.

44. Boyer LA, Plath K, Zeitlinger J, Brambrink T, Medeiros LA, Lee TI, et al. Polycomb complexes repress developmental regulators in murine embryonic stem cells. Nature. 2006;441(7091):349–53.

45. Dodge JE, Kang YK, Beppu H, Lei H, Li E. Histone H3-K9 methyltransferase ESET is essential for early development. Mol Cell Biol. 2004;24(6):2478–86.

46. Loh YH, Zhang W, Chen X, George J, Ng HH. Jmjd1a and Jmjd2c histone H3 Lys 9 demethylases regulate self-renewal in embryonic stem cells. Genes Dev. 2007;21(20):2545–57. doi:10.1101/gad.1588207.

47. O'Carroll D, Erhardt S, Pagani M, Barton SC, Surani MA, Jenuwein T. The polycomb-group gene Ezh2 is required for early mouse development. Mol Cell Biol. 2001;21(13):4330–6. doi:10.1128/MCB.21.13.4330-4336.2001.

48. Tachibana M, Sugimoto K, Nozaki M, Ueda J, Ohta T, Ohki M, et al. G9a histone methyltransferase plays a dominant role in euchromatic histone H3 lysine 9 methylation and is essential for early embryogenesis. Genes Dev. 2002;16(14):1779–91. doi:10.1101/gad.989402.

49. Nicetto D, Hahn M, Jung J, Schneider TD, Straub T, David R, et al. Suv4-20 h histone methyltransferases promote neuroectodermal differentiation by silencing the pluripotency-associated Oct-25 gene. PLoS Genet. 2013;9(1):e1003188. doi:10.1371/journal.pgen.1003188.

50. Allis CD, Jenuwein T, Reinberg D. Epigenetics. Cold Spring Harbor, NY: Cold Spring Harbor Laboratory Press; 2007.

51. Platero JS, Hartnett T, Eissenberg JC. Functional analysis of the chromo domain of HP1. EMBO J. 1995;14(16):3977–86.

52. Fischle W, Wang Y, Jacobs SA, Kim Y, Allis CD, Khorasanizadeh S. Molecular basis for the discrimination of repressive methyl-lysine marks in histone H3 by Polycomb and HP1 chromodomains. Genes Dev. 2003;17(15):1870–81. doi:10.1101/gad.1110503.

53. Stewart MD, Li J, Wong J. Relationship between histone H3 lysine 9 methylation, transcription repression, and heterochromatin protein 1 recruitment. Mol Cell Biol. 2005;25(7):2525–38. doi:10.1128/MCB.25.7.2525-2538.2005.

54. Thiru A, Nietlispach D, Mott HR, Okuwaki M, Lyon D, Nielsen PR, et al. Structural basis of HP1/PXVXL motif peptide interactions and HP1 localisation to heterochromatin. EMBO J. 2004;23(3):489–99. doi:10.1038/sj.emboj.7600088.

55. Yokochi T, Poduch K, Ryba T, Lu J, Hiratani I, Tachibana M, et al. G9a selectively represses a class of late-replicating genes at the nuclear

periphery. Proc Natl Acad Sci USA. 2009;106(46):19363–8. doi:10.1073/pnas.0906142106.

56. Maksakova IA, Thompson PJ, Goyal P, Jones SJ, Singh PB, Karimi MM, et al. Distinct roles of KAP1, HP1 and G9a/GLP in silencing of the two-cell-specific retrotransposon MERVL in mouse ES cells. Epigenetics Chromatin. 2013;6(1):15. doi:10.1186/1756-8935-6-15.

57. Kidder BL, Hu G, Zhao K. ChIP-Seq: technical considerations for obtaining high-quality data. Nat Immunol. 2011;12(10):918–22. doi:10.1038/ni.2117.

58. Langmead B, Salzberg SL. Fast gapped-read alignment with Bowtie 2. Nat Methods. 2012;9(4):357–9. doi:10.1038/nmeth.1923.

59. Xu S, Grullon S, Ge K, Peng W. Spatial clustering for identification of ChIP-enriched regions (SICER) to map regions of histone methylation patterns in embryonic stem cells. Methods Mol Biol. 2014;1150:97–111. doi:10.1007/978-1-4939-0512-6_5.

60. Zang C, Schones DE, Zeng C, Cui K, Zhao K, Peng W. A clustering approach for identification of enriched domains from histone modification ChIP-Seq data. Bioinformatics. 2009;25(15):1952–8. doi:10.1093/bioinformatics/btp340.

61. Zhang Y, Liu T, Meyer CA, Eeckhoute J, Johnson DS, Bernstein BE, et al. Model-based analysis of ChIP-Seq (MACS). Genome Biol. 2008;9(9):R137. doi:10.1186/gb-2008-9-9-r137.

62. Mortazavi A, Williams BA, McCue K, Schaeffer L, Wold B. Mapping and quantifying mammalian transcriptomes by RNA-Seq. Nat Methods. 2008;5(7):621–8. doi:10.1038/nmeth.1226.

63. Robinson MD, McCarthy DJ, Smyth GK. edgeR: a bioconductor package for differential expression analysis of digital gene expression data. Bioinformatics. 2009;26(1):139–40. doi:10.1093/bioinformatics/btp616.

64. Dignam JD, Lebovitz RM, Roeder RG. Accurate transcription initiation by RNA polymerase II in a soluble extract from isolated mammalian nuclei. Nucleic Acids Res. 1983;11(5):1475–89.

65. Wysocka J. Identifying novel proteins recognizing histone modifications using peptide pull-down assay. Methods. 2006;40(4):339–43. doi:10.1016/j.ymeth.2006.05.028.

66. Tan CM, Chen EY, Dannenfelser R, Clark NR, Ma'ayan A. Network2Canvas: network visualization on a canvas with enrichment analysis. Bioinformatics. 2013;29(15):1872–8. doi:10.1093/bioinformatics/btt319.

67. Whyte WA, Bilodeau S, Orlando DA, Hoke HA, Frampton GM, Foster CT, et al. Enhancer decommissioning by LSD1 during embryonic stem cell differentiation. Nature. 2012;482(7384):221–5. doi:10.1038/nature10805.

68. Newman AM, Cooper JB. AutoSOME: a clustering method for identifying gene expression modules without prior knowledge of cluster number. BMC Bioinformatics. 2010;11:117. doi:10.1186/1471-2105-11-117.

69. Heinz S, Benner C, Spann N, Bertolino E, Lin YC, Laslo P, et al. Simple combinations of lineage-determining transcription factors prime cis-regulatory elements required for macrophage and B cell identities. Mol Cell. 2010;38(4):576–89. doi:10.1016/j.molcel.2010.05.004.

13

Decoupling the downstream effects of germline nuclear RNAi reveals that H3K9me3 is dispensable for heritable RNAi and the maintenance of endogenous siRNA-mediated transcriptional silencing in *Caenorhabditis elegans*

Natallia Kalinava[1], Julie Zhouli Ni[1], Kimberly Peterman[1], Esteban Chen[1] and Sam Guoping Gu[1,2*]

Abstract

Background: Germline nuclear RNAi in *C. elegans* is a transgenerational gene-silencing pathway that leads to H3K9 trimethylation (H3K9me3) and transcriptional silencing at the target genes. H3K9me3 induced by either exogenous double-stranded RNA (dsRNA) or endogenous siRNA (endo-siRNA) is highly specific to the target loci and transgenerationally heritable. Despite these features, the role of H3K9me3 in siRNA-mediated transcriptional silencing and inheritance of the silencing state at native target genes is unclear. In this study, we took combined genetic and whole-genome approaches to address this question.

Results: Here we demonstrate that siRNA-mediated H3K9me3 requires combined activities of three H3K9 histone methyltransferases: MET-2, SET-25, and SET-32. *set-32* single, *met-2 set-25* double, and *met-2 set-25;set-32* triple mutant adult animals all exhibit prominent reductions in H3K9me3 throughout the genome, with *met-2 set-25;set-32* mutant worms losing all detectable H3K9me3 signals. Surprisingly, loss of high-magnitude H3K9me3 at the native nuclear RNAi targets has no effect on the transcriptional silencing state. In addition, the exogenous dsRNA-induced transcriptional silencing and heritable RNAi at *oma-1*, a well-established nuclear RNAi reporter gene, are completely resistant to the loss of H3K9me3.

Conclusions: Nuclear RNAi-mediated H3K9me3 in *C. elegans* requires multiple histone methyltransferases, including MET-2, SET-25, and SET-32. H3K9me3 is not essential for dsRNA-induced heritable RNAi or the maintenance of endo-siRNA-mediated transcriptional silencing in *C. elegans*. We propose that siRNA-mediated transcriptional silencing in *C. elegans* can be maintained by an H3K9me3-independent mechanism.

Background

Following the initial discovery of RNAi [1, 2], a variety of small RNA-mediated silencing phenomena have been uncovered. There is a considerable diversity in the biogenesis of small RNA, biochemical function of the Argonaute (AGO) protein, as well as downstream effects among different silencing pathways that involve small RNA. In addition to the posttranscriptional gene silencing (PTGS) mechanism, in which the AGO-siRNA complexes, also referred to as RNA-induced silencing complex (RISC), degrade target mRNA [3–6], RNAi can also induce heterochromatin and (co-)transcriptional gene silencing at the target locus (reviewed in [7–11]). These so-called nuclear RNAi effects, initially

*Correspondence: ggu@dls.rutgers.edu
[2] Nelson Labs A125, 604 Allison Road, Piscataway, NJ 08854, USA
Full list of author information is available at the end of the article

discovered in plants and *Schizosaccharomyces pombe*, provide one of the first examples that RNA can function as a sequence-specific guide to regulate chromatin structure. Subsequent studies in these systems revealed that nuclear RNAi plays critical roles in gene regulation, heterochromatin assembly, genome surveillance and chromosome stability. Recent studies have shown that the nuclear RNAi pathway and its role in genome surveillance are conserved in animals as well. In *Drosophila melanogaster*, PIWI protein and the PIWI-interacting RNA (piRNA) silence transposons through both PTGS and transcriptional gene silencing (TGS) mechanisms [12–15]. piRNA-dependent heterochromatin formation at transposable elements was also shown to occur in mammalian germ cells [16].

In *Caenorhabditis elegans*, nuclear RNAi is required for H3K9me3 and transcriptional silencing in a distinct set of genomic loci that have high levels expression of endo-siRNA (more on the native targets later in Background). Besides endo-siRNA, exogenous dsRNA can also trigger highly specific nuclear RNAi effects at native genes or transgenes [17–21]. The dsRNA-induced H3K9me3 in *C. elegans* can last for at least three generations after the initial dsRNA exposure has been removed [18].

Several NRDE (nuclear RNAi-defective) proteins [19, 20, 22] and a germline nuclear Argonaute protein, HRDE-1 [17, 21, 23], are essential for nuclear RNAi in *C. elegans*, but not PTGS. Previous studies have indicated that dsRNA-triggered gene silencing can persist for multiple generations in *C. elegans* [24–26]. Mutant *C. elegans* strains lacking nuclear RNAi components (e.g., HRDE-1, NRDE-1, or NRDE-2) are defective in heritable RNAi induced by either dsRNA or piRNA [17, 21, 23, 24] and exhibit other transgenerational defects, such as the mortal germline (Mrt) phenotype [17, 21] and heat-induced progressive activation of native target genes [27]. These features make *C. elegans* a uniquely attractive system to study the mechanisms of RNA-mediated chromatin regulation and transgenerational epigenetics, as well as their roles in germline development.

Methylation of histone H3 at lysine 9 (H3K9me), the hallmark of constitutive heterochromatin, is an evolutionarily conserved response of nuclear RNAi [9, 28]. Studies in *S. pombe* have indicated a complex role of H3K9me2/3. Tethering H3K9 methyltransferase (HMT), Clr4, to a target gene leads to transcriptional silencing [29, 30]. H3K9 methylation is also required for stable interaction between RNAi machineries and chromatin [9], which convolutes the determination of the direct cause of transcriptional silencing—whether it being heterochromatin, RNAi, or both. The complexity of the system is further evidenced by the role of heterochromatin in promoting co-transcriptional silencing [31].

In *C. elegans*, H3K9me3-marked regions form large domains, enriched in the arms of autosomes, particularly the ones associated with meiotic pairing center, and the left tip of the sex chromosome [32, 33]. Previous studies have shown that H3K9 methylation is involved in a diverse set of functions: germline development and immortality [12, 34–36], chromosome nuclear localization [37], dosage compensation [38], RNAi [18, 21, 22, 36], and maintaining the genome stability [39].

We previously showed that germline nuclear RNAi-dependent H3K9me3 is highly enriched in long terminal repeat (LTR) retrotransposons [40]. Even though these regions account for only a small fraction of the total H3K9me3-enriched regions in *C. elegans* genome, H3K9me3 profiles at these native germline nuclear RNAi targets are prominent and defined [17, 40]. By performing whole-genome analyses using the *hrde-1* loss-of-function mutant, we identified loci with the germline nuclear RNAi-dependent heterochromatin (GRH) and loci with germline nuclear RNAi-dependent transcriptional silencing (GRTS) in the *C. elegans* genome [40]. Interestingly, GRTS and GRH loci only partially overlap. GRTS loci tend to have much less H3K9me3 defects than the GRH loci in *hrde-1* mutants. Conversely, many GRH loci show little changes in transcriptional repression in *hrde-1* mutants. These results highlight the complexity of germline nuclear RNAi in *C. elegans*, suggesting that the two germline nuclear RNAi effects, H3K9me3 and transcriptional silencing, may not be causally linked.

In this study, we combined genetic and whole-genome approaches to characterize the requirement of H3K9me3 for transcriptional silencing at nuclear RNAi targets. *C. elegans* has 38 putative histone methyltransferases (HMTs) [41, 42]. It is unclear which of them are required for the H3K9me3 response associated with nuclear RNAi. MET-2 (a H3K9 mono- and dimethylation HMT) [43] and SET-25 (a H3K9 trimethylation HMT) are required for all detectable H3K9me3 at the embryonic stage, as shown by mass spectrometry analysis [37, 44]. A recent study also showed a complete loss of H3K9me3 in adult germline of *met-2 set-25* mutants by immunofluorescence (IF) analysis and de-silencing in the *met-2 set-25* mutants leads to increased genome-instability and mutation [39]. Interestingly, many H3K9me3-enriched loci, including LTR retrotransposons, remain silenced in *met-2 set-25* mutant worms [39]. Despite the prominent loss of H3K9me3 in *met-2 set-25* mutants, SET-32 and SET-26 were also shown to be H3K9 HMTs by IF [38] and in vitro HMT assay [42], respectively. A candidate screen using a transgene reporter showed that SET-25 is required for exogenous dsRNA-induced heritable RNAi and SET-32 is required for piRNA-induced gene silencing [21]. The H3K9me3 status at these nuclear RNAi reporter transgenes is unclear. In addition, the requirement of H3K9me3 for nuclear RNAi-mediated transcriptional silencing at native genes has not been tested.

Results

MET-2, SET-25, and SET-32 are required for exogenous dsRNA-triggered H3K9me3

We first performed H3K9me3 ChIP-seq in *met-2 set-25* double mutant worms to examine the requirement of these two HMTs for exogenous dsRNA-triggered H3K9me3. We chose a germline-specific gene *oma-1* as the target gene as it is sensitive to dsRNA-induced nuclear RNAi [17, 45]. Despite the popular usage of *oma-1* in nuclear RNAi and heritable silencing studies

(e.g., [17, 24, 45, 46]), a high-resolution profile of dsRNA-induced H3K9me3 at this locus has not been reported before. A combination of control samples, including wild-type and *hrde-1* mutant animals, both with *oma-1* RNAi, as well as wild-type animals with control RNAi (*gfp*), were used to indicate the full extent of HRDE-1-dependent H3K9me3 response at the *oma-1* locus. Synchronized young adult animals, of which over 50% of total cells are germline [47], were used throughout this study. (Additional file 1: Table S1 lists all samples and sequencing libraries used in this study.)

The *oma-1* dsRNA-induced H3K9me3 response was analyzed using coverage plot (Fig. 1a) and whole-genome 1-kb-resolution scatter plot (Fig. 1b–f). In WT, *oma-1*

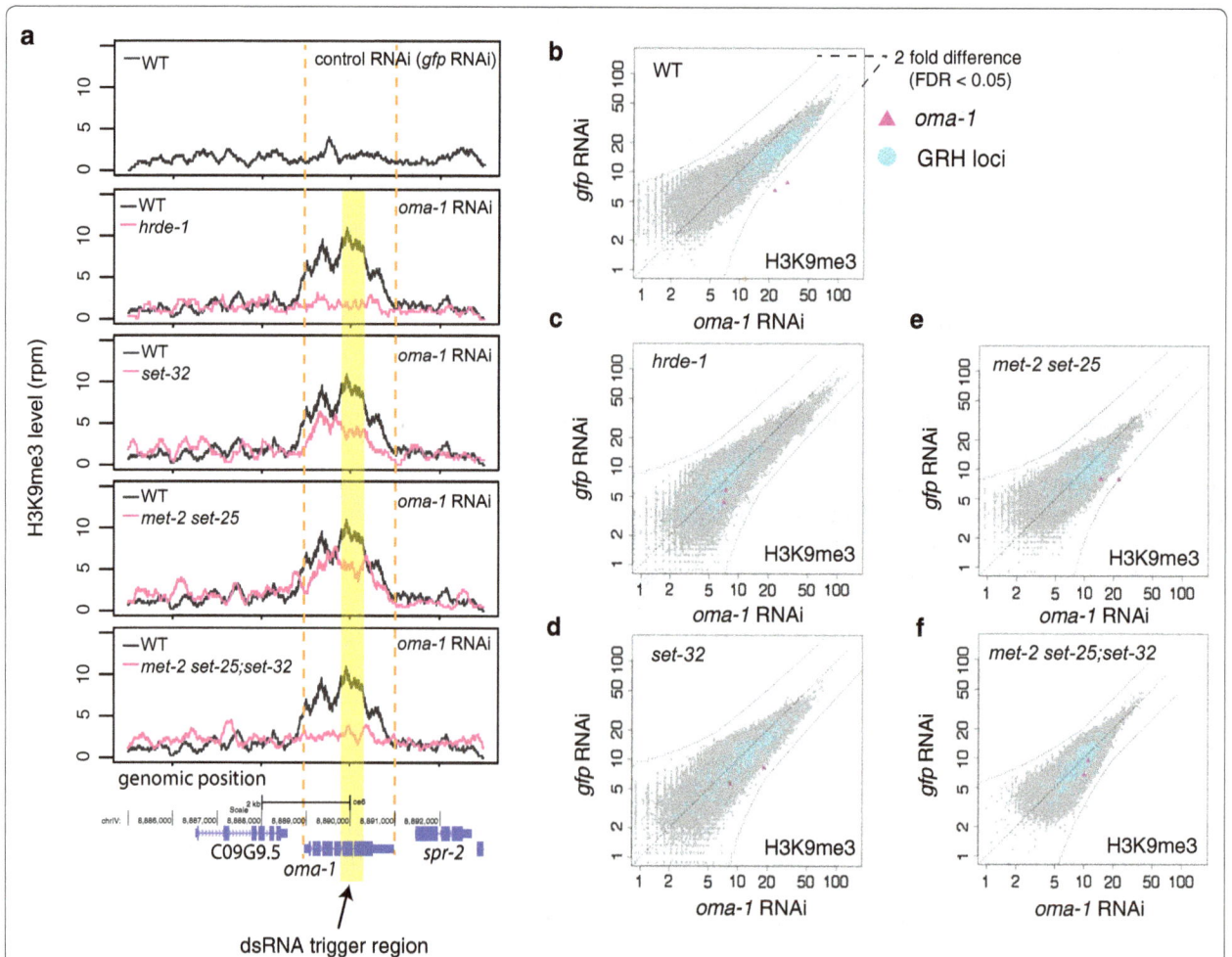

Fig. 1 MET-2, SET-25, and SET-32 are required for the dsRNA-triggered H3K9me3 response at *oma-1*. **a** H3K9me3 levels in different samples are plotted as a function of position along the *oma-1* locus. The WT response (*oma-1* RNAi) is shown in all panels, except the *top one* (WT with *gfp* RNAi), to facilitate comparison with mutant ones. A *yellow block* indicates the region targeted by dsRNA. **b–f** Scatter plots that compare whole-genome H3K9me3 levels in *oma-1* RNAi and *gfp* RNAi samples at 1 kb resolution for WT and various mutant strains. *oma-1* regions (2 kb) and GRH loci (215 kb) are *highlighted*. *Curved dotted lines* indicate the twofold change (FDR < 0.05). Synchronized young adult animals (19 °C) were used throughout this study

dsRNA-induced H3K9me3 peaked at the dsRNA trigger region (0.52 kb) (Fig. 1a) and was limited to *oma-1* (Fig. 1b), a degree of specificity similar to the ones observed for other targets in our previous study [40]. The H3K9me3 response spread throughout the *oma-1* gene (~2 kb) and dropped to the background level around the putative *oma-1* promoter and the polyadenylation site, without spreading into either of the adjacent genes, both of which are less than 0.5 kb away (Fig. 1a). The H3K9me3 response was absent in the *hrde-1* mutant (*oma-1* RNAi) or WT (control RNAi) worms (Fig. 1a–c), as expected [17, 45].

Compared to WT, *met-2 set-25* mutants showed only a partial defect in the H3K9me3 response at the *oma-1* locus (Fig. 1a–e), suggesting that additional H3K9 HMT(s) are involved. This was somewhat surprising because of the more severe H3K9me3 defect at the whole-genome level observed in the same mutant sample (Fig. 4a and see later section). In addition, previous studies showed that *met-2 set-25* double mutants were devoid of H3K9me3 at the embryo stage [37, 44], as well as in adult germline [39].

We then performed *oma-1* RNAi and H3K9me3 ChIP-qPCR to screen additional H3K9 HMT mutants (Additional file 2: Fig. S1), among which only *set-32* mutant and *met-2 set-25;set-32* triple mutant worms showed defects in the H3K9me3 response at the *oma-1* locus (Additional file 2: Fig. S1). *met-2* or *set-25* single mutants did not show any defect in the H3K9me3 response (Additional file 2: Fig. S1). We further confirmed the requirement of *set-32* by H3K9me3 ChIP-seq analysis (Fig. 1a). Compared to *met-2 set-25* mutants (*oma-1* RNAi), *set-32* mutants (*oma-1* RNAi) showed a stronger defect in the H3K9me3 response at the *oma-1* locus (ΔH3K9me3$_{oma-1}$

[*set-32*/WT] = 0.32 and ΔH3K9me3$_{oma-1}$ [*met-2 set-25*/WT] = 0.55) (Fig. 1a, d, e), indicating that SET-32 plays a more prominent role than MET-2 SET-25 in dsRNA-induced H3K9me3. Similar to *hrde-1* mutants (*oma-1* RNAi), *met-2 set-25;set-32* mutants (*oma-1* RNAi) showed a background level of H3K9me3 ChIP-seq signal at the *oma-1* locus (Fig. 1a, c, f), suggesting that MET-2, SET-25, and SET-32, in combination, contribute to the full H3K9me3 response induced by exogenous dsRNA.

H3K9me3 is not required for exogenous dsRNA-induced transcriptional silencing and heritable RNAi

To investigate the role of H3K9me3 in nuclear RNAi and heritable silencing, we performed a set of heritable RNAi experiments using wild-type, *hrde-1*, *met-2 set-25*, *set-32*, and *met-2 set-25;set-32* mutant strains with *oma-1* or *gfp* RNAi (Fig. 2a). qRT-PCR analyses of *oma-1* mRNA and pre-mRNA were performed for the dsRNA-exposed animals (P0 generation) and their descendants (F1, F2, and F3) that were cultured without dsRNA exposure.

In the WT animals, *oma-1* RNAi caused heritable silencing of the target gene in F1, F2, and F3 at both mRNA and pre-mRNA levels (Fig. 2b, c). The heritable silencing was dependent on HRDE-1 (Fig. 2b, c), as expected [17, 24].

All three HMT mutant strains (*met-2 set-25*, *set-32*, and *met-2 set-25;set-32*) showed wild-type-like multigenerational profiles of *oma-1* pre-mRNA (Fig. 2c), despite various degrees of H3K9me3 defect at the *oma-1* locus. These results indicate that heritable transcriptional silencing can occur in the absence of the H3K9me3 response. We note that all three HMT mutant strains had a modest but consistently enhanced heritable RNAi at the mRNA level (Fig. 2b).

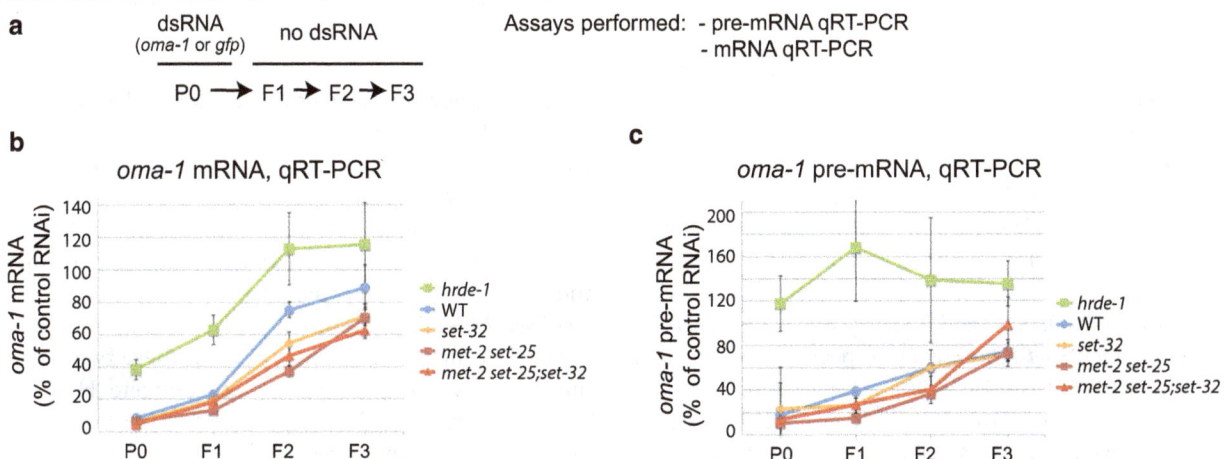

Fig. 2 Requirement of H3K9me3 in heritable RNAi and transcriptional silencing of *oma-1*. **a** Experimental scheme. **b, c** mRNA and pre-mRNA levels of *oma-1* RNAi samples, normalized to the control RNAi, are plotted as a function of generations for WT or various mutant strains

To further examine nuclear RNAi-mediated transcriptional silencing at *oma-1*, we performed RNA polymerase II (Pol II) ChIP-seq for the P0 generation samples, using an antibody against the phosphorylated C-terminal domain (CTD) repeat YSPTSPS at the S2 position (S2P), a modification associated with the elongating Pol II. Our scatter plot analysis showed that, compared to *gfp* RNAi, *oma-1* RNAi did not change the overall Pol II levels at the *oma-1* locus in wild-type, *set-32*, or *met-2 set-25;set-32* mutant worms (Fig. 3a, c, d) [a modest reduction in the Pol II level was observed in the *oma-1* RNAi sample for *met-2 set-25* mutants (Fig. 3b)]. However, coverage plot analysis showed that *oma-1* RNAi changed the Pol II profile at the target gene. Specifically, it led to a reduction in the Pol II level at the 3′ end (for WT and all three HMT mutant strains) and an increase in the gene body (all samples except *met-2 set-25* mutants) (Fig. 3e). Such Pol II shift is consistent with a model in which exogenous dsRNA-induced transcriptional silencing occurs at the elongation step, as previously suggested [19].

Taken together, these results indicate that MET-2, SET-25, and SET-32 in combination are responsible for the full H3K9me3 response for exogenous dsRNA-triggered H3K9me3. However, these HMTs are not required for dsRNA-induced transcription silencing or heritable silencing.

MET-2, SET-25, and SET-32 in combination contribute to all of the detectable H3K9me3 at the native nuclear RNAi targets in adult animals

When examined at the global level, *set-32*, *met-2 set-25*, and *met-2 set-25;set-32* mutant worms all showed significant reduction in the H3K9me3 ChIP-seq signal compared to WT (Fig. 4a and Additional file 3: Fig. S2). *met-2 set-25;set-32* triple mutants had the most severe global H3K9me3 loss among the three HMT mutant strains and showed H3K9me3 depletion throughout each of the six chromosomes (Fig. 4a and Additional file 3: Fig. S2). *met-2 set-25* double mutants ranked the second and had a severity of H3K9me3 loss similar to *met-2 set-25;set-32* mutants. *set-32* mutants showed the least global H3K9me3 loss among the three HMT mutant strains. To quantify the loss of H3K9me3, we defined the top 5 percentile regions of H3K9me3 in WT as the H3K9me3(+) regions (Additional file 4: Table S3). The median values of ΔH3K9me3[mutant/WT] for these H3K9me3(+) regions were 0.30, 0.25, and 0.17 for *set-32*, *met-2;set-25*, and *met-2;set-25;set-32* mutants, respectively (Fig. 5a and Additional file 6: Fig. S3a, all *p* values $<2.2 \times 10^{-16}$).

By using ΔH3K9me3[mutant/WT] ≤0.5 (FDR < 0.05) as the cutoff, we identified 816 kb regions with SET-32-dependent H3K9me3, which were much less than the ones that required MET-2 SET-25 (2677 kb) or all three HMTs

(3290 kb) (Fig. 4b, Additional file 5: Fig. S4a, b, and Additional file 4: Table S3). Regions with SET-32-dependent H3K9me3 largely overlapped with the MET-2 SET-25-dependent ones (Fig. 4b). Taken together, these results indicate that MET-2, SET-25, and SET-32, in combination, contribute to all detectable level of H3K9me3 ChIP-seq signal in adult *C. elegans*. Interestingly, both *set-32* and *met-2 set-25* mutants had more than 50% H3K9me3 loss in many regions throughout the genome compared to wild-type animals, suggesting a synergistic relationship between SET-32 and MET-2 SET-25.

We then limited our analysis to the native targets of germline nuclear RNAi. Consistent with our previous work [40], *hrde-1* mutation led to a much stronger loss of H3K9me3 in GRH than GRTS loci, as shown in exemplary targets in Fig. 6a, b. Similar results were obtained when GRTS and GRH loci were analyzed as groups: the median values of ΔH3K9me3[*hrde-1*/WT] were 0.17 for GRH and 0.38 for GRTS loci (both *p* values $<2.2 \times 10^{-16}$, Fig. 5a and Additional file 6: Fig. S3a). The partial H3K9me3 loss in *hrde-1* mutants indicates that GRTS carry both HRDE-1-dependent and HRDE-1-independent H3K9me3.

met-2 set-25;set-32 triple mutants showed a background level of H3K9me3 at both the GRTS and GRH nuclear RNAi targets (Fig. 6a, b). This is expected because of the aforementioned genome-wide depletion of H3K9me3 in the same mutant worms. *met-2 set-25* double mutant and *set-32* mutant worms both showed strong H3K9me3 loss at both GRTS and GRH loci (Fig. 6a, b). The median values of ΔH3K9me3[mutant/WT] for *set-32*, *met-2 set-25*, and *met-2 set-25;set-32* mutants were 0.21, 0.25 and 0.15 in GRH loci (0.23, 0.20, and 0.13 in the GRTS loci, respectively) (Fig. 5a and Additional file 6: Fig. S3a, all *p* values $<2.2 \times 10^{-16}$). Therefore, these HMTs together are required for HRDE-1-dependent and HRDE-1-independent H3K9me3 at the native germline nuclear RNAi targets.

H3K9me3 is dispensable for transcriptional silencing at the native nuclear RNAi targets

To determine the impact of H3K9me3 loss on the transcriptional silencing in the native germline nuclear RNAi targets, we performed coverage analyses of Pol II ChIP-seq, and pre-mRNA-seq for two exemplary native targets and a control region (Fig. 6a–c). *set-32*, *met-2 set-25*, and *met2 set-25;set-32* mutants all showed background levels of Pol II occupancy and pre-mRNA at these two targets, similar to WT, despite partial or complete H3K9me3 loss in the mutants. In contrast, *hrde-1* mutant worms showed dramatic increases in both Pol II occupancy and pre-mRNA at these two targets, even though it had only partial H3K9me3 loss at *cer8*, a GRTS locus (Fig. 6a).

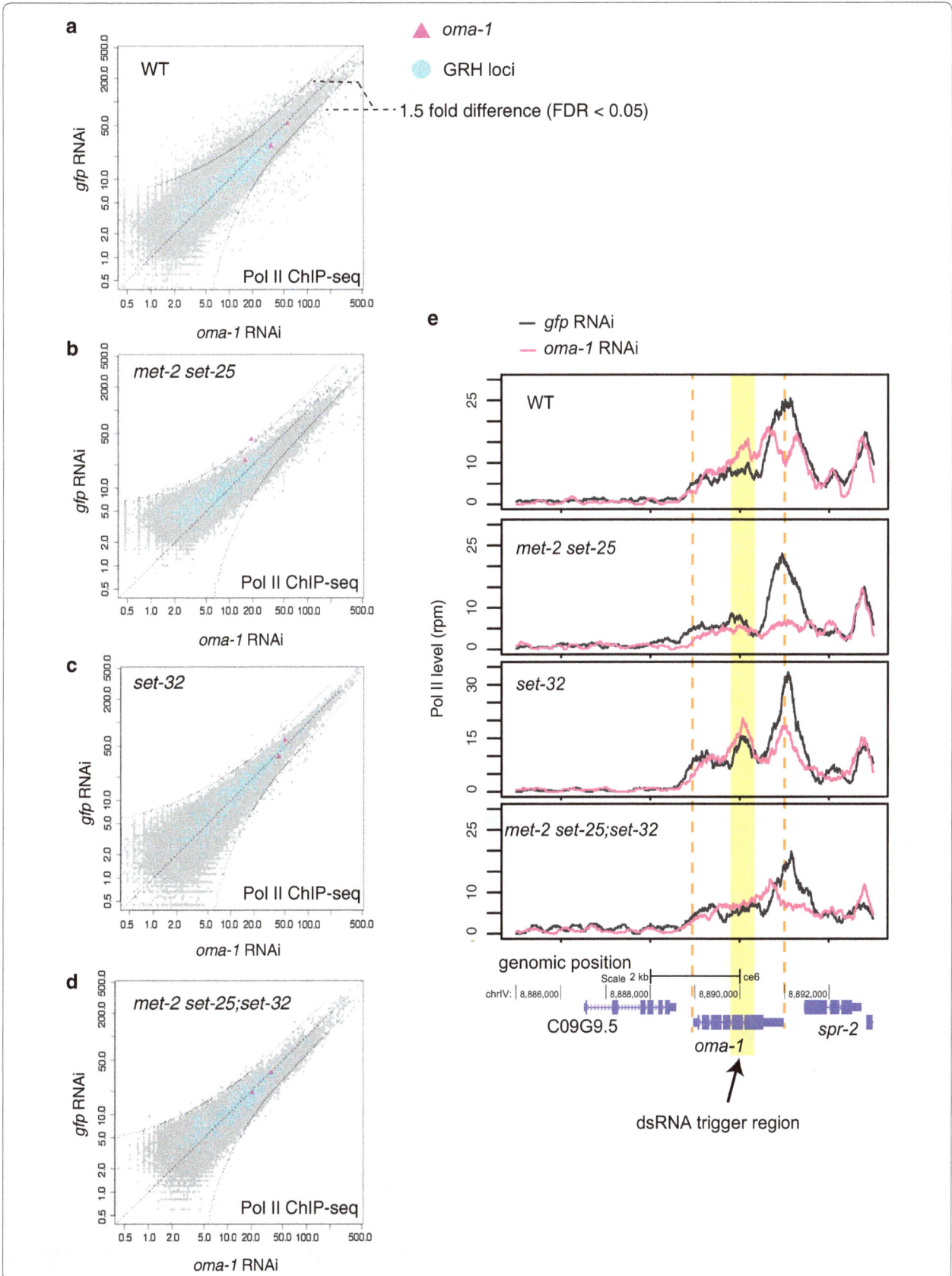

(See figure on previous page.)

Fig. 3 Impact of RNAi on Pol II profile at the *oma-1* locus. **a–d** Scatter plots that compare whole-genome Pol II levels in *oma-1* RNAi and *gfp* RNAi samples at 1 kb resolution for WT and various mutant strains. *oma-1* regions (2 kb) and GRH loci (215 kb) are *highlighted*. *Curved dotted lines* indicate the 1.5-fold change (FDR < 0.05). **e** Pol II levels are plotted as a function of position along the *oma-1* locus

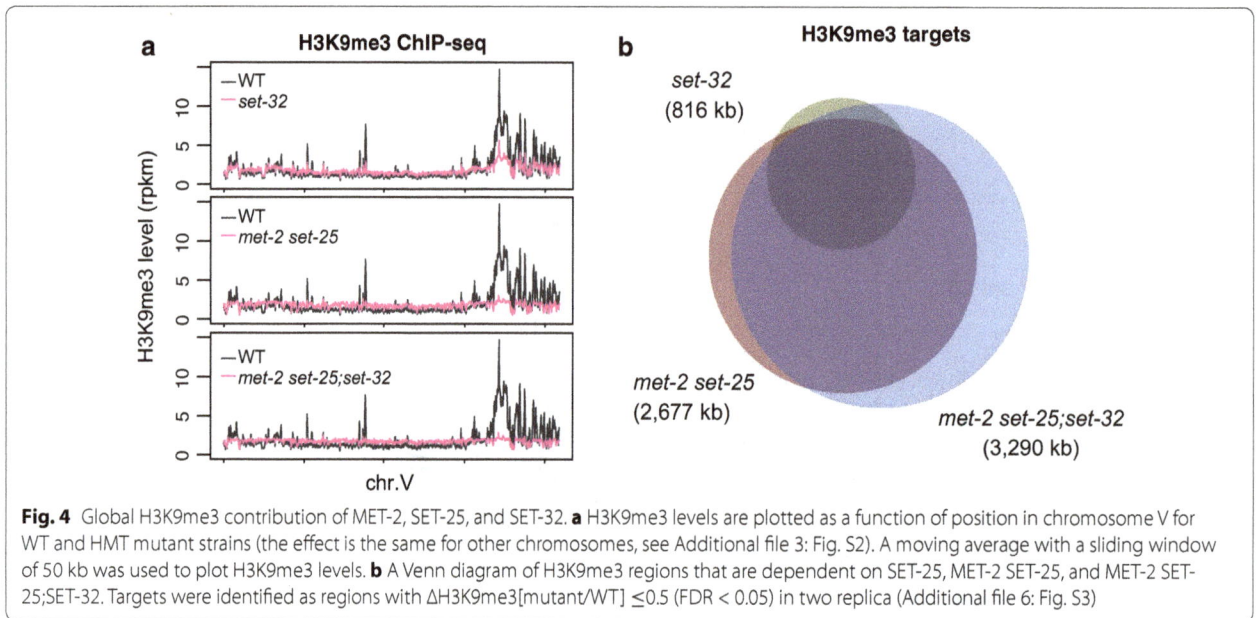

Fig. 4 Global H3K9me3 contribution of MET-2, SET-25, and SET-32. **a** H3K9me3 levels are plotted as a function of position in chromosome V for WT and HMT mutant strains (the effect is the same for other chromosomes, see Additional file 3: Fig. S2). A moving average with a sliding window of 50 kb was used to plot H3K9me3 levels. **b** A Venn diagram of H3K9me3 regions that are dependent on SET-25, MET-2 SET-25, and MET-2 SET-25;SET-32. Targets were identified as regions with ΔH3K9me3[mutant/WT] ≤0.5 (FDR < 0.05) in two replica (Additional file 6: Fig. S3)

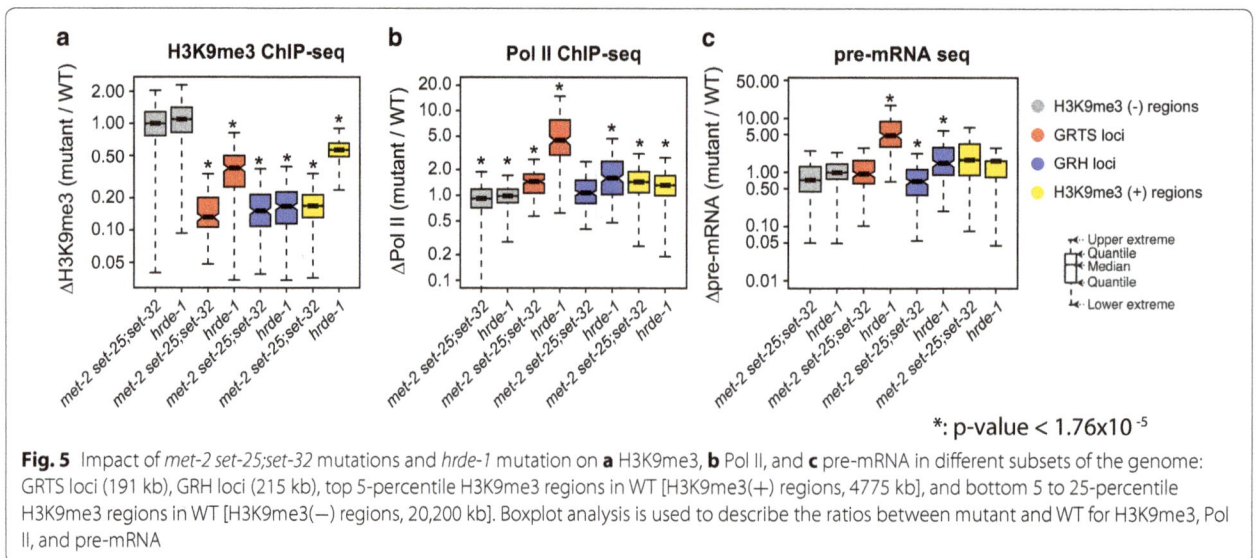

Fig. 5 Impact of *met-2 set-25;set-32* mutations and *hrde-1* mutation on **a** H3K9me3, **b** Pol II, and **c** pre-mRNA in different subsets of the genome: GRTS loci (191 kb), GRH loci (215 kb), top 5-percentile H3K9me3 regions in WT [H3K9me3(+) regions, 4775 kb], and bottom 5 to 25-percentile H3K9me3 regions in WT [H3K9me3(−) regions, 20,200 kb]. Boxplot analysis is used to describe the ratios between mutant and WT for H3K9me3, Pol II, and pre-mRNA

Consistent with our previous study, *hrde-1* mutant worms showed strong transcriptional de-silencing in GRTS loci: the median values of ΔPol II [*hrde-1*/WT] and Δpre-mRNA [*hrde-1*/WT] were 4.52 and 4.85, respectively (p values <3.42 × 10^{-15}) (Fig. 5b, c). In contrast,

met-2 set-25;set-32 mutants showed only a modest increase in Pol II occupancy at GRH loci overall: the median value of ΔPol II [*met2 set-25;set-32*/WT] was 1.46 (p value <3.01 × 10^{-13}). Importantly, *met-2 set-25;set-32* mutants did not show any increase in pre-mRNA at GRTS

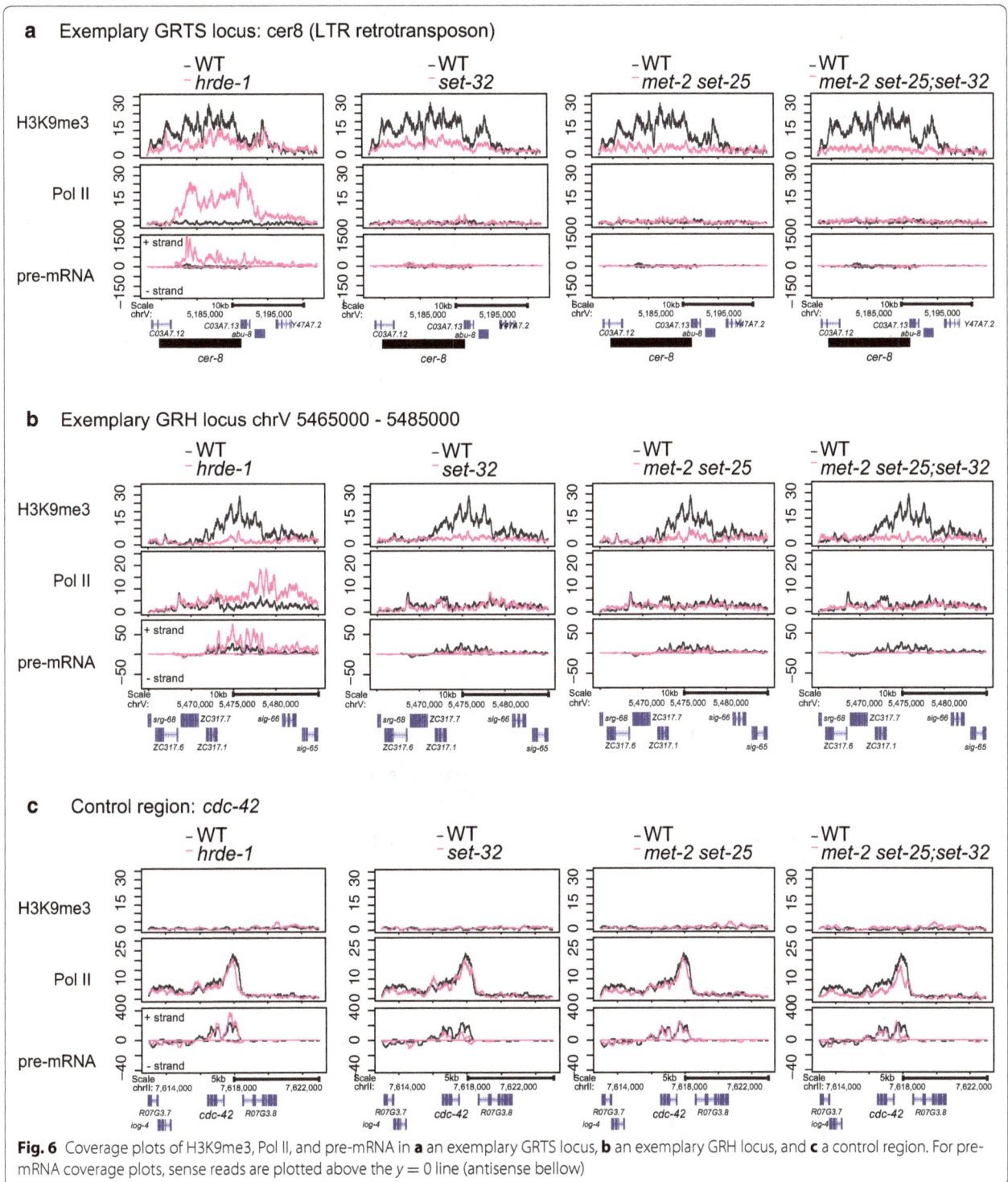

Fig. 6 Coverage plots of H3K9me3, Pol II, and pre-mRNA in **a** an exemplary GRTS locus, **b** an exemplary GRH locus, and **c** a control region. For pre-mRNA coverage plots, sense reads are plotted above the $y = 0$ line (antisense bellow)

loci: the median value of Δpre-mRNA [*met2 set-25;set-32/* WT] was 0.93 (p value = 0.062). In addition, GRH loci showed unchanged Pol II occupancy in *met-2 set-25;set-32* mutants (the median value of ΔPol II [*met2 set-25;set-32/*WT] = 1.07, p value = 0.16) and an unexpected

decrease in pre-mRNA (the median value of Δpre-mRNA [*met2 set-25;set-32/*WT] was 0.68, p value = 7.59×10^{-10}) (Fig. 5b, c and Additional file 6: Fig. S3b-c).

We note that the H3K9me3(+) regions in *met-2 set-25;set-32* mutants, on average, showed a modest increase in the Pol II level over WT: the median value of ΔPol II [*met2 set-25;set-32*/WT] was 1.44 (*p* value = 1.47×10^{-11}) (Fig. 5b, c and Additional file 6: Fig. S3b, c). However, the pre-mRNA increase in these regions was not statistically significant: the median value of Δpre-mRNA [*met2 set-25;set-32*/WT] was 1.70 (*p* value = 0.63). Scatter plot analyses of H3K9me3 ChIP-seq, Pol II ChIP-seq, pre-mRNA-seq, and mRNA-seq confirmed that the loss of H3K9me3 was not associated with transcriptional de-silencing in most of the H3K9me3(+) regions (Additional file 7: Fig. S5). Furthermore, the overall germline chromosome morphology was

similar between *met-2 set-25;set-32* mutant and wild-type worms (Additional file 8: Fig. S6). These data suggest that H3K9me3 plays, at most, a very limited role in germline chromatin condensation or transcription silencing at the global level.

Germline nuclear RNAi-mediated H3K9me3 is accompanied with H3K27me at the native targets

To characterize the other known germline nuclear RNAi-mediated heterochromatin mark in *C. elegans*, H3K27me3 [45], at the whole-genome level, we performed H3K27me3 ChIP-seq in the wild-type and *hrde-1*

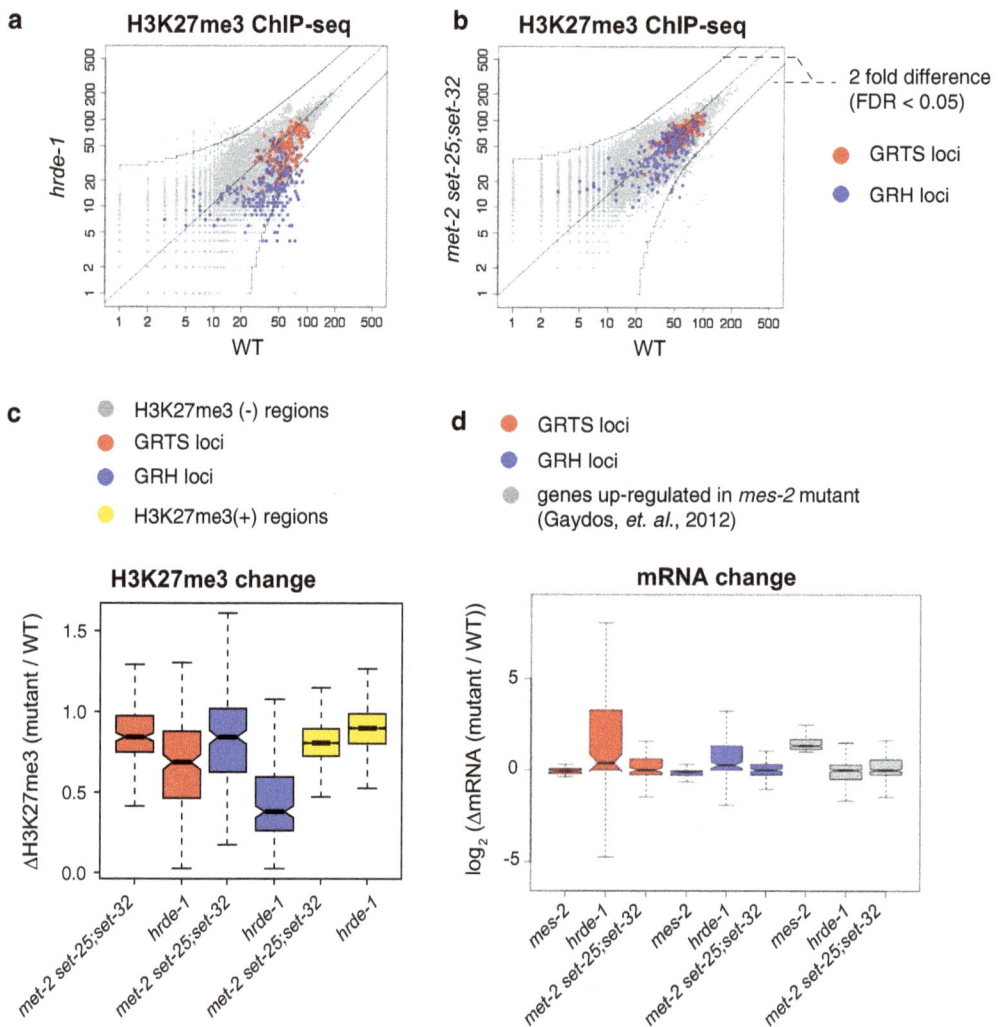

Fig. 7 H3K27me3 is associated with native germline nuclear RNAi targets and not affected in *met-2 set-25;set-32* mutant worms. **a**, **b** Scatter plots comparing whole-genome H3K27me3 levels in WT and *hrde-1* or *met-2 set-25;set-32* mutants at 1 kb resolution. **c** Boxplot analysis showing H3K27me3 changes between WT and *met-2 set-25;set-32* or *hrde-1* mutant worms in different subsets of the genome: GRTS loci (191 kb), GRH loci (215 kb), top 5-percentile H3K27me3 regions in WT (H3K27me3 + regions, 4966 kb). **d** Boxplot analysis comparing mRNA expression between WT and *mes-2*, *hrde-1*, or *met-2 set-25;set-32* mutants in different subsets of the genome: GRTS loci (121 genes), GRH loci (142 genes), genes that are up-regulated in *mes-2* mutant worms (355 genes, ≥2 fold, FDR < 0.05). Published microarray data [48] were used for mRNA expression changes between *mes-2* mutant and WT worms

mutant worms. Consistent with a previous study [45], we found that H3K27me3 was associated with both GRH and GRTS loci and was dependent on HRDE-1 (Fig. 7a, c). The median values for ΔH3K27me3 [hrde-1/WT] at GRTS and GRH loci were 0.68 and 0.38, respectively (all p values $<2.2 \times 10^{-16}$).

Just like H3K9me3, the correlation between HRDE-1-dependent-H3K27me3 and transcriptional silencing was different between the GRTS and GRH loci: GRTS loci showed a much weaker H3K27me3 loss than GRH loci in hrde-1 mutants (Fig. 7a, c), comparing with a much stronger transcriptional de-silencing in GRTS than GRH loci (Fig. 5b, c). In addition, GRTS and GRH loci did not show de-silencing at mRNA level in mes-2 mutants by using a published microarray data [48] (Fig. 7d), nor in met-2 set-25;set-32 mutants by mRNA-seq analysis in this study (Fig. 7d). In contrast, these targets were activated in hrde-1 mutants (Fig. 7d). Taken together, these results suggest that transcriptional silencing at HRDE-1 targets is likely to be independent of H3K27me3 as well.

We note that, by performing H3K27me3 ChIP-seq in met-2 set-25;set-32 mutant worms, we found no evidence of any H3K27me3-based compensating mechanism for the H3K9me3 loss—the triple HMT mutant worms did not show increased H3K27me3 in GRTS and GRH loci (Fig. 7b, c).

Discussion

In plants, S. pombe, as well as Drosophila, nuclear RNAi-mediated H3K9me3 is essential to maintain the transcriptional silencing state at target loci. Here we show that H3K9me3 is completely dispensable for the maintenance of endo-siRNA-mediated transcriptional silencing at the whole-genome level in C. elegans. In addition, H3K9me3 is not required for exogenous dsRNA-triggered transcriptional silencing and heritable RNAi in a well-established native gene target. These findings shift the paradigm of the C. elegans nuclear RNAi pathway, a key model system of studying RNA-mediated transcriptional silencing and transgenerational epigenetics.

The H3K9me3 contributions of MET-2, SET-25, and SET-32 in HRDE-1-dependent regions and elsewhere in the genome

C. elegans has several H3K9 HMTs. Before this study, it was unclear which of these HMTs are required for siRNA-mediated H3K9me3. Here, we provide experimental evidence supporting that germline nuclear RNAi-dependent H3K9me3 requires MET-2, SET-25, and SET-32.

Interestingly, the relative H3K9me3 contributions of these three HMTs are not the same at different targets.

Their activities appear to be synergistic at the native targets (endo-siRNA-targeted), as each of the two mutant strains, met-2 set-25 and set-32, showed >50% of H3K9me3 loss compared to the wild type. Such synergy may be the underlying mechanism that allows the high level of H3K9me3 at the native targets. The synergistic relationship is also evident for essentially all H3K9me3-enriched regions in the genome. The underlying mechanism for such apparent synergy is currently unknown. We also note that MET-2 SET-25 and SET-32 activities are not entirely mutually dependent, as the triple HMT mutant worms showed greater H3K9me3 loss than met-2 set-25 or set-32 mutants. For exogenous dsRNA-induced H3K9me3, MET-2 SET-25-dependent H3K9me3 and SET-32-dependent H3K9me3 are additive, suggesting that the two are independently triggered by dsRNA.

RNAi-mediated transcriptional (or co-transcriptional) silencing at different types of targets

Our ChIP-seq analyses show that HRDE-1-dependent silencing at native targets (endo-siRNA targeted) leads to depletion of Pol II throughout the regions, suggesting that transcription initiation is prevented. In contrast, exogenous dsRNA-triggered RNAi does not significantly change the overall Pol II level at the target gene, but causes a shift of the Pol II profile instead (a reduction at the 3' end and a gain in the gene body). This suggests that exogenous dsRNA-triggered RNAi does not block transcription initiation, but rather occurs during transcription elongation (also suggested previously by [19]) or co-transcriptionally. It is conceivable that the actual effect of RNAi on transcriptional silencing (and perhaps the degree of heritable silencing) is dependent on various features associated with the target gene (e.g., promoter, chromatin landscape, and siRNA).

The role of H3K9me3 in C. elegans germline nuclear RNAi

Our findings are consistent with two models. Model 1: H3K9me3 is not involved in siRNA-mediated transcription silencing and its function lies in another not yet identified aspect of germline nuclear RNAi. Model 2: H3K9me3 is involved in siRNA-mediated transcription silencing, together with an H3K9me3-independent silencing mechanism. Although these two models are mutually exclusive at any given target locus, they may both occur in C. elegans and used by different loci/silencing mechanisms that involve siRNA. A previous study showed that SET-32 and SET-25 are required for HRDE-1-dependent silencing using a GFP transgene as the piRNA reporter or exogenous dsRNA-induced heritable RNAi reporter [21]. In our study, these HMTs are not required for heritable silencing at a native reporter gene, *oma-1*. Future studies are required to explain this difference. Transgene reporter is a powerful tool in studying RNAi. However, the efficacy of small RNA-mediated silencing on a transgene reporter appears to be dependent on a variety of factors, such as transgene structure [23], epigenetic history [49, 50], and DNA sequence [51]. On the other hand, detailed characterization of additional native targets is needed to address whether heritable RNAi is regulated in a target-specific, context-dependent manner.

Conclusions

We found that three H3K9 HMTs, MET-2, SET-25, and SET-32, are required for germline nuclear RNAi-mediated H3K9me3 in *C. elegans*. Loss of the prominent H3K9me3 response in *met-2 set-25;set-32* mutant worms is not associated with any defect in germline nuclear RNAi-mediated transcriptional silencing at the native targets. Therefore, a high level of H3K9me3 is dispensable for the maintenance of the HRDE-1-dependent transcriptional silencing in *C. elegans* germline. In addition, we found that dsRNA-induced H3K9me3 is not required for transgenerational silencing at *oma-1*. We propose that transcriptional and heritable silencing of germline nuclear RNAi pathway in *C. elegans* can be maintained in an H3K9me3-independent manner. Our discovery that H3K9me3 can be decoupled from transcriptional silencing in *C. elegans* provides a unique opportunity to study any H3K9me3-independent silencing mechanism, particularly the direct biochemical effect of AGO-siRNA complex on transcription and co-transcriptional processes.

Methods

Strains

Bristol strain N2 was used as a standard wild-type strain. This study used the following mutations: Chr I: *set-32(ok1457)*; Chr II: *set-13(ok2697)*; Chr III: *set-25(ok5021)*, *hrde-1(tm1200)*; Chr IV: *set-21(ok2327)*, *set-26(tm3526)*, and *set-9(red8)*; Chr V: *met-2(n4256)*. The *set-9(red8)* mutation has a stop codon and a frame shift in the first exon, which was generated in this study using the CRISPR-cas9-mediated genome editing [52, 53]. Genotyping primers and other relevant sequences are listed in Additional file 9: Table S2. All strains were cultured at 19 °C.

Worm grind preparation

Worms were cultured on NGM plates with OP50 *E.coli* as a food source [54]. Synchronized young adult hermaphrodite animals were obtained by first using the bleaching method to collect worm embryos, which were

hatched in M9 buffer without food, and then L1 larvae were released onto NGM plates with OP50 *E. coli*. Young adult worms were ground by mortar and pestle in liquid nitrogen and stored at −80 °C. Worm grind from ~5000 young adult worms was used for each assay in this study.

Multigenerational heritable RNAi

Heritable RNAi experiments were conducted as previously described [18]. *oma-1* and *gfp* RNAi trigger sequences are reported in Additional file 9: Table S2. For *oma-1* RNAi, single nucleotide mismatch at every 30 nt was used to distinguish the trigger sequence from the native *oma-1* gene. All animals at P0 (with dsRNA feeding) and F1, F2, F3 (without dsRNA feeding) generations were collected at young adult stage and ground in liquid nitrogen.

H3K9me3 ChIP-qPCR

Worm grinds of P0 generation of *oma-1* RNAi and *gfp* RNAi samples at young adult stage were used for crosslinking, sonication and ChIP according to the protocol described in [27]. H3K9me3 ChIP from ~5000 worms per grind yield 5–10 ng of DNA. 1 ng of ChIP DNA was used per qPCR reaction. qPCR was set up using KAPA SYBR FAST Universal 2× PCR Master Mix (KAPA Biosystems) on a Mastercycler EP Realplex real-time PCR system (Eppendorf) according to the manufacturer's instructions. qPCR primers are listed in Additional file 9: Table S2. Each sample was processed in triplicate. Reported values for the *oma-1* RNAi H3K9me3 fold change were calculated using $\Delta\Delta CT$ analysis.

mRNA and pre-mRNA qRT-PCR

Worm grinds of P0 (RNAi+) and F1, F2, F3 (RNAi−) generations were used for total RNA extraction with Trizol reagent (Life Technologies), followed by DNase I (NEB) treatment.

mRNA reverse transcription (RT). 1 μg of total RNA was used for the first-strand cDNA synthesis with SuperScript III RT kit (Life Technologies) and oligo(dT)$_{20}$ primer (to enrich for mRNA).

Pre-mRNA RT. 2 μg of total RNA was used for the first-strand cDNA synthesis with SuperScript III RT kit (Life Technologies) and random hexamer primer mix (to capture pre-mRNA).

qRT-PCR. qRT-PCR was performed using KAPA SYBR FAST Universal 2× PCR Master Mix (KAPA Biosystems) on a Mastercycler EP Realplex real-time PCR system (Eppendorf) according to the manufacturer's instructions. qPCR primers are listed in Additional file 9: Table S2. Each sample was processed in triplicate. Reported values for the fold change of mRNA and pre-mRNA at *oma-1* gene were calculated using $\Delta\Delta CT$ analysis.

High-throughput sequencing

Pre-mRNA-seq. Pre-mRNA library was prepared as described in [40]. Anti-RNA Pol II S2 (ab5095, Abcam) antibodies were used for Pol II IP, followed by RNA isolation and library preparation.

mRNA-seq. Worm grinds (~5000 young adult worms) were used for total RNA extraction with Trizol reagent (Life Technologies). mRNA was enriched using the Poly(A) Purist MAG kit (Life Technologies) according to the manufacturer's instructions. 0.5–1 µg of mRNA was used for mRNA-seq library preparation as described in [40]. A mixture of four different 4-mer barcodes was used for the 5′-end ligation in both mRNA and pre-mRNA-seq as indicated Additional file 1: Table S1.

All libraries were sequenced using Illumina HiSeq 2500 platform, with 50-nt single-end run and dedicated index sequencing. Dedicated 6-mer indexes were used to de-multiplex DNA ChIP-seq and RNA-seq libraries for different samples.

Data availability: De-multiplexed raw sequencing data in fastq format for all libraries were deposited in NCBI (GEO accession number: GSE86517).

Data analysis

Sequencing reads were aligned to *C. elegans* genome (WS190 version) by using Bowtie (0.12.7) [55]. Only perfect alignments were used for data analysis. If a read was aligned to N different loci, it was counted as 1/N. Normalization based on sequencing depth of each library was used for all data analysis. In Fig. 5a and Additional file 6: Fig. S3a, besides sequencing depth, we also used the median values of H3K9me3(−) regions for data normalization. Otherwise, the background H3K9me3 levels in HMT mutants are artificially higher than the WT one, due to a high degree of H3K9me3 loss in a large fraction of the genome in the HMT mutants (e.g., Fig. 4a). The three-region Venn diagram was generated using a web-based software (http://www.benfrederickson.com/venn-diagrams-with-d3.js/). Custom R and python scripts were used in this study.

Curves that indicate twofold or 1.5-fold changes (FDR < 0.05) in all scatter plots were calculated using a script from [56]. Welch two-sample t test was used to calculate all p values.

Additional files

Additional file 1: Table S1. List of experiments, libraries, sequencing depth and barcodes used in this study.

Additional file 2: Figure S1. H3K9me3 ChIP-qPCR analysis of *oma-1* dsRNA-triggered H3K9me3 response in WT and various mutants.

Additional file 3: Figure S2. H3K9me3 levels are plotted as a function of position on all chromosomes for WT and HMT mutants (the panels for chromosome V are also shown in Fig. 4a). A moving average with a sliding window of 50 kb was used to plot H3K9me3 levels.

Additional file 4: Table S3. RPKM values of 1-kb windows for all ChIP-seq libraries used in this study. GRTS loci, GRH loci, SET-32-dependent H3K9me3, MET-2 SET-25-dependent H3K9me3, and MET-2 SET-25;SET-32-dependent H3K9me3 are indicated.

Additional file 5: Figure S4. Identification of H3K9me3 that is dependent on SET-32, MET-2 SET-25, or MET-2 SET-25;SET-32. (a) Whole-genome H3K9me3 scatter plots (1 kb resolution) that compare various HMT mutants with WT. (b) Venn diagram analysis between two biological repeats for H3K9me3 that is dependent on SET-32, MET-2 SET-25, or MET-2 SET-25;SET-32.

Additional file 6: Figure S3. Boxplot analysis for the ratios between mutant and WT for H3K9me3, Pol II, and pre-mRNA in different subsets of the genome: GRTS loci (191 kb), GRH loci (215 kb), top 5-percentile H3K9me3 regions in WT (H3K9me3(+) regions, 4775 kb), and bottom 5–25-percentile H3K9me3 regions in WT (H3K9me3(−) regions, 20,200 kb).

Additional file 7: Figure S5. Whole-genome scatter plots (1 kb resolution) that compare various mutant with WT for (a) H3K9me3 ChIP-seq, (b) Pol II ChIP-seq, and (c) pre-mRNA and (d) mRNA-seq (gene-based).

Additional file 8: Figure S6. DAPI staining of germline nuclei from WT and *met-2 set-25;set-32* mutant adult hermaphrodite. Scale bar: 10 µm.

Additional file 9: Table S2. Oligonucleotides and other sequences used in this study.

Additional file 10: Table S4. RPKM values of all annotated genes for all mRNA-seq libraries used in this study. GRTS loci, GRH loci, and H3K9me3-enriched genes are indicated.

Additional file 11: Table S5. RPKM values of 1-kb windows for all pre-mRNA-seq libraries used in this study. GRTS loci, GRH loci, SET-32-dependent H3K9me3, MET-2 SET-25-dependent H3K9me3, and MET-2 SET-25;SET-32-dependent H3K9me3 are indicated.

Abbreviations
HMT: histone methyltransferases; ChIP-seq: chromatin immunoprecipitation—high-throughput sequencing; RNAi: RNA interference; Endo-siRNA: endogenous small interfering RNA; dsRNA: double-stranded RNA; GRH: germline nuclear RNAi-mediated heterochromatin; GRTS: germline nuclear RNAi-mediated transcriptional silencing; FDR: false discovery rate.

Authors' contributions
NK and SG designed the experiments. NK, EC, JN, and KP collected the samples and performed the experiments. NK and SG performed data analysis and wrote the manuscripts. All authors read and approved the final manuscript.

Author details
[1] Department of Molecular Biology and Biochemistry, Rutgers the State University of New Jersey, Piscataway, NJ 08854, USA. [2] Nelson Labs A125, 604 Allison Road, Piscataway, NJ 08854, USA.

Acknowledgements
We thank Zoran Gajic, Julia Hong, Matthew Kim, Max Cabrera, Elaine Gavin, Shobhna Patel and Monica Driscoll for help, suggestions, and support. Some strains were provided by the CGC, which is funded by NIH Office of Research Infrastructure Programs (P40 OD010440).

Competing interests
The authors declare that they have no competing interests.

Funding
Research reported in this publication was supported by the Busch Biomedical Grant and the National Institute of General Medical Sciences of the National Institutes of Health under award number R01GM111752. The content is solely the responsibility of the authors and does not necessarily represent the official views of the National Institutes of Health.

References

1. Fire A, Xu S, Montgomery MK, Kostas SA, Driver SE, Mello CC. Potent and specific genetic interference by double-stranded RNA in *Caenorhabditis elegans*. Nature. 1998;391(6669):806–11. doi:10.1038/35888.

2. Kennerdell JR, Carthew RW. Use of dsRNA-mediated genetic interference to demonstrate that frizzled and frizzled 2 act in the wingless pathway. Cell. 1998;95(7):1017–26.

3. Elbashir SM, Lendeckel W, Tuschl T. RNA interference is mediated by 21- and 22-nucleotide RNAs. Genes Dev. 2001;15(2):188–200.

4. Hammond SM, Bernstein E, Beach D, Hannon GJ. An RNA-directed nuclease mediates post-transcriptional gene silencing in Drosophila cells. Nature. 2000;404(6775):293–6. doi:10.1038/35005107.

5. Hammond SM, Boettcher S, Caudy AA, Kobayashi R, Hannon GJ. Argonaute2, a link between genetic and biochemical analyses of RNAi. Science. 2001;293(5532):1146–50. doi:10.1126/science.1064023.

6. Tuschl T, Zamore PD, Lehmann R, Bartel DP, Sharp PA. Targeted mRNA degradation by double-stranded RNA in vitro. Genes Dev. 1999;13(24):3191–7.

7. Castel SE, Martienssen RA. RNA interference in the nucleus: roles for small RNAs in transcription, epigenetics and beyond. Nat Rev Genet. 2013;14(2):100–12. doi:10.1038/nrg3355.

8. Herr AJ, Baulcombe DC. RNA silencing pathways in plants. Cold Spring Harb Symp Quant Biol. 2004;69:363–70. doi:10.1101/sqb.2004.69.363.

9. Martienssen R, Moazed D. RNAi and heterochromatin assembly. Cold Spring Harb Perspect Biol. 2015;7(8):a019323. doi:10.1101/cshperspect. a019323.

10. Moazed D. Small RNAs in transcriptional gene silencing and genome defence. Nature. 2009;457(7228):413–20. doi:10.1038/nature07756.

11. Wassenegger M. RNA-directed DNA methylation. Plant Mol Biol. 2000;43(2–3):203–20.

12. Sienski G, Batki J, Senti KA, Donertas D, Tirian L, Meixner K, et al. Silencio/ CG9754 connects the Piwi–piRNA complex to the cellular heterochromatin machinery. Genes Dev. 2015;29(21):2258–71. doi:10.1101/gad.271908.115.

13. Sienski G, Donertas D, Brennecke J. Transcriptional silencing of transposons by Piwi and maelstrom and its impact on chromatin state and gene expression. Cell. 2012;151(5):964–80. doi:10.1016/j.cell.2012.10.040.

14. Siomi MC, Sato K, Pezic D, Aravin AA. PIWI-interacting small RNAs: the vanguard of genome defence. Nat Rev Mol Cell Biol. 2011;12(4):246–58. doi:10.1038/nrm3089.

15. Yu Y, Gu J, Jin Y, Luo Y, Preall JB, Ma J, et al. Panoramix enforces piRNA-dependent cotranscriptional silencing. Science. 2015;350(6258):339–42. doi:10.1126/science.aab0700.

16. Pezic D, Manakov SA, Sachidanandam R, Aravin AA. piRNA pathway targets active LINE1 elements to establish the repressive H3K9me3 mark in germ cells. Genes Dev. 2014;28(13):1410–28. doi:10.1101/gad.240895.114.

17. Buckley BA, Burkhart KB, Gu SG, Spracklin G, Kershner A, Fritz H, et al. A nuclear Argonaute promotes multigenerational epigenetic inheritance and germline immortality. Nature. 2012;489(7416):447–51. doi:10.1038/nature11352.

18. Gu SG, Pak J, Guang S, Maniar JM, Kennedy S, Fire A. Amplification of siRNA in *Caenorhabditis elegans* generates a transgenerational sequence-targeted histone H3 lysine 9 methylation footprint. Nat Genet. 2012;44(2):157–64. doi:10.1038/ng.1039.

19. Guang S, Bochner AF, Burkhart KB, Burton N, Pavelec DM, Kennedy S. Small regulatory RNAs inhibit RNA polymerase II during the elongation phase of transcription. Nature. 2010;465(7301):1097–101. doi:10.1038/nature09095.

20. Guang S, Bochner AF, Pavelec DM, Burkhart KB, Harding S, Lachowiec J, et al. An Argonaute transports siRNAs from the cytoplasm to the nucleus. Science. 2008;321(5888):537–41. doi:10.1126/science.1157647.

21. Ashe A, Sapetschnig A, Weick EM, Mitchell J, Bagijn MP, Cording AC, et al. piRNAs can trigger a multigenerational epigenetic memory in the germline of *C. elegans*. Cell. 2012;150(1):88–99. doi:10.1016/j.cell.2012.06.018.

22. Burkhart KB, Guang S, Buckley BA, Wong L, Bochner AF, Kennedy S. A pre-mRNA-associating factor links endogenous siRNAs to chromatin regulation. PLoS Genet. 2011;7(8):e1002249. doi:10.1371/journal. pgen.1002249.

23. Shirayama M, Seth M, Lee HC, Gu W, Ishidate T, Conte D Jr, et al. piRNAs initiate an epigenetic memory of nonself RNA in the *C. elegans* germline. Cell. 2012;150(1):65–77. doi:10.1016/j.cell.2012.06.015.

24. Alcazar RM, Lin R, Fire AZ. Transmission dynamics of heritable silencing induced by double-stranded RNA in *Caenorhabditis elegans*. Genetics. 2008;180(3):1275–88. doi:10.1534/genetics.108.089433.

25. Grishok A, Tabara H, Mello CC. Genetic requirements for inheritance of RNAi in *C. elegans*. Science. 2000;287(5462):2494–7.

26. Vastenhouw NL, Brunschwig K, Okihara KL, Muller F, Tijsterman M, Plasterk RH. Gene expression: long-term gene silencing by RNAi. Nature. 2006;442(7105):882. doi:10.1038/442882a.

27. Ni JZ, Kalinava N, Chen E, Huang A, Trinh T, Gu SG. A transgenerational role of the germline nuclear RNAi pathway in repressing heat stress-induced transcriptional activation in *C. elegans*. Epigenetics Chromatin. 2016;9:3. doi:10.1186/s13072-016-0052-x.

28. Grewal SI. RNAi-dependent formation of heterochromatin and its diverse functions. Curr Opin Genet Dev. 2010;20(2):134–41. doi:10.1016/j. gde.2010.02.003.

29. Audergon PN, Catania S, Kagansky A, Tong P, Shukla M, Pidoux AL, et al. Epigenetics. Restricted epigenetic inheritance of H3K9 methylation. Science. 2015;348(6230):132–5. doi:10.1126/science.1260638.

30. Ragunathan K, Jih G, Moazed D. Epigenetics. Epigenetic inheritance uncoupled from sequence-specific recruitment. Science. 2015;348(6230):1258699. doi:10.1126/science.1258699.

31. Buhler M, Verdel A, Moazed D. Tethering RITS to a nascent transcript initiates RNAi- and heterochromatin-dependent gene silencing. Cell. 2006;125(5):873–86. doi:10.1016/j.cell.2006.04.025.

32. Gu SG, Fire A. Partitioning the *C. elegans* genome by nucleosome modification, occupancy, and positioning. Chromosoma. 2010;119(1):73–87. doi:10.1007/s00412-009-0235-3.

33. Liu T, Rechtsteiner A, Egelhofer TA, Vielle A, Latorre I, Cheung MS, et al. Broad chromosomal domains of histone modification patterns in *C. elegans*. Genome Res. 2011;21(2):227–36. doi:10.1101/gr.115519.110.

34. Kelly WG, Schaner CE, Dernburg AF, Lee MH, Kim SK, Villeneuve AM, et al. X-chromosome silencing in the germline of *C. elegans*. Development. 2002;129(2):479–92.

35. Kerr SC, Ruppersburg CC, Francis JW, Katz DJ. SPR-5 and MET-2 function cooperatively to reestablish an epigenetic ground state during passage through the germ line. Proc Natl Acad Sci USA. 2014;111(26):9509–14. doi:10.1073/pnas.1321843111.

36. She X, Xu X, Fedotov A, Kelly WG, Maine EM. Regulation of heterochromatin assembly on unpaired chromosomes during *Caenorhabditis elegans* meiosis by components of a small RNA-mediated pathway. PLoS Genet. 2009;5(8):e1000624. doi:10.1371/journal.pgen.1000624.

37. Towbin BD, Gonzalez-Aguilera C, Sack R, Gaidatzis D, Kalck V, Meister P, et al. Step-wise methylation of histone H3K9 positions heterochromatin at the nuclear periphery. Cell. 2012;150(5):934–47. doi:10.1016/j. cell.2012.06.051.

38. Snyder MJ, Lau AC, Brouhard EA, Davis MB, Jiang J, Sifuentes MH, et al. Anchoring of heterochromatin to the nuclear lamina reinforces dosage compensation-mediated gene repression. PLoS Genet. 2016;12(9):e1006341. doi:10.1371/journal.pgen.1006341.

39. Zeller P, Padeken J, van Schendel R, Kalck V, Tijsterman M, Gasser SM. Histone H3K9 methylation is dispensable for *Caenorhabditis elegans* development but suppresses RNA:DNA hybrid-associated repeat instability. Nat Genet. 2016;48(11):1385–95. doi:10.1038/ng.3672.

40. Ni JZ, Chen E, Gu SG. Complex coding of endogenous siRNA, transcriptional silencing and H3K9 methylation on native targets of germline nuclear RNAi in *C. elegans*. BMC Genom. 2014;15:1157. doi:10.1186/1471-2164-15-1157.

41. Andersen EC, Horvitz HR. Two *C. elegans* histone methyltransferases repress lin-3 EGF transcription to inhibit vulval development. Development. 2007;134(16):2991–9. doi:10.1242/dev.009373.

42. Greer EL, Beese-Sims SE, Brookes E, Spadafora R, Zhu Y, Rothbart SB, et al. A histone methylation network regulates transgenerational epigenetic memory in *C. elegans*. Cell Rep. 2014;7(1):113–26. doi:10.1016/j.celrep.2014.02.044.

43. Bessler JB, Andersen EC, Villeneuve AM. Differential localization and independent acquisition of the H3K9me2 and H3K9me3 chromatin modifications in the *Caenorhabditis elegans* adult germ line. PLoS Genet. 2010;6(1):e1000830. doi:10.1371/journal.pgen.1000830.

44. Garrigues JM, Sidoli S, Garcia BA, Strome S. Defining heterochromatin in *C. elegans* through genome-wide analysis of the heterochromatin protein 1 homolog HPL-2. Genome Res. 2015;25(1):76–88. doi:10.1101/gr.180489.114.

45. Mao H, Zhu C, Zong D, Weng C, Yang X, Huang H, et al. The Nrde pathway mediates small-RNA-directed histone H3 lysine 27 trimethylation in *Caenorhabditis elegans*. Curr Biol. 2015;25(18):2398–403. doi:10.1016/j.cub.2015.07.051.

46. Houri-Ze'evi L, Korem Y, Sheftel H, Faigenbloom L, Toker IA, Dagan Y, et al. A tunable mechanism determines the duration of the transgenerational small RNA inheritance in *C. elegans*. Cell. 2016;165(1):88–99. doi:10.1016/j.cell.2016.02.057.

47. Hirsh D, Oppenheim D, Klass M. Development of the reproductive system of *Caenorhabditis elegans*. Dev Biol. 1976;49(1):200–19.

48. Gaydos LJ, Rechtsteiner A, Egelhofer TA, Carroll CR, Strome S. Antagonism between MES-4 and polycomb repressive complex 2 promotes appropriate gene expression in *C. elegans* germ cells. Cell Rep. 2012;2(5):1169–77. doi:10.1016/j.celrep.2012.09.019.

49. Seth M, Shirayama M, Gu W, Ishidate T, Conte D Jr, Mello CC. The *C. elegans* CSR-1 argonaute pathway counteracts epigenetic silencing to promote germline gene expression. Dev Cell. 2013;27(6):656–63. doi:10.1016/j.devcel.2013.11.014.

50. Leopold LE, Heestand BN, Seong S, Shtessel L, Ahmed S. Lack of pairing during meiosis triggers multigenerational transgene silencing in *Caenorhabditis elegans*. Proc Natl Acad Sci USA. 2015;112(20):E2667–76. doi:10.1073/pnas.1501979112.

51. Bagijn MP, Goldstein LD, Sapetschnig A, Weick EM, Bouasker S, Lehrbach NJ, et al. Function, targets, and evolution of *Caenorhabditis elegans* piRNAs. Science. 2012;337(6094):574–8. doi:10.1126/science.1220952.

52. Arribere JA, Bell RT, Fu BX, Artiles KL, Hartman PS, Fire AZ. Efficient marker-free recovery of custom genetic modifications with CRISPR/Cas9 in *Caenorhabditis elegans*. Genetics. 2014;198(3):837–46. doi:10.1534/genetics.114.169730.

53. Paix A, Folkmann A, Rasoloson D, Seydoux G. High efficiency, homology-directed genome editing in *Caenorhabditis elegans* using CRISPR-Cas9 ribonucleoprotein complexes. Genetics. 2015;201(1):47–54. doi:10.1534/genetics.115.179382.

54. Brenner S. The genetics of *Caenorhabditis elegans*. Genetics. 1974;77(1):71–94.

55. Langmead B, Trapnell C, Pop M, Salzberg SL. Ultrafast and memory-efficient alignment of short DNA sequences to the human genome. Genome Biol. 2009;10(3):R25. doi:10.1186/gb-2009-10-3-r25.

56. Maniar JM, Fire AZ. EGO-1, a *C. elegans* RdRP, modulates gene expression via production of mRNA-templated short antisense RNAs. Curr Biol. 2011;21(6):449–59. doi:10.1016/j.cub.2011.02.019.

Permissions

All chapters in this book were first published in E&C, by BioMed Central; hereby published with permission under the Creative Commons Attribution License or equivalent. Every chapter published in this book has been scrutinized by our experts. Their significance has been extensively debated. The topics covered herein carry significant findings which will fuel the growth of the discipline. They may even be implemented as practical applications or may be referred to as a beginning point for another development.

The contributors of this book come from diverse backgrounds, making this book a truly international effort. This book will bring forth new frontiers with its revolutionizing research information and detailed analysis of the nascent developments around the world.

We would like to thank all the contributing authors for lending their expertise to make the book truly unique. They have played a crucial role in the development of this book. Without their invaluable contributions this book wouldn't have been possible. They have made vital efforts to compile up to date information on the varied aspects of this subject to make this book a valuable addition to the collection of many professionals and students.

This book was conceptualized with the vision of imparting up-to-date information and advanced data in this field. To ensure the same, a matchless editorial board was set up. Every individual on the board went through rigorous rounds of assessment to prove their worth. After which they invested a large part of their time researching and compiling the most relevant data for our readers.

The editorial board has been involved in producing this book since its inception. They have spent rigorous hours researching and exploring the diverse topics which have resulted in the successful publishing of this book. They have passed on their knowledge of decades through this book. To expedite this challenging task, the publisher supported the team at every step. A small team of assistant editors was also appointed to further simplify the editing procedure and attain best results for the readers.

Apart from the editorial board, the designing team has also invested a significant amount of their time in understanding the subject and creating the most relevant covers. They scrutinized every image to scout for the most suitable representation of the subject and create an appropriate cover for the book.

The publishing team has been an ardent support to the editorial, designing and production team. Their endless efforts to recruit the best for this project, has resulted in the accomplishment of this book. They are a veteran in the field of academics and their pool of knowledge is as vast as their experience in printing. Their expertise and guidance has proved useful at every step. Their uncompromising quality standards have made this book an exceptional effort. Their encouragement from time to time has been an inspiration for everyone.

The publisher and the editorial board hope that this book will prove to be a valuable piece of knowledge for researchers, students, practitioners and scholars across the globe.

List of Contributors

Pavla Navratilova and Gemma Barbara Danks
Sars International Centre for Marine Molecular Biology, University of Bergen, 5008 Bergen, Norway

Abby Long, Stephen Butcher and John Robert Manak
Departments of Biology and Pediatrics and the Roy J. Carver Center for Genomics, 459 Biology Building, University of Iowa, Iowa City, IA 52242, USA

Eric M. Thompson
Sars International Centre for Marine Molecular Biology, University of Bergen, 5008 Bergen, Norway
Department of Biology, University of Bergen, 5020 Bergen, Norway

Arnab Nayak and Stefan Müller
Institute of Biochemistry II, Goethe University Medical School, University Hospital Building 75, Theodor-Stern-Kai 7, 60590 Frankfurt am Main, Germany

Anja Reck and Christian Morsczeck
Department of Oral and Maxillofacial Surgery, University of Regensburg, 93042 Regensburg, Germany

Ksenia Skvortsova, Elena Zotenko, Phuc-Loi Luu and Cathryn M. Gould
Epigenetics Research Laboratory, Genomics and Epigenetics Division, Garvan Institute of Medical Research, 384 Victoria Street, Darlinghurst, Sydney, NSW 2010, Australia

Shalima S. Nair, Susan J. Clark and Clare Stirzaker
Epigenetics Research Laboratory, Genomics and Epigenetics Division, Garvan Institute of Medical Research, 384 Victoria Street, Darlinghurst, Sydney, NSW 2010, Australia
St Vincent's Clinical School, UNSW Australia, Sydney, NSW 2010, Australia

David Latrasse, Natalia Y. Rodriguez-Granados, Abdelhafid Bendahmane, Celine Camps, vivien Sommard, Cécile Raynaud and Adnane Boualem
Institute of Plant Sciences Paris-Saclay (IPS2), CNRS, INRA, University Paris-Sud, University of Evry, University Paris-Diderot, Sorbonne Paris-Cite, University of Paris-Saclay, Batiment 630, 91405 Orsay, France

Alaguraj Veluchamy and Kiruthiga Gayathri Mariappan
Division of Biological and Environmental Sciences and Engineering, King Abdullah University of Science and Technology, Thuwal 23955-6900, Kingdom of Saudi Arabia

Claudia Bevilacqua and Nicolas Crapart
UMR 1313 Génétique Animale et Biologie Intégrative, Institut National de la Recherche Agronomique, 78350 Jouy-en-Josas, France

Catherine Dogimont
UR 1052, Unité de Génétique et d'Amélioration des Fruits et Légumes, INRA, BP94, 84143 Montfavet, France

Moussa Benhamed
Institute of Plant Sciences Paris-Saclay (IPS2), CNRS, INRA, University Paris-Sud, University of Evry, University Paris-Diderot, Sorbonne Paris-Cite, University of Paris-Saclay, Batiment 630, 91405 Orsay, France
Division of Biological and Environmental Sciences and Engineering, King Abdullah University of Science and Technology, Thuwal 23955-6900, Kingdom of Saudi Arabia

Paul R. Hillman
Department of Veterinary Pathobiology, College of Veterinary Medicine and Biomedical Sciences, Texas A&M University, College Station, TX 77845, USA
Department of Molecular and Cellular Medicine, College of Medicine, Texas A&M Health Science Center, College Station, TX 77845, USA

Sarah G. B. Christian
Department of Veterinary Pathobiology, College of Veterinary Medicine and Biomedical Sciences, Texas A&M University, College Station, TX 77845, USA

Ryan Doan
Department of Veterinary Pathobiology, College of Veterinary Medicine and Biomedical Sciences, Texas A&M University, College Station, TX 77845, USA

Interdisciplinary Genetics Program, College of Agriculture and Life Sciences, Texas A&M University, College Station, TX 77845, USA

Noah D. Cohen
Department of Large Animal Clinical Sciences, College of Veterinary Medicine and Biomedical Sciences, Texas A&M University, College Station, TX, USA

Kranti Konganti
Institute for Genome Science and Society, Texas A&M University, College Station, TX 77845, USA

Kory Douglas
Department of Large Animal Clinical Sciences, College of Veterinary Medicine and Biomedical Sciences, Texas A&M University, College Station, TX, USA
Department of Veterinary Integrative Biosciences, College of Veterinary Medicine and Biomedical Sciences, Texas A&M University, College Station, TX 77843, USA

Xu Wang
Department of Molecular Biology and Genetics, Cornell University,Ithaca, NY 14853, USA

Paul B. Samollow
Department of Veterinary Integrative Biosciences, College of Veterinary Medicine and Biomedical Sciences, Texas A&M University, College Station, TX 77843, USA

Scott V. Dindot
Department of Veterinary Pathobiology, College of Veterinary Medicine and Biomedical Sciences, Texas A&M University, College Station, TX 77845, USA
Department of Molecular and Cellular Medicine, College of Medicine, Texas A&M Health Science Center, College Station, TX 77845, USA
Department of Veterinary Pathobiology, College of Veterinary Medicine and Biomedical Sciences, Texas A&M University, 4467 TAMU, College Station, TX 77843, USA

Sergey V. Ulianov and Sergey V. Razin
Institute of Gene Biology of the Russian Academy of Sciences, Moscow, Russia 119334. 2 Faculty of Biology, M.V. Lomonosov Moscow State University, Moscow, Russia 119992

Faculty of Biology, M.V. Lomonosov Moscow State University, Moscow, Russia 119992

Aleksandra A. Galitsyna
Institute of Gene Biology of the Russian Academy of Sciences, Moscow,Russia 119334
Faculty of Bioengineering and Bioinformatics, M.V.Lomonosov Moscow State University, Moscow, Russia 119992
Institute for Information Transmission Problems (the Kharkevich Institute) of the Russian Academy of Sciences, Moscow, Russia 127051

Ilya M. Flyamer
Institute of Gene Biology of the Russian Academy of Sciences, Moscow,Russia 119334
Faculty of Biology, M.V. Lomonosov Moscow State University, Moscow, Russia 119992
MRC Human Genetics Unit, Institute of Genetics and Molecular Medicine, University of Edinburgh, Edinburgh, UK

Arkadiy K. Golov and Alexey A. Gavrilov
Institute of Gene Biology of the Russian Academy of Sciences, Moscow, Russia 119334

Ekaterina E. Khrameeva
Skolkovo Institute of Science and Technology, Skolkovo, Russia 143026
Institute for Information Transmission Problems (the Kharkevich Institute) of the Russian Academy of Sciences, Moscow, Russia 127051

Maxim V. Imakaev
Department of Physics,Massachusetts Institute of Technology, Cambridge, MA 02139, USA

Nezar A. Abdennur
Computational and Systems Biology Graduate Program, Massachusetts Institute of Technology, Cambridge, MA, USA

Mikhail S. Gelfand
Faculty of Bioengineering and Bioinformatics, M.V. Lomonosov Moscow State University, Moscow, Russia 119992
Skolkovo Institute of Science and Technology, Skolkovo, Russia 143026
Institute for Information Transmission Problems (the Kharkevich Institute) of the Russian Academy of Sciences, Moscow, Russia 127051
Faculty of Computer Science, Higher School of Economics, Moscow, Russia 125319

Marion Cremer, Ines Hellmann, Andreas Maiser, Heinrich Leonhardt and Thomas Cremer
LMU Biocenter, Department Biology II, Ludwig Maximilians-Universität (LMU Munich), Grosshadernerstr

Volker J. Schmid
BioImaging Group, Department of Statistics, Ludwig Maximilians-Universität (LMU Munich), Munich, Germany

Felix Kraus
LMU Biocenter, Department Biology II, Ludwig Maximilians-Universität (LMU Munich), Grosshadernerstr
Department of Biochemistry and Molecular Biology, Monash Biomedicine Discovery Institute, Monash University, Melbourne 3800, Australia

Yolanda Markaki
LMU Biocenter, Department Biology II, Ludwig Maximilians-Universität (LMU Munich), Grosshadernerstr.
Department of Biological Chemistry, David Geffen School of Medicine at UCLA, Los Angeles, CA, USA

Sam John
Department of Genome Sciences, University of Washington, Seattle, WA, USA
Center for Cancer Research, National Cancer Institute, Bethesda, MD, USA

John Stamatoyannopoulos
Department of Genome Sciences, University of Washington, Seattle, WA, USA. Center for Cancer Research, National Cancer Institute, Bethesda, MD, USA

Archana P. Gupta, Lei Zhu, Jaishree Tripathi, Michal Kucharski, Alok Patra and Zbynek Bozdech
School of Biological Sciences, Nanyang Technological University, 60 Nanyang Drive, Singapore 637551, Singapore

Hironao Wakabayashi, Christopher Tucker, Jeffrey J. Hayes and Elena Rustchenko
Department of Biochemistry and Biophysics, University of Rochester Medical Center, Rochester, NY, USA

Gabor Bethlendy
Roche Diagnostics Corporation, Indianapolis, IN, USA

Parabase Genomics, Dorchester, MA, USA

Anatoliy Kravets
Department of Biochemistry and Biophysics, University of Rochester Medical Center, Rochester, NY, USA

Stephen L. Welle
Department of Medicine, University of Rochester Medical Center, Rochester, NY, USA
Department of Pediatrics, Center for Pediatric Biochemical Research, University of Rochester Medical Center, Rochester, NY, USA

Michael Bulger
Department of Pediatrics, Center for Pediatric Biochemical Research,University of Rochester Medical Center, Rochester, NY, USA

Sonia Heras, Katrien Smits, Catharina De Schauwer and Ann Van Soom
Department of Reproduction, Obstetrics and Herd Health, Faculty of Veterinary Medicine, Ghent University, 9820 Merelbeke, Belgium

Erin K. Shanle
Department of Biological and Environmental Sciences, Longwood University, Farmville, VA 23909, USA.

Stephen A. Shinsky
Department of Biochemistry and Biophysics,The University of North Carolina, Chapel Hill, NC 27599, USA
Lineberger Comprehensive Cancer Center, The University of North Carolina School of Medicine, Chapel Hill, NC 27599, USA
Roy and Diana Vagelos Labora-tories, Department of Chemistry, University of Pennsylvania, Philadelphia, PA, USA

Joseph B. Bridgers and Krzysztof Krajewski
Department of Biochemistry and Biophysics, The University of North Carolina, Chapel Hill, NC 27599, USA.

Narkhyun Bae, Cari Sagum and Mark T. Bedford
Department of Epigenetics and Molecular Carcinogenesis, The University of Texas MD Anderson Cancer Center, Smithville, TX 78957, USA.

Scott B. Rothbart
Center for Epigenetics, Van Andel Research Institute, Grand Rapids, MI 49503, USA

Brian D. Strahl
Department of Biochemistry and Biophysics, The University of North Carolina, Chapel Hill, NC 27599, USA
Lineberger Comprehensive Cancer Center, The University of North Carolina School of Medicine, Chapel Hill, NC 27599, USA

Benjamin L. Kidder
Department of Oncology, Wayne State University School of Medicine, Detroit, MI, USA.
Karmanos Cancer Institute, Wayne State University School of Medicine, Detroit, MI, USA
Systems Biology Center, National Heart, Lung and Blood Institute, National Institutes of Health, Bethesda, MD, USA

Gangqing Hu, Kairong Cui and Keji Zhao
Systems Biology Center, National Heart, Lung and Blood Institute, National Institutes of Health, Bethesda, MD, USA.

Natallia Kalinava, Julie Zhouli Ni, Kimberly Peterman and Esteban Chen
Department of Molecular Biology and Biochemistry, Rutgers the State University of New Jersey, Piscataway, NJ 08854, USA

Sam Guoping Gu
Department of Molecular Biology and Biochemistry, Rutgers the State University of New Jersey, Piscataway, NJ 08854, USA
Nelson Labs A125, 604 Allison Road, Piscataway, NJ 08854, USA

Index

www.ingramcontent.com/pod-product-compliance
Lightning Source LLC
Chambersburg PA
CBHW080244230326
41458CB00097B/3337

9 781641 161268